WITHDRAWN

Biology of Extracellular Matrix

For further volumes:
www.springer.com/series/8422

Robert P. Mecham
Editor

The Extracellular Matrix: an Overview

Springer

Editor
Professor Robert P. Mecham
Washington University School of Medicine
Department of Cell Biology & Physiology
660 South Euclid Ave., Campus Box 8228
St. Louis, MO 63110
USA
bmecham@wustl.edu

ISSN 0887-3224   e-ISSN 2191-1959
ISBN 978-3-642-16554-2   e-ISBN 978-3-642-16555-9
DOI 10.1007/978-3-642-16555-9
Springer Heidelberg Dordrecht London New York

Library of Congress Control Number: 2011920814

© Springer-Verlag Berlin Heidelberg 2011
This work is subject to copyright. All rights are reserved, whether the whole or part of the material is concerned, specifically the rights of translation, reprinting, reuse of illustrations, recitation, broadcasting, reproduction on microfilm or in any other way, and storage in data banks. Duplication of this publication or parts thereof is permitted only under the provisions of the German Copyright Law of September 9, 1965, in its current version, and permission for use must always be obtained from Springer. Violations are liable to prosecution under the German Copyright Law.
The use of general descriptive names, registered names, trademarks, etc. in this publication does not imply, even in the absence of a specific statement, that such names are exempt from the relevant protective laws and regulations and therefore free for general use.

Cover design: deblik, Berlin

Printed on acid-free paper

Springer is part of Springer Science+Business Media (www.springer.com)

Preface to the Series

The first volume of the *Biology of Extracellular Matrix* series was published in 1986 and was titled "Regulation of Matrix Accumulation." Twelve volumes in the series were published over a period of 12 years and each volume provided timely reviews on current topics of ECM biology. With the contraction of the publishing industry in the late 1990s, Academic Press, the former Series publisher, was purchased by Elsevier and they decided to discontinue most of their monograph series, including the *Biology of Extracellular Matrix*. I was able to retain the rights to the series title and was pleased when Springer agreed to resume publication. The volume "The Extracellular Matrix: An Overview," Robert P. Mecham (Editor), is the first under the new publisher. It should also be noted that the series is being published in collaboration with the American Society for Matrix Biology.

The Study of Extracellular Matrix Biology

Over the years, our understanding of extracellular matrix (ECM) function has evolved from the early concept of a static "connective tissue" that ties everything together to one of a dynamic biomaterial that provides strength and elasticity, interacts with cell-surface receptors, and controls the availability of growth factors. There is now no question that ECM is an important part of cell biology, and to understand cellular differentiation, tissue development, and tissue remodeling requires an in-depth consideration of the ECM components that are produced by the cell. As we look back through the relatively short history of ECM biology, we find that the field was first dominated by biochemistry (mostly chemistry!) where investigators were trying to isolate and identify the individual ECM components. The proteins that were identified were, indeed, unique in their structure, composition, and function and were unlike other proteins in living cells. The ECM is designed to function as homo- and heteropolymers that are generally insoluble in their mature state. They also have relatively long half-lives compared with other proteins in the body. Some contain unique cross-links, some have high amounts of sulfated polysaccharides, some are designed to be "sticky" in terms of interacting with cells, and others form complex adhesion surfaces and diffusion barriers between different cell layers. In all cases, however, each class of ECM molecule is designed to interact with another to produce the unique physical and signaling properties that support tissue structure, growth, and function.

This brings us to the field today, where the tools of cell and molecular biology together with the power of model organism genetics allow us to focus on the functional complexities of the ECM biopolymer. Constructing a complex, mechanically appropriate matrix requires the cell to know the instructions for assembly, to have knowledge of the available building materials, and be able to interpret information about the stresses that the final material will have to endure. In this regard, it is clear that cells are adept at reading the instructive signals from the microenvironment, and changing the mix of matrix proteins needs to be added at any particular instance. While there is still a need to use biochemistry to characterize the individual ECM components, to fully understand the ECM requires a fundamental knowledge of cell biology. We need to understand, for example, the cellular mechanisms that lead to coordinated expression, both temporally and spatially, of complex sets of genes that encode ECM proteins as well as the enzymes responsible for their secretion and assembly. Building a functional collagen fiber, for example, involves activating and regulating genes for collagen alpha chains, hydroxylating enzymes, proteases to process propeptide regions, lysyl oxidases for cross-linking, and other chaperones and assembly proteins. Similar complexities are involved in the processing and assembly of most ECM networks, including basement membranes, elastic fibers, and large proteoglycan matrices. Virtually, all fields of animal and plant biology are concerned with questions of extracellular matrix in some manner. It is my hope that this series will prove helpful to all those seeking an introduction to EMC biology as well as experienced ECM investigators who are interested in greater insight into ECM function. In the preface to volume one of this series over 2 decades ago, I pointed out that the series cannot thrive without a large measure of enthusiasm and active participation from the ECM community. I welcome your suggestions of topics for future volumes and look forward to your feedback as we explore the extracellular matrix.

St. Louis MO, USA   Robert P. Mecham

Preface to the Volume

The objective of this overview volume in the "Biology of Extracellular Matrix" series is to update and build upon topics discussed in previous volumes in this series as well as in classic ECM review texts, such as Betty Hay's *Cell Biology of Extracellular Matrix*. The first chapter by Jürgen Engel and Matthias Chiquet is the ideal introduction to ECM biology. It provides an overview of ECM structure and function in a creative and insightful interpretation of the ECM as a complex "machine." The authors outline the basic features of the major classes of ECM components and describe how their multidomain structure allows multiple functions to be combined in one, often large molecule that is engineered to undergo multimeric assembly and extended multimolecular networks. They show how ECM components together with their cell surface receptors can be viewed as intricate nanodevices that allow cells to physically organize their 3D environment, as well as to sense and respond to various types of mechanical stress. They also make the point that metazoan evolution would not have been possible without the concomitant expansion of ECM complexity. Using examples of phylogenetically "old" versus "young" ECM protein families, they review the evidence that today's incredible diversity of ECM components arose from the recombination of preexisting protein modules by exon shuffling during evolution.

The second chapter focuses on fibronectin and other glycoproteins that mediate cell adhesion through interactions with integrins. Jielin Xu and Deane Mosher use an in-depth analysis of fibronectin as a prototype to illustrate how the domain organization of adhesive glycoproteins is structured to bridge interactions between cells (integrin binding domains) and other components of the ECM, including collagen, heparan, and fibrin. They also discuss fibronectin assembly and the importance of integrins and the cellular cytoskeleton in this process. Other glycoproteins discussed in this chapter include vitronectin, the laminins, thrombospondins, tenascins, entactins, nephronectin, and fibrinogen. A short section on integrin signaling is also included.

Of all of the ECM proteins, few are as old as collagen. In early parazoa (like sponges), cells are embedded in an ECM consisting mainly of fibrillar collagens not unlike those of higher animals. In Chap. 3, David Birk and Peter Brückner bring us up-to-date on collagen types and collagen fibril assembly. There are 28 different types of collagen in vertebrates (many more in invertebrates) that assemble into a variety of supramolecular structures including fibrils, microfibrils, and network-like structures. This chapter begins with a general discussion of collagen molecules and their supramolecular structure, assembly, and function within extracellular matrices. One of the more interesting aspects of collagen biology as outlined in this chapter is the description of mechanistic principles involved in the assembly of collagen-containing suprastructures. This includes the characterization of tissue-specific collagen fibrillogenesis, which serves to generate the diversity in extracellular matrix structures and functions required for individual tissue function.

As multicellular organisms evolved and grew more complex, there arose a need during development to separate polarized epithelial cells from underlying mesenchymal cells. This separation process, i.e., gastrulation, would not be possible without the appearance of the basement membrane – a unique ECM structure that combines the structural rigidity and unique basket-weave-forming properties of collagen type IV with cell-adhesive proteins (e.g., laminins) and charged proteoglycans (e.g., perlecan and agrin). The chapter on basement membranes by Jeffrey Miner summarizes our current knowledge about the basement membrane components and their receptors on cells. Basement membrane assembly is also discussed along with a number of human genetic diseases caused by mutations that affect basement membrane components.

The discussion of proteoglycans is separated into two chapters. The first, Chap. 5, authored by Thomas Wight, Bryan Toole, and Vincent Hascall, focuses on hyaluronan and the large aggregating proteoglycans. This family includes aggrecan, versican, neurocan, and brevican. These proteoglycans form macromolecular complexes with hyaluronan and contribute to the structural and mechanical stability of different tissues. Considerable evidence suggests that the large hydrodynamic space occupied by glycosaminoglycan chains influences tissue turgidity and viscoelasticity. In addition, recent data point to a prominent role for these ECM structures in direct cell signaling as well as an ability to bind and sequester growth factors and morphogens that are important for cell movement and differentiation. The chapter also contains a description of new functions mapped to the proteoglycan core protein.

The small leucine-rich proteoglycans (SLRPs) are discussed in Chap. 6 by Renato Iozzo, Silvia Goldoni, Agnes Berendsen, and Marian Young. SLRPs serve as tissue organizers by orienting and ordering various collagenous matrices during ontogeny, wound repair, and cancer. They also interact with a number of surface receptors and growth factors thereby regulating cell behavior. The focus of this chapter is on novel conceptual and functional advances in our understanding of SLRP biology with special emphasis on genetic diseases, cancer growth, fibrosis, osteoporosis, and other biological processes where these proteoglycans play a central role.

One of the newest ECM structures to be described and characterized, but among the oldest ECM structures in evolution, is the microfibril. The core elements of these 10–15 nm filaments are the fibrillins – large cysteine-rich proteins that can be found as far back in evolution as the placozoans and, perhaps, parazoans. First described as components of elastic fibers, microfibrils are now known to be important regulators of growth factor signaling through their ability to bind and sequester growth factors, particularly TGF-β family members. In Chap. 7, Dirk Hubmacher and Dieter Reinhardt provide an overview of the structure, assembly, and functions of fibrillins and microfibrils as well as the pathobiology associated with genetic aberrations in the microfibril system.

Vertebrate evolution would not have been as successful as it was without elastin. As the name implies, elastin imparts elasticity to tissues, particularly large blood vessels and the lung. Without elastic vessels, it would not be possible to evolve an efficient closed, pulsatile circulatory system that supports efficient distal perfusion and body growth. Similarly, the mechanical function of the vertebrate lung would not be possible without elastin. Beth Kozel, Robert Mecham, and Joel Rosenbloom discuss this unique, highly cross-linked protein in Chap. 8. Emphasis is given to how the protein works as an elastomer and why damage to elastic fibers is so detrimental to tissue integrity and overall longevity. Diseases linked to mutations in the elastin gene are discussed, as are animal models of these diseases.

Collagen and elastin function is a polymer where individual chains are cross-linked one to another via modified lysine residues. The enzyme responsible for initiating the cross-linking reaction is one or more members of the lysyl oxidase family. These copper-requiring enzymes catalyze the oxidative removal of lysine epsilon-amino groups to form a reactive aldehyde, the cross-link precursor. There are five known members of this amine oxidase family (lysyl oxidase and 4 lysyl oxidase-like enzymes), and in Chap. 9, Herbert Kagan and Faina Ryfkin provide a detailed analysis of the amino oxidase mechanism of lysyl oxidase and bring us up-to-date on the known functions of the individual family members. They also review evidence showing that LOX can function both as an anti-oncogenic agent as well as an enhancer of malignancy in selected cancerous conditions.

Fibulins are a family of proteins that share a common architectural signature, namely a series of epidermal growth factor (EGF)-like modules followed by a carboxy terminal fibulin-type module. Over the last few years, the biological role of the fibulins has become clearer as new members of the family were identified and knockout mice provided insight into fibulin function. In Chap. 10, Marion Cooley and Scott Argraves review the current understanding of structure–function relationships for the fibulins, particularly with regards to elastogenesis. They also discuss the role that fibulins play in diseases such as cancer, cardiovascular disease, and eye disease.

In the final chapter of the volume (Chap. 10), David Roberts and Lester Lau provide an extensive review of a class of extracellular matrix components referred to as "matricellular proteins." These proteins, in general, share a complex modular structure that enables them to interact with specific components of the matrix while engaging specific cell surface receptors through which they control cell behavior. Matricellular proteins, including the thrombospondins, some thrombospondin-repeat superfamily members, tenascins, SPARC, CCN proteins, and SIBLING proteins, are increasingly recognized to play important roles in inherited disorders, responses to injury and stress, and the pathogenesis of several chronic diseases of aging.

What Is Not Included and Plans for the Future

Trying to review the entirety of extracellular matrix in one volume is an impossible task. For this reason, I have chosen to focus this first volume on the major molecules that make up the ECM. Subsequent volumes that are either in production or in the planning stages include ECM turnover, glycoprotein biology, integrins and receptors for ECM, and volumes devoted to topics such as ECM in development and the role of ECM in specific diseases. It is hoped that this "overview" volume will be used as a basis of reference as we explore ECM function more deeply in subsequent publications.

St. Louis MO, USA   Robert P. Mecham

Contents

1 **An Overview of Extracellular Matrix Structure and Function** 1
Jürgen Engel and Matthias Chiquet

2 **Fibronectin and Other Adhesive Glycoproteins** 41
Jielin Xu and Deane Mosher

3 **Collagens, Suprastructures, and Collagen Fibril Assembly** 77
David E. Birk and Peter Brückner

4 **Basement Membranes** 117
Jeffrey H. Miner

5 **Hyaluronan and the Aggregating Proteoglycans** 147
Thomas N. Wight, Bryan P. Toole, and Vincent C. Hascall

6 **Small Leucine-Rich Proteoglycans** 197
Renato V. Iozzo, Silvia Goldoni, Agnes Berendsen,
and Marian F. Young

7 **Microfibrils and Fibrillin** 233
Dirk Hubmacher and Dieter P. Reinhardt

8 **Elastin** 267
Beth A. Kozel, Robert P. Mecham, and Joel Rosenbloom

9 **Lysyl Oxidase and Lysyl Oxidase-Like Enzymes** 303
Herbert M. Kagan and Faina Ryvkin

10	**The Fibulins**	337
	Marion A. Cooley and W. Scott Argraves	
11	**Matricellular Proteins**	369
	David D. Roberts and Lester F. Lau	

Index ... 415

Contributors

W. Scott Argraves Department of Regenerative Medicine and Cell Biology, Medical University of South Carolina, 173 Ashley Avenue, Charleston, SC 29425, USA, argraves@musc.edu

Agnes Berendsen Craniofacial and Skeletal Diseases Branch, NIDCR, National Institutes of Health, Bethesda, MD, USA

David E. Birk Department of Pathology and Cell Biology, University of South Florida, College of Medicine, 12901 Bruce B. Downs Boulevard, MDC11, Tampa, FL 33612-4799, USA, dbirk@health.usf.edu

Peter Brückner Institute for Physiological Chemistry and Pathobiochemistry, University Hospital of Münster, Waldeyerstr. 15, 48149 Münster, Germany, peter.bruckner@uni-muenster.de

Matthias Chiquet Department of Orthodontics and Dentofacial Orthopedics, University of Bern Freiburgstrasse 7, CH-3010, Bern, Switzerland, matthias.chiquet@zmk.unibe.ch

Marion A. Cooley Department of Regenerative Medicine and Cell Biology, Medical University of South Carolina, 173 Ashley Avenue, Charleston, SC 29425, USA

Jürgen Engel Department of Biophysical Chemistry, Biocenter, University of Basel, Klingelbergstrasse 50/70, CH-4056 Basel, Switzerland, Juergen.engel@unibas.ch

Silvia Goldoni Department of Pathology, Anatomy and Cell Biology and the Cancer Cell Biology and Signaling Program, Kimmel Cancer Center, Thomas Jefferson University, Room 249 JAH, 1020 Locust Street, Philadelphia, PA 19107, USA

Vincent C. Hascall Department of Biomedical Engineering, Cleveland Clinic Foundation, Cleveland, OH, USA

Dirk Hubmacher Department of Anatomy and Cell Biology, Faculty of Medicine, McGill University Montreal, Montreal, QC, Canada H3A 2B2

Renato V. Iozzo Department of Pathology, Anatomy and Cell Biology and the Cancer Cell Biology and Signaling Program, Kimmel Cancer Center, Thomas Jefferson University, Room 249 JAH, 1020 Locust Street, Philadelphia, PA 19107, USA, iozzo@mail.jci.tju.edu

Herbert M. Kagan Department of Biochemistry, Boston University School of Medicine, 715 Albany Street, Boston, MA 02118, USA, kagan@biochem.bumc.bu.edu

Beth A. Kozel Department of Pediatrics, Washington University School of Medicine, 660 South Euclid Avenue, St. Louis, MO 63110, USA

Robert P. Mecham Department of Cell Biology and Physiology, Washington University School of Medicine, 660 South Euclid Avenue, St. Louis, MO 63110, USA, bmecham@wustl.edu

Jeffrey H. Miner Renal Division, Washington University School of Medicine, Box 8126, 660 S. Euclid Avenue, St. Louis, MO 63110, USA, minerj@wustl.edu

Deane Mosher Department of Biomolecular Chemistry, University of Wisconsin, 4285 MSC, 1300 University Avenue, Madison, WI 53706, USA, dfm1@medicine.wisc.edu

Dieter P. Reinhardt Department of Anatomy and Cell Biology, Faculty of Medicine, McGill University Montreal, Montreal, QC, Canada H3A 2B2, dieter.reinhardt@mcgill.ca

Joel Rosenbloom Thomas Jefferson University, Philadelphia, PA, USA

Faina Ryvkin Emmanuel College, 400 The Fenway, Boston, MA 02115, USA, ryvkin@emmanuel.edu

Bryan P. Toole Department of Regenerative Medicine and Cell Biology, Medical University of South Carolina, Charleston, SC, USA

Thomas N. Wight The Hope Heart Matrix Biology Program, Benaroya Research Institute at Virginia Mason, 1201 Ninth Avenue, Seattle, WA, USA, twight@benaroyaresearch.org

Jielin Xu Department of Biomolecular Chemistry, University of Wisconsin, 4285 MSC, 1300 University Avenue, Madison, WI 53706, USA

Marian Young Craniofacial and Skeletal Diseases Branch, NIDCR, National Institutes of Health, Bethesda, MD, USA

Chapter 1
An Overview of Extracellular Matrix Structure and Function

Jürgen Engel and Matthias Chiquet

Abstract Extracellular matrix (ECM) not only provides a stable framework for maintaining the shape of multicellular organisms under physical load and during locomotion, but it is also essential for their morphogenesis and regenerative capacity. In this introductory chapter, we describe the basic features of the major classes of ECM components, namely, collagens, glycoproteins, and proteoglycans. We emphasize their multidomain structure that allows multiple functions to be combined in one, often large molecule. Of the many types of protein modules found in ECM components, some are devoted to multimeric assembly, and hence, for their crucial ability to form extended multimolecular networks or matrices. We argue that ECM components together with integrin receptors on the cell surface can be viewed as intricate nanodevices that allow cells to physically organize their 3D environment, as well as to sense and respond to various types of mechanical stress. In addition, ECM functions as part of a cell-controlled machinery to store and activate growth factors during development. We also make the point that metazoan evolution would not have been possible without the concomitant expansion of ECM complexity. Using examples of phylogenetically "old" versus "young" ECM protein families, we review the evidence that today's incredible diversity of ECM components arose from the recombination of preexisting protein modules by exon shuffling during evolution.

J. Engel
Department of Biophysical Chemistry, Biocenter, University of Basel, Klingelbergstrasse 50/70, CH-4056 Basel, Switzerland
e-mail: Juergen.engel@unibas.ch

M. Chiquet (✉)
Department of Orthodontics and Dentofacial Orthopedics, University of Bern Freiburgstrasse 7, CH-3010, Bern, Switzerland
e-mail: matthias.chiquet@zmk.unibe.ch

1.1 Introduction: No Metazoans Without Extracellular Matrix

> Were the various types of cells to lose their stickiness for one another and for the supporting extracellular white fibers, reticuli, etc., our bodies would at once disintegrate and flow off into the ground in a mixed stream of ectodermal, muscle, mesenchyme, endothelial, liver, pancreatic, and many other types of cells. (W. H. Lewis, 1922)

Extracellular matrix (ECM) is the term for the organic matter that is found between most cells in plants (Dhugga 2001) and animals (Hay 1991). ECM glues together the cells and organs of multicellular organisms. It is also essential for their morphogenesis, and later provides a stable framework for tissues that is required to maintain shape under gravity and other physical loads (Alberts et al. 2002). Without ECM, multicellular life could not exist, plants would not grow tall, and animals could neither swim, nor walk or fly. Given the importance of ECM for the integrity of complex organisms, it is not surprising that the evolution of its components is closely linked to the phylogeny of plants and animals themselves (Exposito et al. 2002; Hynes and Zhao 2000; Muller et al. 2004; Adair and Mecham 1990). Life on earth started about 4 billion years ago, and after at least 3 billion years of single cell life forms, multicellularity presumably evolved gradually from colony-forming unicellular organisms (Furusawa and Kaneko 2002; Muller et al. 2004). It is difficult to date the dawn of true multicellular organisms. An important step in evolution was certainly the emergence within the same cell colony of various types of specialized, mortal somatic cells next to the potentially immortal stem and germ line cells (Sanchez Alvarado and Kang 2005). Eventually, this division of labor resulted in higher order organisms whose isolated parts could no longer survive on their own. Multicellular individuals with their own identity evolved that were born, propagated, and died. In their body plan, some of the first multicellular species might have resembled today's most primitive green algae, e.g., Volvox globator (Hallmann 2003). Volvox is a small sphere of translucent ECM with identical, ciliated somatic cells on its surface, and with centrally located germ cells. The ECM of Volvox is well structured but very divergent from that of today's plants and animals, and the same might have been the case for the first multicellular species during evolution.

The plant and animal kingdoms have been separated about a billion years ago, and this event is likely to coincide with the birth of two types of "modern" ECM. The sessile algae, fungi, and plants acquired a comparatively uniform ECM in their cell walls (Dhugga 2001) that consists primarily of various long-chain polysaccharides (e.g., cellulose in plants) with few, albeit important, proteins (Brownlee 2002). This contrasts with the more sophisticated and diverse ECM of animals, which is the theme of this book. About 450 million years ago, in an evolutionarily short time called the "Cambrian explosion" (Couso 2008; Cummings 2006), essentially all of the animal phyla appeared on earth that live today, together with many more that have long vanished again and for which only fossil records are available. Evolution of metazoan animals, for which motility is a way of life, rapidly generated an incredible diversity of body plans. This was possible because entire gene families became devoted to intercellular adhesion and communication, as well as to the construction of very intricate ECM networks with diverse structure and function (Hynes and Zhao 2000; Whittaker et al. 2006). Obviously, the engagement of many genes allows for more complex regulation of ECM structure in metazoans than that found in modern plants. An elaborate gene network controls assembly and remodeling, as well as physical organization and functional properties of ECM in the various animal tissues. Although all metazoan ECM has highly preserved structural features and is made from the same classes of molecules (Exposito et al. 2002; Whittaker et al. 2006), namely, collagens, glycoproteins, and proteoglycans (see Sect. 1.2), it can have the material properties of a soft gel (e.g., in the vitreous body of our eyes), a polymer fiber rope (as in tendons and ligaments), or a rock-hard composite (e.g., in our bones) (Alberts et al. 2002). In terms of functions, animal ECM covers an entire spectrum, from maintaining body shape, to sustaining large mechanical stresses during motions, to acting as an instructive environment for the adhesion, growth, and differentiation of cells and organs (Adams and Watt 1993). By providing a mechanically stable yet permanently reconstructing framework, metazoan ECM is indispensable for both embryonic development and tissue remodeling in the adult.

1.2 Building Blocks of Extracellular Matrix

1.2.1 Basic Features of Extracellular Matrix Proteins

The ECM is formed by a large variety of proteins with different structures and functions but some common features are apparent. Many proteins of the ECM are very large. To their size contributes an often extensive glycosylation, which is on average 35 weight%, and in the case of proteoglycans the covalent attachment of glycosaminoglycan (GAG) chains. Molar masses of 100–1,000 kDa are frequent and even larger proteins are known. In general, ECM proteins are highly asymmetric in shape.

All ECM proteins are multidomain proteins, in which different or equal domains are arranged in a specific domain organization. Domains are defined as homologous units. The homology follows from amino acid sequence comparisons. In many cases, structures of domains are known at atomic resolution, which provides a more sensitive detection of structural homology. Individual domains may have distinct functions even after fragmentation from the intact protein. Even homologous domains may have sufficiently large sequential and structural differences to show rather different functions. The combination of different domains leads to a multifunctionality of essentially all ECM proteins. Commonly, several domains in a protein act in a concerted fashion. It was also observed that domains interact with each other in a given multidomain protein and form new functional entities in this way. The multifunctionality and the expanded shapes provide the potential for lateral interactions, favoring the formation of fibers and other supramolecular assemblies of ECM proteins.

ECM proteins are normally grouped as glycoproteins, proteoglycans, and collagens. Proteoglycans contain long, charged glycosaminoglycan chains covalently attached to serines or threonines of the core protein. Some GAG chains are also found unconnected to a protein, e.g., hyaluronan. Collagens are defined as glycoproteins or proteoglycans with one or more collagenous domains. The latter consist of segments with a repeating $(GXY)_n$ sequence. Three chains with such sequences combine to a collagen triple helix.

1.2.2 Domains in Extracellular Matrix Proteins

As mentioned earlier, domains are defined as homologous protein units. Homology means that the domains have a common precursor. This is a yes or no decision and the often used phrase of a percentage of homology is therefore meaningless. It should be replaced by percentage of identity.

It is often difficult to define a homologous group. Comparing sequences, the range of 25% sequence identity and below is a twilight zone, in which it is difficult to decide on the existence of homology (Doolittle 1992). A comparison of 3D-structures is a more sensitive way to detect homology. Definitions of domains are done with some ambiguity. In particular, differently defined domains may have a distant common origin and may belong to the same homology group. It is mentioned in section 1.4.1 that the creation of completely new folds was a rare event in evolution. Consequently, the number of homologous groups without a common origin should be small.

In spite of the mentioned ambiguities, the domain concept has a large practical value for grouping and comparing different ECM proteins. Peer Bork and Amos Bairoch, the inventors of the SwissProt database, pioneered the domain concept (reviewed in Bork et al. 1996). Today large lists of domains can be found in databases like PROSITE, SMART, CDART also called CDSEARCH, PFAM, and others. These databases are continuously updated. In the last issue of the SMART database, about 850 domains are listed, of which 250 are found in extracellular proteins. Numbers of domains are even larger in PFAM and CDSEARCH. The databases can be used to display the domain organization of a protein of known sequence (Adams and Engel 2007). It is also possible to list all proteins, which have similar organizations or which contain a given set of domains. In addition, the databases guide to the three-dimensional structures of domains, in case such structures were solved by crystallography or by NMR-spectroscopy. A list of important domains frequently found in ECM proteins is given in Table 1.1.

Most domains are of globular shape and have a defined size. For many of the domains, the three-dimensional structure at atomic resolution was elucidated by X-ray crystallography or NMR-spectroscopy. Domain structures are reviewed

1 An Overview of Extracellular Matrix Structure and Function

Table 1.1 Domains occuring in ECM Proteins

Letter code	Full name	Size (aa)	ECM proteins with homologues	Frequent function
CA	Cadherin domain	110–130	E-cadherin, N-cadherin, desmoglein, many other adhesion proteins	Homo association
C4	Collagen IV carboxy terminal domain	Trimer (3 × 110)	Collagen IV	Hexamer formation
CUB	CUB	About 110	BMP-1, Tolloid, neuropilin, many complement components	
EGF	Epidermal growth factor domain	About 50	Agrin, BMP-1, CASP, CMP, TSP-1 to -5, tenascins, many others	One of the most abundant domains with many functions
EF	EF-hand domain	12 Flanked by α-helices	BM-40/SPARC and very many cytosolic proteins	Binds Ca^{2+} and other divalent ions
KU	Kunitz inhibitor domain	About 60	α3 chain collagen VI, α1 chain collagen VII and many protease inhibitors	No protease inhibition in the collagens
F1	Fibronectin type 1 domain	About 40	Fibronectin and many coagulation factors	
F2	Fibronectin type 2 domain	About 60	Fibronectin and many coagulation factors	Involved in collagen binding
F3	Fibronectin type 3 domain	About 90	Fibronectin, tenascin, α1-chain of collagen I	RGD-loop binds to several integrins
FBG	Fibrinogen C-terminal domain	About 225	Tenascin, ficolin, angiopoitin, fibrinogen	
Ig	IG-like domain	70–100	NCAM, FGFR most abundant in IgGs and MHC	
LE	Laminin-type EGF-like domain	About 60	Laminin α-, β-, γ-chain, agrin, perlecan, unc-52, netrin	Specialized domains bind entactin/nidogen
LN	Laminin N-terminal domain		Laminin α-, β-, γ-chain, netrin	Involved in basement membrane assembly
LG	Laminin G-like domain	About 190	Laminin α-chain, agrin, neurexin, slit protein	In some cases binding to integrin α6β1
TB	TGF-beta binding domain		TGF-beta binding protein, fibrillin-1,-2,-3, follistatin	

(*continued*)

Table 1.1 (continued)

Letter code	Full name	Size (aa)	ECM proteins with homologues	Frequent function
TSP1	Thrombospondin type 1 domain	About 55	Thrombospondin-1,-2,-3,-4 ADAMTS, properdin	
TSP2	Synonymous to EGF			
TSP3	Thrombospondin type 3 domain	Composed of many EF-hand like repeats	Thrombospondins 1–5	Binding of Ca-ions
TSPC	C-terminal L-lectin-like thrombospondin domain		Highly conserved domain of all thrombospondins	
VWFA	Von Willebrand factor A domain, I-domain of integrin	About 190	Present in 22 ECM proteins including VWF, collagens VI, VII, XII, XIV, matrilin, integrin α-chain	Binding of specific sites of some collagens
VWFC	Von Willebrand factor C domain	About 70	VWF, thrombospondins-1 and -2, chordin	
VWFD	Von Willebrand factor D domain		VWF, BMP-binding regulator protein	

by Bork et al. (1996) and Hohenester and Engel (2002). Detailed information is contained in the RCSB Protein Data Bank.

Many types of folds can be distinguished and the three-dimensional structures reveal interesting details. For example, the VWFA domain has six α-helices connected by parallel β-strands. The C-terminal and N-terminal are very close, which directs the repeat of several domains in a polypeptide chain. The collagen-binding site was localized by cocrystallization of an inhibiting Fab–antibody fragment (Romijn et al. 2001). The F3 domain is a β-sheath protein of an immunoglobulin-like fold. Its N- and C-termini are at opposite ends of the molecule making it suitable for a linear arrangement of repeating domains. The RGD site of the cell-binding domain of fibronectin, which binds to integrin α5β1, is located in a flexible surface exposed loop. In general, binding sites are located at the surface of domains. Erroneous assignments of binding sites were frequently corrected by the elucidation of 3D-structures.

Three-dimensional structures were also obtained for fragments containing several globular domains. As an example, the structure of a fibronectin fragment FN1-FN2-FN2 revealed noncovalent interactions between the first and third domain (Pickford et al. 2001) demonstrating that the traditional "pearls on a string" representation of multidomain proteins is woefully inadequate. Interactions between domains are important for many recognition processes and may even force the polypeptide to fold back on itself.

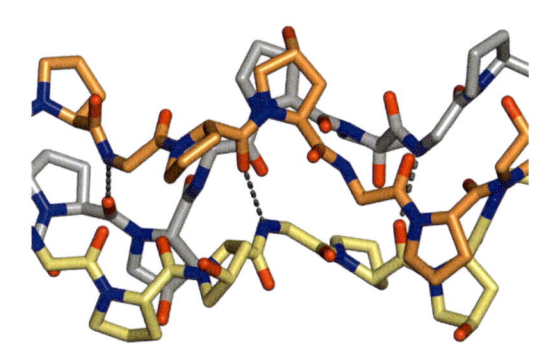

1 An Overview of Extracellular Matrix Structure and Function

Fig. 1.1 Structure of the collagen triple helix as demonstrated for the model peptide (Gly-Pro-Hyp). Each polypeptide chain forms a left-handed polyproline-II-type structure. These are not stable on their own but three chains (colored *orange*, *yellow*, and *gray*) wind to a right-handed superhelix. The entire structure is stabilized by hydrogen bonds between the NH groups of glycines and the CO groups of hydroxyproline in a neighboring chain. N and O are marked in *blue* and *red*, respectively. An important stabilization originates from the frozen N–C bonds in the imino acid residues, favoring the polyproline-II-helix and reducing the stability of the unfolded chains. In addition the OH of hydroxyproline has a stabilizing function, probably by inductive effects of its dipole on ring conformation. Note that all side chains [in case of (Gly-Pro-Hyp)$_n$ the proline rings] are pointing out of the triple helix and can therefore interact with neighboring molecules in collagen fibrils and other assembly forms. For details on structure and stabilization see Bächinger and Engel (2005)

In addition to the globular domains discussed so far, multidomain proteins also contain structures of a different type with an elongated rod-like shape. Examples are the collagen triple helix (Fig. 1.1) and the α-helical coiled-coil structure (Fig. 1.2).

The collagen triple helix is formed by three polypeptide chains with the repeating sequence (Gly-X-Y)$_n$, in which proline occurs frequently in the X-position and 4(R)hydroxyproline in the Y-position. Each chain forms a left-handed polyproline type II helix and the three helices intertwine to form a right-handed super helix. The translation per residue is about 0.29 nm. The triple helical structure is stabilized by hydrogen bonds between Gly on one chain and Pro in the X-position of a neighboring chain and by the sterical constrains of the proline rings. For the formation of a regular triple helix, it is essential that Gly residues should repeat in every third position. Only these residues fit into the center of the triple helix and any side chain larger than H would destabilize it (Fig. 1.1). 4-(R)hydroxyproline in the Y-position causes a strong additional stabilization, which originates from the inductive effect of the OH-group on ring puckering (Vitagliano et al. 2001). Inductive effects are well known in organic chemistry and cause electron withdrawals in the ring structure. Importantly, the chains in the collagen triple helix are staggered by one residue. This gives rise to two stereoisomers A-B-C and A-C-B in which the B chain is either staggered by one or by two residues against the A chain in the circular arrangement of the three chains. Many collagens have two or three different chains and in these cases, the surface of the triple helix is very different in the

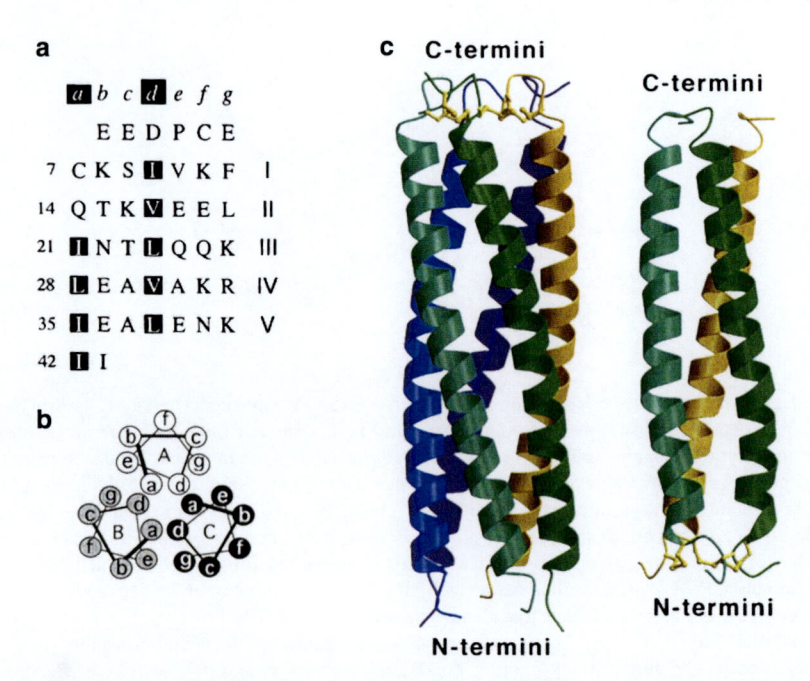

Fig. 1.2 Schematic representation of α-helical coiled coil structures. In (**a**) the sequence of the coiled-coil domain of matrilin-1 (also called cartilage matrix protein) is broken up in five successive heptad repeats *abcdefg* in which residues with hydrophobic side chains occur predominantly in the positions *a* and *d*. This sequence region also processes a high potential for α-helix formation. The α-helix has 3.6 residues per turn, which implies that residues in the positions *a* and *d* get very close to each other thus forming a hydrophobic boundary that runs almost parallel to the helix axis. This is shown in (**b**) in which single heptad repeats in α-helical conformation are shown in cross section; the three-stranded structure of matrilin-1 is depicted. The thickness of the lines connecting the residues indicates the position perpendicular to the paper plane. Thick lines are nearer to the viewer than thin lines. In all coiled-coil structures several α-helices are assembled and stabilized by interactions between the hydrophobic boundaries of residues in *a* and *d* positions. Electrostatic interaction between residues in *e* and *g* positions may also play a role. Depending on special features of the sequence two-stranded, three- and five-stranded coiled-coils are possible. In (**c**), the side views of the five-stranded coiled-coil domain of thrombospondin-5 (also called cartilage oligomeric matrix protein or COMP) and the three-stranded structure of matrilin-1 (cartilage matrix protein) are shown as ribbon diagrams. These structures are five to six heptads long but other coiled-coil domains can be shorter (e.g., in tenascins) or much longer (e.g., in laminins). Original data for matrilin-1, Dames et al. (1998) and for thrombospondin-5, Malashkevich et al. (1996)

stereoisomers. For most collagens, it is not known in which isomeric form they exist and whether two forms may coexist (Bächinger and Engel 2005).

Like the collagen helix, the α-helical coiled-coil structure is extended and its translation per residue is about 0.15 nm. It exists with different numbers of polypeptide chains. The most common coiled-coil structures in ECM contain three chains, but five-stranded structures are also known. For unknown reasons, the many cytosolic coiled-coil structures are only two-stranded. The coiled-coil structures are stabilized by hydrophobic interactions between repeated hydrophobic side chains. The most common repeat is the heptad repeat, *abcdefg*, in which residues in positions *a* and *d* carry hydrophobic side chains (Parry et al. 2008). For α-helices, these residues are located at a surface boundary, which runs almost parallel to the helix axis. The hydrophobic boundaries combine and form a three- or five-stranded structure (Fig. 1.2). Some deviations are found from the regular heptad repeat, which also lead to suitable interaction boundaries. In contrast to the collagen triple helix, the chains in the α-helical coiled-coil structure are not staggered.

The collagen triple helix and the α-helical coiled-coil structure occur as domains in many ECM proteins. In contrast to the globular domains, their size varies over a broad range. Collagen triple helices may be only 14 nm long in minicollagens of jellyfish and 800 nm long in annelid cuticle collagens (Engel 1997). The coiled-coil structures are also found in different lengths. In matrilins and thrombospondins, only three to five heptad repeats form short rods of 3–5 nm, whereas the long arm of laminin is about 75 nm long. In laminin, short sequence regions of a different type (Engel 1992) frequently interrupt the repeats. Interruptions of the regular GXY-repeats are also found in many collagens. They may serve to introduce flexible kinks. For collagen IV, such flexible kinks were demonstrated by electron microscopy (Hofmann et al. 1984). Uninterrupted collagen triple helices and coiled-coil structures possess a high stiffness, which give them the appearance of rods with only gradual bending in electron micrographs. The diameter of the collagen triple helix is about 1.2 nm, whereas the coiled-coil structure is somewhat thicker (see Figs. 1.1 and 1.2).

Membrane spanning domains occur only in a limited number of ECM proteins. They have been predicted from N-terminal hydrophobic sequences in collagen XIII, XVII, XXIII, and XXV. These collagens are classified as cell surface receptors. With their collagenous ectodomains, they are involved in cell attachment. Similarly, membrane spanning domains of the α- and β-subunits of integrins link cells to ECM components (Hynes 2002). The integrins are major receptors for many ECM proteins and will be introduced below (Sect. 1.3.3). Their ectodomains have a multidomain organization similar to that of ECM proteins and they are intimately connected to the ECM.

1.2.3 Do Homologous Domains Have Related Functions?

It is the hope of computational biologists that the function of proteins can be predicted from their sequence. A frequently discussed concept is the assignment of hypothetical functions to multidomain proteins with the assumption that homologous domains will have similar function (Friedberg et al. 2006). As a general principle, this assumption does not hold and only for rather basic functions like membrane spanning and oligomerization may the concept be successful.

A large number of experimental investigations show that homologous domains adopted a variety of functions. They should be looked at as related folds in which specific regions have been adopted for different functions. One of the many examples is the Kunitz inhibitor domain, which is found in the γ-chains of collagen VI. The active Kunitz inhibitor forms a strong complex with trypsin, α-chymotrypsin, or related proteases and its specificity depends on the structure of its active site. An extensive search for a related function of the Kunitz domain in collagen VI was not successful (Mayer et al. 1994). In addition, the three-dimensional structure of the collagen VI domain was solved, and it was realized that an inhibitor site is missing in the Kunitz domain of collagen VI and that this domain may be involved in a different function (Kohfeldt et al. 1996). This function is still ill defined, but the domain participates in the linkage of several collagen VI molecules during assembly of beaded filaments. The Ig domains in perlecan provide a second example. Only one of the Ig binds to nidogen/entactin, whereas the other Ig domains have no binding function and are classified as spacer elements (Kvansakul et al. 2001).

The issue is further complicated by the experimental observation that many domains exhibit their functions only when properly glycosylated or after another type of posttranslational modification. A frequently occurring modification is limited proteolytic cleavage by one of the many matrix proteinases. Again these changes do not depend on the type the domain alone but on its specific structure, tissue environment, and other factors.

Only a few and in part trivial functional predictions are possible on the basis of the amino acid sequence alone. It is possible to predict binding sites for bivalent cations in the EGFCa or EF-hand domains. It is also possible to predict the oligomerization potential of coiled-coil domains and collagen triple helices. More detailed predictions are possible with the help of the three-dimensional structure by which potential interaction sites can be explored. The most valid approach is still experimental investigation. Isolated domains of interest may be recombinantly expressed for functional studies. For ECM-domains, expression in mammalian cells is preferred over expression in *Escherichia coli* because of the need for proper disulfide linkage and glycosylation. Experiments teach us that single domains often show only incomplete functions. Elucidation of the full function requires a concerted action of many different domains, which are arranged in a machine-like, exactly defined spatial arrangement (Engel 2007).

1.2.4 Domain Organization

The polypeptide chains of ECM proteins often consist of a large number of individual domains. The linear representation of domains is called domain organization. The domain organization of the typical adhesive ECM glycoprotein fibronectin is shown in Fig. 1.3.

As mentioned earlier, the domain organization of any protein of known sequence may be obtained from databases like CDART, SMART, and SPAM. Clearly, only domains will be identified that are already entered in the database. Some of the sequence regions for which no domains are displayed may contain novel domains.

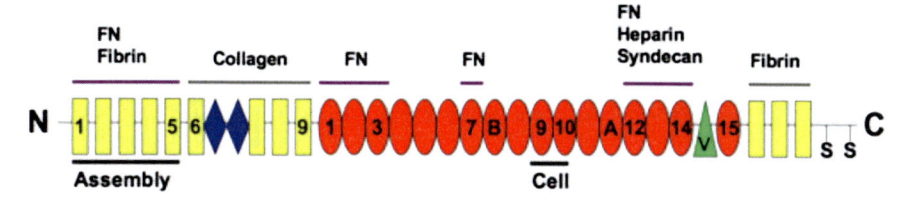

1 An Overview of Extracellular Matrix Structure and Function 11

Fig. 1.3 Domain organization and functional regions of fibronectin. Internally homologous domains are shown in a linear arrangement for one of the two subunits of fibronectin. A second chain (not shown) is connected by disulfide bonds (S–S) at the C-terminus. Rectangles stand for F1-, chevrons for F2-, ellipsoids for F3-domains, and a triangle for a variable domain. Regions with binding potential for other matrix components are indicated on top of the structure and other functional regions below the structure. The cell-binding F3-domain 10 contains an exposed Arg-Gly-Asp (RGD)-site to which the fibronectin-specific cellular receptor integrin α5β1 binds. For more information see: Hynes (1985); Mao and Schwarzbauer (2005); Potts and Campbell (1996)

The curators of the databases ask researchers for suggestions for potentially new domains.

In single molecule electron micrographs, many ECM proteins show an extended shape (Fig. 1.4). The observed structures can often be matched with the domain organization (see Fig. 1.5). By electron microscopy in combination with structural studies of individual domains, a representation of the total structure of a complex protein may be obtained (Engel 1994).

It should be noted that under the experimental conditions employed for electron microscopy, noncovalent interdomain interactions might be disrupted. Interactions between domains within the same protein have been frequently demonstrated and interactions between distant domains are also possible. Such interactions may lead to large changes in the global conformation of the protein. An important example is fibronectin, which exists in a condensed and extended form (Markovic et al. 1983). Only the latter is able to polymerize into fibronectin fibrils (Mao and Schwarzbauer 2005).

The domain organization of many ECM proteins may also vary due to the existence of splice variants. The variants contain novel domains and frequently a repertoire of proteins with slightly different domain organization is found. The expression of the splice variants is highly regulated by splicing factors and is frequently tissue specific. The functions of splice variants often differ significantly from those of the parent protein. An example is agrin for which only a specific splice variant interacts with the acetylcholine receptor (Gesemann et al. 1995).

1.2.5 Multimerization of Several Polypeptide Chains

The complexity of ECM proteins is further increased by the fact that several identical or different polypeptide chains associate into large oligomers. Examples of such proteins are shown in Fig. 1.5.

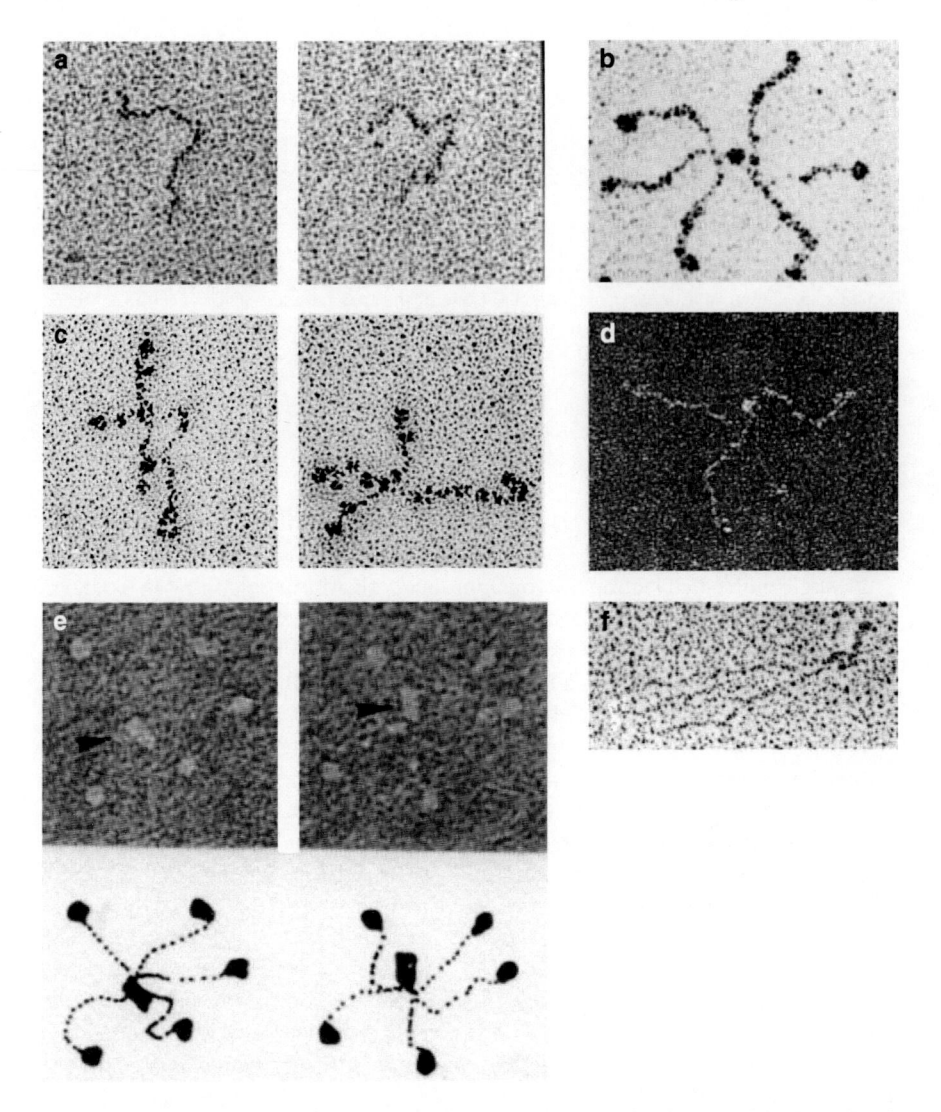

Fig. 1.4 Electron microscopic images of (**a**) fibronectin, (**b**) tenascin-C, (**c**) laminin-111, (**d**) collagen XII, (**e**) thrombospondin-5 (COMP), and (**f**) perlecan. Images were taken after rotary shadowing (**a–c, f**) or negative staining (**d, e**). For fibronectin (**a**), the two subunits of about 70 nm in length are connected with an average angle of 70°. Bending of the chains indicates a flexibility of the structural organization. Individual domains (see Fig. 1.3) cannot be distinguished at the resolution of the electron microscope. The six arms of tenascin-C (**b**) are about 75–90 nm long, depending on species and splice variant. Three arms are linked by a short coiled-coil domain, which is linked face to face to a second coiled-coil domain connecting the three other arms. This assembly structure appears as a dense cluster of platinum crystallites formed during rotary shadowing. Also the C-terminal FBG-domains can be recognized as globules, whereas other domains cannot be distinguished. Laminin-111 (**c**) has the shape of a Greek cross with three short arms of 35 nm and a long arm of 77 nm in length. Several of the domains in laminin (see Fig. 1.5) can be distinguished and the entire structure has a high degree of flexibility. Collagen XII

Association is mediated by oligomerization domains, of which the collagen triple helix and the α-helical coiled-coil domains are the most common. All collagen helix-containing proteins are trimers and all noncollagenous domains of these proteins also exist in three copies. By convention, these proteins are normally called collagens, although their content of triple helix may be small and other domains may predominate. For example, collagen XII (Fig. 1.5) consists mainly of VWFA and F3 domains and its triple helix is small (Koch et al. 1992). FACIT collagens contain several short triple helices. To facilitate proper chain association and to avoid slippage of chains, these and other collagens contain short coiled-coil regions (McAlinden et al. 2003) or other globular domains with a strong potential for trimerization (Boudko et al. 2009). Coiled-coil domains form trimers (see laminin in Fig. 1.5) and in some cases (thrombospondin 5, cartilage oligomeric matrix protein) pentamers. Tenascins are hexamers in which two three-stranded coiled-coil structures are arranged in an antiparallel way (Fig. 1.5). Several other globular domains also have a potential for oligomerization. An example is the POZ/BTB-domain in Mac-2 binding protein (Muller et al. 1999). This domain exists only as a dimer and is unstable in monomeric form. Domains with this property are called obligatory oligomerization domains. An obligatory trimer is the C-terminal NC1 domain of collagen IV (C4 in Table 2.1), which leads to a strong and specific antiparallel dimerization of this collagen (Khoshnoodi et al. 2008; Than et al. 2002). Fibronectin is normally found as a disulfide linked dimer and the disulfide-containing region at its C-terminus may also be classified as a dimerization domain.

1.2.6 Posttranslational Modifications

Domains of ECM proteins are extensively modified by a large number of chemical modifications. Some of these like the hydroxylation of proline to hydroxyproline occur in the interior of the cells; others like the proteolytic removal of the N-terminal and C-terminal domains of interstitial collagens occur at the cell surface. Many different limited proteolysis steps are mediated by matrix proteases

←

Fig. 1.4 (continued) (**d**) is a homotrimer. The domain organization of the subunits is shown in Fig. 1.5. In the electron micrograph, the noncollagenous domains appear as three flexible arms 90 nm in length. They are joined in the about 50 nm long collagen triple helix that is seen as a thin strand pointing to 5 pm in the picture. In thrombospondin-5 (COMP) (**e**), five about 40 nm long strands are connected by the pentameric coiled-coil domain (see Fig. 1.2). This domain appears as a rectangular body, which is marked by an *arrowhead* in the electron micrograph. Perlecan (**f**) is a low density proteoglycan whose three glycosaminoglycan (GAG)-chains are only faintly seen in the electron micrograph. The GAG chains are of variable length of up to 100 nm, and the core protein to which they are attached has a length of 30–50 nm. Original data for (**a**) and (**c**): Engel et al. (1981), (**b**): Spring et al. (1989), (**d**): Koch et al. (1992), (**e**): Morgelin et al. (1992), and (**f**): Paulsson et al. (1987)

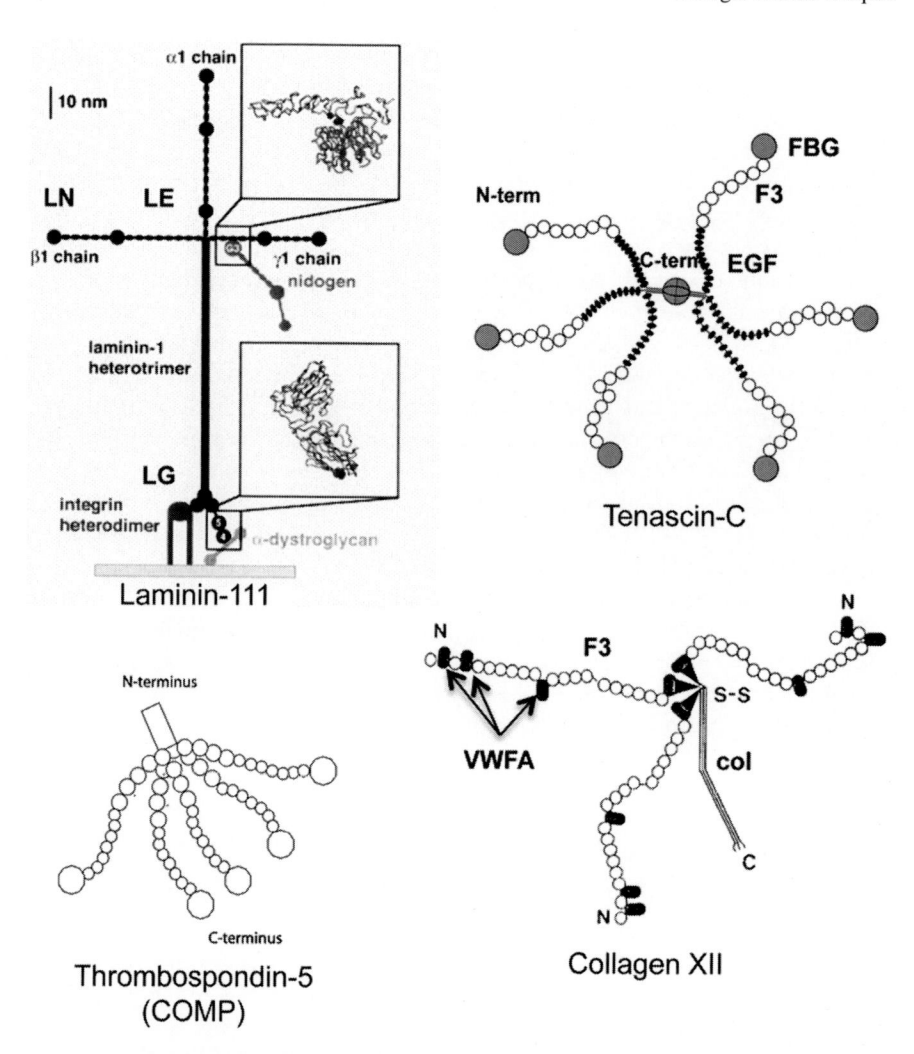

Fig. 1.5 Domain organization of multimeric ECM proteins laminin-111, tenascin-C, thrombospondin-5 (COMP), and collagen XII. Laminin-111 is a heterotrimer of an α1-, a β1-, and a γ1-chain, which are connected by a 77 nm long coiled-coil domain. At the C-terminus of the α-chain, 5 LG-domains are located, which are binding to integrin α6β1 and α-dystroglycan. The crystal structure of two LG domains was solved and is indicated in a *box*. The short arms are composed of many EGF-like (LE) domains and are terminated by LN domains involved in assembly. Two EGF-domains of the γ1-chain constitute the binding site for nidogen/entactin, which forms a very tight complex with laminin-111. The complex of the laminin-binding domain of nidogen with three EGF-domains was solved at atomic resolution and is shown in a box. Figure is taken from Sasaki et al. (2004).

Tenascin-C is a homo-hexamer whose 75 nm long subunits (for the avian 190 kDa splice variant) start with a FBG domain at the N-terminus (*gray circles*), followed by 8 F3 modules (*open circles*), and 11 EGF domains (*black ovals*). Three arms are joined by a short C-terminal three-stranded coiled-coil domain that is connected face to face to an identical trimeric assembly, giving rise to a six-armed ("hexabrachion") molecule with bilateral symmetry (Chiquet-Ehrismann et al. 1988; Spring et al. 1989).

in the extracellular space. All these events may lead to an activation of latent domains, large changes of function, and a degradation of ECM proteins. Proteolytic modifications are mediated by several hundred matrix proteases of the MMP, ADAM, ADAMTS, and other families (Tang and Hong 1999). Important for functional modifications and also for solubilization and structural stabilization are the N-glycosylation of asparagines and the O-glycosylation of serines and threonines. A special case is the attachment of long glycosaminoglycan (GAG) chains to the core protein of some ECM proteins. These proteins then exist as proteoglycans. Frequently two forms, one with and one without GAG chains are found. Furthermore, GAG chains are of variable length depending on age and tissue. All glycosylation occurs in the Golgi (exception: hyaluronan synthesis at the cell surface), and a large repertoire of enzymes is required.

It is not possible to deal with the large field of protein modifications in this chapter. However, it should be stressed that the building blocks of the ECM are dynamically remodeled by modifications and that large temporal and spatial variations exist. This leads to dramatic changes of functions and to a large increase of complexity. Examples are the degradation of collagens during pregnancy, the remodeling of bone, and the activation of thrombospondins. A static view on the matrix and its functions is therefore dangerous and functional prediction cannot be made without a detailed knowledge of the modification state.

1.2.7 Calcium Binding and Mineralization

Calcium and other divalent cations are essential for the regulation of many events inside cells. Here their concentration is in the nanomolar range and highly variable. Calcium-binding regulatory proteins such as calmodulin are changing their conformational state and function in response to the change in calcium concentration. In contrast, the calcium concentration is high in the ECM and rather uniform

---

Fig. 1.5 (continued) Thrombospondin-5, also called cartilage oligomeric matrix protein (COMP), has 5 about 35 nm long arms that are connected by a five-stranded coiled-coil domain (*box*, see Fig. 1.2). Each arm consists of 4 EGF domains (*medium size circles*) and 7 TSP3 domains (*small circles*) and is terminated by a TSPC domain (*large circle*). Crystallography showed that the TSPC forms a joint structure with the TSP3 domains (Kvansakul et al. 2004).
Collagen XII is composed of three identical subunits with very large noncollagenous "arms" about 70 nm in length, which are joined by a 50 nm long C-terminal collagen (col) triple helix with an interruption (visible as kink by EM; cf. Fig. 1.4). The N-terminal noncollagenous parts of the subunits consist of many F3 modules (*open circles*), interspersed with VWFA domains (*black ovals*), and a thrombospondin N-terminal like domain (TSPN; *black triangles*). The VWFA modules are recognizable in EM micrographs as globules protruding from the tandem F3 domains (cf. Fig. 1.4d). The scheme depicts the large variant of collagen XII (320 kDa per subunit), which is also a proteoglycan since it carries chondroitin sulfate chains at the N-terminus. Alternative splicing gives rise to a small variant (220 kDa) that lacks half of the noncollagenous domain and chondroitin sulfate (Koch et al. 1992; Koch et al. 1995)

because the matrix is not compartmentalized by membranes. Small changes may be caused by Donnan effects in the vicinity of charged groups of, for example, glycosaminoglycans. It is usually assumed that the concentration of free calcium ions is approximately equal to the concentration in blood plasma, which is about 2 mM. Many domains within ECM proteins are stabilized by calcium ions. Examples are the EGF-Ca domains of fibrillin, and the EF-hand domains of SPARC/osteonectin. The affinity for calcium of most of these binding domains in ECM proteins is high and they are therefore always saturated at the high free calcium concentrations in the matrix. The small changes of calcium concentrations can therefore not cause conformational changes and lead to regulation of protein function. Importantly, certain specialized components of the ECM may act as catalysts or inhibitors for calcification of tissues. Templates of the matrix nucleate crystal formation and promote mineralization of bone, teeth, and other calcified tissues. This complex field is only partially understood because it requires a concerted action of several proteins and domains in an exact spatial order.

1.3 Extracellular Matrix Assemblies as "Nanodevices"

1.3.1 Self-Organization of Extracellular Matrix

As a rule, ECM components do not function as isolated soluble molecules, but mainly in the context of macroscopic assemblies of complex composition. ECM molecules have an intrinsic property to self-associate and to form ordered assemblies with other ECM proteins. These properties arise from their modular structure in which domains of different binding potential are arranged in defined patterns. Heteromeric structures have been compared with alloys of metals, which exhibit new properties (see Chap. 3). Most interactions between ECM components are noncovalent, but nonetheless very stable due to multivalency and cooperative binding. In addition, specific enzymes can form irreversible covalent crosslinks. Examples are elastin and the various forms of assembly of collagens including fibrils, sheath-like structures, and beaded filaments (Chaps. 3 and 8). The laminins autonomously form extended networks in the presence of calcium ions, and interact with nidogen/entactin, α-dystroglycan, and other proteins. The assembly processes will finally lead to complex structures like the basement membranes, which consist of many proteins, including laminins and collagen IV (Chap. 4).

Important basic steps in the formation of supramolecular structures are self-assembly processes driven by interaction potentials between domains. Self-assembly is combined with the influence of cellular receptors and forces originating from the cellular surrounding. The assembly process is highly regulated by the influence of cells and by protein modifications. The resulting supramolecular structures exhibit a high degree of sophistication. Their biological functions are normally the result of a concerted action of many components, which cooperate in a network. The view that an isolated ECM-domain has a defined function is usually misleading. In reality, several domains of the same protein and of other proteins in the network act cooperatively. Examples are the network between elastin and fibrillin (Ramirez and Dietz 2009) (Chap. 8), the presentation and activation of bone morphogenic protein by domains in fibrillin (Sengle et al. 2008a), and the activation of integrins by cytosolic, transmembrane, and extracellular proteins (Coller and Shattil 2008; Hynes 2002).

The increasing power of structural and dynamic methods starts to reveal details of the mechanisms of interactions and conformational transitions in ECM-networks. At a nanoscale level, parts of the structure cooperate like the elements of a machine. Molecular machines are widely known in biology, and parts of the ECM might be regarded as such (Engel 2007).

1.3.2 Cellular Control of Extracellular Matrix Assembly

However, self-assembly is not the sole principle governing ECM organization. In many instances, ECM composition and structure are under tight control of the embedded cells (Kadler et al. (2008). Various types of connective tissue cells "spin" collagen fibrils that exhibit tissue-specific compositions, diameters, and orientations (Canty et al. 2006). Cells arrange collagen fibrils in plywood structures (e.g., in dermis), in ropes (e.g., in tendon) (Canty et al. 2006), or in extremely ordered and transparent layers (in the cornea) (Kao and Liu 2002). Although both laminin and collagen IV form autonomous but irregular 3D assemblies in vitro, epithelial cells mold these molecules into very refined suprastructures, i.e., sheet-like basement membranes of intertwined laminin and collagen IV meshworks that are connected by nidogen and linked to perlecan (Yurchenco et al. 2004). How cells determine precise topographic parameters of ECM assembly is not understood in detail. However, an important mechanism is that cells bind secreted ECM components on their surface and pull on them to arrange them in space. This is best known for collagen fibril rearrangement (Meshel et al. 2005), and for the assembly of fibronectin into a pericellular fibril meshwork (Mao and Schwarzbauer 2005). Single fibronectin molecules secreted by fibroblasts and other cells are retained on the cell surface by ECM receptors of the integrin family. The structure of integrins and their role in cell adhesion and signaling will be described in more detail in the next paragraph. Here, it suffices to say that cells can use integrins to "grab" ECM components in their surroundings. Internally, integrins are connected to the cytoskeleton. By applying actomyosin-generated traction force to the cytoplasmic tail of integrins, cells can pull on surface-bound fibronectin dimers, stretching them from a globule into an extended conformation. This favors the lateral assembly of fibronectin into a pericellular meshwork of fibrils (Cho and Mosher 2006; Mao and Schwarzbauer 2005). By a similar mechanism, cells use cytoskeletal force and integrins to pull on collagen fibrils to which they are attached, reorganizing them in space (Meshel et al. 2005) (Fig. 1.6). This is thought to be an important principle, e.g., in the morphogenesis of tendons and ligaments (Stopak and Harris 1982). A similar cell-mediated process occurs in the formation of elastic fibers (Czirok et al. 2006).

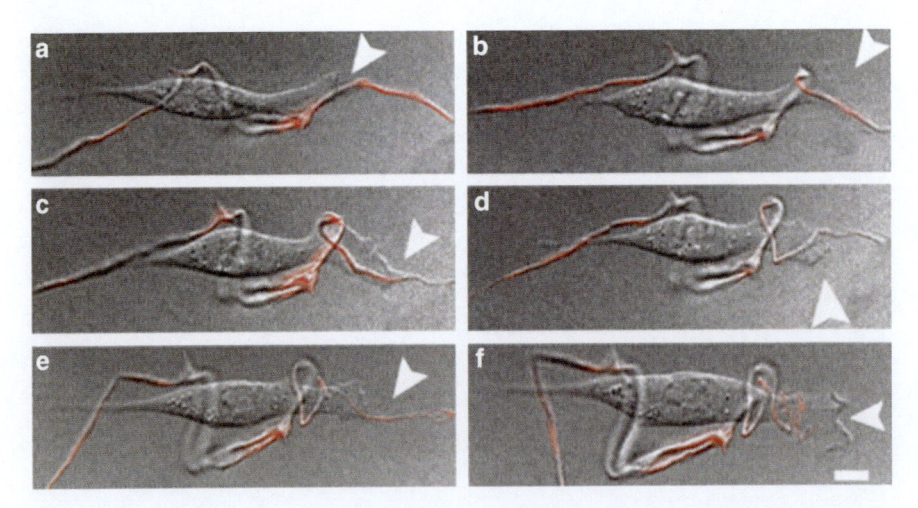

Fig. 1.6 Mechanical reorganization of fibrillar ECM by connective tissue cells. In this experiment, cultured fibroblasts were incubated with fluorescently labeled collagen fibrils, some of which happened to be deposited on their upper cell surface. The six frames from a time lapse movie show how a fibroblast grabs a single collagen fibril and pulls on it. By cycles of expansion and contraction of the lamellipodium at its front end (*arrowheads*), the cell moves and distorts the fibril that became attached to its upper surface via integrin receptors, in a way resembling the hand-over-hand reeling of a rope. By this mechanism, fibroblasts can reorganize and align three-dimensional ECM networks. Reprinted by permission from Macmillan Publishers Ltd. from Meshel et al. (2005)

1.3.3 *Integration of Cells with Extracellular Matrix*

As mentioned in the last paragraph, cells adhere to the surrounding ECM by means of specific membrane receptors. Of these, the integrins are the most abundant and essential, but alternative ECM receptors such as cell surface proteoglycans (e.g., syndecans) and glycoproteins (e.g., discoidin domain receptors) have important regulatory functions as well (Hohenester et al. 2006; Klass et al. 2000; Midwood et al. 2004; Leitinger and Hohenester 2007). This book mainly focuses on the ECM components themselves; the function of ECM cannot be understood without considering its interaction with cells, however. Therefore, the basic facts of integrin structure and function will be introduced here; the reader is referred to recent excellent reviews for further information (Arnaout et al. 2007; Hynes 2002; Takagi 2007). Integrins comprise a large family of heterodimeric cell surface proteins present abundantly on all cells (except erythrocytes). Each integrin subunit (called α and β) has a short cytoplasmic tail, a single transmembrane, and a large extracellular domain. Both extracellular domains of an integrin heteodimer together form the binding site for an ECM ligand (Arnaout et al. 2007). Ligand specificity is determined by the combination of a distinct α with a β subunit. In mammals, 18 α and 8 β chains combine to at least 24 different integrins. Many of these bind to short exposed peptide loops in their ECM ligands, such as to an Arg-Gly-Asp (RGD) sequence found in fibronectin (see Chap. 2) and many other ECM proteins. Other integrins (e.g., some of those binding to laminins and fibrillar collagens) recognize complex epitopes to which several ligand subunits contribute (Takagi 2007). Integrins physically connect the cell's interior cytoskeleton to the exterior ECM environment (Delon and Brown 2007). In conjunction with a multitude of adaptor and signaling components linked to their cytoplasmic tails, they form various types of matrix adhesion complexes (Zaidel-Bar et al. 2004), intricate "nanomachines" whose importance in animal biology cannot be overestimated.

Together with their intra- and extracellular ligands, integrins are responsible not only for cell adhesion to ECM, but also in general, for the transmission of forces across the cell membrane (Bershadsky et al. 2006; Katsumi et al. 2004). Hence, they are essential for processes such as cell and tissue movements during morphogenesis (Brakebusch et al. 1997) or muscle function in the adult (Mayer et al. 1997b). Conformational changes cause integrins to switch between inactive and active (ligand-binding) states (Hynes 2002) (Fig. 1.7). Up to now, for two (out of about 24) integrins, the atomic details of these changes have been demonstrated by X-ray crystallography, electron microscopy, and other structural methods (Xiao et al. 2004; Xiong et al. 2009; Xiong et al. 2001). The extracellular "heads" of both integrin chains have a multidomain structure with some modules that are familiar from ECM proteins, such as VWFA and EGF-like domains. The integrin heads are in a bent conformation close to the cell surface in the inactive state, and switch into an upright position upon activation, which also involves separation of the two cytoplasmic tails and binding of the adaptor protein talin. Activation, ligand binding, and subsequent signal transduction by integrins are triggered by signals from within the cell (Hynes 2002), but are also controlled by extracellular ligand binding (Ginsberg et al. 2005) and by the applied force associated with bound ligand (Kong et al. 2009). Hence, integrins are also at the center of mechanotransduction, i.e., of the process by which mechanical stimuli are converted into chemical signals within the cell (Katsumi et al. 2004). Moreover, integrin-containing matrix adhesions are assembly platforms for a multitude of signaling pathways that control cell growth, differentiation, and death (Zaidel-Bar et al. 2007). These adhesion structures mediate the synergy between growth factor- and integrin-dependent signaling, which is responsible for the anchorage-dependent growth of most normal cells (Frisch and Ruoslahti 1997; Reddig and Juliano 2005) (Fig. 1.8).

Clearly, the complex network of ECM components with their receptors and linked signaling pathways comprise a sophisticated machinery of many components operating on a nanometer scale. It has been pointed out by Hynes (2002) that not only are integrins nanomachines by themselves as outlined above, but they also form supramolecular complexes with their extracellular and cytosolic partners. These networks are most likely highly ordered and perform their distinct functions similar to the different components of a multienzyme complex, or like the various parts of a sophisticated man-designed machine.

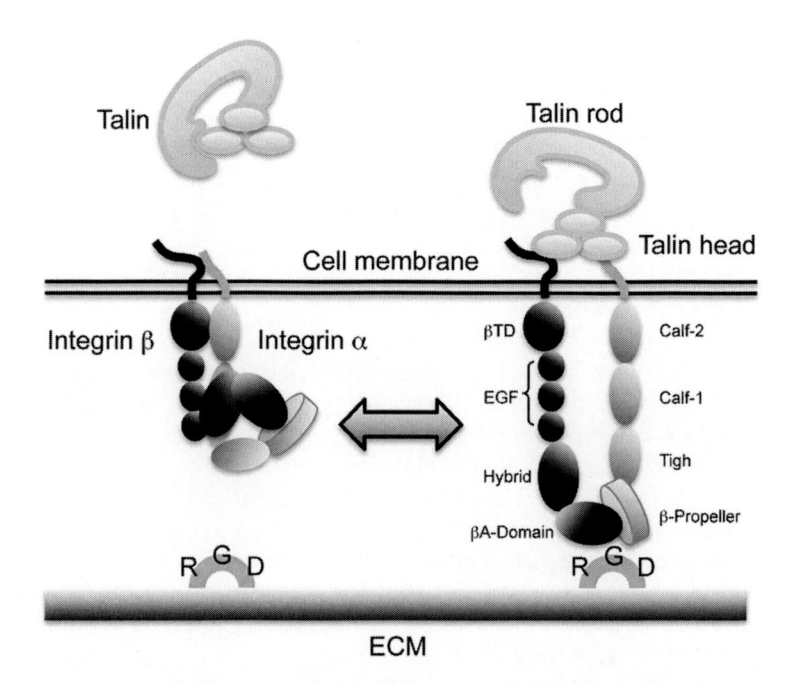

Fig. 1.7 Mechanism of integrin function. Integrins are heterodimeric cell membrane proteins; like their ECM ligands, they are built of multiple domains. The various modules of integrin α and β chains are indicated for the molecule at right. Many integrins are not constitutively active, but expressed on cell surfaces in an inactive state that neither binds extracellular ligand nor is linked intracellularly to the cytoskeleton (*left side* of scheme). Structural analysis of certain integrins (αvβ3 and αIIbβ3) revealed that in the inactive form, their extracellular domains assume a "bent" (globular) conformation that forces the transmembrane and cytoplasmic regions of the α and β chains into close apposition. Integrin activation is accompanied by extensive conformational changes that lead to the unclasping of the extracellular domains and to the separation of the cytoplasmic tails. In the extended (activated) form, integrins cluster and avidly bind to, e.g., an Arg-Gly-Asp (RGD) peptide motif of an extracellular matrix component, as well as to intracellular adaptor proteins such as talin that link them to the cytoskeleton (*right side* of scheme). Integrins can be activated in two ways. In adherent cells, inactive molecules can bind with low affinity to insoluble, multimeric ECM ligands and thereby become activated and clustered, which largely increases their avidity ("outside-in signaling"). In suspended cells such as leukocytes or blood platelets, signals from within the cell can release a self-inhibitory function of talin, which consequently binds to and separates the cytoplasmic tails of inactive integrins. The conformational change within the integrin tails is propagated through the cell membrane to the extracellular domains, which become activated and are now able to bind avidly to immobilized as well as soluble ECM ligands ("inside-out signaling"). Scheme modified from Saltel et al. (2009). For further details, see Ginsberg et al. (2005); Hynes (2002)

1.3.4 Presentation and Activation of Growth Factors via Extracellular Matrix

Twenty years ago pioneers in modern ECM research (Bernfield and Sanderson 1990) described the function of the ECM in development and tissue formation as follows:

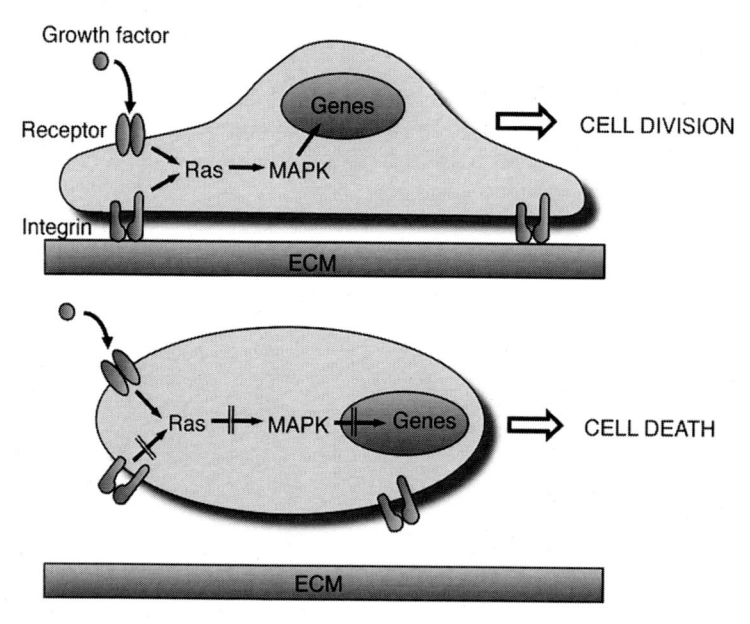

1 An Overview of Extracellular Matrix Structure and Function

Fig. 1.8 Cooperativity between growth factor receptor and integrin signaling, which is the basis for the phenomenon called "anchorage dependence of growth" of normal cells. Many growth factors (e.g., epidermal growth factor EGF or platelet derived growth factor PDGF) signal to mitogen-activated protein kinases (MAPK) via their receptor tyrosine kinases and the small GTPase Ras. In adherent cells, however, additional signals arise from ligand-bound (activated) integrins and converge on the Ras-MAPK pathway. The synergism between growth factor- and integrin-derived signals is essential for full activation of MAPK that supports the growth of normal adherent cells. If normal (but not cancer) cells lose contact with their ECM substrate, even excess growth factor is unable to stimulate MAPK signaling efficiently, and the cells will undergo apoptosis. For further reading, see Ruoslahti et al. (1992)

"Cellular behavior during development is dictated by the insoluble extracellular matrix and the soluble growth factors, the major molecules responsible for integrating cells into morphologically and functionally defined groups." Already at this time it was recognized that, contrary to earlier textbook knowledge, many growth factors are not freely diffusible through tissues, but are more or less immobilized in the ECM. This can occur through either direct interaction with specific ECM components, or indirectly via binding proteins. For proteoglycans such as syndecans (Bernfield and Sanderson 1990) and others (Kresse and Schonherr 2001), binding potential for several growth factors was demonstrated. Syndecans are coreceptors in fibroblast growth factor (FGF) signaling. In many cases, immobilized growth factors are present in ECM in a latent form and need to be activated for function. Activation often requires proteolytic processing, but recent evidence indicates that certain growth factors can also be activated by cellular traction force. Deposition of latent growth factors within the ECM is a mechanism by which cells can achieve precise spatial and temporal specificity in signaling to their neighbors.

The presentation of TGF-β and the bone morphogenic proteins (BMP-2 to BMP-10) by the ECM is of particular interest, because of the prominent role played by these growth factors in development and in the regulation of ECM expression (Butzow et al. 1993; Kahari et al. 1991a, b; Ruoslahti et al. 1992). The BMPs are expressed in many tissues and are essential for the embryonic development of skeletal structures, kidney, eye, and other organs. A binding protein for TGF-β was found by Butzow et al. (1993). Latent TGF-β was found in a disulfide-bridged complex with a cysteine rich peptide named latent TGF-β binding protein (LTBP) (Miyazono et al. 1992; Saharinen and Keski-Oja 2000). A domain family with homology within the Cys-rich peptide was defined and named TB (see Table 2.1). The latent form of TGF-β and BMPs is a complex of the growth factor dimer with its propeptide. The propeptide is called latency-associated peptide (LAP) as it confers latency either by blocking the binding to the receptor or by altering the conformation of the growth factor (De Crescenzo et al. 2001).

In the ternary complex of TGF-β, LAP, and LTBP (referred to as large latent complex), the peptide bond between TGF-β and its propeptide is already enzymatically cleaved but the latent growth factor is completely inactive until further activation. The TGF-β large latent complex and pro-BMPs are targeted to the ECM by binding to the microfibrillar system of fibrillin (Gregory et al. 2005; Sengle et al. 2008a). The growth factor binding site is located near the N-terminus of fibrillin-1. A model for growth factor activation, namely a competition of the prodomain (LAP) of BMP with the type-II-receptor of TGF-β has been developed (Sengle et al. 2008b). It was further shown that this mechanism applies to a number of TGF-β family members, namely BMP-2, -4, -7, and -10. This mechanism does not require enzymatic activation and therefore offers the possibility of reversible regulation by interfering factors. In addition, the attachment of the latent complex to the microfibrillar system implies a spatial control and a susceptibility to mechanical forces and reorientation. Not all details of the mechanism have been elucidated, but it is clear that the presentation, activation, and regulation of TGF-β family members are performed by a sophisticated nanomachine, which is composed of many components (Ramirez and Rifkin 2009).

It is of interest to add a side remark to this paragraph. Some confusion arose from the presence of TB domains in fibrillin (see Table 1.1). These domains are homologous but are not very closely related to the TB module of LTBP in the secreted complex with TGF-β. Nevertheless, it was believed that the TB domains of fibrillin are involved in the binding of the growth factor. The experimental data show that this is not the case. Referring to Sect. 1.2.3, this is an additional example for the incorrectness of deriving functions from sequence homology.

1.3.5 *Extracellular Matrix, Mechanical Stress, and Mechanotransduction*

As mentioned at the beginning, the evolution of large multicellular organisms would not have been possible without the development of ECM. Although gravitational forces acting on single cells are very small (in the nanonewton range), mechanical stresses can reach enormous values (megapascals) in the joints and tendons of a running animal, due to leverage action of the tissue masses involved (Magnusson et al. 2008). It is an important function of ECM to withstand these stresses, and to shield embedded cells from adverse effects (Guilak et al. 2006). Not surprisingly, specialized types of ECM have evolved that deal with distinct modes of mechanical stress, such as tension, compression, and shear. The parallel collagen bundles of tendons are perfectly adapted to bear tensile stress (Magnusson et al. 2008). They are physically linked to the contractile apparatus of muscle fibers via cell-ECM adhesions at the myotendinous junction (MTJ) (Fig.1.9), and thereby transmit muscle-generated forces to the skeleton. Cartilage consists of a collagen meshwork with encaged proteoglycans that act like a water cushion; it is best suited to bear compressive stress (Guilak et al. 2006). Bone is a "compromise" material in the sense that it is a stiff composite that can tolerate compression, tension, and shear without suffering much strain (Carpenter and Carter 2007). On the macroscopic scale, these various ECMs form large structures; on the micro- and nanometer scale, they are very anisotropic and adapted to local stresses, with a sophisticated nanoarchitecture that is not matched by any man-made material.

Bones, muscles, tendons, or the circulatory system all depend on mechanical stimulation for their homeostasis; decreased mechanical load leads to atrophy and overload to hypertrophy (Kjaer 2004; Wakatsuki et al. 2004). Cells adhering to ECM respond to mechanical stresses (or rather to the resulting strains, i.e., deformations) by changes in the expression of specific genes among them for ECM components themselves (Chiquet et al. 2009). This biological feedback leads to the adaptation of ECM to altered load patterns as, e.g., manifested by bone trabeculae that remodel according to a changed pattern of tension and compression lines (Wolff 1892). Modern finite element analysis of stresses and strains in loaded skeletal elements

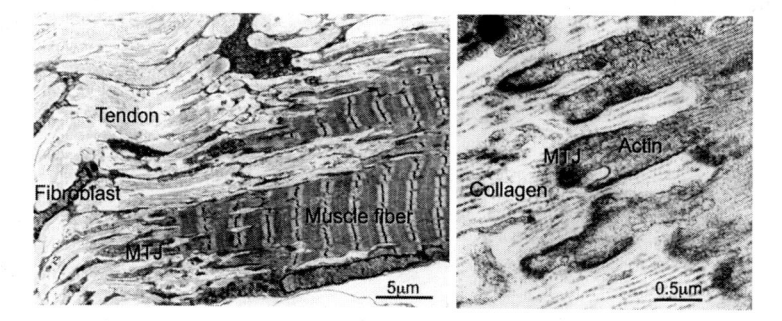

Fig. 1.9 The myotendinous junction (MTJ) as seen in the electron microscope. The cross-striated skeletal muscle fiber (at *right* of both images) inserts into the parallel collagen bundles of the tendon (*left side* of both images). The MTJ is characterized by invaginations of the muscle fiber sarcolemma. These are occupied by individual tendon collagen fibrils that are attached to the muscle basement membrane (*right*: enlarged image of MTJ). Note the colinear arrangement of intracellular muscle actin and extracellular tendon collagen fibrils. The sarcolemmal densities at the tips of muscle fiber processes are equivalent to focal adhesions of cultured cells and, like those, contain integrins and adaptor proteins that link the muscle cytoskeleton to basement membrane and tendon ECM. Rat soleus muscle; courtesy of V. Gaschen, University of Bern

can predict precisely areas of bone resorption (high compressive stress), bone formation (high tensile stress), and cartilage formation (high hydrostatic pressure), respectively (Carpenter and Carter 2007). This clearly shows that different modes of mechanical stress affect cell differentiation and shape skeletal structures. How do cells attached to ECM sense stresses and strains in their environment? As said earlier, cells adhere to the ECM via integrin-containing matrix adhesion sites on their surface; these structures link the ECM to the cytoskeleton both in cultured cells and in vivo (as exemplified by the MTJ, Fig. 1.9). Tissues with their cells embedded in ECM have been compared to "tensegrity" structures in which forces are propagated in all directions (Ingber 2006). Cell-matrix adhesions are well suited to transduce mechanical forces into chemical signals because of their strategic location at the cell surface (Chen 2008). Many intracellular signaling pathways are known to be triggered in an integrin-dependent way; some of them by pulling on integrins from the outside or the inside (Choquet et al. 1997; Riveline et al. 2001; Schmidt et al. 1998). Recent evidence suggests that certain integrin-associated components in matrix adhesion sites might act as "strain gauges." For example, the focal adhesion adaptor protein p130Cas appears to be physically stretched in response to applied force, thereby exposing hidden phosphorylation sites (Sawada et al. 2006). When modified by Src family kinases, these sites can trigger binding of SH2-domain containing signaling molecules and downstream responses (Tamada et al. 2004). This is one of the several examples of how mechanical cues could be converted into chemical signals intracellularly in an integrin-dependent fashion (Chen 2008; Engler et al. 2006). In addition, we have seen earlier that ECM proteins such as fibronectin can themselves undergo conformational changes upon mechanical stress, which influences their assembly (Mao and Schwarzbauer 2005). Since integrins on cell surfaces might interact differently with fibronectin in different conformational states, mechanical information contained in the ECM might be converted into distinct ligand binding (i.e., chemical information) already at the extracellular face of a cell membrane (Vogel and Sheetz 2006). In conclusion, the ECM is the major source of mechanical stimuli sensed by most cells, and a biological feedback loop ensures that the ECM is in turn constantly adapted to changes in load that act on the animal body. This feedback is controlled by cell-matrix adhesions, intricate "nanodevices" responsible for mechano-sensing and -transduction.

1.4 Evolution of Metazoan Extracellular Matrix

1.4.1 Diversity of Extracellular Matrix by Evolvement of New Multidomain Proteins

At the end of this chapter, we will return to the question mentioned at the very beginning, namely how the ECM coevolved together with the metazoan phyla, and what we can learn from its phylogeny to better understand the incredible diversity of modern animals and their adaptive capability to occupy every possible ecological niche. According to current estimates, the first multicellular metazoans appeared between 700 and 600 million years ago (Conway Morris 2006; Couso 2008). Depending on their reproduction time, from hundreds of million to several billion generations contributed to the phylogenetic diversity that we observe today. The first ECM building blocks must have evolved long before the appearance of metazoa, and may have been rather different from the present. They were continuously remodeled by single nucleotide mutations, and also by mechanisms on the gene level (see below), all under the influence of selection pressure where only useful changes survived. The probability to create a new protein ternary structure by random mutations was very low during early evolution, and only between 200 (Dorit and Gilbert 1991) and less than 1,000 (Han et al. 2007; Orengo and Thornton 2005) unique new structures evolved. However, many variations originated from the same ancestor fold, and several thousand protein domains are described in today's databases (although there is considerable redundancy). As mentioned earlier, sequence homology between domains does not necessarily indicate a shared function: Homologous domains often adopted different functions by modification of structure, binding sites, and enzymatic activity.

Although still debated, the "exon theory of genes" assumes that the most ancient and simple proteins consisted of just one or two stably folding protein modules, each encoded by a single small exon (Gilbert et al. 1997). To generate today's ECM and other multidomain proteins, it is thought that first a limited number of primordial genes encoding small proteins duplicated and fused to larger genes with a more complex exon–intron structure, and with tandem arrays of certain domains. By nonhomologous intronic recombination (exon shuffling), different types of domains were then swapped between genes. By these genetic mechanisms, certain proteins acquired novel domains whereas others lost some; this finally generated the many patchwork genes and proteins that we observe in modern species (Patthy 2003; Vogel et al. 2004). Exon (i.e., domain) shuffling between genes can only work if participating exons contain an integer number of codons, and if their splice sites are all in the same phase. For example, the majority of FN3 and other domains in different ECM proteins is encoded by a single exon with intron/exon junctions on both sides in phase 1, i.e., interrupting a codon after the first nucleotide (in contrast to modules in ancient proteins encoded by exons preferentially in phase zero) (Kaessmann et al. 2002). This means that complete FN3 modules can be duplicated or deleted within an ECM gene, or inserted into another gene having exons in phase without disturbing the open reading frame. These events must occur frequently in evolution because paralogs and orthologs of ECM proteins in different species often vary in their number of FN3 domains (Tucker et al. 2006). Obviously, this mechanism provided a means to rapidly duplicate a function within an ECM component, or to transfer it to another (even unrelated) protein. It is important to stress again, however, that no function can be predicted a priori from the presence of a specific domain in a protein (see Sect. 1.2.3). FN3 and other domains seem to be convenient "shuttles" for a plethora of different sites of interactions both with other ECM components and with cellular receptors. In addition, it is possible that many of these modules serve no other function than filling space as "bricks" of a large ECM framework, or as adaptors between two functional domains.

A further method of diversification for many ECM proteins is the occurrences of splice variants that exhibit slightly different expression patterns and functions (Astrof and Hynes 2009; Chiquet-Ehrismann et al. 1991; Koch et al. 1995). Often, differentially spliced domains are "young," i.e., have been duplicated recently in evolution (Tucker et al. 2006). It is conceivable that some of the older constant domains were originally spliced but became fixed because they lost their splice site. Thus, splicing might be another mechanism favoring multidomain protein evolution (Xing and Lee 2006). In addition, the glycosylation pattern on one and the same ECM protein can vary depending on the tissue, and can change during pathogenesis (Prabhakar et al. 2009; Vynios et al. 2008). Despite a large effort in recent years, the additional information hidden in the differential glycosylation of ECM proteins is still barely understood and appreciated.

1.4.2 Conserved Versus Variable Genetic Setups in Extracellular Matrix Components

In evolutionary terms, ECM components can be divided into two classes. One class comprises proteins for which orthologs are found in most of today's animal phyla; they are highly conserved in structure and function. The other class is evolutionarily younger, more variable in structure and phylum-specific. Already in parazoa, like sponges, cells are embedded in an ECM consisting mainly of fibrillar collagens not unlike those of higher animals (Exposito et al. 2008). With the evolution of gastrulation in eumetazoan animals, polarized epithelial cells became separated from mesenchymal (interstitial) cells by specialized, sheet-like ECMs called basement membranes. These structures are composed of a distinct set of highly conserved lattice-forming ECM components consisting of collagen type IV and laminins. Together with their cellular integrin receptors, fibrillar collagens and basement membrane components comprise the basic cell-ECM adhesion "tool kit" common to all eumetazoan species from hydra to man (Hynes and Zhao 2000; Zhang et al. 2007). ECM components belonging to the second class are phylum-specific expansions and usually fulfill specialized functions (Whittaker et al. 2006). Among the latter ECM components are, for example, the bone-specific proteins of vertebrates, or the cuticle collagens of nematode worms. The genes for the ancient conserved ECM components vary considerably in structure from those for the "younger," more variable components. In this context, a "young" ECM protein family is less than 450 million years old and evolved after the divergence of modern animal phyla in the so-called Cambrian explosion (Conway-Morris 2003). This difference is exemplified in the discussion that follows.

Many of the ancient "toolkit" components of the ECM are extremely well conserved both on the gene and the protein level, and one essential common feature is their capability to assemble into oligomers and higher order structures, such as fibrils or networks. The fibril-forming interstitial collagens found from sponges to vertebrates are one prominent example. When comparing the gene structure of sponge collagen with that of fibrillar vertebrate collagens (types I, III, XI, and most of all type V), striking similarities are found. For one, the ~300 nm long triple-helical region of all these collagens is encoded by many small exons of which most have a defined length of 45 or 54 nucleotides, which translate to either 5 or 6 Gly-X-Y triplets (Exposito et al. 2000; Yamada et al. 1980). Biophysical measurements show that this is about the minimal size for a collagenous peptide to assemble into a stable triple helix with two partners at ambient temperature (Persikov et al. 2005). It is therefore assumed that the primordial collagen gene of primitive animals encoded a single small oligomerization domain of about this size, which during evolution grew into larger triple helices by duplications of the first mini-gene, followed by exon shuffling and fusion (Yamada et al. 1980). According to the "exon theory of genes" (Gilbert 1987), this evolutionary process required ancient exons to be in phase zero, i.e., to start and end with a complete codon, and indeed, all exons of fibrillar collagen genes start with a codon for glycine and end with a codon for a Y-position. However, the evolution of a large fibril-forming collagen gene probably occurred long before the appearance of animals ~700 million years ago, because a full-size collagen must already have existed at the time when eumetazoa separated from parazoa. The reason becomes obvious from considering other similarities between sponge and vertebrate fibrillar collagen genes. In the region coding for the triple helix, not only the size of exons, but, amazingly enough, also their number and arrangement is perfectly conserved (Exposito et al. 2000). Moreover, fibrillar collagens from all animal species are practically identical in length, and they all possess C-terminal noncollagenous domains with significant homology and with a shared function in collagen helix assembly. Thus, the capability to form collagenous ECM networks presumably goes back to the dawn of the animal kingdom.

As stated above, basement membranes and their components are not found in sponges but are present in all eumetazoa, and the latter are again highly conserved in structure and function. For example, laminins in all eumetazoan species have three different subunits (α, β, γ) that are joined in a long rod-like domain, giving the individual molecules their typical cross- (or T-) like shape. Laminins assemble via a triple α-helical coiled-coil domain. In addition, laminin subunits exhibit extended N-terminal arms with EGF-like and laminin-specific modules (see Sect. 1.2), which are responsible for Ca^{2+}-dependent assembly of the protein into large networks. Moreover, the α-chains of all laminins possess a highly conserved, globular C-terminal "G" domain that contains binding sites for integrin receptors, thus mediating the contact with cell surfaces (Scheele et al. 2007). Laminin can be considered the ancient prototype of an oligomeric, modular, and multifunctional ECM glycoprotein since each of its subunits has separate domains for oligomerization, self-assembly into networks, binding to other ECM components, and interaction with cells. Laminins are essential substrates for all cell types that adhere to basement membranes (e.g., epithelial, endothelial, muscle, and fat cells), and they are well known for mediating the growth of axons during development and regeneration (Chiquet et al. 1988; Manthorpe et al. 1983). Thus, not only the overall domain structure, but also decisive functions are conserved in laminins of all modern species from jellyfish to *Drosophila* to man, indicating the ancient origin of this ECM molecule. However, a shared function per se does not necessarily mean that binding sites are identical between vertebrate and invertebrate laminins: Leech laminin allows the growth of leech but not mouse axons and vice versa, meaning that the mutual binding sites between laminins and their integrin partners must have coevolved within each animal phylum, but diverged between phyla during evolution (Chiquet et al. 1988). On the other hand, the binding site of laminin to its ECM partner nidogen seems to be conserved functionally as well as structurally from insects to mammals, since *Drosophila* laminin binds to mouse and human nidogen with high affinity (Mayer et al. 1997a).

After having discussed several of the ancient and highly conserved ECM components, we will turn to examples of matrix proteins that are more variable in structure, evolutionarily younger, and phylum-specific. Many of these components have specialized functions related to differences in body plan and physiology between different phyla. For example, worms and insects rely on an exoskeleton for maintaining body shape, and not surprisingly, their tough cuticle contains many ECM components that are not found in vertebrates. Although vertebrates possess around 40 collagen genes, more than 170 have been identified in the genome of *C. elegans* (Myllyharju and Kivirikko 2004). Conversely, in vertebrates two main features evolved in their body plan that required specialized ECM. One is the endoskeleton made of bones, cartilage, ligaments, and tendons. The other is the closed circulatory system with vessel walls that have to sustain high blood pressures, and with a blood clotting system providing a provisionary matrix in case of vessel damage. Consequently, certain blood clotting factors and components of resilient and elastic connective tissues are only found in vertebrates (Huxley-Jones et al. 2007; Wagenseil and Mecham, 2009).

Another typical example of a "young," more variable ECM component is fibronectin, only present in the chordate/vertebrate lineage. From textbooks and many publications, fibronectin is probably the most well-known adhesive ECM glycoprotein of vertebrates (Astrof and Hynes 2009). Fibronectin is a minor component of adult ECM but is abundant in blood plasma where it circulates as a soluble protein. In the case of injury, plasma fibronectin exits blood vessels and, due to its ability to self-aggregate and to interact with collagen and fibrin, anchors the blood clot to the interstitial ECM in wounds. Given its major function in blood clotting and wound healing, it is not surprising that fibronectin is found in all animals with a closed circulatory system, i.e., vertebrates. The only invertebrate for which a fibronectin-like gene has been identified is the sea squirt, Ciona, a primitive chordate and distant relative of vertebrates (Tucker and Chiquet-Ehrismann 2009). The Ciona protein has the same overall domain structure as vertebrate fibronectin, indicating that it indeed represents a distant ortholog of the vertebrate protein, and that there was a common precursor. However, notable differences suggest that fibronectin-like proteins diverged considerably after separation of the chordate and the vertebrate subphyla. Ciona fibronectin has only one type I (F1) domain at the N-terminus; vertebrates have two clusters of six and three. Since multiple F1 domains are required for fibrin binding, they might be an acquisition of vertebrates or might have been lost in chordates, which do not possess fibrinogen. Ciona fibronectin has no RGD sequence, and it is unknown whether it can function as an integrin ligand. Conversely, it includes modules not found in vertebrate fibronectins, namely three Ig domains interspersed with the FN3 repeats. Tandem arrays of Ig with F3 domains occur in several other protein families involved in cell adhesion, most notably Ig-CAMs (Hynes and Zhao 2000). Thus, Ciona "fibronectin" might more closely resemble an ancient precursor ECM protein than vertebrate fibronectins, but the opposite cannot be excluded.

Fibronectin presents just one of many examples of phylum-specific and compared to the ancient ECM components, less conserved and more variable proteins. There is only one fibronectin gene in higher vertebrates, compared to the several orthologs found for many other vertebrate ECM genes. This might indicate that the fibronectin gene was assembled from preexisting protein modules relatively late in evolution, perhaps after the one or two genome duplications that are presumed to have occurred early in the chordate/vertebrate lineage. Typical for most phylum-specific ECM components are certain types of protein modules that on the genome level have features of mobile elements. Especially, the fibronectin type III (F3) domain seems to be a module whose complete sequence within a gene is easily duplicated to large tandem arrays, as well as exchanged between related and unrelated genes (Kenny et al. 1999; Tucker et al. 2006). Like the Ig domain, the F3 module expanded and spread a great deal in the genomes of eumetazoa (Vogel et al. 2005). The intracellular giant muscle protein titin boosts over a hundred F3 repeats in tandem with Ig domains, and among the many ECM proteins containing such repeats, FACIT collagen XII and tenascin-X hold the record with 18 and 32 per subunit, respectively.

1.4.3 Phylogeny of Conserved Domains in Extracellular Matrix Protein Families

As explained in the last paragraph, the modular structure of ECM proteins facilitated rapid genetic changes during evolution, allowing an incredible variety of components to be generated hand-in-hand with the divergence of metazoan phyla with their different body plans. Thus, to understand the structural and functional complexity of ECM in various modern animals, it is informative to study the phylogeny of ECM components (or rather of their most conserved domains; see later) from simple metazoa to higher vertebrates. In simple animals, an ECM protein family usually consists of only one or two members, whereas in vertebrates with their complex tissue structure, the same family has expanded to four or more homologous proteins with distinct expression patterns and slightly different functions (Adams et al. 2003; Tucker et al. 2006). Such proteins are called "paralogs" to indicate that they diverged from a common ancestor before the separation of related species (in contrast, "ortholog" is the term for the closest relative of a protein in a different species). As a rule, most or all paralogs of an ECM protein family are found in all modern vertebrate species. Typical examples are the thrombospondin and the tenascin gene families, which both consist of large oligomeric ECM glycoproteins with largely (but not entirely) conserved modular structure. In the following, we will describe how the paralogs of these two ECM protein families might have evolved. We will see that it is not possible to construct meaningful phylogenetic trees for entire large ECM proteins [phylogenomics is a more reliable approach to identifying paralogous proteins between genomes (McKenzie et al. 2006)]. Rather, phylogenetic trees can only be obtained for the most conserved single domains (or a few adjacent domains) of a large protein. In some cases, phylogenetic trees constructed from separate domains of one ECM protein are consistent, in other cases they are not.

Thrombospondins. Thrombospondins (TSPs) are calcium-binding ECM glycoproteins with three to five large extended and identical subunits. They support and modulate cell attachment, regulate cell shape and migration, and are important for angiogenesis (see Chap. 11). In vertebrates, the TSP family has five members, of which TSP-1 and -2 are trimeric and the others pentameric. An α-helical coiled-coil domain adjacent to the globular N-terminal domain (except in TSP-5) mediates oligomerization of subunits. The homologous "arms" of the molecule consist of 3–4 EGF-like repeats, a Ca-binding TSP3 repeat composed of seven EF hand-like repeats, and a highly conserved TSPC domain related to L-type-lectins. TSP-1 and -2 possess an additional cassette consisting of a procollagen and three TSP1 domains related to properdin-like modules (Adams and Lawler 2004) (Table 2.1). By multiple alignment of sequences retrieved from available invertebrate genome databases using the CLUSTALW program, two thrombospondin-like genes could be identified in the urochordate *C. intestinalis*, and a single gene in *D. melanogaster* and two other insects; none was found in the nematode *C. elegans*, however (Adams et al. 2003). As stated above, it is not possible to perform multiple sequence alignments with entire large genes, but only with small conserved regions usually comprising one or a few adjacent protein modules. To identify the unknown genes unequivocally as invertebrate thrombospondins, other criteria had to be considered, the most important being a conserved domain structure. After putative novel members of the protein family had been identified in a BLAST search, their domain structure was explored by computer programs such as CDART or SMART. Using this approach, "bona fide" TSPs were selected from the many newly discovered invertebrate proteins with TSP domains. Only the N-terminal domain of the insect TSP1 is not found in vertebrate TSPs (Adams et al. 2003). As mentioned, phylogenetic trees of an ECM gene family again can be constructed by comparing small conserved regions separately. The result (obtained with computer programs such as PROTPARS or SATCHMO) are so-called "unrooted" trees where the length of the branches is proportional to the percent amino acid change between two nodes, without defining a single evolutionary origin (or root) of the tree (Adams and Engel 2007). To obtain a rooted phylogenetic tree, one has to assume that the number of amino acid exchanges between two related sequences is proportional to their phylogenetic distance, i.e., to time. However, it is well known that mutation rates can differ greatly between different proteins, and even between different domains of a single protein, thereby distorting the time axis (Thorne 2007). In the case of TSPs, sequence analysis of three distinct conserved gene regions produced consistent unrooted phylogenetic trees (Adams et al. 2003), suggesting synchronous evolution of the entire TSP genes. A rooted phylogenetic tree was generated that with some confidence could be calibrated in time against the vertebrate fossil record (Lawler et al. 1993). In the case of other ECM protein families, this can prove difficult or impossible, as we will see in the following section.

The phylogenetic analysis of the TSP gene family (Fig. 1.10) indicates that a single TSP gene of early metazoans was duplicated once early in chordate/vertebrate evolution, giving rise to two TSPs as found today in Ciona. Since Ciona TSP-A forms a clade with vertebrate TSP-1 and -2, whereas Ciona TSP-B groups with vertebrate TSP-3, -4, and -5 (Adams et al. 2003), it is assumed that a vertebrate-specific duplication of part or whole of the genome generated the additional TSPs. This notion is supported by the organization of vertebrate genomes: consecutive stretches containing many genes on one chromosome that are homologous to an array of related genes on a different chromosome (paralogous genomic regions) are taken as evidence for probably two genome duplications early in the chordate/vertebrate lineage. Indeed, all vertebrate TSP genes are found in paralogous regions

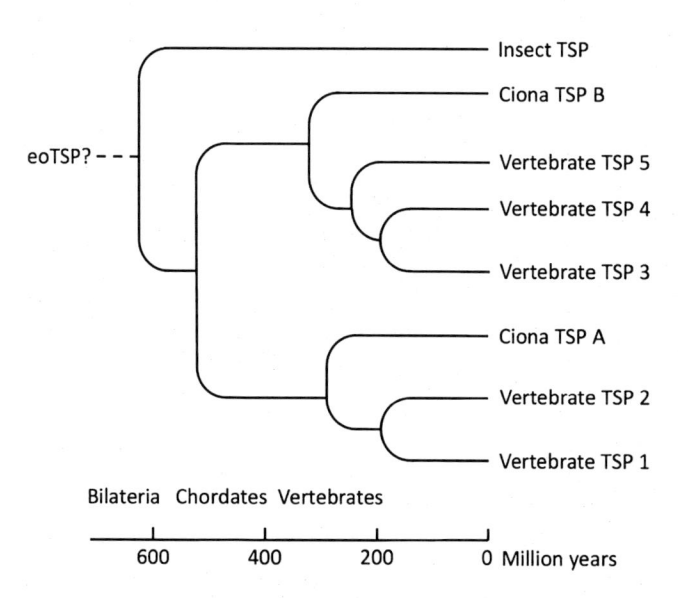

Fig. 1.10 Rooted phylogenetic tree of the thrombospondin gene family, with approximate time scale (simplified; after Lawler et al. 1993 and Adams et al. 2003). Invertebrates possess only one thrombospondin gene. A gene duplication early in the chordate lineage is thought to have generated clades A and B, which in the urochordate Ciona still consist of only one thrombospondin gene each. In vertebrates, one or two additional gene duplications are presumed to have generated the further members of each clade

of different chromosomes, confirming their ancient evolutionary origin (McKenzie et al. 2006).

Tenascins. Like the TSPs, tenascins (TNs) are typical examples of large, oligomeric ECM glycoproteins with conserved modular structure. Tenascins interact with many partners both in the ECM and on cell surfaces, and are known to modulate ECM assembly and cell adhesion (Chiquet-Ehrismann and Chiquet 2003). The subunits of all four vertebrate TN paralogs (TN-C, -R, -W, and -X) have an oligomerization domain at the C-terminus responsible for the formation of trimers or hexamers ("hexabrachions"), followed by a number of EGF repeats, tandem FN3 modules, and a fibrinogen-related FBG domain at the C-terminus (see Fig. 1.5). The conserved domain structure has been used to identify proteins in the entire chordate lineage that are undoubtedly tenascins. However, the number especially of the FN3 repeats in TNs can vary greatly even between orthologs from different vertebrate species. This has led to much confusion in the literature as to whether two TNs in two species are paralogs or orthologs (Tucker et al. 2006). However, the FBG domain is the part of the tenascins that is most conserved, and a meaningful phylogenetic tree of the gene family could be reconstructed from comparing just this small region in related genes. The close similarity between FBG domains provided one reason for concluding that avian "TN-Y" is in fact the ortholog of mammalian TN-X. Like for TSPs, another compelling argument for orthology was shared synteny: on their chromosomes, avian "TN-Y" and mammalian TN-X gene are both flanked by the same neighboring genes (Tucker et al. 2006). The phylogenetic tree of the tenascin gene family indicates that tenascins are an "invention" of the chordate/vertebrate lineage. The only invertebrates known to possess a (single) tenascin-C gene are the cephalochordate Amphioxus (lancelet) and the urochordate Ciona (sea squirt) (Tucker and Chiquet-Ehrismann 2009). All vertebrates have four paralogous tenascin genes; TN-W, -C, and -R group in one clade, whereas TN-X forms a separate one. Tenascin-C and -R are the most closely related paralogs and probably arose by a recent gene duplication. Thus, the phylogenetic tree of TNs clearly differs from that of TSPs with only one family member in primitive chordates and more recent duplications in vertebrates.

Although the phylogeny of tenascins based on similarity between FBG domains makes sense in terms of vertebrate evolution, analysis of the large arrays of FNIII repeats is all but straightforward. It seems that this central region of tenascin subunits has been (and probably still is) a playground for rapid genetic change. Interestingly, based on genomic organization there seem to be different types of FN3 repeats in tenascins. The most N- and C-terminal repeats are encoded by two exons each and are highly conserved between species. The central FNIII domains are all encoded by a single exon with class 1 intron–exon junctions, usually occur in tandem, and are homologous to each other (>50%, in some cases 90–100% identity), but divergent from the corresponding repeats even in orthologous tenascins. It is likely that this type of FN3 repeats arose by exon duplication long after the separation of species, in some cases only a few million years ago (Tucker et al. 2006). Remarkably, some of these domains show greatest homology to an unrelated protein such as fibronectin or collagen XII, rather than to a paralogous TN. One might speculate that they were imported from another protein and inserted into an existing TN by exon shuffling late in phylogeny, and independently from the evolution of the "constant" part of the molecule.

The remarkable hypervariability of tandem FN3 domains is a structural feature not only of tenascins, but also of other ECM proteins with such repeats, such as FACIT collagens and fibronectin itself [the fact that additional variability arises from alternative splicing of FN3 repeats (Astrof and Hynes 2009; Chiquet-Ehrismann and Chiquet 2003) has not even been discussed here]. All in all, these evolutionary mechanisms at the gene level have enabled an incredible diversity in the structure of ECM proteins even within the same family. The functional consequences of rapid structural changes in orthologous ECM proteins are not clear, however. In the case of TN subunits, the N- and C-terminal "business ends" are conserved whereas the extended middle part is variable. This affects the spacing between other functional domains, or might promote novel interactions. However, why should TN-X be twice as large in mammals than in birds, or why should human (and chick!) TN-C have a functional RGD motif but not the mouse ortholog? Similar puzzles exist in many other ECM protein families. Clearly, much still needs to be learned about the amazing structural, functional, and evolutionary adaptability of metazoan ECM components.

Acknowledgement We thank Dr. Josephine C. Adams for critical reading of the manuscript.

References

Adair WS, Mecham RP (1990) Organization and assembly of plant and animal extracellular matrix. Academic, San Diego

Adams JC, Engel J (2007) Bioinformatic analysis of adhesion proteins. Methods Mol Biol 370:147–172

Adams JC, Lawler J (2004) The thrombospondins. Int J Biochem Cell Biol 36:961–968

Adams JC, Watt FM (1993) Regulation of development and differentiation by the extracellular matrix. Development 117:1183–1198

Adams JC, Monk R, Taylor AL, Ozbek S, Fascetti N, Baumgartner S, Engel J (2003) Characterisation of *Drosophila* thrombospondin defines an early origin of pentameric thrombospondins. J Mol Biol 328:479–494

Alberts B, Johnson A, Lewis J, Raff M, Roberts K, Walter P (2002) Molecular biology of the cell, 4th edn. Garland Science, New York, Chapter 19

Arnaout MA, Goodman SL, Xiong JP (2007) Structure and mechanics of integrin-based cell adhesion. Curr Opin Cell Biol 19:495–507

Astrof S, Hynes RO (2009) Fibronectins in vascular morphogenesis. Angiogenesis 12:165–175

Bächinger HP, Engel J (eds) (2005) Protein folding handbook. Wiley-VHC, Hoboken, NJ

Bernfield M, Sanderson RD (1990) Syndecan, a developmentally regulated cell surface proteoglycan that binds extracellular matrix and growth factors. Philos Trans R Soc Lond B Biol Sci 327:171–186

Bershadsky A, Kozlov M, Geiger B (2006) Adhesion-mediated mechanosensitivity: a time to experiment, and a time to theorize. Curr Opin Cell Biol 18:472–481

Bork P, Downing AK, Kieffer B, Campbell ID (1996) Structure and distribution of modules in extracellular proteins. Q Rev Biophys 29:119–167

Boudko SP, Sasaki T, Engel J, Lerch TF, Nix J, Chapman MS, Bachinger HP (2009) Crystal structure of human collagen XVIII trimerization domain: a novel collagen trimerization fold. J Mol Biol 392:787–802

Brakebusch C, Hirsch E, Potocnik A, Fassler R (1997) Genetic analysis of beta1 integrin function: confirmed, new and revised roles for a crucial family of cell adhesion molecules. J Cell Sci 110 (Pt 23):2895–2904

Brownlee C (2002) Role of the extracellular matrix in cell-cell signalling: paracrine paradigms. Curr Opin Plant Biol 5:396–401

Butzow R, Fukushima D, Twardzik DR, Ruoslahti E (1993) A 60-kD protein mediates the binding of transforming growth factor-beta to cell surface and extracellular matrix proteoglycans. J Cell Biol 122:721–727

Canty EG, Starborg T, Lu Y, Humphries SM, Holmes DF, Meadows RS, Huffman A, O'Toole ET, Kadler KE (2006) Actin filaments are required for fibripositor-mediated collagen fibril alignment in tendon. J Biol Chem 281:38592–38598

Carpenter RD, Carter DR (2007) The mechanobiological effects of periosteal surface loads. Biomech Model Mechanobiol 7:227–242

Chen CS (2008) Mechanotransduction – a field pulling together? J Cell Sci 121:3285–3292

Chiquet M, Masuda-Nakagawa L, Beck K (1988) Attachment to an endogenous laminin-like protein initiates sprouting by leech neurons. J Cell Biol 107:1189–1198

Chiquet M, Gelman M, Lutz R, Maier S (2009) From mechanotransduction to extracellular matrix gene expression in fibroblasts. Biochim Biophys Acta 1793:911–920

Chiquet-Ehrismann R, Chiquet M (2003) Tenascins: regulation and putative functions during pathological stress. J Pathol 200:488–499

Chiquet-Ehrismann R, Kalla P, Pearson CA, Beck K, Chiquet M (1988) Tenascin interferes with fibronectin action. Cell 53:383–390

Chiquet-Ehrismann R, Matsuoka Y, Hofer U, Spring J, Bernasconi C, Chiquet M (1991) Tenascin variants: differential binding to fibronectin and distinct distribution in cell cultures and tissues. Cell Regul 2:927–938

Cho J, Mosher DF (2006) Role of fibronectin assembly in platelet thrombus formation. J Thromb Haemost 4:1461–1469

Choquet D, Felsenfeld DP, Sheetz MP (1997) Extracellular matrix rigidity causes strengthening of integrin-cytoskeleton linkages. Cell 88:39–48

Coller BS, Shattil SJ (2008) The GPIIb/IIIa (integrin alphaIIbbeta3) odyssey: a technology-driven saga of a receptor with twists, turns, and even a bend. Blood 112:3011–3025

Conway Morris S (2006) Darwin's dilemma: the realities of the Cambrian 'explosion'. Philos Trans R Soc Lond B Biol Sci 361:1069–1083

Conway-Morris S (2003) The Cambrian "explosion" of metazoans and molecular biology: would Darwin be satisfied? Int J Dev Biol 47:505–515

Couso JP (2008) Segmentation, metamerism and the Cambrian explosion. Int J Dev Biol 53:1305–1316

Cummings FW (2006) On the origin of pattern and form in early Metazoans. Int J Dev Biol 50:193–208

Czirok A, Zach J, Kozel BA, Mecham RP, Davis EC, Rongish BJ (2006) Elastic fiber macro-assembly is a hierarchial, cell motion-mediated process. J Cell Physiol 207:97–106

Dames SA, Kammerer RA, Wiltscheck R, Engel J, Alexandrescu AT (1998) NMR structure of a parallel homotrimeric coiled coil. Nat Struct Biol 5:687–691

De Crescenzo G, Grothe S, Zwaagstra J, Tsang M, O'Connor-McCourt MD (2001) Real-time monitoring of the interactions of transforming growth factor-beta (TGF-beta) isoforms with latency-associated protein and the ectodomains of the TGF-beta type II and III receptors reveals different kinetic models and stoichiometries of binding. J Biol Chem 276:29632–29643

Delon I, Brown NH (2007) Integrins and the actin cytoskeleton. Curr Opin Cell Biol 19:43–50

Dhugga KS (2001) Building the wall: genes and enzyme complexes for polysaccharide synthases. Curr Opin Plant Biol 4:488–493

Doolittle RF (1992) Stein and Moore Award address. Reconstructing history with amino acid sequences. Protein Sci 1:191–200

Dorit RL, Gilbert W (1991) The limited universe of exons. Curr Opin Genet Dev 1:464–469

Engel J (1992) Laminins and other strange proteins. Biochemistry 31:10643–10651

Engel J (1994) Electron microscopy of extracellular matrix components. Methods Enzymol 245:469–488

Engel J (1997) Versatile collagens in invertebrates. Science 277:1785–1786

Engel J (2007) Visions for novel biophysical elucidations of extracellular matrix networks. Int J Biochem Cell Biol 39:311–318

Engel J, Odermatt E, Engel A, Madri JA, Furthmayr H, Rohde H, Timpl R (1981) Shapes, domain organizations and flexibility of laminin and fibronectin, two multifunctional proteins of the extracellular matrix. J Mol Biol 150:97–120

Engler AJ, Sen S, Sweeney HL, Discher DE (2006) Matrix elasticity directs stem cell lineage specification. Cell 126:677–689

Exposito J, Cluzel C, Lethias C, Garrone R (2000) Tracing the evolution of vertebrate fibrillar collagens from an ancestral alpha chain. Matrix Biol 19:275–279

Exposito JY, Cluzel C, Garrone R, Lethias C (2002) Evolution of collagens. Anat Rec 268:302–316

Exposito JY, Larroux C, Cluzel C, Valcourt U, Lethias C, Degnan BM (2008) Demosponge and sea anemone fibrillar collagen diversity reveals the early emergence of A/C clades and the maintenance of the modular structure of type V/XI collagens from sponge to human. J Biol Chem 283:28226–28235

Friedberg I, Jambon M, Godzik A (2006) New avenues in protein function prediction. Protein Sci 15:1527–1529

Frisch SM, Ruoslahti E (1997) Integrins and anoikis. Curr Opin Cell Biol 9:701–706

Furusawa C, Kaneko K (2002) Origin of multicellular organisms as an inevitable consequence of dynamical systems. Anat Rec 268:327–342

Gesemann M, Denzer AJ, Ruegg MA (1995) Acetylcholine receptor-aggregating activity of agrin isoforms and mapping of the active site. J Cell Biol 128:625–636

Gilbert W (1987) The exon theory of genes. Cold Spring Harb Symp Quant Biol 52:901–905

Gilbert W, de Souza SJ, Long M (1997) Origin of genes. Proc Natl Acad Sci USA 94:7698–7703

Ginsberg MH, Partridge A, Shattil SJ (2005) Integrin regulation. Curr Opin Cell Biol 17:509–516

Gregory KE, Ono RN, Charbonneau NL, Kuo CL, Keene DR, Bachinger HP, Sakai LY (2005) The prodomain of BMP-7 targets the BMP-7 complex to the extracellular matrix. J Biol Chem 280:27970–27980

Guilak F, Alexopoulos LG, Upton ML, Youn I, Choi JB, Cao L, Setton LA, Haider MA (2006) The pericellular matrix as a transducer of biomechanical and biochemical signals in articular cartilage. Ann NY Acad Sci 1068:498–512

Hallmann A (2003) Extracellular matrix and sex-inducing pheromone in Volvox. Int Rev Cytol 227:131–182

Han JH, Batey S, Nickson AA, Teichmann SA, Clarke J (2007) The folding and evolution of multidomain proteins. Nat Rev Mol Cell Biol 8:319–330

Hay ED (ed) (1991) Cell biology of extracellular matrix, 2nd edn. Plenum, New York and London

Hofmann H, Voss T, Kuhn K, Engel J (1984) Localization of flexible sites in thread-like molecules from electron micrographs. Comparison of interstitial, basement membrane and intima collagens. J Mol Biol 172:325–343

Hohenester E, Engel J (2002) Domain structure and organisation in extracellular matrix proteins. Matrix Biol 21:115–128

Hohenester E, Hussain S, Howitt JA (2006) Interaction of the guidance molecule slit with cellular receptors. Biochem Soc Trans 34:418–421

Huxley-Jones J, Robertson DL, Boot-Handford RP (2007) On the origins of the extracellular matrix in vertebrates. Matrix Biol 26:2–11

Hynes R (1985) Molecular biology of fibronectin. Annu Rev Cell Biol 1:67–90

Hynes RO (2002) Integrins: bidirectional, allosteric signaling machines. Cell 110:673–687
Hynes RO, Zhao Q (2000) The evolution of cell adhesion. J Cell Biol 150:89–96
Ingber DE (2006) Mechanical control of tissue morphogenesis during embryological development. Int J Dev Biol 50:255–266
Kadler KE, Hill A, Canty-Laird EG (2008) Collagen fibrillogenesis: fibronectin, integrins, and minor collagens as organizers and nucleators. Curr Opin Cell Biol 20:495–501
Kaessmann H, Zollner S, Nekrutenko A, Li WH (2002) Signatures of domain shuffling in the human genome. Genome Res 12:1642–1650
Kahari VM, Larjava H, Uitto J (1991a) Differential regulation of extracellular matrix proteoglycan (PG) gene expression. Transforming growth factor-beta 1 up-regulates biglycan (PGI), and versican (large fibroblast PG) but down-regulates decorin (PGII) mRNA levels in human fibroblasts in culture. J Biol Chem 266:10608–10615
Kahari VM, Peltonen J, Chen YQ, Uitto J (1991b) Differential modulation of basement membrane gene expression in human fibrosarcoma HT-1080 cells by transforming growth factor-beta 1. Enhanced type IV collagen and fibronectin gene expression correlates with altered culture phenotype of the cells. Lab Invest 64:807–818
Kao WW, Liu CY (2002) Roles of lumican and keratocan on corneal transparency. Glycoconj J 19:275–285
Katsumi A, Orr AW, Tzima E, Schwartz MA (2004) Integrins in mechanotransduction. J Biol Chem 279:12001–12004
Kenny PA, Liston EM, Higgins DG (1999) Molecular evolution of immunoglobulin and fibronectin domains in titin and related muscle proteins. Gene 232:11–23
Khoshnoodi J, Pedchenko V, Hudson BG (2008) Mammalian collagen IV. Microsc Res Tech 71:357–370
Kjaer M (2004) Role of extracellular matrix in adaptation of tendon and skeletal muscle to mechanical loading. Physiol Rev 84:649–698
Klass CM, Couchman JR, Woods A (2000) Control of extracellular matrix assembly by syndecan-2 proteoglycan. J Cell Sci 113(Pt 3):493–506
Koch M, Bernasconi C, Chiquet M (1992) A major oligomeric fibroblast proteoglycan identified as a novel large form of type-XII collagen. Eur J Biochem 207:847–856
Koch M, Bohrmann B, Matthison M, Hagios C, Trueb B, Chiquet M (1995) Large and small splice variants of collagen XII: differential expression and ligand binding. J Cell Biol 130:1005–1014
Kohfeldt E, Gohring W, Mayer U, Zweckstetter M, Holak TA, Chu ML, Timpl R (1996) Conversion of the Kunitz-type module of collagen VI into a highly active trypsin inhibitor by site-directed mutagenesis. Eur J Biochem 238:333–340
Kong F, Garcia AJ, Mould AP, Humphries MJ, Zhu C (2009) Demonstration of catch bonds between an integrin and its ligand. J Cell Biol 185:1275–1284
Kresse H, Schonherr E (2001) Proteoglycans of the extracellular matrix and growth control. J Cell Physiol 189:266–274
Kvansakul M, Hopf M, Ries A, Timpl R, Hohenester E (2001) Structural basis for the high-affinity interaction of nidogen-1 with immunoglobulin-like domain 3 of perlecan. EMBO J 20:5342–5346
Kvansakul M, Adams JC, Hohenester E (2004) Structure of a thrombospondin C-terminal fragment reveals a novel calcium core in the type 3 repeats. EMBO J 23:1223–1233
Lawler J, Duquette M, Urry L, McHenry K, Smith TF (1993) The evolution of the thrombospondin gene family. J Mol Evol 36:509–516
Leitinger B, Hohenester E (2007) Mammalian collagen receptors. Matrix Biol 26:146–155
Magnusson SP, Narici MV, Maganaris CN, Kjaer M (2008) Human tendon behaviour and adaptation, in vivo. J Physiol 586:71–81
Malashkevich VN, Kammerer RA, Efimov VP, Schulthess T, Engel J (1996) The crystal structure of a five-stranded coiled coil in COMP: a prototype ion channel? Science 274:761–765
Manthorpe M, Engvall E, Ruoslahti E, Longo FM, Davis GE, Varon S (1983) Laminin promotes neuritic regeneration from cultured peripheral and central neurons. J Cell Biol 97:1882–1890

Mao Y, Schwarzbauer JE (2005) Fibronectin fibrillogenesis, a cell-mediated matrix assembly process. Matrix Biol 24:389–399

Markovic Z, Lustig A, Engel J, Richter H, Hormann H (1983) Shape and stability of fibronectin in solutions of different pH and ionic strength. Hoppe Seylers Z Physiol Chem 364:1795–1804

Mayer U, Poschl E, Nischt R, Specks U, Pan TC, Chu ML, Timpl R (1994) Recombinant expression and properties of the Kunitz-type protease-inhibitor module from human type VI collagen alpha 3(VI) chain. Eur J Biochem 225:573–580

Mayer U, Mann K, Fessler LI, Fessler JH, Timpl R (1997a) *Drosophila* laminin binds to mammalian nidogen and to heparan sulfate proteoglycan. Eur J Biochem 245:745–750

Mayer U, Saher G, Fassler R, Bornemann A, Echtermeyer F, von der Mark H, Miosge N, Poschl E, von der Mark K (1997b) Absence of integrin alpha 7 causes a novel form of muscular dystrophy. Nat Genet 17:318–323

McAlinden A, Smith TA, Sandell LJ, Ficheux D, Parry DA, Hulmes DJ (2003) Alpha-helical coiled-coil oligomerization domains are almost ubiquitous in the collagen superfamily. J Biol Chem 278:42200–42207

McKenzie P, Chadalavada SC, Bohrer J, Adams JC (2006) Phylogenomic analysis of vertebrate thrombospondins reveals fish-specific paralogues, ancestral gene relationships and a tetrapod innovation. BMC Evol Biol 6:33

Meshel AS, Wei Q, Adelstein RS, Sheetz MP (2005) Basic mechanism of three-dimensional collagen fibre transport by fibroblasts. Nat Cell Biol 7:157–164

Midwood KS, Valenick LV, Hsia HC, Schwarzbauer JE (2004) Coregulation of fibronectin signaling and matrix contraction by tenascin-C and syndecan-4. Mol Biol Cell 15:5670–5677

Miyazono K, Thyberg J, Heldin CH (1992) Retention of the transforming growth factor-beta 1 precursor in the Golgi complex in a latent endoglycosidase H-sensitive form. J Biol Chem 267:5668–5675

Morgelin M, Heinegard D, Engel J, Paulsson M (1992) Electron microscopy of native cartilage oligomeric matrix protein purified from the Swarm rat chondrosarcoma reveals a five-armed structure. J Biol Chem 267:6137–6141

Muller SA, Sasaki T, Bork P, Wolpensinger B, Schulthess T, Timpl R, Engel A, Engel J (1999) Domain organization of Mac-2 binding protein and its oligomerization to linear and ring-like structures. J Mol Biol 291:801–813

Muller WE, Wiens M, Adell T, Gamulin V, Schroder HC, Muller IM (2004) Bauplan of urmetazoa: basis for genetic complexity of metazoa. Int Rev Cytol 235:53–92

Myllyharju J, Kivirikko KI (2004) Collagens, modifying enzymes and their mutations in humans, flies and worms. Trends Genet 20:33–43

Orengo CA, Thornton JM (2005) Protein families and their evolution-a structural perspective. Annu Rev Biochem 74:867–900

Parry DA, Fraser RD, Squire JM (2008) Fifty years of coiled-coils and alpha-helical bundles: a close relationship between sequence and structure. J Struct Biol 163:258–269

Patthy L (2003) Modular assembly of genes and the evolution of new functions. Genetica 118:217–231

Paulsson M, Yurchenco PD, Ruben GC, Engel J, Timpl R (1987) Structure of low density heparan sulfate proteoglycan isolated from a mouse tumor basement membrane. J Mol Biol 197:297–313

Persikov AV, Ramshaw JA, Brodsky B (2005) Prediction of collagen stability from amino acid sequence. J Biol Chem 280:19343–19349

Pickford AR, Smith SP, Staunton D, Boyd J, Campbell ID (2001) The hairpin structure of the (6)F1(1)F2(2)F2 fragment from human fibronectin enhances gelatin binding. EMBO J 20:1519–1529

Potts JR, Campbell ID (1996) Structure and function of fibronectin modules. Matrix Biol 15:313–320

Prabhakar V, Capila I, Sasisekharan R (2009) Glycosaminoglycan characterization methodologies: probing biomolecular interactions. Methods Mol Biol 534:331–340

Ramirez F, Dietz HC (2009) Extracellular microfibrils in vertebrate development and disease processes. J Biol Chem 284:14677–14681

Ramirez F, Rifkin DB (2009) Extracellular microfibrils: contextual platforms for TGFbeta and BMP signaling. Curr Opin Cell Biol 21:616–622

Reddig PJ, Juliano RL (2005) Clinging to life: cell to matrix adhesion and cell survival. Cancer Metastasis Rev 24:425–439

Riveline D, Zamir E, Balaban NQ, Schwarz US, Ishizaki T, Narumiya S, Kam Z, Geiger B, Bershadsky AD (2001) Focal contacts as mechanosensors: externally applied local mechanical force induces growth of focal contacts by an mDia1-dependent and ROCK-independent mechanism. J Cell Biol 153:1175–1186

Romijn RA, Bouma B, Wuyster W, Gros P, Kroon J, Sixma JJ, Huizinga EG (2001) Identification of the collagen-binding site of the von Willebrand factor A3-domain. J Biol Chem 276:9985–9991

Ruoslahti E, Yamaguchi Y, Hildebrand A, Border WA (1992) Extracellular matrix/growth factor interactions. Cold Spring Harb Symp Quant Biol 57:309–315

Saharinen J, Keski-Oja J (2000) Specific sequence motif of 8-Cys repeats of TGF-beta binding proteins, LTBPs, creates a hydrophobic interaction surface for binding of small latent TGF-beta. Mol Biol Cell 11:2691–2704

Saltel F, Mortier E, Hytonen VP, Jacquier MC, Zimmermann P, Vogel V, Liu W, Wehrle-Haller B (2009) New PI(4, 5)P2- and membrane proximal integrin-binding motifs in the talin head control beta3-integrin clustering. J Cell Biol 187:715–731

Sanchez Alvarado A, Kang H (2005) Multicellularity, stem cells, and the neoblasts of the planarian Schmidtea mediterranea. Exp Cell Res 306:299–308

Sasaki T, Fassler R, Hohenester E (2004) Laminin: the crux of basement membrane assembly. J Cell Biol 164:959–963

Sawada Y, Tamada M, Dubin-Thaler BJ, Cherniavskaya O, Sakai R, Tanaka S, Sheetz MP (2006) Force sensing by mechanical extension of the Src family kinase substrate p130Cas. Cell 127:1015–1026

Scheele S, Nystrom A, Durbeej M, Talts JF, Ekblom M, Ekblom P (2007) Laminin isoforms in development and disease. J Mol Med 85:825–836

Schmidt C, Pommerenke H, Durr F, Nebe B, Rychly J (1998) Mechanical stressing of integrin receptors induces enhanced tyrosine phosphorylation of cytoskeletally anchored proteins. J Biol Chem 273:5081–5085

Sengle G, Charbonneau NL, Ono RN, Sasaki T, Alvarez J, Keene DR, Bachinger HP, Sakai LY (2008a) Targeting of bone morphogenetic protein growth factor complexes to fibrillin. J Biol Chem 283:13874–13888

Sengle G, Ono RN, Lyons KM, Bachinger HP, Sakai LY (2008b) A new model for growth factor activation: type II receptors compete with the prodomain for BMP-7. J Mol Biol 381:1025–1039

Spring J, Beck K, Chiquet-Ehrismann R (1989) Two contrary functions of tenascin: dissection of the active sites by recombinant tenascin fragments. Cell 59:325–334

Stopak D, Harris AK (1982) Connective tissue morphogenesis by fibroblast traction. I. Tissue culture observations. Dev Biol 90:383–398

Takagi J (2007) Structural basis for ligand recognition by integrins. Curr Opin Cell Biol 19:557–564

Tamada M, Sheetz MP, Sawada Y (2004) Activation of a signaling cascade by cytoskeleton stretch. Dev Cell 7:709–718

Tang BL, Hong W (1999) ADAMTS: a novel family of proteases with an ADAM protease domain and thrombospondin 1 repeats. FEBS Lett 445:223–225

Than ME, Henrich S, Huber R, Ries A, Mann K, Kuhn K, Timpl R, Bourenkov GP, Bartunik HD, Bode W (2002) The 1.9-A crystal structure of the noncollagenous (NC1) domain of human placenta collagen IV shows stabilization via a novel type of covalent Met-Lys cross-link. Proc Natl Acad Sci USA 99:6607–6612

Thorne JL (2007) Protein evolution constraints and model-based techniques to study them. Curr Opin Struct Biol 17:337–341
Tucker RP, Chiquet-Ehrismann R (2009) Evidence for the evolution of tenascin and fibronectin early in the chordate lineage. Int J Biochem Cell Biol 41:424–434
Tucker RP, Drabikowski K, Hess JF, Ferralli J, Chiquet-Ehrismann R, Adams JC (2006) Phylogenetic analysis of the tenascin gene family: evidence of origin early in the chordate lineage. BMC Evol Biol 6:60
Vitagliano L, Berisio R, Mastrangelo A, Mazzarella L, Zagari A (2001) Preferred proline puckerings in cis and trans peptide groups: implications for collagen stability. Protein Sci 10:2627–2632
Vogel V, Sheetz M (2006) Local force and geometry sensing regulate cell functions. Nat Rev Mol Cell Biol 7:265–275
Vogel C, Bashton M, Kerrison ND, Chothia C, Teichmann SA (2004) Structure, function and evolution of multidomain proteins. Curr Opin Struct Biol 14:208–216
Vogel C, Teichmann SA, Pereira-Leal J (2005) The relationship between domain duplication and recombination. J Mol Biol 346:355–365
Vynios DH, Theocharis DA, Papageorgakopoulou N, Papadas TA, Mastronikolis NS, Goumas PD, Stylianou M, Skandalis SS (2008) Biochemical changes of extracellular proteoglycans in squamous cell laryngeal carcinoma. Connect Tissue Res 49:239–243
Wagenseil JE, Mecham RP (2009) Vascular extracellular matrix and arteria mechanics. Physiol Rev 89:957–989
Wakatsuki T, Schlessinger J, Elson EL (2004) The biochemical response of the heart to hypertension and exercise. Trends Biochem Sci 29:609–617
Whittaker CA, Bergeron KF, Whittle J, Brandhorst BP, Burke RD, Hynes RO (2006) The echinoderm adhesome. Dev Biol 300:252–266
Wolff J (1892) The law of transformation of bones. Verlag August Hirschwald, Berlin
Xiao T, Takagi J, Coller BS, Wang JH, Springer TA (2004) Structural basis for allostery in integrins and binding to fibrinogen-mimetic therapeutics. Nature 432:59–67
Xing Y, Lee C (2006) Alternative splicing and RNA selection pressure–evolutionary consequences for eukaryotic genomes. Nat Rev Genet 7:499–509
Xiong JP, Stehle T, Diefenbach B, Zhang R, Dunker R, Scott DL, Joachimiak A, Goodman SL, Arnaout MA (2001) Crystal structure of the extracellular segment of integrin alpha Vbeta3. Science 294:339–345
Xiong JP, Mahalingham B, Alonso JL, Borrelli LA, Rui X, Anand S, Hyman BT, Rysiok T, Muller-Pompalla D, Goodman SL, Arnaout MA (2009) Crystal structure of the complete integrin alphaVbeta3 ectodomain plus an alpha/beta transmembrane fragment. J Cell Biol 186:589–600
Yamada Y, Avvedimento VE, Mudryj M, Ohkubo H, Vogeli G, Irani M, Pastan I, de Crombrugghe B (1980) The collagen gene: evidence for its evolutionary assembly by amplification of a DNA segment containing an exon of 54 bp. Cell 22:887–892
Yurchenco PD, Amenta PS, Patton BL (2004) Basement membrane assembly, stability and activities observed through a developmental lens. Matrix Biol 22:521–538
Zaidel-Bar R, Cohen M, Addadi L, Geiger B (2004) Hierarchical assembly of cell-matrix adhesion complexes. Biochem Soc Trans 32:416–420
Zaidel-Bar R, Itzkovitz S, Ma'ayan A, Iyengar R, Geiger B (2007) Functional atlas of the integrin adhesome. Nat Cell Biol 9:858–867
Zhang X, Boot-Handford RP, Huxley-Jones J, Forse LN, Mould AP, Robertson DL, Lili AM, Sarras MP Jr (2007) The collagens of hydra provide insight into the evolution of metazoan extracellular matrices. J Biol Chem 282:6792–6802

Chapter 2
Fibronectin and Other Adhesive Glycoproteins

Jielin Xu and Deane Mosher

Abstract Cells adhere to the extracellular matrix through interaction with adhesive extracellular matrix glycoproteins, including fibronectin, laminins, vitronectin, thrombospondins, tenascins, entactins (or nidogens), nephronectin, fibrinogen, and others. Most adhesive glycoproteins bind cells through cell surface integrin receptors in conjunction with other cell surface receptors, such as dystroglycans and syndecans, and interact with other extracellular matrix proteins to form an intensive matrix network. Interactions between cells and the extracellular matrix may mediate many cellular responses, such as cell migration, growth, differentiation, and survival. Cells receive and respond to signals from surrounding extracellular matrix, and in turn, modulate surrounding extracellular matrix through control of matrix assembly. This chapter discusses the adhesive glycoproteins and focuses on the interaction between integrins and adhesive glycoproteins.

2.1 Introduction

The interaction between cells and glycoproteins of the extracellular matrix mediates cell adhesion, migration, growth, differentiation, and survival of adherent cells. Each of these glycoproteins has distinct functional domains or polypeptide sequences to bind specific cell surface receptors, such as the integrins, dystroglycan, and syndecans; or to interact with other extracellular matrix proteins such as collagens.

Integrins are arguably the most important cell surface receptors that anchor cells to the extracellular matrix. We focus on the interaction between integrins and adhesive glycoproteins in this chapter. We concentrate on two aspects: first, the integrin-binding sequences, especially the dominant integrin-binding residue – aspartate – of each adhesive glycoprotein; second, the relationships between the

J. Xu and D. Mosher (✉)
Department of Biomolecular Chemistry, University of Wisconsin, 4285 MSC, 1300 University Avenue, Madison, WI 53706, USA
e-mail: dfm1@medicine.wisc.edu

ligand–integrin interaction and the deposition of the ligand in extracellular matrix. The major cell adhesion protein, fibronectin, which interacts with more than ten different integrin receptors, is considered in the greatest detail. Other adhesive glycoproteins, including laminins, vitronectin, thrombospondins, tenascins, entactins, nephronectin, and fibrinogen, are discussed.

2.2 Fibronectin

2.2.1 Overview

Fibronectin was first discovered in 1948 as a contaminant of plasma fibrinogen with insolubility at low temperature and was termed "cold-insoluble globulin" (Morrison et al. 1948; Mosesson and Umfleet 1970). Fibronectin is a high molecular weight dimeric glycoprotein (~450 kDa per dimer) widely expressed by a wide variety of cells in embryos and adult tissues (Hynes 1990; Mosher 1989). Plasma fibronectin is synthesized in the liver by hepatocytes and present in a soluble form in blood plasma at a concentration of around 300 µg/ml. Cellular fibronectin is secreted by fibroblasts and multiple other cell types and is organized into fibrils contributing to the insoluble extracellular matrix. The name "fibronectin" is derived from the Latin word *fibra*, meaning fiber, and *nectere*, meaning to bind. Fibronectin is crucial for vertebrate development, presumably by mediating a variety of adhesive and migratory events. Targeted inactivation of the fibronectin gene is lethal at embryonic day 8.5 in embryos homozygous for the disruption (George et al. 1993). Plasma fibronectin is also important for thrombosis. Conditional fibronectin knockout mice with plasma fibronectin levels reduced to less than 2% of normal have a delay in thrombus formation after vascular injury and defects in thrombus growth and stability (Ni et al. 2003). Fibronectin is organized into a fibrillar network on the cell surface through interaction with cell surface receptors and regulates cell functions, such as cell adhesion, migration, growth, and differentiation (Hynes 1990; Mosher 1989).

2.2.2 Structure of Fibronectin

2.2.2.1 Basic Structure

Visualization of soluble fibronectin by rotary shadowing electron microscopy in the early 1980s revealed two identical and apparently flexible strands (Engel et al. 1981; Erickson et al. 1981). Fibronectin mainly exists as a dimeric glycoprotein, with two similar ~240-kDa subunits covalently linked through a pair of disulfide bonds near the C-terminus. There are three types of repeating modules in each fibronectin subunit: 12 type I (termed FN1), 2 type II (termed FN2), and 15–17 type III repeats (termed FN3) (Fig. 2.1); accounting for 90% of the sequence. The remaining sequences include a connector between modules [5]FN1 and [6]FN1, a short connector between [1]FN3 and [2]FN3, and a variable (V) sequence that is not homologous to other parts of fibronectin.

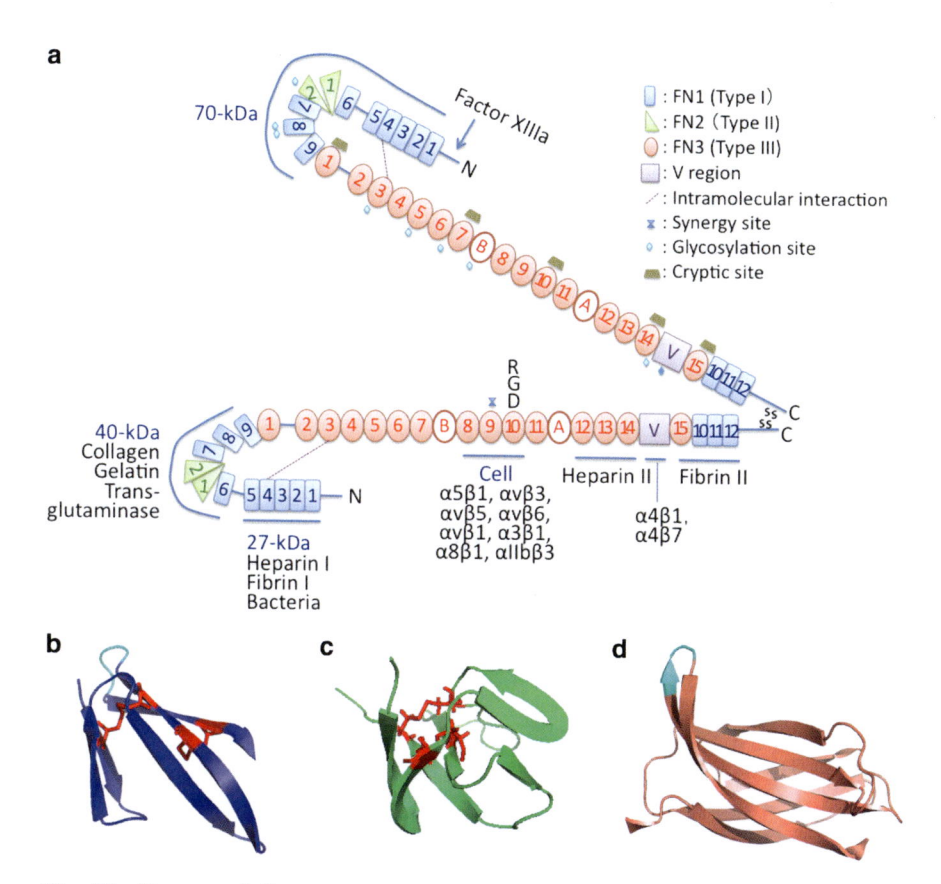

Fig. 2.1 *Diagram of fibronectin modular structure and structures of fibronectin modules.* (**a**) Diagram of the modular structure of fibronectin. Each fibronectin dimer is composed of two monomers linked at the C-terminus by a pair of disulfide bonds. 12 type I modules (*blue rectangles*) termed FN1, 2 type II modules (*green triangles*) termed FN2, and 15–17 type III modules (*salmon ovals*) termed FN3. The number of FN3 modules varies due to the presence of [A]FN3 (EDA) and [B]FN3 (EDB) based on alternative splicing. The alternatively spliced V region is shown as a purple square. Proteolytic 27-kDa, 40-kDa, and 70-kDa N-terminal fragments and the protein-binding sites on fibronectin are *underlined* with receptors listed. (**b**) The ribbon structure of [5]FN1 is drawn using PyMOL of structure PDB: 2RKY (Bingham et al. 2008). The cysteine residues and disulfide bonds are shown in *red*, with other residues shown in *blue* to match panel **a**. (**c**) The ribbon structure of [2]FN2 is drawn using PyMOL of solution structure PDB: 1E8B (Pickford et al. 2001). The cysteine residues and disulfide bonds are shown in *red*. (**d**) The ribbon structure of [10]FN3 is drawn using PyMOL of solution structure PDB: 1FNF (Leahy et al. 1996). The Arg-Gly-Asp (RGD) residues are shown in cyan

The FN1 module is found only in chordates (Tucker and Chiquet-Ehrismann 2009) (see Chap. 1). It has been noted that the N-terminal sub-domain of the VWF type C module of α2 procollagen shows a structural similarity with the fibronectin FN1 module (O'Leary et al. 2004) and suggested that the VWF type C module, which has been found in a large number of proteins of flies and worms, may be the precursor of the fibronectin FN1 module. Each FN1 module is about 45 amino acid residues long and contains two intrachain disulfide bonds (shown in red in Fig. 2.1b). NMR spectroscopy showed that the FN1 module has compact stacked antiparallel β-sheets enclosing a hydrophobic core with conserved aromatic residues (Baron et al. 1990; Potts et al. 1999). One sheet has two strands (A and B), and the other has three strands (C, D, and E). One disulfide bond, which links two nonadjacent β-strands, connects the first and third cysteines. The other disulfide bond, connecting the second and fourth cysteines, links adjacent β-strands D and E.

FN2 modules are rare and are similar to the kringle domains, which are present in lower organisms besides vertebrates (Ozhogina et al. 2001). Interestingly, FN2 modules are found in matrix metallo-proteinases (Collier et al. 1988). Each FN2 module is approximately 60 residues long with two intrachain disulfide bonds in each repeat. NMR spectroscopy shows that the solution structure of FN2 module consists of several highly conserved aromatic residues, two double-stranded antiparallel β-sheets perpendicular to each other, and four cysteines that form two disulfide bonds connecting cysteines 1–3 and 2–4 (Constantine et al. 1992; Pickford et al. 1997) (Fig. 2.1c). NMR studies identified an interaction between ^{6}FN1 and ^{2}FN2 (Pickford et al. 2001), and thus the FN2 modules are thought to cause a departure from a "head-to-tail" arrangement of FN modules (Fig. 2.1).

The FN3 module is found in multiple copies in many other extracellular matrix proteins, cell surface receptors, and cytoskeletal proteins of vertebrates and non-vertebrates (Bork and Doolittle 1992). Each FN3 module is about 90 residues long and lacks disulfide bonds. It consists of two antiparallel β-sheets formed from seven β-strands similar to Ig domains without disulfide bonds (Fig. 2.1d). One β-sheet is formed by four β-strands (G, F, C, and C') and the other β-sheet is formed by three β-strands (A, B, and E), arranged as a β sandwich to enclose a hydrophobic core (Dickinson et al. 1994a; Dickinson et al. 1994b; Leahy et al. 1996; Main et al. 1992). The β-strands are connected by flexible loops. The main integrin-binding motif Arg-Gly-Asp (RGD) (shown in cyan in Fig. 2.1d) is in one of the flexible loops connecting two β-strands (Dickinson et al. 1994b).

2.2.2.2 Alternative Splicing

One large single gene (~50 kb for human fibronectin) encodes fibronectin in most species (Hirano et al. 1983). Alternative pre-mRNA splicing and various posttranslational modifications result in heterogeneity of fibronectin, with up to 20 variants in human fibronectin (ffrench-Constant 1995; Kosmehl et al. 1996). There are two alternatively spliced segments in fibronectin due to alternative exon usage: extra domain A (EDA) located between the 11th and 12th FN3 modules, and extra domain

B (EDB) between the seventh and eighth FN3 modules (Fig. 2.1a). The nonhomologous variable (V) region between the 14th and 15th FN3 modules, which is subject to exon subdivisions, resulting in five different V region variants in human fibronectin (V0, V64, V89, V95, and V120, with the number standing for the number of amino acid residues in each variant). There is a special type of cartilage-specific splicing [termed $(V + C)^-$], with fibronectin lacking in the entire V region through the ^{10}FN1 module (Burton-Wurster et al. 1999; MacLeod et al. 1996).

Alternative splicing of fibronectin is regulated by cell type, stage of development, and age (ffrench-Constant 1995; Kornblihtt et al. 1996). Fibronectin isolated from plasma tends to have a lower molecular weight than fibronectin isolated from cell culture, which has resulted in the terms, plasma fibronectin and cellular fibronectin. Plasma fibronectin generally lacks EDA and EDB sequences, and contains a subunit that is V0. Cellular fibronectin is a more heterogeneous group of splice variants with variable presence of EDA, EDB, and V region isoforms. Certain isoforms of fibronectin, especially those containing EDA and EDB modules, are upregulated after wounding, and in malignant cells (ffrench-Constant 1995).

The EDA module of fibronectin mediates cell differentiation (Jarnagin et al. 1994). Fibronectin containing the EDA module is much better at promoting cell adhesion and spreading than fibronectin lacking the EDA module (Manabe et al. 1997). The presence of EDA module in fibronectin enhances fibronectin-$\alpha 5\beta 1$ integrin interaction and promotes cell adhesion (Manabe et al. 1997). A direct interaction between EDA and $\alpha 9\beta 1$ integrins, however, is critical for lymphatic valve morphogenesis through regulation of fibronectin assembly (Bazigou et al. 2009). Genetically manipulated mice that lacked EDA developed normally, but with a shorter life span, abnormal wound healing, and edematous granulation tissue (Muro et al. 2003), suggesting that EDA is not required for embryonic development but is important for a normal life span and emphasizing the role of fibronectin in organization of the granulation tissue and in wound healing. EDB knockout mice developed normally as well, but with reduced fibronectin matrix assembly (Fukuda et al. 2002). The presence of the EDB module exposes a cryptic binding site in the ^{7}FN3 module (Carnemolla et al. 1992). EDB-containing fibronectins are concentrated in tumors and are found at low levels in plasma (Menrad and Menssen 2005). For this reason, tumor therapy research has focused on developing antibodies specific to the EDB module of fibronectin.

2.2.2.3 Posttranslational Modifications

In addition to alternative pre-mRNA splicing, various posttranslational modifications that occur intracellularly during trafficking through the endoplasmic reticulum and Golgi contribute to the heterogeneity of fibronectin. Fibronectin can be glycosylated, phosphorylated, and sulfated (Paul and Hynes 1984). The intrachain and intramodule disulfide bonds of FN1 and FN2 modules are formed in this step as well.

There are seven N-linked carbohydrate chains and one or two O-linked carbohydrate chains per fibronectin subunit (Mosher 1989). Generally, fibronectin contains about 5% carbohydrate although higher levels of glycosylation occur in some tissues (Mosher 1989; Ruoslahti et al. 1981). Nonglycosylated fibronectin is more sensitive to proteolysis than glycosylated fibronectin and has an altered binding affinity to proteins such as collagen, suggesting that carbohydrates stabilize fibronectin against degradation and regulate its affinity to some substrates (Bernard et al. 1982; Jones et al. 1986; Olden et al. 1979). The 40-kDa gelatin-binding domain contains three N-linked glycosylation sites (Skorstengaard et al. 1984), with two sites, Asn497 and Asn511, present in the ^{8}FN1 module. Nonglycosylated ^{8}FN1 has decreased thermal stability and decreased gelatin-binding activity compared with glycosylated ^{8}FN1 (Ingham et al. 1995; Millard et al. 2005).

O-phosphoserine was identified at a concentration of two residues per molecule in human plasma fibronectin (Etheredge et al. 1985). Phosphorylation has also been identified in the carboxyl-terminal region of bovine plasma fibronectin (Skorstengaard et al. 1982). Most of the sulfation of fibronectin occurs at tyrosine residues as tyrosine-O–SO$_4$, probably in the V region (Liu and Lipmann 1985; Paul and Hynes 1984). It should be noted that the referenced analyses are somewhat dated; application of new mass spectrometric techniques should allow localization of modifications to specific residues and may reveal additional sites of modification.

2.2.3 Functional Domains

Fibronectin has important roles in mediating a variety of cell adhesive and migratory activities. Fibronectin binds to cells through cell surface receptors (integrins) and specifically interacts with other proteins, including collagen, fibrin, and heparin/heparan sulfate. The functional domains of fibronectin have been defined by studies of proteolytic fragments and recombinant constructs (Pankov and Yamada 2002).

2.2.3.1 Integrin Interaction Domains

Two major sites of fibronectin that mediate cell adhesion are the "cell-binding domain" (^{9}FN3–^{10}FN3) and the alternatively spliced V region (Fig. 2.1a). Fibronectin interacts with many integrins. For example, α3β1, α5β1, α8β1, αvβ1, αIIbβ3, αvβ3, αvβ5, and αvβ6 integrins interact with the Arg-Gly-Asp (RGD) sequence in the central cell-binding domain. Integrins α4β1 and α4β7, in contrast, recognize the Leu-Asp-Val (LDV) sequence in the V region (Humphries et al. 2006; Leiss et al. 2008). The integrin-binding motifs all contain a critical residue Asp (D), which interacts with a metal in the metal-ion dependent adhesion site (MIDAS) in the integrins (Fig. 2.2). Additional integrin-binding sites are also available in the EDA module, which binds α4β1 or α9β1 integrin (Liao et al.

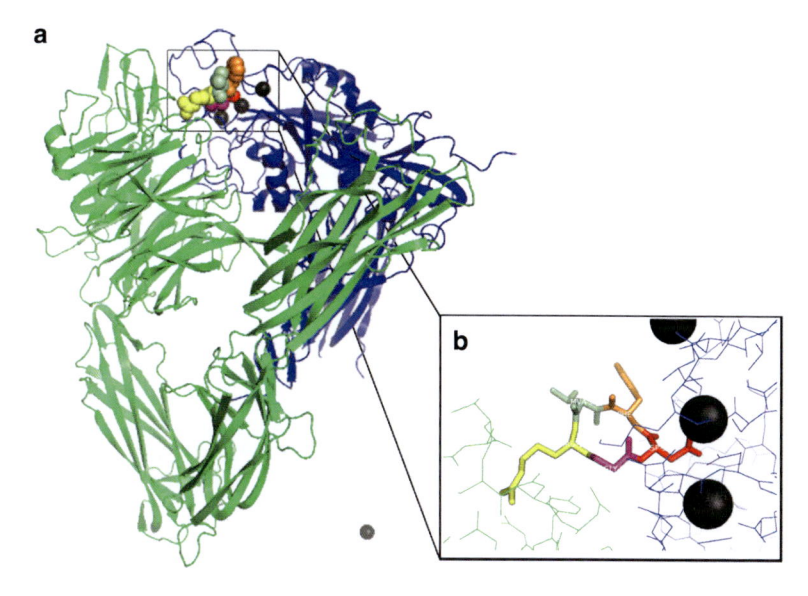

Fig. 2.2 *Structure of RGD – αvβ3 integrin interaction,* (**a**) The ribbon structure of clyco(RGDf-N V) – αvβ3 integrin interaction in the presence of Mn PDB: 1L5G (Xiong et al. 2002). (**b**) Ball-and-stick representation of RGD-integrin interaction. αv is shown in *green*, and β3 is shown in *blue*. The three Mn^{2+} ions are shown in *black*. The peptide residues R, G, D, f, and MVA are shown in *yellow, magenta, red, orange,* and *pale green,* respectively. The aspartate side chain binds the Mn^{2+} in the middle – the MIDAS site

2002); ^{14}FN3 module, which binds α4β1 integrin through the IDAPS sequence (Pankov and Yamada 2002); and ^{5}FN3, which binds activated α4β1 or α4β7 integrin through the KLDAPT sequence (Moyano et al. 1997). Recently, it was demonstrated that the ^{3}FN3 module may mediate cell spreading and migration through interaction with β1 integrin(s) combined with specific although unresolved α-subunit(s) in an RGD-dependent manner (Obara et al. 2010). Also, *iso*-Asp-Gly-Arg (*iso*-DGR), spontaneously converted from Asn-Gly-Arg (NGR) by deamidation of asparagine with isomerization of the backbone linkage to the β-position, in ^{5}FN1 and ^{7}FN1 repeats may interact with αvβ3 integrins (Curnis et al. 2006).

Major Cell-Binding Domain

The RGD motif, which mediates cell adhesion through interaction with cell surface integrin receptors and widely exists in adhesive glycoproteins such as vitronectin and von Willebrand factor, was first identified in fibronectin in 1983 (Pierschbacher and Ruoslahti 1984). The interesting history of the discovery of RGD motif can be found in an essay written by Ruoslahti (2003).

The RGD motif is essential for development. Site-directed mutagenesis to substitute a Glu (E) for Asp (D) in the RGD motif caused a >95% loss of cell-adhesive ability (Obara et al. 1988). Mouse embryos in which the RGD motif was replaced with inactive RGE died at embryonic day 10 with shortened posterior trunk and severe vascular defects (Takahashi et al. 2007). Interestingly, changing Arg (R) to Lys (K) to generate peptides with a KGD sequence caused loss of interaction with α5β1 but the interaction with αIIbβ3 was not affected (Scarborough et al. 1993).

In fibronectin, the RGD motif localizes in a flexible loop connecting two β-strands of the ^{10}FN3 module, protruding out of the protein structure (Dickinson et al. 1994b) (Fig. 2.1d). The high-affinity interaction of α5β1 integrin with fibronectin RGD motif requires the synergy site Pro-His-Ser-Arg-Asn (PHSRN) in the ^{9}FN3 repeat (Aota et al. 1994). Crystal structure of a fibronectin fragment of ^{7}FN3–^{10}FN3 revealed that the RGD loop in the ^{10}FN3 module and the "synergy" site in the ^{9}FN3 module are on the same face of ^{7}FN3–^{10}FN3, presumably enabling simultaneous interaction of both sites with a single integrin molecule (Leahy et al. 1996). Antibody blocking- and epitope-mapping studies with α5β1 integrin and fibronectin cell-binding domain fragments suggested that the synergy site primarily binds to the α subunit of integrin while the RGD motif binds to the β subunit of integrin (Mould et al. 1997). Mechanical studies showed that there are two forms of α5β1-fibronectin bonds: relaxed bonds and tensioned bonds, with the tensioned bonds being required for phosphorylation of focal adhesion kinase (Friedland et al. 2009). It was found that the relaxed bonds only involve the RGD sequence and the tensioned bonds require both RGD and the synergy site. Another recent study using purified integrins found that activated αvβ3 integrin could not bind soluble fibronectin, while α5β1 integrin binds soluble fibronectin efficiently, suggesting that the RGD sequence in soluble fibronectin is not exposed, and that α5 integrin binds to the synergy site first and causes a conformational change, which exposes the RGD sequence for β1 integrin (Huveneers et al. 2008). The idea that RGD sequence in soluble fibronectin is cryptic is also supported by studies showing that binding of the functional upstream domain (FUD) of a bacterial adhesin protein to the N-terminal portion of soluble fibronectin causes fibronectin to undergo conformational changes and expose the epitope for a monoclonal antibody that recognizes the ^{10}FN3 module (Ensenberger et al. 2004).

Alternatively Spliced Cell-Binding Domains

α4β1 and α4β7 integrins recognize the LDV sequence in the alternatively spliced V region of fibronectin (Guan and Hynes 1990; Mould et al. 1991; Wayner et al. 1989). It is hypothesized that LDV binds integrins at the junction between the α and β subunits similar to the way RGD binds (Humphries et al. 2006). The interaction between α4β1 and the V region may mediate lymphocyte adhesion under inflammatory conditions (Elices et al. 1994).

α4β1 and α9β1 integrins recognize the adjacent D (Asp) and G (Gly) residues in the C–C′ loop of the EDA module (Shinde et al. 2008). EDA – α9β1 integrin interaction regulates fibronectin assembly in lymphatic cells and mediates lymphatic valve morphogenesis (Bazigou et al. 2009).

Effects of Fibronectin–Integrin Interactions

Integrins are heterodimeric transmembrane receptors (with α and β subunits) that interact with extracellular matrix glycoproteins, connect to the cytoskeleton inside the cell through their cytoplasmic tails, and regulate intracellular signal transduction pathways utilizing signals from extracellular ligands (Hynes 2002). Ligand–integrin interaction mediates cell adhesion, induces integrin clustering, and regulates cell shape, proliferation, differentiation, and apoptosis (Ginsberg et al. 2005). Interestingly, many of the integrin-triggered signaling pathways are similar to the growth factor-triggered signaling pathways, and most of these pathways require cells to be adherent (Hynes 2002; Schwartz and Assoian 2001). Many integrin-associated proteins, such as Src family protein tyrosine kinases, integrin-linked kinase, and protein kinase C may interact with integrins and mediate the intracellular signaling pathways (Ginsberg et al. 2005).

Fibronectin–integrin interaction may induce cytoskeleton reorganization, focal adhesion formation, actin microfilament bundle assembly, and importantly, cell-generated tension to unfold cryptic fibronectin, which is critical for fibronectin matrix assembly (Geiger et al. 2001; Hynes 1990; Mosher 1989). $\alpha 5\beta 1$ integrin binds soluble fibronectin and supports the focal adhesion distribution, Rho activation, and fibronectin assembly (Huveneers et al. 2008). The roles of integrin in fibronectin matrix assembly are discussed in details in Sect. 2.2.4.

2.2.3.2 Collagen-Binding Domains

The collagen-binding domain of fibronectin is identified as ^{6}FN1–^{9}FN1 including the ^{1}FN2–^{2}FN2 modules (Fig. 2.1a). Fibronectin binds denatured collagen (gelatin) more effectively than native collagen (Engvall et al. 1978). Collagens denature locally at physiological temperatures and unfold their triple helices (Leikina et al. 2002), enabling fibronectin to interact with native collagen in vivo. Fibronectin–collagen interaction may mediate cell adhesion to denatured collagen, form noncovalent crosslinking of fibronectin and collagen in migratory pathways, and regulate the removal of denatured collagenous materials from blood and tissue (Mosher 1989; Pankov and Yamada 2002). Two segments of the gelatin-binding domain ^{6}FN1–^{7}FN1 (including ^{1}FN2–^{2}FN2) and ^{8}FN1–^{9}FN1 bind the same sequence of collagen $\alpha 1$ (Erat et al. 2009; Pickford et al. 2001).

2.2.3.3 Fibrin-Binding Domains

There are three fibrin-binding sites in fibronectin. The first and the major fibrin-binding site is in the N-terminal ^{4}FN1–^{5}FN1 (Williams et al. 1994). The second binding site is ^{10}FN1–^{12}FN1 close to the C-terminus. The third binding site appears following chymotrypsin digestion of fibronectin, and is immediately adjacent to the collagen-binding domain (Mosher 1989). At physiological temperatures, the fibronectin–fibrin interaction is very weak. Covalent crosslinking of fibrin and fibronectin mediated by Factor XIII transglutaminase at a Gln residue close to the N-terminus stabilizes this interaction, helps incorporate fibronectin into the fibrin-clot, stimulates platelet thrombus growth on fibrin, and has the potential to modulate cell adhesion or migration into fibronectin–fibrin clots upon wound healing (Cho and Mosher 2006; Magnusson and Mosher 1998).

2.2.3.4 Heparin-Binding Domains

Fibronectin contains at least two heparin-binding domains that interact mainly with heparan sulfate proteoglycans. The first and strongest site localizes to ^{12}FN3–^{14}FN3 modules in the C-terminus. The crystal structure of ^{12}FN3–^{14}FN3 modules and other related studies revealed the heparin-binding site to be a group of six positively charged residues in ^{13}FN3 and a minor heparin-binding site in ^{14}FN3 (Barkalow and Schwarzbauer 1991; Ingham et al. 1990; Sharma et al. 1999). The second and weaker site is in the N-terminal ^{1}FN1–^{5}FN1 modules. Fibronectin and heparin interact with high affinity, with at least two sets of affinities with $K_d = 10^{-7}$ to 4×10^{-9} M (Hynes 1990; Mosher 1989; Yamada et al. 1980). Other novel heparin-binding domains have been identified in ^{5}FN3 module and in the alternatively spliced V region (Mostafavi-Pour et al. 2001; Moyano et al. 1999).

Heparin-binding domains may cooperate with cell-binding domain of fibronectin and potentiate cell adhesion, cell spreading, and formation of actin microfilament bundles on fibronectin for certain cell types (Beyth and Culp 1984; Izzard et al. 1986; Lark et al. 1985; Laterra et al. 1983a; Laterra et al. 1983b; Woods et al. 1986).

2.2.3.5 Bacteria-Binding Domains

Besides heparin and fibrin, the N-terminal ^{1}FN1–^{5}FN1 can bind several types of bacteria, such as *Staphylococus aureus* or *Streptococcus pyogenes* (Mosher 1989). Recently, much attention has been paid to the bacterial fibronectin-binding proteins (FnBPs) that mediate cell adhesion and induce entry of bacteria into nonphagocytic host cells using fibronectin (Schwarz-Linek et al. 2004). Crystal and NMR studies revealed that the FnBPs are disordered in their unbound state and upon interactions with fibronectin become ordered through an unusual and distinctive tandem β-zipper mechanism (Bingham et al. 2008) (Fig. 2.3).

2.2.4 Fibronectin Matrix Assembly

Fibronectin is important for many activities including cell migration and tissue morphogenesis (Dzamba et al. 2009; Zhou et al. 2008). These activities require

2   Fibronectin and Other Adhesive Glycoproteins   51

[Figure: Ribbon and stick structure showing ^{4}FN1 and ^{5}FN1 modules with labeled N and C termini]

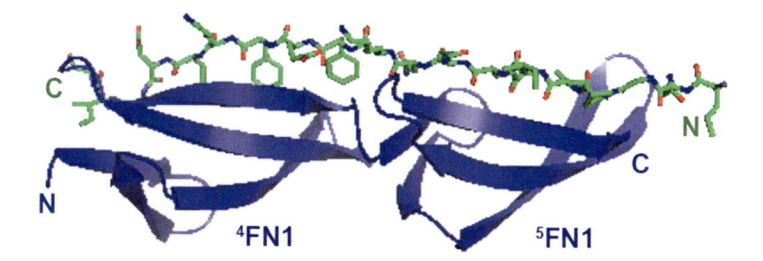

^{4}FN1 ^{5}FN1

Fig. 2.3 *FnBP-1 in complex with ^{4}FN1–^{5}FN1 modules.* Ribbon and stick structure of ^{4}FN1–^{5}FN1/FnBP-1 from PBD: 2RKY (Bingham et al. 2008). Fibronectin modules are shown as ribbon in *blue*, and FnBP-1 forms a fourth β-strand in the major β-sheet and is shown in sticks with carbon, nitrogen, and oxygen atoms shown in *green, blue, and red*, respectively

fibronectin to be assembled into fibronectin fibrils, which are one of the earliest components of extracellular matrix, and provide scaffolding for deposition of the fibronectin-interacting proteins such as collagen and heparan sulfate proteoglycans in the extracellular matrix (Hynes 2009). Inhibition of fibronectin fibril formation causes delay in embryonic development (Darribere et al. 1990). Unlike assembly of collagen or laminin, fibronectin fibrillogenesis does not occur spontaneously at physiological salt concentrations and pH. It requires the presence of assembly-competent cells. The rules for fibronectin assembly seems to be the same for plasma fibronectin and cellular fibronectins (Bae et al. 2004).

2.2.4.1   Steps of Fibronectin Matrix Assembly

Soluble compact fibronectin needs to be assembled to its fibrillar matrix form in a cell-mediated, stepwise manner. Fibronectin assembly is initiated by binding of soluble fibronectin to cell surface receptors that induce conformational changes that expose cryptic binding sites in bound fibronectin. These changes facilitate fibronectin–fibronectin interactions, forming fibronectin fibrils, fibronectin fibril elongation through cell-generated tension mediated by integrins, and the formation of an insoluble fibrillar network (Fig. 2.4).

One hypothesis is that fibronectin assembly begins by interactions of the fibronectin cell-binding domain (RGD motif in ^{10}FN3) with cell surface integrin receptors (Mao and Schwarzbauer 2005). Dimeric fibronectin induces integrin clustering by binding two integrins with its two cell-binding domains. Clustered integrins become activated, cause actin filament rearrangement, facilitate the extension of fibronectin that exposes cryptic binding sites, enable interactions of the N-terminal 70K region (^{1}FN1–^{9}FN1, termed 70K) with other parts of fibronectin, and cause irreversible association of fibronectin to a fibrillar matrix. However, Coussen et al. found that neither monomeric nor dimeric ^{7}FN3–^{10}FN3 binds integrins stably; a trimer is required (Coussen et al. 2002), suggesting that an interacting fibronectin dimer is not sufficient to cause clustering of integrins. An alternative hypothesis is that fibronectin assembly is initiated by interaction between the N-terminal 70-kDa

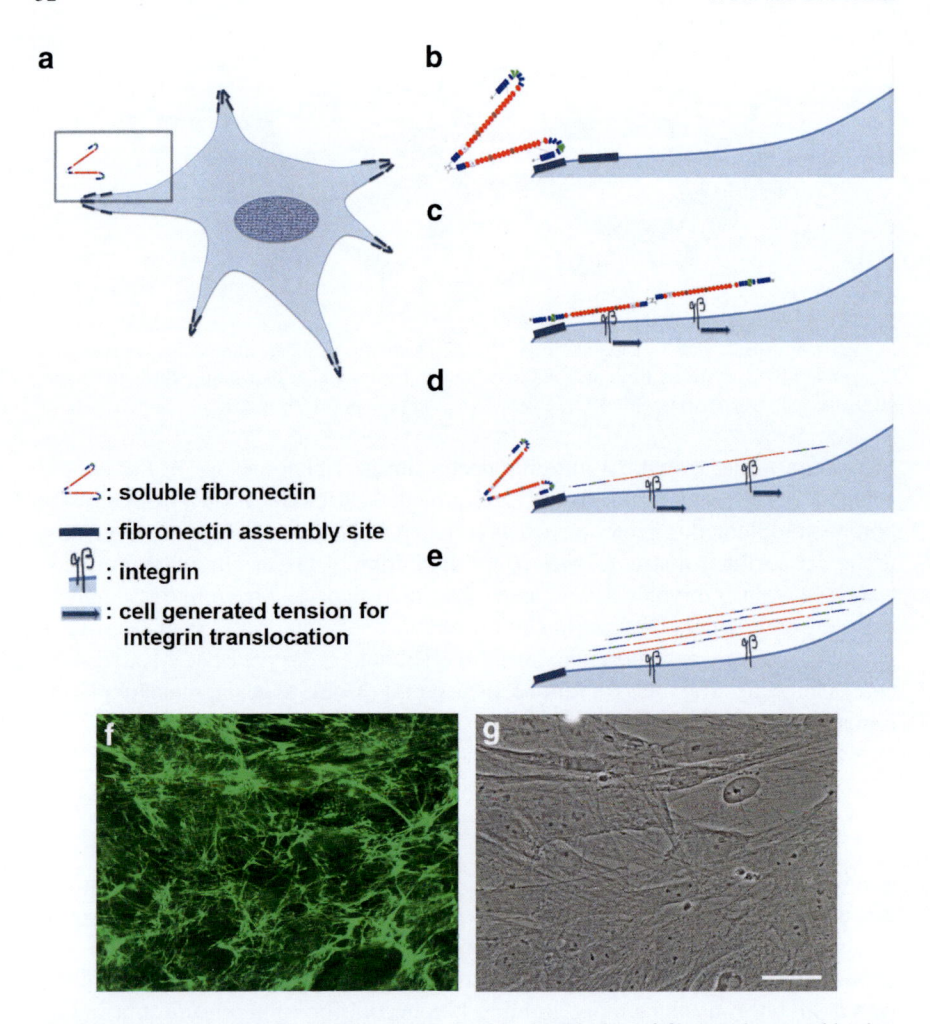

Fig. 2.4 *Hypothetical model of fibronectin assembly.* (**a**) Display of fibronectin assembly sites (*dark blue* strips at the focal adhesions) on the cell surface is controlled by the adherent substrates to which cells are attached. (**b–e**) show the enlarged *boxed area* of (**a**). (**b**) Soluble fibronectin dimer binds to linearly arrayed fibronectin assembly sites through the N-terminal 70-kDa region of fibronectin (70 K). (**c**) The binding of 70 K to the cell surface fibronectin assembly receptors induces unfolding of fibronectin, which exposes the RGD sequence in ^{10}FN3. (**d**) The RGD-integrin (integrins are shown as "αβ" on the cell surface) interaction activates Rho, and stretches fibronectin through tension generated from integrins and cytoskeleton contractility. Besides causing elongation of fibronectin, translocation of integrins toward the center of the cells also frees the peripheral fibronectin assembly sites for the second soluble fibronectin, and more soluble fibronectin follows. (**e**) Such elongation of fibronectin exposes more cryptic fibronectin–fibronectin interacting sites, leading to the formation of insoluble fibronectin fibrils through fibronectin–fibronectin interactions. (**f–g**) show immunofluorescence staining of fibronectin matrix. AH1F human foreskin fibroblasts were incubated in serum-containing medium for 24 h, and stained for their assembled fibronectin fibrils with an anti-human fibronectin monoclonal antibody followed by FITC-conjugated secondary antibody; an extensive network of fibrils is seen (**f**). Cells are shown by phase microscopy (**g**). *Scale bar* = 20 μm

region (^{1}FN1–^{9}FN1, termed 70K) and its cell surface receptors. 70K is able to bind to fibronectin assembly sites on the cell surface without the presence of intact fibronectin (Tomasini-Johansson et al. 2006) and inhibit assembly of intact fibronectin (McKeown-Longo and Mosher 1985). In the alternative hypothesis, binding of the 70K region to cell surface receptors unfolds fibronectin, which exposes the integrin-binding site RGD to interact with cell surface integrins followed by elongation of bound fibronectin, exposing cryptic fibronectin–fibronectin interaction sites forming fibronectin fibrils.

2.2.4.2 Essential Domains for Fibronectin Matrix Assembly

The fibronectin assembly initiation site located in the N-terminal 70-kDa region (70K) is essential for fibronectin matrix assembly, especially ^{1}FN1–^{5}FN1. A recombinant fibronectin construct including ^{1}FN1–^{5}FN1 followed by ^{8}FN3 to the C-terminus undergoes fibrillogenesis whereas removal of ^{5}FN1 module from the same construct caused loss of fibrillogenesis ability (Schwarzbauer 1991). The five FN1 modules, ^{1}FN1–^{5}FN1, likely work as a functional unit in interacting with other proteins. Removal of any of the five modules or mutation of conserved Tyr residues in individual modules results in decreased affinity (Magnusson and Mosher 1998; Sottile et al. 1991). 70K binds to the cell surface with the same affinity and at the same binding sites as intact fibronectin, but is not assembled into insoluble matrix (McKeown-Longo and Mosher 1985; Tomasini-Johansson et al. 2006). Although 70K is not assembled into insoluble matrix, 70K blocks the binding and assembly of fibronectin efficiently (McKeown-Longo and Mosher 1985).

Controversy exists as to the exact role of the cell-binding domain, especially RGD and the synergy site PHSRN. Antibodies binding to fibronectin's cell-binding domain or a fragment containing the cell-binding domain inhibited fibronectin matrix assembly in vitro (McDonald et al. 1987). Fibronectin lacking the synergy site showed reduced matrix assembly, which could be rescued by Mn^{2+}, suggesting a modulatory role of the synergy site on integrin function (Sechler et al. 1997). Mouse embryos in which the RGD sequence was replaced with inactive RGE die at embryonic day 10. However, RGE-FN is assembled in fibrils in vivo (Takahashi et al. 2007). Fibronectin lacking the RGD sequence can be assembled using $\alpha 4\beta 1$ integrins (Schwarzbauer 1991; Sechler et al. 2000). Our unpublished observation with RGE-FN and fibronectin-null cells suggest that the RGD motif is not required for initial binding of soluble fibronectin, but may mediate cell adhesion, activate cells to become assembly competent and, most importantly, mediate elongation of fibronectin (unpublished data of Xu and Mosher 2009).

Besides the 70K and the cell-binding domain, there are several other regions that are essential for fibronectin matrix assembly (Pankov and Yamada 2002). Fibronectin needs to be dimeric to be assembled. Removal of the cysteines at the C-terminus of fibronectin that form the interchain disulfide bonds generates monomeric fibronectin that does not assemble. In contrast, a recombinant fibronectin construct lacking ^{1}FN3–^{7}FN3, that can still dimerize, is competent for fibrillogenesis (Schwarzbauer 1991).

The ability of adherent fibronectin-null fibroblasts to assemble exogenous fibronectin is dependent on the adherent substrate: cells adherent to vitronectin could not assemble exogenous fibronectin, while cells adherent to collagen, laminin, or fibronectin are competent for fibronectin assembly (Bae et al. 2004). In identification of smaller fragments in fibronectin that account for the supportive activity, the ^{1}FN3 module and the C-terminal modules are found to be required for activation of adherent cells to be optimally competent for fibronectin assembly (Xu et al. 2009). The mechanism of how vitronectin suppresses or how fibronectin, collagen, or laminin supports adherent cells for fibronectin assembly is obscure. Vitronectin mainly interacts with αvβ3 integrin, while collagen, laminin, or fibronectin mainly interacts with β1 integrins. β3 integrin recycles through an endosomal "short-loop" recycling pathway, and β1 integrin recycles through a perinuclear "long-loop" recycling pathway (White et al. 2007). It is found that the recycling of αvβ3 integrin may inhibit the return of internalized α5β1 integrin back to the plasma membrane (White et al. 2007). Therefore, we hypothesize that for cells adherent to vitronectin, αvβ3 integrin recycles rapidly and inhibits the recycling of α5β1 integrin, which is important for fibronectin assembly.

2.2.4.3 Role of Integrins and Cytoskeletal Contractility in Fibronectin Assembly

α5β1 integrins are widespread. Monoclonal antibodies to α5 or β1 integrin subunits inhibited fibronectin assembly and 70K binding (Akiyama et al. 1989; Fogerty et al. 1990). Elevated levels of α5β1 integrin in Chinese hamster ovary (CHO) cells resulted in enhanced fibronectin assembly (Giancotti and Ruoslahti 1990). Recent studies found that the binding of α5β1 integrin by soluble fibronectin causes Rho activation and fibronectin assembly independent of syndecan-4 (Huveneers et al. 2008). Besides α5β1, other integrins like α4β1, αvβ3, and α9β1 have been reported to be able to support fibronectin assembly (Akiyama et al. 1989; Bazigou et al. 2009; Sechler et al. 2000; Wennerberg et al. 1996; Yang and Hynes 1996), although other studies have also shown that αvβ3 integrin could not bind soluble fibronectin and is not able to support fibronectin assembly in the absence of α5β1 integrin (Huveneers et al. 2008).

Fibronectin requires conformational changes to expose its cryptic sites for fibronectin–fibronectin interactions. Besides the conformational change caused by direct interaction between fibronectin and integrins, cell-driven integrin movement along the cell surface may stretch fibronectin and cause further exposure of cryptic self-association sites. Loss of cell contractility by blockage of Rho, myosin light chain kinase, or actin–myosin interaction inhibits fibronectin matrix formation (Halliday and Tomasek 1995; Wu et al. 1995b; Zhang et al. 1994; Zhang et al. 1997; Zhong et al. 1998). The majority of cryptic fibronectin–fibronectin interaction sites are in the FN3 modules (Geiger et al. 2001). The lack of disulfide bonds in these modules is thought to facilitate the stretched-induced exposure of cryptic sites (Ohashi and Erickson 2005).

2.2.4.4 Future Prospects

Fibronectin is a late addition to the repertoire of molecules that mediate cell-extracellular matrix adhesion (discussed in Chap. 1). It can be thought of as an amalgam of FN3 modules with sites of cell adhesion and unique and distinctive FN1 modules that mediate assembly. However, how the amalgam works is still not known. A number of important questions remain unanswered. How do different adherent substrates differentially mediate adherent cells to assemble soluble fibronectin? What are the cell surface binding sites for the N-terminal 70-kDa region of fibronectin that initiates fibronectin assembly? How does fibronectin convert from soluble dimer to multimers? Which cryptic sites are required for fibronectin assembly? What are the requirements of integrins in fibronectin assembly? A better appreciation of such issues would better define the assembly of fibronectin and may be of considerable value to manipulate assembly of fibronectin matrix.

2.3 Laminin

2.3.1 *Introduction*

Laminins, which are present in worms and flies and are among the first extracellular matrix proteins produced during embryogenesis, are the major cell adhesive proteins of the basement membrane (Yurchenco and Wadsworth 2004) (see Chap. 4). Compared with fibronectin, which is found only in chordates, laminins are evolutionarily ancient and conserved, with sequence similarities with a laminin gene found in *Hydra vulgaris* (Tzu and Marinkovich 2008). Laminins bind cell surface receptors and thereby connect basement membrane with adjacent cell layers. Laminins are large (400–900 kDa) heterotrimeric glycoproteins of three different polypeptide chains: α, β, and γ (Fig. 2.5). Unlike fibronectin, which is encoded by a single gene and generates variants through alternative splicing, multiple genes encode each of the three laminin subunits, which can assemble in different combinations of laminin variants.

Laminins undergo self-polymerization and form filaments and layered sheets, which initiate basement membrane assembly. Interestingly, laminin sheets are generally mixtures of multiple laminins instead of separate networks of each laminin (Scheele et al. 2007). When laminin polymerization is inhibited, basement membrane assembly seems to be disrupted even in the presence of other major constituents such as entactin, type IV collagen, and perlecan (Li et al. 2002). Laminin binds cell surface receptors like heparin, integrins, and α-dystroglycan,

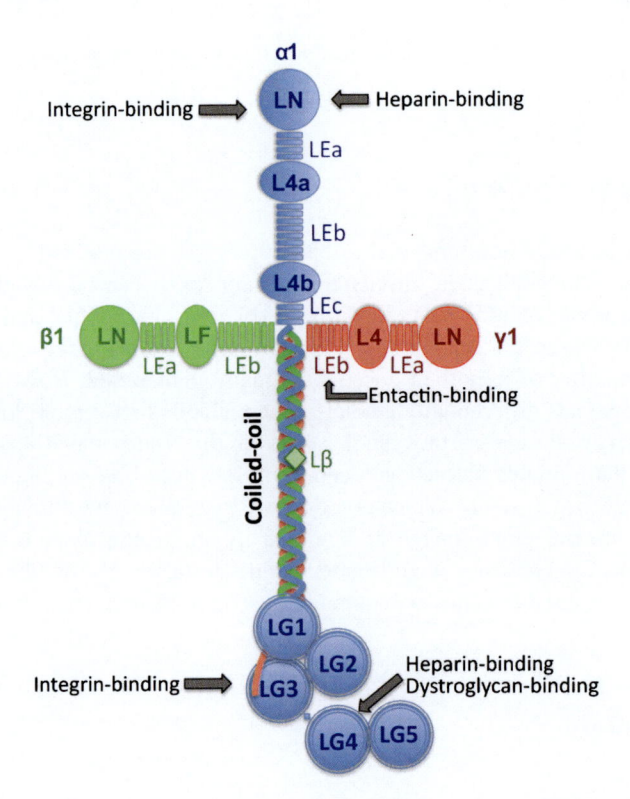

Fig. 2.5 *A schematic model of the laminin-111 domain structure*. Laminins are heterotrimers composed of α (*blue*), β (*green*), and γ (*red*) chains linked together at the coiled-coil region. The abbreviations are: *LN* laminin N-terminal domain, *LE* laminin epidermal growth factor-like domain, *L4* laminin 4 domain, *LF* laminin four domain, *LG* laminin globular domain. Binding sites for heparin, dystroglycan, integrin, and entactin are indicated in the figure

which make laminin the central adhesive protein of basement membranes. Laminins mediate cell adhesion (Nomizu et al. 1998), proliferation (Kubota et al. 1992), migration (Colucci et al. 1996), and differentiation (Rozzo et al. 1993) through interaction with cell surface receptors and also play a role in neurite outgrowth (Weeks et al. 1990; Weeks et al. 1991), metastasis (Colognato and Yurchenco 2000; Malinda and Kleinman 1996), and angiogenesis (Kibbey et al. 1992). The roles of different laminins in development and disease was reviewed recently by Scheele et al. (2007), and in Chap. 4.

2.3.2 Laminin-Interacting Proteins

Laminins interact with other laminins via their N-terminal globular LN domains to self-polymerize and initiate basement membrane assembly. There are also many protein-binding sites on laminins for extracellular matrix proteins, such as entactin (or nidogen), and for cells surface receptors, such as syndecans, integrins, and α-dystroglycan. Interestingly, most of the noncellular extracellular matrix protein-binding sites are in the short arms of the three chains, whereas most of the cell surface receptor-binding sites are in the N- and C- terminus of laminin α chains, especially in the LG domain (Timpl et al. 2000).

A major class of laminin receptor for linking cells with the basement membrane is the integrins. Laminin–integrin interaction activates a series of intracellular signaling pathways involving focal adhesion kinases (FAK), small rho GTPases, mitogen-activated protein kinases (MAPK), phosphatases, and cytoskeleton components, and therefore mediates cell adhesion, migration, proliferation, differentiation, and survival (Belkin and Stepp 2000; Givant-Horwitz et al. 2005; Gonzales et al. 1999; Hintermann and Quaranta 2004; Watt 2002).

Of the 24 different known integrin heterodimers, α1β1, α2β1, α3β1, α6β1, α6β4, α7β1, α9β1, and αvβ3 integrins have been reported to bind laminins (Nishiuchi et al. 2006; Patarroyo et al. 2002). As stated above, integrins mostly recognize laminins through the C-terminal globular LG domains of the α chains, with some integrin-binding activity at the N-terminus of α chains. The β and γ chains can be recognized by integrins as well (Patarroyo et al. 2002). Therefore, unlike the RGD and LDV sequences that define major and minor integrin-binding sites for fibronectin, the integrin-binding sites in laminins vary (Patarroyo et al. 2002). Studies of the major integrin-binding site in laminin-511 showed that deletion of the LG3 domain caused loss of its integrin-binding abilities, suggesting LG3 domain is required for integrin binding (Ido et al. 2004) (see Chap. 4 for laminin nomenclature and structure). However, recombinant LG1-3 domains do not bind integrin (Ido et al. 2006). Further studies by the same group found that Glu-1607 of the γ1 chain and the homologous Glu residue of the γ2 chain are critical for integrin binding, although Glu-1607 is not directly involved in integrin binding (Ido et al. 2007). Surprisingly, the γ3 chain lacks such a Glu residue, and laminin-113 or laminin-213 is not able to bind integrins (Ido et al. 2008). When the C-terminal four residues of the γ1 chain, including the conserved Glu residue, were swapped to the γ3 chain, the chimeric laminin-213 regained its integrin-binding activity (Ido et al. 2008). The above results suggest that integrins bind laminin through a combination of the C-terminal conserved Glu residue of the γ chain and the LG3 domain of the α chain, although the exact binding pattern is not known. Integrins may either bind to the LG3 domain and use the Glu residue of the γ chain as an auxiliary site, or bind to a cryptic integrin-binding site in the LG3 domain exposed only upon the interaction between the LG3 domain of the α chain and the Glu residue of the γ chain (Fig. 2.5).

In addition to integrins, laminins also interact with collagen, sulfatides, heparan sulfate proteoglycans, 67-kDa laminin receptor, and α-dystroglycan (Givant-Horwitz et al. 2005; Miner and Yurchenco 2004). The LG4 domain contains a heparin-binding site that is critical for basement membrane assembly (Li et al. 2002). Other binding sites include a single entactin (or nidogen) binding site, which locates to a loop of a LEb3 domain of the ϒ1 chain (Stetefeld et al. 1996). Such interaction between laminin and entactin (or nidogen) serve to bridge laminin with the collagen IV network and has significant developmental importance (Mayer et al. 1998; Yurchenco and Schittny 1990).

2.4 Other Adhesive Glycoproteins

2.4.1 Vitronectin

Vitronectin is a 75-kDa glycoprotein present in blood plasma at a concentration of 200–400 µg/ml (2.5–5.0 µM). It is also present in other body fluids such as amniotic fluid and urine, and in the extracellular matrix of many tissues (Preissner 1991; Tomasini and Mosher 1991). Vitronectin was independently studied under the names "serum spreading factor," "epibolin," and "S protein (site-specific protein)" in the late 1970s and early 1980s until investigators realized their findings relate to the same protein, vitronectin, named for its ability to bind glass. Human vitronectin is a protein of 459 amino acids mainly synthesized in the liver (Seiffert et al. 1994). In human blood, it exists in two forms: one is a single chain 75-kDa form, and the other is a two-chain form cleaved after Arg^{379} generating 65 and 10-kDa chains connected by a disulfide bridge (Cys^{274}–Cys^{453}) (Schvartz et al. 1999).

Vitronectin has many important protein-binding domains (Fig. 2.6). A somatomedin B domain is located at the N-terminus (amino acids 1–44) and binds plasminogen activator inhibitor-1 (Zhou et al. 2003) and interacts with the urokinase receptor (Wei et al. 1994). Immediately following the somatomedin B domain is an RGD cell adhesion sequence (residues 45–47), which is the major integrin-binding site in the protein. Adjacent to the RGD is a binding domain (amino acids 53–64) for thrombin–antithrombin complex and collagen (Schvartz et al. 1999). The core of vitronectin (residues 132–459) is homologous to hemopexin. At the C terminus, there are a plaminogen-binding site (residues 332–348) (Kost et al. 1992), two heparin-binding sites (residues 347–352 and 354–362) (Cardin and Weintraub 1989), and another plasminogen activator inhibitor-1 (PAI-1) binding site (residues

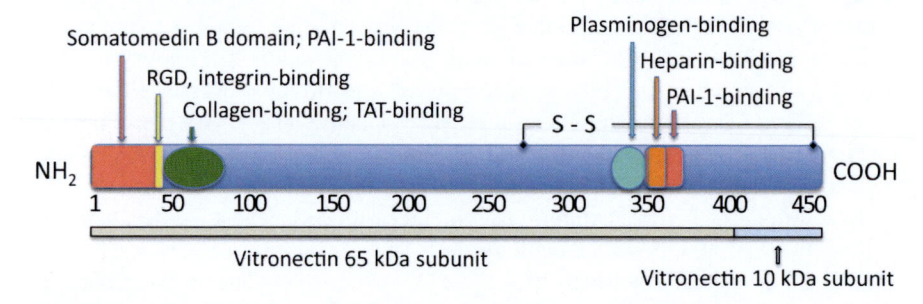

Fig. 2.6 *Protein-binding domains of vitronectin*. Vitronectin's major ligand-binding sites are indicated, with the major integrin-binding site RGD locates in residues 45–47. The disulfide bond connects the 65 and 10 kDa subunits of the cleaved form of vitronectin. PAI-1, plasminogen activator inhibitor-1; TAT, thrombin-antithrombin

348–370) (Gechtman et al. 1997). Vitronectin assumes different conformational states upon binding to ligands such as thrombin–antithrombin complex (Tomasini and Mosher 1991).

Vitronectin interacts with the extracellular matrix through its collagen- and heparin-binding domains, and with cells through its RGD integrin-binding sequence. Integrins αIIbβ3, αvβ1, αvβ3, αvβ5, αvβ8, and α8β1 recognize the RGD motif of vitronectin (Brooks et al. 1994; Marshall et al. 1995; Nishimura et al. 1994; Schnapp et al. 1995; Smith et al. 1990; Thiagarajan and Kelly 1988). α5β1, the major integrin receptor for fibronectin, does not recognize the RGD of vitronectin. Vitronectin–integrin interaction activates intracellular signaling pathways, induces protein phosphorylation, activates MAP kinase pathways, and mediates cell adhesion, spreading, migration, cell growth, differentiation, proliferation, and apoptosis (Felding-Habermann and Cheresh 1993; Meredith et al. 1996; Savill et al. 1990; Schvartz et al. 1999).

Vitronectin functions in wound healing, viral infection, and tumor growth and metastasis (Felding-Habermann and Cheresh 1993; Schvartz et al. 1999). Interestingly, vitronectin knockout mice developed normally with no major defects (Zheng et al. 1995), suggesting either vitronectin is dispensable or other molecule might play a rescue role in the absence of vitronectin.

2.4.2 Thrombospondins

Thrombospondins are a family of structurally related multifunctional, multimodular calcium-binding extracellular matrix glycoproteins encoded by separate genes. Five thrombospondins have been identified so far and can be divided into two groups: group A with thrombospondin-1 and -2 forming homotrimers, and group B with thrombospondin-3, -4, and -5 (also known as cartilage oligomeric matrix protein) forming homopentamers (Lawler 2000) (see Chap. 11). A single thrombospondin gene is present in *Drosophila* (Adams et al. 2003).

Thrombospondins have been shown to bind to cells, platelets, calcium, and various substances such as heparin, integrins, fibronectin, collagen, laminin, fibrinogen, plasminogen, osteonectin, and transforming growth factor-β; and are important for cell adhesion and spreading, platelet aggregation, angiogenesis, neurite outgrowth, and apoptosis (Adams 1997; Esemuede et al. 2004; Frazier 1991; Mosher 1990). Various functions of thrombospondins have been mapped to different structural domains. The N-terminal domain has a high affinity heparin-binding site with roles in platelet aggregation and endocytosis of thrombospondin-1. Besides various cell-binding sites in type I repeats and the C-terminal domain of thrombospondins, there is a RGD sequence in the type III calcium-binding repeats of thrombospondin-1, -2, and -5. The RGD sequence is not conserved in all thrombospondins as it exists in thrombospondin-4 and -5 of some species but is not found in any species of thrombospondin-3. The RGD cell-adhesive motif, which is found in repeat 12 of TSP-1 and TSP-2, makes these proteins potential ligands for $\alpha V\beta 3$, $\alpha IIb\beta 3$, $\alpha 5\beta 1$, and other RGD-recognizing integrins. Main-chain and side-chain coordination of calcium by RGD, however, forces it into a conformation that would not be expected to interact with integrins (Carlson et al. 2005; Kvansakul et al. 2004). Cell adhesion and biochemical experiments suggest that the sequence becomes active at low calcium concentrations (Chen et al. 1994; Kvansakul et al. 2004; Lawler and Hynes 1989; Lawler et al. 1988) or after disulfide reduction (Sun et al. 1992). Thus, this may be an example of an RGD sequence that is conditionally active.

Thrombospondin-1 can inhibit endothelial cell proliferation and migration, inhibit neovascularization, and promote growth and migration of smooth muscle cells and fibroblasts (Bagavandoss and Wilks 1990; Esemuede et al. 2004; Majack et al. 1988; Vogel et al. 1993). The medical focus of thrombospondin is on the role of thrombospondin in angiogenesis and tumor therapy.

2.4.3 Tenascins

Tenascins are a family of extracellular matrix glycoproteins including tenascin-C, tenascin-R, tenascin-W, tenascin-X, and tenascin-Y (Jones and Jones 2000). Tenascin-C was the first tenascin identified and is mainly synthesized by the nervous system and connective tissues. Tenascin-R is found in the nervous system. Tenascin-X and tenascin-Y are found primarily in muscle connective tissues. Tenascin-W is found in kidney and developing bone with a KGD sequence that interacts with integrins (Meloty-Kapella et al. 2008). The basic structure of tenascins is variable numbers of epidermal growth factor-like repeats followed by alternatively spliced fibronectin type III modules and a fibrinogen-like globular C-terminal domain (see Chap. 11).

Like thrombospondin-1, tenascin-C contains an RGD motif and is recognized by diverse integrins, yet is classified as an antiadhesive or adhesion-modulatory protein (Orend and Chiquet-Ehrismann 2000). $\alpha 8\beta 1$, $\alpha v\beta 3$, and $\alpha v\beta 6$ integrins all bind to the RGD motif in the third fibronectin type III repeat of tenascin-C, and $\alpha 9\beta 1$ integrin binds to the same module but to a different motif: the IDG motif in sequence EIDGIELT (Joshi et al. 1993; Prieto et al. 1993; Schnapp et al. 1995; Sriramarao et al. 1993; Yokosaki et al. 1998). Similar to $\alpha 9\beta 1$, $\alpha 7\beta 1$ integrin interacts with a VFDNFVLK sequence in the alternately spliced fibronectin type-III repeat D, which corresponds to the EIDGIELT sequence for $\alpha 9\beta 1$ integrin, and both $\alpha 7\beta 1$ and $\alpha 9\beta 1$ integrins promote neurite outgrowth (Andrews et al. 2009; Mercado et al. 2004). Human umbilical vein endothelial cells adhere to tenascin-C and partially spread through $\alpha 2\beta 1$ and $\alpha v\beta 3$ integrins (Sriramarao et al. 1993).

Cell adhesion to tenascin is weak, with adherent cells being elongated instead of flattened. Adhesion usually does not result in a rearranged actin cytoskeleton as is the case with cells adherent to fibronectin (Lotz et al. 1989; Sriramarao et al. 1993). Tenascin-C causes cells adherent to fibronectin to detach through direct interaction of tenascin-C with the ^{13}FN3 module of FN, which inhibit the binding of syndecan-4 to fibronectin followed by suppression of focal adhesion kinase and RhoA activity (Huang et al. 2001; Midwood and Schwarzbauer 2002). The metalloprotease meprin cleaves human tenascin-C at the seventh fibronectin type III repeats and destroys the antiadhesive ability of tenascin-C by removing the C-terminal anti-adhesion domain (Ambort et al. 2010).

2.4.4 Entactins (or Nidogens)

Entactins, also known as nidogens, are ubiquitous basement membrane glycoproteins (Timpl 1989). Two entactins expressed by distinct genes have been identified in vertebrates, named entactin-1 (~150 kDa) and entactin-2 (~200 kDa) (or nidogen-1 and nidogen-2) (Kohfeldt et al. 1998). Each isoform contains three globular domains with two in the N-terminus (named G1 and G2) and the third in the C-terminus (G3). A rod-like connecting domain composed of cysteine-rich epidermal growth factor-like repeats, which include the RGD integrin-binding sequence and a thyroglobulin-like repeat, connects the N- and C-terminal globules (see Chap. 4).

Entacin-1 binds strongly to both the laminin γ1 chain through globular domain G3 and to collagen IV through G2 (Fox et al. 1991; Poschl et al. 1996; Reinhardt et al. 1993), and serves as a link between self-assembled laminin and collagen IV to stabilize basement membrane (Timpl and Brown 1996) and integrate other extracellular matrix proteins. Entactin-1 also binds fibronectin, perlecan, and fibulins through its G2 and G3 domains (Hsieh et al. 1994; Kvansakul et al. 2001; Reinhardt et al. 1993; Sasaki et al. 1995).

The RGD integrin-binding sequence localizes to the second epidermal growth factor-like repeat in the rod-like domain. Entactin-1 mediates cell adhesion through $\alpha v \beta 3$ integrin recognizing the RGD sequence and $\alpha 3 \beta 1$ integrin recognizing a cysteine-rich epidermal growth factor repeat in the G2 globular domain (Dong et al. 1995; Gresham et al. 1996; Wu et al. 1995a; Yi et al. 1998). Mouse entactin-2 also contains a RGD sequence, but the RGD is changed to YGD in human entactin-2 (Kohfeldt et al. 1998). Mouse entactin-2 is found to mediate cell adhesion mainly through $\alpha 3 \beta 1$ and $\alpha 6 \beta 1$ integrins from antibody inhibition studies, although the GRGDS peptide only showed low inhibition suggesting that the RGD sequence in the mouse entactin-2 is not the major integrin-binding site (Salmivirta et al. 2002). While human entactin-2 can promote cell adhesion of many different cell lines, the receptor for cell adhesion has not been identified (Kohfeldt et al. 1998).

2.4.5 Nephronectin

Nephronectin is an extracellular matrix glycoprotein identified as a novel ligand for $\alpha 8 \beta 1$ integrins, an interaction that is essential for kidney development as demonstrated in mice lacking $\alpha 8 \beta 1$ integrin or nephronectin (Brandenberger et al. 2001).

Nephronectin is 70–90 kDa, with five epidermal growth factor like-repeats (residues 57–250), an RGD-containing linker domain (residues 382–384), and a C-terminal domain with sequence homology to meprin-A5 protein-receptor protein-tyrosine phosphatase μ, named the MAM domain (residues 417–561) (Brandenberger et al. 2001). Nephronectin is widely expressed in kidney, lung, brain, uterus, placenta, thyroid gland, and blood vessels (Huang and Lee 2005) with a similar distribution as α8β1 integrins (Brandenberger et al. 2001; Manabe et al. 2008; Wagner et al. 2003). Mice deficient in nephronectin have a similar phenotype as α8β1 knockout mice with kidney agenesis and hypoplasia (Linton et al. 2007; Muller et al. 1997).

To date, α8β1 integrin is the only identified receptor for nephronectin. Nephronectin interacts with α8β1 integrin through its RGD sequence in the linker domain and a synergetic LFEIFEIER sequence on the C-terminal side of the RGD motif (Sato et al. 2009). A synthetic peptide containing both RGD and LFEIFEIER sequence binds α8β1 integrin ~2,000 fold better than a peptide with only the RGD motif (Sato et al. 2009). The high affinity binding of nephronectin to α8β1 integrin partly answers why other α8β1 ligands with lower affinities such as fibronectin, vitronectin, or tenascin-C are not able to compensate for the deficiency of nephronectin in kidney development.

2.4.6 Fibrinogen

The interaction of the C-terminal tail of the γ-chain of fibrinogen with αIIbβ3 integrins on platelets has been subjected to extensive study because of its importance in platelet aggregation and thrombus formation. These studies have revealed a distinctive recognition motif (Springer et al. 2008). The sequence of the tail is ...GAKQAGDV in human. By crystallography, this sequence is unstructured in fibrinogen. Crystal structures of γC peptides bound to αIIbβ3 revealed that the peptide binds over an extended region with interaction of carboxyl groups of the penultimate aspartate and the C-terminal valine with metals in the integrin.

2.5 Integrin Signaling Pathways

Integrins are heterodimeric transmembrane proteins composed of two subunits, α and β, each with a large extracellular domain, a transmembrane domain, and a cytoplasmic domain. It is clear from the above descriptions that integrins are the major protein cells used to both bind and respond to the adhesive glycoproteins, linking the extracellular matrix to the intracellular cytoskeleton with a bidirectional signaling pathway across the plasma membrane, with integrin extracellular domains interacting with extracellular matrix, and the integrin cytoplasmic domains linking to the cytoskeleton and signal transduction pathways (Harburger and Calderwood 2009).

Integrin activation is controlled by "inside-out" signals to achieve high-affinity binding between integrins and adhesive glycoproteins (Banno and Ginsberg 2008). And, in turn, the ligation of adhesive glycoproteins with integrins activates the "outside-in" signaling pathways regulating cell responses such as migration, survival, differentiation, and proliferation (Hynes 2002). Many integrins are expressed in an inactive state (Hynes 2002). The binding of the phosphotyrosine-binding (PTB) like domain of talin and kindlin to the cytoplasmic domain of integrin-β subunit triggers the "inside-out" integrin activation, likely through disruption of a connection between the cytoplasmic domains of the α and β subunits of integrin, which leads to tail separation and conformational changes of integrin's extracellular domains, allowing the high affinity ligand-binding of integrins (Ginsberg et al. 2005). The reinforced ligation of integrins and adhesive glycoproteins triggers "outside-in" signals and induces integrin microclustering, conformational changes of integrin cytoplasmic domains, and recruitment of additional intracellular proteins to the integrin cytoplasmic domains forming a dynamic integrin "adhesome," including focal adhesion kinase (FAK), Src-family protein tyrosine kinases (SFKs), Ras and Rho GTPases, integrin-linked kinase (ILK), paxillin, vinculin, and others (Ginsberg et al. 2005; Zaidel-Bar et al. 2007). Such dynamic multiprotein complex can be assembled or disassembled by altering the associated proteins through integrin phosphorylation, competitor binding, or mechanical stresses, and therefore mediate cellular responses such as adhesion disassembly and cell migration (Harburger and Calderwood 2009).

2.6 Concluding Remarks

Adhesive glycoproteins have multiple cell receptor binding sites (Table 2.1) that interact with different integrins and regulate various cell functions, including cell adhesion, migration, differentiation, growth, neurite outgrowth, apoptosis, and tumor metastases. For example, fibronectin has the RGD in ^{10}FN3, LDV in the V region, IDAPS in ^{14}FN3, KLDAPT in ^{5}FN3, and probably more sites that remain to be discovered. These different integrin-binding sites interact with their own sets of integrins. Thus, glycoproteins may use different integrin-binding sites to bind different cells, and cells may use different integrins to adhere to different glycoproteins. For example, fibronectin uses the EDA module to bind $\alpha 9\beta 1$ integrin of endothelial cells of the lymphatic valve, and uses RGD in ^{10}FN3 module to bind $\alpha 5\beta 1$ integrin of fibroblasts, while fibroblasts use $\alpha 6\beta 1$ instead of $\alpha 5\beta 1$ to bind laminins.

There are several different ways of binding integrins. The major integrin-binding sequence is the RGD sequence, which is first discovered in fibronectin and found in more than 100 other proteins, including laminin, vitronectin, thrombospondins, tenascins, collagen, entactins, and nephronectin. Other similar sequences include LDV, iso-DGR, IDAPS, and KLDAPT in fibronectin; IGD and VFDNFVLK in tenascin-C; and EGD in entactins. All of those sequences include a major residue, aspartate (D), to bind integrins. Some RGD sequences are accompanied by a synergy

Table 2.1 Summary of glycoprotein–integrin interactions

Glycoprotein	Integrin-recognition sites	Integrins
Fibronectin	RGD in ^{10}FN3	α3β1, α5β1, α8β1, αvβ1, αIIbβ3, αvβ3, αvβ5, αvβ6
	LDV in V region	α4β1, α4β7
	^{A}FN3 (EDA)	α4β1, α9β1
	IDAPS in ^{14}FN3	α4β1
	KLDAPT in ^{5}FN3	α4β1, α4β7
	^{3}FN3	αβ1 (unknown α chain)
	Iso-DGR (spontaneously converted from NGR in ^{5}FN1 or ^{7}FN1)	αvβ3
Laminin	Combination of C-terminal conserved Glu residue of the γ subunit and the LG3 domain of the α subunit	α1β1, α2β1, α3β1, α6β1, α6β4, α7β1, α9β1, αvβ3
Vitronectin	RGD	αIIbβ3, αvβ1, αvβ3, αvβ5, αvβ8, α8β1
TSP-1	RGD	αvβ3, αIIbβ3, α5β1
TSP-2	RGD	αvβ3, αIIbβ3, α5β1
Tenascin-C	RGD	α8β1, αvβ3, αvβ6
	IDG (in EIDGELT)	α9β1
	VFDNFVLK	α7β1
Entactin-1	RGD	αvβ3
	EGF repeat in G2	α3β1
Nephronectin	RGD	α8β1
Fibrinogen	…GAKQAGDV	αIIbβ3
Collagen IV	GFOGER	α2β1
		α1β1

site, such as the PHSRN sequence in fibronectin and LEFIFEIER in nephronectin. The synergy site supports high-affinity integrin-RGD binding and is required for the formation of tensioned α5β1-fibronectin bonds. An important question for the future is whether synergy sites exist more widely in integrin-interacting proteins. Other integrin–glycoprotein interactions, including a critical GFOGER motif within a triple helical collagen peptide that binds to the I domain of α2β1 (Zhang et al. 2003), use Glu as the critical cation-coordinating residue. α6β1 integrin binds laminin-111 through a combination of the C-terminal conserved Glu residue of the γ subunit and the LG3 domain of the α subunit. Finally, in the case of fibrinogen, the C-terminal carboxyl group is recognized by αIIbβ3. These variations upon the RGD paradigm indicate that much more needs to be learned about such fine details of ligand–integrin interactions.

References

Adams JC (1997) Thrombospondin-1. Int J Biochem Cell Biol 29:861–865
Adams JC, Monk R, Taylor AL, Ozbek S, Fascetti N, Baumgartner S, Engel J (2003) Characterisation of *Drosophila* thrombospondin defines an early origin of pentameric thrombospondins. J Mol Biol 328:479–494

Akiyama SK, Yamada SS, Chen WT, Yamada KM (1989) Analysis of fibronectin receptor function with monoclonal antibodies: roles in cell adhesion, migration, matrix assembly, and cytoskeletal organization. J Cell Biol 109:863–875

Ambort D, Brellier F, Becker-Pauly C, Stocker W, Andrejevic-Blant S, Chiquet M, Sterchi EE (2010) Specific processing of tenascin-C by the metalloprotease meprinbeta neutralizes its inhibition of cell spreading. Matrix Biol 29:31–42

Andrews MR, Czvitkovich S, Dassie E, Vogelaar CF, Faissner A, Blits B, Gage FH, Ffrench-Constant C, Fawcett JW (2009) Alpha9 integrin promotes neurite outgrowth on tenascin-C and enhances sensory axon regeneration. J Neurosci 29:5546–5557

Aota S, Nomizu M, Yamada KM (1994) The short amino acid sequence Pro-His-Ser-Arg-Asn in human fibronectin enhances cell-adhesive function. J Biol Chem 269:24756–24761

Bae E, Sakai T, Mosher DF (2004) Assembly of exogenous fibronectin by fibronectin-null cells is dependent on the adhesive substrate. J Biol Chem 279:35749–35759

Bagavandoss P, Wilks JW (1990) Specific inhibition of endothelial cell proliferation by thrombospondin. Biochem Biophys Res Commun 170:867–872

Banno A, Ginsberg MH (2008) Integrin activation. Biochem Soc Trans 36:229–234

Barkalow FJ, Schwarzbauer JE (1991) Localization of the major heparin-binding site in fibronectin. J Biol Chem 266:7812–7818

Baron M, Norman D, Willis A, Campbell ID (1990) Structure of the fibronectin type 1 module. Nature 345:642–646

Bazigou E, Xie S, Chen C, Weston A, Miura N, Sorokin L, Adams R, Muro AF, Sheppard D, Makinen T (2009) Integrin-alpha9 is required for fibronectin matrix assembly during lymphatic valve morphogenesis. Dev Cell 17:175–186

Belkin AM, Stepp MA (2000) Integrins as receptors for laminins. Microsc Res Tech 51:280–301

Bernard BA, Yamada KM, Olden K (1982) Carbohydrates selectively protect a specific domain of fibronectin against proteases. J Biol Chem 257:8549–8554

Beyth RJ, Culp LA (1984) Complementary adhesive responses of human skin fibroblasts to the cell-binding domain of fibronectin and the heparan sulfate-binding protein, platelet factor-4. Exp Cell Res 155:537–548

Bingham RJ, Rudino-Pinera E, Meenan NA, Schwarz-Linek U, Turkenburg JP, Hook M, Garman EF, Potts JR (2008) Crystal structures of fibronectin-binding sites from Staphylococcus aureus FnBPA in complex with fibronectin domains. Proc Natl Acad Sci USA 105:12254–12258

Bork P, Doolittle RF (1992) Proposed acquisition of an animal protein domain by bacteria. Proc Natl Acad Sci USA 89:8990–8994

Brandenberger R, Schmidt A, Linton J, Wang D, Backus C, Denda S, Muller U, Reichardt LF (2001) Identification and characterization of a novel extracellular matrix protein nephronectin that is associated with integrin alpha8beta1 in the embryonic kidney. J Cell Biol 154:447–458

Brooks PC, Clark RA, Cheresh DA (1994) Requirement of vascular integrin alpha v beta 3 for angiogenesis. Science 264:569–571

Burton-Wurster N, Gendelman R, Chen H, Gu DN, Tetreault JW, Lust G, Schwarzbauer JE, MacLeod JN (1999) The cartilage-specific (V + C)- fibronectin isoform exists primarily in homodimeric and monomeric configurations. Biochem J 341(Pt 3):555–561

Cardin AD, Weintraub HJ (1989) Molecular modeling of protein-glycosaminoglycan interactions. Arteriosclerosis 9:21–32

Carlson CB, Bernstein DA, Annis DS, Misenheimer TM, Hannah BL, Mosher DF, Keck JL (2005) Structure of the calcium-rich signature domain of human thrombospondin-2. Nat Struct Mol Biol 12:910–914

Carnemolla B, Leprini A, Allemanni G, Saginati M, Zardi L (1992) The inclusion of the type III repeat ED-B in the fibronectin molecule generates conformational modifications that unmask a cryptic sequence. J Biol Chem 267:24689–24692

Chen H, Sottile J, O'Rourke KM, Dixit VM, Mosher DF (1994) Properties of recombinant mouse thrombospondin 2 expressed in Spodoptera cells. J Biol Chem 269:32226–32232

Cho J, Mosher DF (2006) Enhancement of thrombogenesis by plasma fibronectin cross-linked to fibrin and assembled in platelet thrombi. Blood 107:3555–3563

Collier IE, Wilhelm SM, Eisen AZ, Marmer BL, Grant GA, Seltzer JL, Kronberger A, He CS, Bauer EA, Goldberg GI (1988) H-ras oncogene-transformed human bronchial epithelial cells (TBE-1) secrete a single metalloprotease capable of degrading basement membrane collagen. J Biol Chem 263:6579–6587

Colognato H, Yurchenco PD (2000) Form and function: the laminin family of heterotrimers. Dev Dyn 218:213–234

Colucci S, Giannelli G, Grano M, Faccio R, Quaranta V, Zallone AZ (1996) Human osteoclast-like cells selectively recognize laminin isoforms, an event that induces migration and activates Ca^{2+} mediated signals. J Cell Sci 109(Pt 6):1527–1535

Constantine KL, Brew SA, Ingham KC, Llinas M (1992) 1H-n.m.r. studies of the fibronectin 13 kDa collagen-binding fragment. Evidence for autonomous conserved type I and type II domain folds. Biochem J 283(Pt 1):247–254

Coussen F, Choquet D, Sheetz MP, Erickson HP (2002) Trimers of the fibronectin cell adhesion domain localize to actin filament bundles and undergo rearward translocation. J Cell Sci 115:2581–2590

Curnis F, Longhi R, Crippa L, Cattaneo A, Dondossola E, Bachi A, Corti A (2006) Spontaneous formation of L-isoaspartate and gain of function in fibronectin. J Biol Chem 281:36466–36476

Darribere T, Guida K, Larjava H, Johnson KE, Yamada KM, Thiery JP, Boucaut JC (1990) In vivo analyses of integrin beta 1 subunit function in fibronectin matrix assembly. J Cell Biol 110:1813–1823

Dickinson CD, Gay DA, Parello J, Ruoslahti E, Ely KR (1994a) Crystals of the cell-binding module of fibronectin obtained from a series of recombinant fragments differing in length. J Mol Biol 238:123–127

Dickinson CD, Veerapandian B, Dai XP, Hamlin RC, Xuong NH, Ruoslahti E, Ely KR (1994b) Crystal structure of the tenth type III cell adhesion module of human fibronectin. J Mol Biol 236:1079–1092

Dong LJ, Hsieh JC, Chung AE (1995) Two distinct cell attachment sites in entactin are revealed by amino acid substitutions and deletion of the RGD sequence in the cysteine-rich epidermal growth factor repeat 2. J Biol Chem 270:15838–15843

Dzamba BJ, Jakab KR, Marsden M, Schwartz MA, DeSimone DW (2009) Cadherin adhesion, tissue tension, and noncanonical Wnt signaling regulate fibronectin matrix organization. Dev Cell 16:421–432

Elices MJ, Tsai V, Strahl D, Goel AS, Tollefson V, Arrhenius T, Wayner EA, Gaeta FC, Fikes JD, Firestein GS (1994) Expression and functional significance of alternatively spliced CS1 fibronectin in rheumatoid arthritis microvasculature. J Clin Invest 93:405–416

Engel J, Odermatt E, Engel A, Madri JA, Furthmayr H, Rohde H, Timpl R (1981) Shapes, domain organizations and flexibility of laminin and fibronectin, two multifunctional proteins of the extracellular matrix. J Mol Biol 150:97–120

Engvall E, Ruoslahti E, Miller EJ (1978) Affinity of fibronectin to collagens of different genetic types and to fibrinogen. J Exp Med 147:1584–1595

Ensenberger MG, Annis DS, Mosher DF (2004) Actions of the functional upstream domain of protein F1 of Streptococcus pyogenes on the conformation of fibronectin. Biophys Chem 112:201–207

Erat MC, Slatter DA, Lowe ED, Millard CJ, Farndale RW, Campbell ID, Vakonakis I (2009) Identification and structural analysis of type I collagen sites in complex with fibronectin fragments. Proc Natl Acad Sci USA 106:4195–4200

Erickson HP, Carrell N, McDonagh J (1981) Fibronectin molecule visualized in electron microscopy: a long, thin, flexible strand. J Cell Biol 91:673–678

Esemuede N, Lee T, Pierre-Paul D, Sumpio BE, Gahtan V (2004) The role of thrombospondin-1 in human disease. J Surg Res 122:135–142

Etheredge RE, Han S, Fossel E, Tanzer ML, Glimcher MJ (1985) Identification and quantitation of O-phosphoserine in human plasma fibronectin. FEBS Lett 186:259–262

Felding-Habermann B, Cheresh DA (1993) Vitronectin and its receptors. Curr Opin Cell Biol 5:864–868

ffrench-Constant C (1995) Alternative splicing of fibronectin – many different proteins but few different functions. Exp Cell Res 221:261–271

Fogerty FJ, Akiyama SK, Yamada KM, Mosher DF (1990) Inhibition of binding of fibronectin to matrix assembly sites by anti-integrin (alpha 5 beta 1) antibodies. J Cell Biol 111:699–708

Fox JW, Mayer U, Nischt R, Aumailley M, Reinhardt D, Wiedemann H, Mann K, Timpl R, Krieg T, Engel J et al (1991) Recombinant nidogen consists of three globular domains and mediates binding of laminin to collagen type IV. EMBO J 10:3137–3146

Frazier WA (1991) Thrombospondins. Curr Opin Cell Biol 3:792–799

Friedland JC, Lee MH, Boettiger D (2009) Mechanically activated integrin switch controls alpha5beta1 function. Science 323:642–644

Fukuda T, Yoshida N, Kataoka Y, Manabe R, Mizuno-Horikawa Y, Sato M, Kuriyama K, Yasui N, Sekiguchi K (2002) Mice lacking the EDB segment of fibronectin develop normally but exhibit reduced cell growth and fibronectin matrix assembly in vitro. Cancer Res 62:5603–5610

Gechtman Z, Belleli A, Lechpammer S, Shaltiel S (1997) The cluster of basic amino acids in vitronectin contributes to its binding of plasminogen activator inhibitor-1: evidence from thrombin-, elastase- and plasmin-cleaved vitronectins and anti-peptide antibodies. Biochem J 325(Pt 2):339–349

Geiger B, Bershadsky A, Pankov R, Yamada KM (2001) Transmembrane crosstalk between the extracellular matrix–cytoskeleton crosstalk. Nat Rev Mol Cell Biol 2:793–805

George EL, Georges-Labouesse EN, Patel-King RS, Rayburn H, Hynes RO (1993) Defects in mesoderm, neural tube and vascular development in mouse embryos lacking fibronectin. Development 119:1079–1091

Giancotti FG, Ruoslahti E (1990) Elevated levels of the alpha 5 beta 1 fibronectin receptor suppress the transformed phenotype of Chinese hamster ovary cells. Cell 60:849–859

Ginsberg MH, Partridge A, Shattil SJ (2005) Integrin regulation. Curr Opin Cell Biol 17:509–516

Givant-Horwitz V, Davidson B, Reich R (2005) Laminin-induced signaling in tumor cells. Cancer Lett 223:1–10

Gonzales M, Haan K, Baker SE, Fitchmun M, Todorov I, Weitzman S, Jones JC (1999) A cell signal pathway involving laminin-5, alpha3beta1 integrin, and mitogen-activated protein kinase can regulate epithelial cell proliferation. Mol Biol Cell 10:259–270

Gresham HD, Graham IL, Griffin GL, Hsieh JC, Dong LJ, Chung AE, Senior RM (1996) Domain-specific interactions between entactin and neutrophil integrins. G2 domain ligation of integrin alpha3beta1 and E domain ligation of the leukocyte response integrin signal for different responses. J Biol Chem 271:30587–30594

Guan JL, Hynes RO (1990) Lymphoid cells recognize an alternatively spliced segment of fibronectin via the integrin receptor alpha 4 beta 1. Cell 60:53–61

Halliday NL, Tomasek JJ (1995) Mechanical properties of the extracellular matrix influence fibronectin fibril assembly in vitro. Exp Cell Res 217:109–117

Harburger DS, Calderwood DA (2009) Integrin signalling at a glance. J Cell Sci 122:159–163

Hintermann E, Quaranta V (2004) Epithelial cell motility on laminin-5: regulation by matrix assembly, proteolysis, integrins and erbB receptors. Matrix Biol 23:75–85

Hirano H, Yamada Y, Sullivan M, de Crombrugghe B, Pastan I, Yamada KM (1983) Isolation of genomic DNA clones spanning the entire fibronectin gene. Proc Natl Acad Sci USA 80:46–50

Hsieh JC, Wu C, Chung AE (1994) The binding of fibronectin to entactin is mediated through the 29 kDa amino terminal fragment of fibronectin and the G2 domain of entactin. Biochem Biophys Res Commun 199:1509–1517

Huang JT, Lee V (2005) Identification and characterization of a novel human nephronectin gene in silico. Int J Mol Med 15:719–724

Huang W, Chiquet-Ehrismann R, Moyano JV, Garcia-Pardo A, Orend G (2001) Interference of tenascin-C with syndecan-4 binding to fibronectin blocks cell adhesion and stimulates tumor cell proliferation. Cancer Res 61:8586–8594

Humphries JD, Byron A, Humphries MJ (2006) Integrin ligands at a glance. J Cell Sci 119: 3901–3903

Huveneers S, Truong H, Fassler R, Sonnenberg A, Danen EH (2008) Binding of soluble fibronectin to integrin alpha5 beta1 - link to focal adhesion redistribution and contractile shape. J Cell Sci 121:2452–2462

Hynes R (1990) Fibronectins. Springer, New York

Hynes RO (2002) Integrins: bidirectional, allosteric signaling machines. Cell 110:673–687

Hynes RO (2009) The extracellular matrix: not just pretty fibrils. Science 326:1216–1219

Ido H, Harada K, Futaki S, Hayashi Y, Nishiuchi R, Natsuka Y, Li S, Wada Y, Combs AC, Ervasti JM, Sekiguchi K (2004) Molecular dissection of the alpha-dystroglycan- and integrin-binding sites within the globular domain of human laminin-10. J Biol Chem 279:10946–10954

Ido H, Harada K, Yagi Y, Sekiguchi K (2006) Probing the integrin-binding site within the globular domain of laminin-511 with the function-blocking monoclonal antibody 4C7. Matrix Biol 25:112–117

Ido H, Ito S, Taniguchi Y, Hayashi M, Sato-Nishiuchi R, Sanzen N, Hayashi Y, Futaki S, Sekiguchi K (2008) Laminin isoforms containing the gamma3 chain are unable to bind to integrins due to the absence of the glutamic acid residue conserved in the C-terminal regions of the gamma1 and gamma2 chains. J Biol Chem 283:28149–28157

Ido H, Nakamura A, Kobayashi R, Ito S, Li S, Futaki S, Sekiguchi K (2007) The requirement of the glutamic acid residue at the third position from the carboxyl termini of the laminin gamma chains in integrin binding by laminins. J Biol Chem 282:11144–11154

Ingham KC, Brew SA, Atha DH (1990) Interaction of heparin with fibronectin and isolated fibronectin domains. Biochem J 272:605–611

Ingham KC, Brew SA, Novokhatny VV (1995) Influence of carbohydrate on structure, stability, and function of gelatin-binding fragments of fibronectin. Arch Biochem Biophys 316:235–240

Izzard CS, Radinsky R, Culp LA (1986) Substratum contacts and cytoskeletal reorganization of BALB/c 3T3 cells on a cell-binding fragment and heparin-binding fragments of plasma fibronectin. Exp Cell Res 165:320–336

Jarnagin WR, Rockey DC, Koteliansky VE, Wang SS, Bissell DM (1994) Expression of variant fibronectins in wound healing: cellular source and biological activity of the EIIIA segment in rat hepatic fibrogenesis. J Cell Biol 127:2037–2048

Jones FS, Jones PL (2000) The tenascin family of ECM glycoproteins: structure, function, and regulation during embryonic development and tissue remodeling. Dev Dyn 218:235–259

Jones GE, Arumugham RG, Tanzer ML (1986) Fibronectin glycosylation modulates fibroblast adhesion and spreading. J Cell Biol 103:1663–1670

Joshi P, Chung CY, Aukhil I, Erickson HP (1993) Endothelial cells adhere to the RGD domain and the fibrinogen-like terminal knob of tenascin. J Cell Sci 106(Pt 1):389–400

Kibbey MC, Grant DS, Kleinman HK (1992) Role of the SIKVAV site of laminin in promotion of angiogenesis and tumor growth: an in vivo Matrigel model. J Natl Cancer Inst 84:1633–1638

Kohfeldt E, Sasaki T, Gohring W, Timpl R (1998) Nidogen-2: a new basement membrane protein with diverse binding properties. J Mol Biol 282:99–109

Kornblihtt AR, Pesce CG, Alonso CR, Cramer P, Srebrow A, Werbajh S, Muro AF (1996) The fibronectin gene as a model for splicing and transcription studies. FASEB J 10:248–257

Kosmehl H, Berndt A, Katenkamp D (1996) Molecular variants of fibronectin and laminin: structure, physiological occurrence and histopathological aspects. Virchows Arch 429: 311–322

Kost C, Stuber W, Ehrlich HJ, Pannekoek H, Preissner KT (1992) Mapping of binding sites for heparin, plasminogen activator inhibitor-1, and plasminogen to vitronectin's heparin-binding region reveals a novel vitronectin-dependent feedback mechanism for the control of plasmin formation. J Biol Chem 267:12098–12105

Kubota S, Tashiro K, Yamada Y (1992) Signaling site of laminin with mitogenic activity. J Biol Chem 267:4285–4288

Kvansakul M, Adams JC, Hohenester E (2004) Structure of a thrombospondin C-terminal fragment reveals a novel calcium core in the type 3 repeats. EMBO J 23:1223–1233

Kvansakul M, Hopf M, Ries A, Timpl R, Hohenester E (2001) Structural basis for the high-affinity interaction of nidogen-1 with immunoglobulin-like domain 3 of perlecan. EMBO J 20:5342–5346

Lark MW, Laterra J, Culp LA (1985) Close and focal contact adhesions of fibroblasts to a fibronectin-containing matrix. Fed Proc 44:394–403

Laterra J, Norton EK, Izzard CS, Culp LA (1983a) Contact formation by fibroblasts adhering to heparan sulfate-binding substrata (fibronectin or platelet factor 4). Exp Cell Res 146:15–27

Laterra J, Silbert JE, Culp LA (1983b) Cell surface heparan sulfate mediates some adhesive responses to glycosaminoglycan-binding matrices, including fibronectin. J Cell Biol 96:112–123

Lawler J (2000) The functions of thrombospondin-1 and -2. Curr Opin Cell Biol 12:634–640

Lawler J, Hynes RO (1989) An integrin receptor on normal and thrombasthenic platelets that binds thrombospondin. Blood 74:2022–2027

Lawler J, Weinstein R, Hynes RO (1988) Cell attachment to thrombospondin: the role of ARG-GLY-ASP, calcium, and integrin receptors. J Cell Biol 107:2351–2361

Leahy DJ, Aukhil I, Erickson HP (1996) 2.0 A crystal structure of a four-domain segment of human fibronectin encompassing the RGD loop and synergy region. Cell 84:155–164

Leikina E, Mertts MV, Kuznetsova N, Leikin S (2002) Type I collagen is thermally unstable at body temperature. Proc Natl Acad Sci USA 99:1314–1318

Leiss M, Beckmann K, Giros A, Costell M, Fassler R (2008) The role of integrin binding sites in fibronectin matrix assembly in vivo. Curr Opin Cell Biol 20:502–507

Li S, Harrison D, Carbonetto S, Fassler R, Smyth N, Edgar D, Yurchenco PD (2002) Matrix assembly, regulation, and survival functions of laminin and its receptors in embryonic stem cell differentiation. J Cell Biol 157:1279–1290

Liao YF, Gotwals PJ, Koteliansky VE, Sheppard D, Van De Water L (2002) The EIIIA segment of fibronectin is a ligand for integrins alpha 9beta 1 and alpha 4beta 1 providing a novel mechanism for regulating cell adhesion by alternative splicing. J Biol Chem 277:14467–14474

Linton JM, Martin GR, Reichardt LF (2007) The ECM protein nephronectin promotes kidney development via integrin alpha8beta1-mediated stimulation of Gdnf expression. Development 134:2501–2509

Liu MC, Lipmann F (1985) Isolation of tyrosine-O-sulfate by Pronase hydrolysis from fibronectin secreted by Fujinami sarcoma virus-infected rat fibroblasts. Proc Natl Acad Sci USA 82:34–37

Lotz MM, Burdsal CA, Erickson HP, McClay DR (1989) Cell adhesion to fibronectin and tenascin: quantitative measurements of initial binding and subsequent strengthening response. J Cell Biol 109:1795–1805

MacLeod JN, Burton-Wurster N, Gu DN, Lust G (1996) Fibronectin mRNA splice variant in articular cartilage lacks bases encoding the V, III-15, and I-10 protein segments. J Biol Chem 271:18954–18960

Magnusson MK, Mosher DF (1998) Fibronectin: structure, assembly, and cardiovascular implications. Arterioscler Thromb Vasc Biol 18:1363–1370

Main AL, Harvey TS, Baron M, Boyd J, Campbell ID (1992) The three-dimensional structure of the tenth type III module of fibronectin: an insight into RGD-mediated interactions. Cell 71:671–678

Majack RA, Goodman LV, Dixit VM (1988) Cell surface thrombospondin is functionally essential for vascular smooth muscle cell proliferation. J Cell Biol 106:415–422

Malinda KM, Kleinman HK (1996) The laminins. Int J Biochem Cell Biol 28:957–959

Manabe R, Ohe N, Maeda T, Fukuda T, Sekiguchi K (1997) Modulation of cell-adhesive activity of fibronectin by the alternatively spliced EDA segment. J Cell Biol 139:295–307

Manabe R, Tsutsui K, Yamada T, Kimura M, Nakano I, Shimono C, Sanzen N, Furutani Y, Fukuda T, Oguri Y, Shimamoto K, Kiyozumi D, Sato Y, Sado Y, Senoo H, Yamashina S, Fukuda S, Kawai J, Sugiura N, Kimata K, Hayashizaki Y, Sekiguchi K (2008) Transcriptome-based systematic identification of extracellular matrix proteins. Proc Natl Acad Sci USA 105:12849–12854

Mao Y, Schwarzbauer JE (2005) Fibronectin fibrillogenesis, a cell-mediated matrix assembly process. Matrix Biol 24:389–399

Marshall JF, Rutherford DC, McCartney AC, Mitjans F, Goodman SL, Hart IR (1995) Alpha v beta 1 is a receptor for vitronectin and fibrinogen, and acts with alpha 5 beta 1 to mediate spreading on fibronectin. J Cell Sci 108(Pt 3):1227–1238

Mayer U, Kohfeldt E, Timpl R (1998) Structural and genetic analysis of laminin-nidogen interaction. Ann NY Acad Sci 857:130–142

McDonald JA, Quade BJ, Broekelmann TJ, LaChance R, Forsman K, Hasegawa E, Akiyama S (1987) Fibronectin's cell-adhesive domain and an amino-terminal matrix assembly domain participate in its assembly into fibroblast pericellular matrix. J Biol Chem 262:2957–2967

McKeown-Longo PJ, Mosher DF (1985) Interaction of the 70,000-mol-wt amino-terminal fragment of fibronectin with the matrix-assembly receptor of fibroblasts. J Cell Biol 100:364–374

Meloty-Kapella CV, Degen M, Chiquet-Ehrismann R, Tucker RP (2008) Effects of tenascin-W on osteoblasts in vitro. Cell Tissue Res 334:445–455

Menrad A, Menssen HD (2005) ED-B fibronectin as a target for antibody-based cancer treatments. Expert Opin Ther Targets 9:491–500

Mercado ML, Nur-e-Kamal A, Liu HY, Gross SR, Movahed R, Meiners S (2004) Neurite outgrowth by the alternatively spliced region of human tenascin-C is mediated by neuronal alpha7beta1 integrin. J Neurosci 24:238–247

Meredith JE Jr, Winitz S, Lewis JM, Hess S, Ren XD, Renshaw MW, Schwartz MA (1996) The regulation of growth and intracellular signaling by integrins. Endocr Rev 17:207–220

Midwood KS, Schwarzbauer JE (2002) Tenascin-C modulates matrix contraction via focal adhesion kinase- and Rho-mediated signaling pathways. Mol Biol Cell 13:3601–3613

Millard CJ, Campbell ID, Pickford AR (2005) Gelatin binding to the 8F19F1 module pair of human fibronectin requires site-specific N-glycosylation. FEBS Lett 579:4529–4534

Miner JH, Yurchenco PD (2004) Laminin functions in tissue morphogenesis. Annu Rev Cell Dev Biol 20:255–284

Morrison PR, Edsall JT, Miller SG (1948) Preparation and properties of serum and plasma proteins; the separation of purified fibrinogen from fraction I of human plasma. J Am Chem Soc 70:3103–3108

Mosesson MW, Umfleet RA (1970) The cold-insoluble globulin of human plasma. I. Purification, primary characterization, and relationship to fibrinogen and other cold-insoluble fraction components. J Biol Chem 245:5728–5736

Mosher D (1989) Fibronectin. Academic, San Diego

Mosher DF (1990) Physiology of thrombospondin. Annu Rev Med 41:85–97

Mostafavi-Pour Z, Askari JA, Whittard JD, Humphries MJ (2001) Identification of a novel heparin-binding site in the alternatively spliced IIICS region of fibronectin: roles of integrins and proteoglycans in cell adhesion to fibronectin splice variants. Matrix Biol 20:63–73

Mould AP, Askari JA, Aota S, Yamada KM, Irie A, Takada Y, Mardon HJ, Humphries MJ (1997) Defining the topology of integrin alpha5beta1-fibronectin interactions using inhibitory anti-alpha5 and anti-beta1 monoclonal antibodies. Evidence that the synergy sequence of fibronectin is recognized by the amino-terminal repeats of the alpha5 subunit. J Biol Chem 272: 17283–17292

Mould AP, Komoriya A, Yamada KM, Humphries MJ (1991) The CS5 peptide is a second site in the IIICS region of fibronectin recognized by the integrin alpha 4 beta 1. Inhibition of alpha 4 beta 1 function by RGD peptide homologues. J Biol Chem 266:3579–3585

Moyano JV, Carnemolla B, Albar JP, Leprini A, Gaggero B, Zardi L, Garcia-Pardo A (1999) Cooperative role for activated alpha4 beta1 integrin and chondroitin sulfate proteoglycans in cell adhesion to the heparin III domain of fibronectin. Identification of a novel heparin and cell binding sequence in repeat III5. J Biol Chem 274:135–142

Moyano JV, Carnemolla B, Dominguez-Jimenez C, Garcia-Gila M, Albar JP, Sanchez-Aparicio P, Leprini A, Querze G, Zardi L, Garcia-Pardo A (1997) Fibronectin type III5 repeat contains a novel cell adhesion sequence, KLDAPT, which binds activated alpha4beta1 and alpha4beta7 integrins. J Biol Chem 272:24832–24836

Muller U, Wang D, Denda S, Meneses JJ, Pedersen RA, Reichardt LF (1997) Integrin alpha8beta1 is critically important for epithelial-mesenchymal interactions during kidney morphogenesis. Cell 88:603–613

Muro AF, Chauhan AK, Gajovic S, Iaconcig A, Porro F, Stanta G, Baralle FE (2003) Regulated splicing of the fibronectin EDA exon is essential for proper skin wound healing and normal lifespan. J Cell Biol 162:149–160

Ni H, Yuen PS, Papalia JM, Trevithick JE, Sakai T, Fassler R, Hynes RO, Wagner DD (2003) Plasma fibronectin promotes thrombus growth and stability in injured arterioles. Proc Natl Acad Sci USA 100:2415–2419

Nishimura SL, Sheppard D, Pytela R (1994) Integrin alpha v beta 8. Interaction with vitronectin and functional divergence of the beta 8 cytoplasmic domain. J Biol Chem 269:28708–28715

Nishiuchi R, Takagi J, Hayashi M, Ido H, Yagi Y, Sanzen N, Tsuji T, Yamada M, Sekiguchi K (2006) Ligand-binding specificities of laminin-binding integrins: a comprehensive survey of laminin-integrin interactions using recombinant alpha3beta1, alpha6beta1, alpha7beta1 and alpha6beta4 integrins. Matrix Biol 25:189–197

Nomizu M, Kuratomi Y, Malinda KM, Song SY, Miyoshi K, Otaka A, Powell SK, Hoffman MP, Kleinman HK, Yamada Y (1998) Cell binding sequences in mouse laminin alpha1 chain. J Biol Chem 273:32491–32499

O'Leary JM, Hamilton JM, Deane CM, Valeyev NV, Sandell LJ, Downing AK (2004) Solution structure and dynamics of a prototypical chordin-like cysteine-rich repeat (von Willebrand Factor type C module) from collagen IIA. J Biol Chem 279:53857–53866

Obara M, Kang MS, Yamada KM (1988) Site-directed mutagenesis of the cell-binding domain of human fibronectin: separable, synergistic sites mediate adhesive function. Cell 53:649–657

Obara M, Sakuma T, Fujikawa K (2010) The third type III module of human fibronectin mediates cell adhesion and migration. J Biochem 147:327–335

Ohashi T, Erickson HP (2005) Domain unfolding plays a role in superfibronectin formation. J Biol Chem 280:39143–39151

Olden K, Pratt RM, Yamada KM (1979) Role of carbohydrate in biological function of the adhesive glycoprotein fibronectin. Proc Natl Acad Sci USA 76:3343–3347

Orend G, Chiquet-Ehrismann R (2000) Adhesion modulation by antiadhesive molecules of the extracellular matrix. Exp Cell Res 261:104–110

Ozhogina OA, Trexler M, Banyai L, Llinas M, Patthy L (2001) Origin of fibronectin type II (FN2) modules: structural analyses of distantly-related members of the kringle family idey the kringle domain of neurotrypsin as a potential link between FN2 domains and kringles. Protein Sci 10:2114–2122

Pankov R, Yamada KM (2002) Fibronectin at a glance. J Cell Sci 115:3861–3863

Patarroyo M, Tryggvason K, Virtanen I (2002) Laminin isoforms in tumor invasion, angiogenesis and metastasis. Semin Cancer Biol 12:197–207

Paul JI, Hynes RO (1984) Multiple fibronectin subunits and their post-translational modifications. J Biol Chem 259:13477–13487

Pickford AR, Potts JR, Bright JR, Phan I, Campbell ID (1997) Solution structure of a type 2 module from fibronectin: implications for the structure and function of the gelatin-binding domain. Structure 5:359–370

Pickford AR, Smith SP, Staunton D, Boyd J, Campbell ID (2001) The hairpin structure of the (6)F1(1)F2(2)F2 fragment from human fibronectin enhances gelatin binding. EMBO J 20:1519–1529

Pierschbacher MD, Ruoslahti E (1984) Cell attachment activity of fibronectin can be duplicated by small synthetic fragments of the molecule. Nature 309:30–33

Poschl E, Mayer U, Stetefeld J, Baumgartner R, Holak TA, Huber R, Timpl R (1996) Site-directed mutagenesis and structural interpretation of the nidogen binding site of the laminin gamma1 chain. EMBO J 15:5154–5159

Potts JR, Bright JR, Bolton D, Pickford AR, Campbell ID (1999) Solution structure of the N-terminal F1 module pair from human fibronectin. Biochemistry 38:8304–8312

Preissner KT (1991) Structure and biological role of vitronectin. Annu Rev Cell Biol 7:275–310

Prieto AL, Edelman GM, Crossin KL (1993) Multiple integrins mediate cell attachment to cytotactin/tenascin. Proc Natl Acad Sci USA 90:10154–10158

Reinhardt D, Mann K, Nischt R, Fox JW, Chu ML, Krieg T, Timpl R (1993) Mapping of nidogen binding sites for collagen type IV, heparan sulfate proteoglycan, and zinc. J Biol Chem 268:10881–10887

Rozzo C, Ratti P, Ponzoni M, Cornaglia-Ferraris P (1993) Modulation of alpha 1 beta 1, alpha 2 beta 1 and alpha 3 beta 1 integrin heterodimers during human neuroblastoma cell differentiation. FEBS Lett 332:263–267

Ruoslahti E (2003) The RGD story: a personal account. Matrix Biol 22:459–465

Ruoslahti E, Engvall E, Hayman EG, Spiro RG (1981) Comparative studies on amniotic fluid and plasma fibronectins. Biochem J 193:295–299

Salmivirta K, Talts JF, Olsson M, Sasaki T, Timpl R, Ekblom P (2002) Binding of mouse nidogen-2 to basement membrane components and cells and its expression in embryonic and adult tissues suggest complementary functions of the two nidogens. Exp Cell Res 279:188–201

Sasaki T, Gohring W, Pan TC, Chu ML, Timpl R (1995) Binding of mouse and human fibulin-2 to extracellular matrix ligands. J Mol Biol 254:892–899

Sato Y, Uemura T, Morimitsu K, Sato-Nishiuchi R, Manabe R, Takagi J, Yamada M, Sekiguchi K (2009) Molecular basis of the recognition of nephronectin by integrin alpha8beta1. J Biol Chem 284:14524–14536

Savill J, Dransfield I, Hogg N, Haslett C (1990) Vitronectin receptor-mediated phagocytosis of cells undergoing apoptosis. Nature 343:170–173

Scarborough RM, Naughton MA, Teng W, Rose JW, Phillips DR, Nannizzi L, Arfsten A, Campbell AM, Charo IF (1993) Design of potent and specific integrin antagonists. Peptide antagonists with high specificity for glycoprotein IIb-IIIa. J Biol Chem 268:1066–1073

Scheele S, Nystrom A, Durbeej M, Talts JF, Ekblom M, Ekblom P (2007) Laminin isoforms in development and disease. J Mol Med 85:825–836

Schnapp LM, Hatch N, Ramos DM, Klimanskaya IV, Sheppard D, Pytela R (1995) The human integrin alpha 8 beta 1 functions as a receptor for tenascin, fibronectin, and vitronectin. J Biol Chem 270:23196–23202

Schvartz I, Seger D, Shaltiel S (1999) Vitronectin. Int J Biochem Cell Biol 31:539–544

Schwartz MA, Assoian RK (2001) Integrins and cell proliferation: regulation of cyclin-dependent kinases via cytoplasmic signaling pathways. J Cell Sci 114:2553–2560

Schwarz-Linek U, Hook M, Potts JR (2004) The molecular basis of fibronectin-mediated bacterial adherence to host cells. Mol Microbiol 52:631–641

Schwarzbauer JE (1991) Identification of the fibronectin sequences required for assembly of a fibrillar matrix. J Cell Biol 113:1463–1473

Sechler JL, Corbett SA, Schwarzbauer JE (1997) Modulatory roles for integrin activation and the synergy site of fibronectin during matrix assembly. Mol Biol Cell 8:2563–2573

Sechler JL, Cumiskey AM, Gazzola DM, Schwarzbauer JE (2000) A novel RGD-independent fibronectin assembly pathway initiated by alpha4beta1 integrin binding to the alternatively spliced V region. J Cell Sci 113(Pt 8):1491–1498

Seiffert D, Crain K, Wagner NV, Loskutoff DJ (1994) Vitronectin gene expression in vivo. Evidence for extrahepatic synthesis and acute phase regulation. J Biol Chem 269:19836–19842

Sharma A, Askari JA, Humphries MJ, Jones EY, Stuart DI (1999) Crystal structure of a heparin- and integrin-binding segment of human fibronectin. EMBO J 18:1468–1479

Shinde AV, Bystroff C, Wang C, Vogelezang MG, Vincent PA, Hynes RO, Van De Water L (2008) Identification of the peptide sequences within the EIIIA (EDA) segment of fibronectin that mediate integrin alpha9beta1-dependent cellular activities. J Biol Chem 283:2858–2870

Skorstengaard K, Thogersen HC, Petersen TE (1984) Complete primary structure of the collagen-binding domain of bovine fibronectin. Eur J Biochem 140:235–243

Skorstengaard K, Thogersen HC, Vibe-Pedersen K, Petersen TE, Magnusson S (1982) Purification of twelve cyanogen bromide fragments from bovine plasma fibronectin and the amino acid sequence of eight of them. Overlap evidence aligning two plasmic fragments, internal homology in gelatin-binding region and phosphorylation site near C terminus. Eur J Biochem 128:605–623

Smith JW, Vestal DJ, Irwin SV, Burke TA, Cheresh DA (1990) Purification and functional characterization of integrin alpha v beta 5. An adhesion receptor for vitronectin. J Biol Chem 265:11008–11013

Sottile J, Schwarzbauer J, Selegue J, Mosher DF (1991) Five type I modules of fibronectin form a functional unit that binds to fibroblasts and *Staphylococcus aureus*. J Biol Chem 266: 12840–12843

Springer TA, Zhu J, Xiao T (2008) Structural basis for distinctive recognition of fibrinogen gammaC peptide by the platelet integrin alphaIIbbeta3. J Cell Biol 182:791–800

Sriramarao P, Mendler M, Bourdon MA (1993) Endothelial cell attachment and spreading on human tenascin is mediated by alpha 2 beta 1 and alpha v beta 3 integrins. J Cell Sci 105(Pt 4): 1001–1012

Stetefeld J, Mayer U, Timpl R, Huber R (1996) Crystal structure of three consecutive laminin-type epidermal growth factor-like (LE) modules of laminin gamma1 chain harboring the nidogen binding site. J Mol Biol 257:644–657

Sun X, Skorstengaard K, Mosher DF (1992) Disulfides modulate RGD-inhibitable cell adhesive activity of thrombospondin. J Cell Biol 118:693–701

Takahashi S, Leiss M, Moser M, Ohashi T, Kitao T, Heckmann D, Pfeifer A, Kessler H, Takagi J, Erickson HP, Fassler R (2007) The RGD motif in fibronectin is essential for development but dispensable for fibril assembly. J Cell Biol 178:167–178

Thiagarajan P, Kelly K (1988) Interaction of thrombin-stimulated platelets with vitronectin (S-protein of complement) substrate: inhibition by a monoclonal antibody to glycoprotein IIb–IIIa complex. Thromb Haemost 60:514–517

Timpl R (1989) Structure and biological activity of basement membrane proteins. Eur J Biochem 180:487–502

Timpl R, Brown JC (1996) Supramolecular assembly of basement membranes. Bioessays 18: 123–132

Timpl R, Tisi D, Talts JF, Andac Z, Sasaki T, Hohenester E (2000) Structure and function of laminin LG modules. Matrix Biol 19:309–317

Tomasini BR, Mosher DF (1991) Vitronectin. Prog Hemost Thromb 10:269–305

Tomasini-Johansson BR, Annis DS, Mosher DF (2006) The N-terminal 70-kDa fragment of fibronectin binds to cell surface fibronectin assembly sites in the absence of intact fibronectin. Matrix Biol 25:282–293

Tucker RP, Chiquet-Ehrismann R (2009) Evidence for the evolution of tenascin and fibronectin early in the chordate lineage. Int J Biochem Cell Biol 41:424–434

Tzu J, Marinkovich MP (2008) Bridging structure with function: structural, regulatory, and developmental role of laminins. Int J Biochem Cell Biol 40:199–214

Vogel T, Guo NH, Krutzsch HC, Blake DA, Hartman J, Mendelovitz S, Panet A, Roberts DD (1993) Modulation of endothelial cell proliferation, adhesion, and motility by recombinant heparin-binding domain and synthetic peptides from the type I repeats of thrombospondin. J Cell Biochem 53:74–84

Wagner TE, Frevert CW, Herzog EL, Schnapp LM (2003) Expression of the integrin subunit alpha8 in murine lung development. J Histochem Cytochem 51:1307–1315

Watt FM (2002) Role of integrins in regulating epidermal adhesion, growth and differentiation. EMBO J 21:3919–3926

Wayner EA, Garcia-Pardo A, Humphries MJ, McDonald JA, Carter WG (1989) Identification and characterization of the T lymphocyte adhesion receptor for an alternative cell attachment domain (CS-1) in plasma fibronectin. J Cell Biol 109:1321–1330

Weeks BS, DiSalvo J, Kleinman HK (1990) Laminin-mediated process formation in neuronal cells involves protein dephosphorylation. J Neurosci Res 27:418–426

Weeks BS, Papadopoulos V, Dym M, Kleinman HK (1991) cAMP promotes branching of laminin-induced neuronal processes. J Cell Physiol 147:62–67

Wei Y, Waltz DA, Rao N, Drummond RJ, Rosenberg S, Chapman HA (1994) Identification of the urokinase receptor as an adhesion receptor for vitronectin. J Biol Chem 269:32380–32388

Wennerberg K, Lohikangas L, Gullberg D, Pfaff M, Johansson S, Fassler R (1996) Beta 1 integrin-dependent and -independent polymerization of fibronectin. J Cell Biol 132:227–238

White DP, Caswell PT, Norman JC (2007) alpha v beta3 and alpha5beta1 integrin recycling pathways dictate downstream Rho kinase signaling to regulate persistent cell migration. J Cell Biol 177:515–525

Williams MJ, Phan I, Harvey TS, Rostagno A, Gold LI, Campbell ID (1994) Solution structure of a pair of fibronectin type 1 modules with fibrin binding activity. J Mol Biol 235:1302–1311

Woods A, Couchman JR, Johansson S, Hook M (1986) Adhesion and cytoskeletal organisation of fibroblasts in response to fibronectin fragments. EMBO J 5:665–670

Wu C, Chung AE, McDonald JA (1995a) A novel role for alpha 3 beta 1 integrins in extracellular matrix assembly. J Cell Sci 108(Pt 6):2511–2523

Wu C, Keivens VM, O'Toole TE, McDonald JA, Ginsberg MH (1995b) Integrin activation and cytoskeletal interaction are essential for the assembly of a fibronectin matrix. Cell 83:715–724

Xiong JP, Stehle T, Zhang R, Joachimiak A, Frech M, Goodman SL, Arnaout MA (2002) Crystal structure of the extracellular segment of integrin alpha Vbeta3 in complex with an Arg-Gly-Asp ligand. Science 296:151–155

Xu J, Bae E, Zhang Q, Annis DS, Erickson HP, Mosher DF (2009) Display of cell surface sites for fibronectin assembly is modulated by cell adherence to 1F3 and C-terminal modules of fibronectin. PLoS ONE 4:e4113

Yamada KM, Kennedy DW, Kimata K, Pratt RM (1980) Characterization of fibronectin interactions with glycosaminoglycans and identification of active proteolytic fragments. J Biol Chem 255:6055–6063

Yang JT, Hynes RO (1996) Fibronectin receptor functions in embryonic cells deficient in alpha 5 beta 1 integrin can be replaced by alpha V integrins. Mol Biol Cell 7:1737–1748

Yi XY, Wayner EA, Kim Y, Fish AJ (1998) Adhesion of cultured human kidney mesangial cells to native entactin: role of integrin receptors. Cell Adhes Commun 5:237–248

Yokosaki Y, Matsuura N, Higashiyama S, Murakami I, Obara M, Yamakido M, Shigeto N, Chen J, Sheppard D (1998) Identification of the ligand binding site for the integrin alpha9 beta1 in the third fibronectin type III repeat of tenascin-C. J Biol Chem 273:11423–11428

Yurchenco PD, Schittny JC (1990) Molecular architecture of basement membranes. FASEB J 4:1577–1590

Yurchenco PD, Wadsworth WG (2004) Assembly and tissue functions of early embryonic laminins and netrins. Curr Opin Cell Biol 16:572–579

Zaidel-Bar R, Itzkovitz S, Ma'ayan A, Iyengar R, Geiger B (2007) Functional atlas of the integrin adhesome. Nat Cell Biol 9:858–867

Zhang Q, Checovich WJ, Peters DM, Albrecht RM, Mosher DF (1994) Modulation of cell surface fibronectin assembly sites by lysophosphatidic acid. J Cell Biol 127:1447–1459

Zhang Q, Magnusson MK, Mosher DF (1997) Lysophosphatidic acid and microtubule-destabilizing agents stimulate fibronectin matrix assembly through Rho-dependent actin stress fiber formation and cell contraction. Mol Biol Cell 8:1415–1425

Zhang WM, Kapyla J, Puranen JS, Knight CG, Tiger CF, Pentikainen OT, Johnson MS, Farndale RW, Heino J, Gullberg D (2003) alpha 11beta 1 integrin recognizes the GFOGER sequence in interstitial collagens. J Biol Chem 278:7270–7277

Zheng X, Saunders TL, Camper SA, Samuelson LC, Ginsburg D (1995) Vitronectin is not essential for normal mammalian development and fertility. Proc Natl Acad Sci USA 92: 12426–12430

Zhong C, Chrzanowska-Wodnicka M, Brown J, Shaub A, Belkin AM, Burridge K (1998) Rho-mediated contractility exposes a cryptic site in fibronectin and induces fibronectin matrix assembly. J Cell Biol 141:539–551

Zhou A, Huntington JA, Pannu NS, Carrell RW, Read RJ (2003) How vitronectin binds PAI-1 to modulate fibrinolysis and cell migration. Nat Struct Biol 10:541–544

Zhou X, Rowe RG, Hiraoka N, George JP, Wirtz D, Mosher DF, Virtanen I, Chernousov MA, Weiss SJ (2008) Fibronectin fibrillogenesis regulates three-dimensional neovessel formation. Genes Dev 22:1231–1243

Chapter 3
Collagens, Suprastructures, and Collagen Fibril Assembly

David E. Birk and Peter Brückner

Abstract Extracellular matrices are composed of collagens, proteoglycans, glycosaminoglycans, glycoproteins, and elastin. Extracellular matrix not only serves as structural scaffolds in organs and tissues, but also determines cellular function through cell–matrix interactions. Accordingly, the structures and organization of extracellular matrices are diverse and adapted to tissue-specific function. This chapter focuses on the collagen family. There are 28 different types of collagen that assemble into a variety of supramolecular structures including fibrils, microfibrils, and network-like structures. This chapter begins with a discussion of collagen molecules. This is followed by a definition of the supramolecular structure of different collagen types and their assembly and function within extracellular matrices. A discussion of general mechanistic principles involved in the assembly of collagen-containing suprastructures is presented. Finally, the regulation of tissue-specific collagen fibrillogenesis is used to illustrate how these general principles are applied in different tissues to generate the diversity in extracellular matrix structures and functions observed.

3.1 Introduction

The genomes of vertebrates and higher invertebrates include genes for a family of 28 extracellular matrix glycoproteins, the collagens (Table 3.1). Collagen types are classified based on domain structure homology and are assigned Roman numerals

---

D.E. Birk (✉)
Department of Pathology and Cell Biology, University of South Florida, College of Medicine, 12901 Bruce B. Downs Boulevard, MDC11, Tampa, FL 33612-4799, USA
e-mail: dbirk@health.usf.edu

P. Brückner
Institute for Physiological Chemistry and Pathobiochemistry, University Hospital of Münster, Waldeyerstr. 15, 48149 Münster, Germany
e-mail: peter.bruckner@uni-muenster.de

Table 3.1 Collagens

Collagen type	Genes	Molecular structure	Other
Type I	COL1A1, COL1A2	$\alpha 1(I)_2 \alpha 2(I)$	
		$\alpha 1(I)_3$	
Type II	COL2A1 (A,B)	$\alpha 1(II)_3$	Alternative splicing
Type III	COL3A1	$\alpha 1(III)_3$	
Type IV	COL4A1, COL4A2,	$\alpha 1(IV)_2 \alpha 2(IV)$	
	COL4A3, COL4A4,	$\alpha 3(IV)\alpha 4(IV)\alpha 5(IV)$	
	COL4A5, COL4A6	$\alpha 5(IV)_2 \alpha 6(IV)$	
Type V	COL5A1, COL5A2,	$\alpha 1(V)_2 \alpha 2(V)^a$	Alternative splicing
	COL5A3	$\alpha 1(V)_3$	
		$\alpha 1(V)\alpha 2(V)\alpha 3(V)$	
Type VI	COL6A1, COL6A2,	$\alpha 1(VI)\alpha 2(VI)\alpha 3(VI)$	Alternative splicing
	COL6A3, COL6A4,	$\alpha 1(VI)\alpha 2(VI)\alpha 4(VI)$	
	COL6A5, COL6A6	$\alpha 1(VI)\alpha 2(VI)\alpha 5(VI)$	
		$\alpha 1(VI)\alpha 2(VI)\alpha 6(VI)$	
Type VII	COL7A1	$\alpha 1(VII)_3$	
Type VIII	COL8A1, COL8A2	$\alpha 1(VIII)_2 \alpha 2(VIII)$	
		$\alpha 1(VIII)_3$	
		$\alpha 2(VIII)_3$	
Type IX	COL9A1, COL9A2,	$\alpha 1(IX)\alpha 2(IX)\alpha 3(IX)_3$	Alternative splicing
	COL9A3		
Type X	COL10A1	$\alpha 1(X)_3$	
Type XI	COL11A1(A,B,C),	$\alpha 1(XI)\alpha 2(XI)\alpha 3(XI)$	Alternative splicing
	COL11A2, COL2A1(A)[b]		
Type XII	COL12A1	$\alpha 1(XII)_3$	Alternative splicing
Type XIII	COL13A1	$\alpha 1(XIII)_3$	Alternative splicing
Type XIV	COL14A1	$\alpha 1(XIV)_3$	Alternative splicing
Type XV	COL15A1	$\alpha 1(XV)_3$	
Type XVI	COL16A1	$\alpha 1(XVI)_3$	
Type XVII	COL17A1	$\alpha 1(XVII)_3$	
Type XVIII	COL18A1	$\alpha 1(XVIII)_3$	Alternative splicing
Type XIX	COL19A1	$\alpha 1(XIX)_3$	
Type XX	COL20A1[c]	$\alpha 1(XX)_3$	
Type XXI	COL21A1	$\alpha 1(XXI)_3$	
Type XXII	COL22A1	$\alpha 1(XXII)_3$	
Type XXIII	COL23A1	$\alpha 1(XXIII)_3$	
Type XXIV	COL24A1	$\alpha 1(XXIV)_3$	
Type XXV	COL25A1	$\alpha 1(XXV)_3$	Alternative splicing
Type XXVI	COL26A1	$\alpha 1(XXVI)_3$	
Type XXVII	COL27A1	$\alpha 1(XXVII)_3$	
Type XXVIII	COL28A1	$\alpha 1(XXVIII)_3$	

[a] $\alpha 1(XI)$ and $\alpha 2(V)$ have been shown to form heterotrimers
[b] The alpha chain is known as $\alpha 3(XI)$ when assembled as type XI collagen
[c] Chicken, human has not been described

I–XXVIII based on the chronological order of discovery. There are at least 45 different collagen genes that code for collagen polypeptides, called alpha chains. Genetically distinct alpha chains within the same collagen type also are numbered based on the order of discovery using Arabic numerals. The majority of collagens are homotrimeric being composed of three identical alpha chains, e.g., [α1(II)]₃ for collagen II. However, other collagens can be heterotrimeric, e.g., [α1(I)]₂α2(I) for collagen I or α3(IV),α5(IV),α6(IV) for one of the isoforms of collagen IV. In addition, a single collagen type can have multiple chain compositions, e.g., [α1(V)]₂α2(V), α1(V)α2(V)α3(V), or [α1(V)]₃, for collagen V. The alpha chains of one collagen type are unique from the alpha chains of another collagen type, i.e., they are encoded by different genes and have different primary (domain) structures. For example, the human α1(I) chain is encoded by the COL1A1 gene (human collagen genes are always designated in capitals) and the mouse α1(II) chain by the *Col2a1* gene.

Collagens assemble into suprastructures that are responsible for the structure and function of extracellular matrices. It is striking that one collagen type can predominate overwhelmingly in tissues that are extremely diverse. A good example is collagen I, which is the predominant collagen in striated collagen fibrils that have a vast array of tissue-specific structures and organizations. The most likely explanation for this is that most, if not all, collagen-containing suprastructures have complex macromolecular compositions that include other collagen types as well as noncollagenous components. These additional macromolecules may be substantial or occur in minute quantities. However, invariably the composite structure of collagen suprastructures is a major determinant of tissue-specific architecture and function.

This chapter begins with a discussion of collagen molecules followed by a definition of the supramolecular structure of different collagen types, their assembly, and function within extracellular matrices. General mechanistic principles involved in the assembly of collagen-containing suprastructures are discussed. Finally, the regulation of tissue-specific collagen fibrillogenesis is used as an example of how these general principles can provide a sequence of regulatory interactions to generate the diverse extracellular matrix structures and functions observed.

3.2 Collagen Molecules

All collagen family members are trimers and share common features. They have at least one stiff, rod-like domain of varying length, termed a collagenous or COL domain, defined by a specific unique motif, the collagen triple helix. All collagens have at least one COL domain as well as noncollagenous or NC domains. The number and structure of COL and NC domains is dependent on the specific collagen type.

3.2.1 Structural Hallmark of Collagen Molecules: The Triple Helix

Within a COL domain each of the three alpha chains is coiled into a left-handed helix that lacks intra-chain hydrogen bonds. The alpha chains are supercoiled to form a triple helix. This parallel, right-handed superhelix is stabilized by inter-chain hydrogen bonds that are almost perpendicular to the triple helical axis. As a result of the high content of imino acids, collagen polypeptides assume an elongated polyproline II-like helix with all peptide bonds in the *trans* configuration. The pitch of the polyproline II helix in collagenous polypeptides corresponds to three amino acids, almost exactly. Steric constraints require that only glycine, the smallest amino acid, can occupy the positions at the center of the triple helix. Hence, the triple helical domains have a repeating $(Gly-X-Y)_n$ structure with the X and Y positions being any of the 21 amino acids. However, the X and Y positions are frequently proline and hydroxyproline, respectively, necessary for helix formation and stability (see below). If glycine residues are replaced by other amino acids, the triple helix motif is interrupted, and hence, the rod-like structures experience rigid kinks or flexible hinges (Vogel et al. 1988; Engel and Bachinger 2005; Shoulders and Raines 2009). This situation is normal in many collagen types where there is more than one triple helical domain and provides flexibility. This substitution of glycine residues may also result from missense mutations in the corresponding collagen genes and represents the underlying cause for mild or severe, systemic connective tissue diseases. This subject is covered in several recent reviews (Myllyharju and Kivirikko 2004; Malfait and De 2009).

Collagen polypeptides contain two unique amino acids, hydroxyproline and hydroxylysine, which are important in triple helix stability and glycosylation, respectively. These unique amino acids are introduced during biosynthesis by enzymatic hydroxylation of almost all prolyl and some of the lysyl residues in the Y positions. Hydroxylysyl modification may be followed by O-glycosidic substitution with the resulting addition of one or two monosaccharide units. The first glycosylation reaction is catalyzed by a collagen-specific galactosyl transferase. A limited number of galactosyl-hydroxylsine residues, e.g., one residue per alpha chain of collagen I in skin, are then further glycosylated by a glucose residue with an unusual β-1,2-glycosidic link. The enzymes catalyzing hydroxylation reactions, i.e., the Fe^{2+}-dependent prolyl and lysyl hydroxylases and the glycosyl transferases, act cotranslationally on the nascent alpha chains in the rough endoplasmic reticulum. The enzymes recognize their substrates prior to triple helix formation, but there is no further modification once the triple helix has formed. Therefore, changes in the rate of helix formation, as seen in some point mutations, result in over- or under-modified collagen.

Hydroxyproline residues are important in triple helix stability. By virtue of the inductive effect of the 4-*trans* hydroxyl group, the pucker of free hydroxyproline is directed upwards, similar to that of Y-hydroxyproline in collagen triple helices. In contrast, free proline prefers the downward pucker, which also occurs in X-proline residues within the triple helix (for further details, see reviews by Shoulders and Raines 2009 and Okuyama et al. 2009). Hence, a forced integration of proline into Y positions absorbs more ring deformation energy, which results in a decrease of about 15°K in the denaturation temperature of so-called unhydroxylated protocollagen compared with physiological hydroxylated collagen (Berg and Prockop 1973).

3.2.2 Triple Helix Assembly

A key step common to all collagen types is trimerization, which involves selection and alignment of appropriate alpha chains with subsequent assembly into specific trimeric collagen molecules (Khoshnoodi et al. 2006). This involves specific interactions between noncollagenous (NC) globular domains at their C-terminal end. This process also facilitates initiation of triple helix formation, starting from the C-terminal end only after completion of translation. However, once trimerization is initiated it must be controlled to allow for the important posttranslational hydroxylation and glycosylation steps. Collagenous domains are unusually rich in *cis*-peptide bonds due to their high content of imino acids that favor *cis*-peptide bond formation. Because the free Gibbs enthalpy of isomerization of peptidyl-imino – but not amino acid – bonds is ~8 kJ/mol, the *cis* state is highly populated at physiological temperatures in nascent collagen polypeptides compared with other proteins with much lower proline contents. The Arrhenius activation energy of *cis*-to-*trans* isomerization is about 83 kJ/mol. Thus, the structural incompatibility of *cis*-peptide bonds with the collagen triple helix constitutes a formidable kinetic barrier against triple helix formation. Isomerization of each *cis*-peptide bond encountered is required during the zipper-like process of collagen triple helix, which results in unusually slow folding in comparison with other proteins (Bachinger et al. 1980; Bruckner et al. 1981). At the start of triple helix formation, a variable number of *cis* bonds are distributed throughout the still unfolded procollagen polypeptides. Hence, the folding times required for full-length triple helix formation are heterogeneous because kinetic barriers against folding are encountered more frequently in molecules with more *cis* bonds than in those with less.

The *cis*-to-*trans* isomerization of Gly-Pro-, but not X-Hyp peptide bonds, is catalyzed in fibroblasts by cyclophilin B, a protein that acts as peptidyl prolyl *cis/trans*-isomerase, and is inhibited by the immuno-suppressor cyclosporin A (Steinmann et al. 1991). In addition, cyclophilin B, prolyl-3-hydroxylase, and "cartilage-associated protein (Crtap)" form a ternary complex with high chaperone activity in the endoplasmic reticulum. Prolyl-3-hydroxylase introduces a single 3-Hyp-residue at the C-terminal end of the triple helical domain of nascent fibrillar procollagens. The apparent function of this complex is to direct cyclophilin B activity near the initiation sites of procollagen folding to ensure efficient catalytic isomerization of peptidyl-prolyl *cis* bonds. In support of this notion, null mutations in LEPRE1 and CRTAP, the genes encoding human prolyl-3-hydroxylase and Crtap, respectively, cause severe recessive osteogenesis imperfecta (for review see Marini et al. 2010). The collagens formed by cells from patients with these mutations are posttranslationally over-modified, which is consistent with a slow folding rate and, hence, an excessive substrate availability for lysyl hydroxylases and glycosyl transferases. Over-modified collagen is a general result of slow folding also in collagens with glycine mutations in patients with osteogenesis imperfecta and underlies abnormal fibrillogenesis.

Posttranslational glycosylation of collagens is another factor affecting fibril structure. Covalent modifications occurring after polypeptide synthesis are important in collagens and have an impact on the assembly of fibrils (Torre-Blanco et al. 1992; Batge et al. 1997; Notbohm et al. 1999). The circumference of collagen triple helical domains is affected by the extent of hydroxylation of lysyl residues and their subsequent galactosylation and glucosyl-galactosylation. Intermolecular center-to-center distances correlate with the extent of glycosylation, especially if the posttranslational modifications affect polypeptide parts eventually situated in overlap regions of the fibril. The extent of glycosylation of hydroxylysine can be manipulated by a variety of factors including activity levels of enzymes (Keller et al. 1985) and by disease-causing mutations (Myllyharju and Kivirikko 2004). Collagen mutations can substantially reduce the rates of procollagen triple helix formation in the rough endoplasmic reticulum and cause over-modification, thereby compromising normal fibrillar organization. However, differences in the extent of glycosylation can also be a mode of physiological regulation of fibril organization. Collagens in specific tissues have different extents of glycosylation. Corneal collagen I and cartilage collagen II have high levels of glycosylation relative to other tissue types and this can provide tissue-specific structural differences.

Another important factor influencing the molecular organization of collagen in fibrils is the incorporation of variable amounts of intrafibrillar water. This results in differing intermolecular distances between lateral or longitudinal neighbors (Brodsky et al. 1982; Katz et al. 1986). It is known that drying of fibrils results in a shortening of the D-periodicity of collagen fibrils and also a reduction in intermolecular lateral distances.

Modifiers of structure such as glycosylation influence the incorporation of collagen into fibrils. Structural malleability is a feature of collagen molecules that is important for their mode of incorporation into fibrils. Collagens exhibit a highly elastic flexibility in their triple helical twist. The elasticity of the helical pitch at low energy cost confers the option of azimuthal distortion to collagen molecules in their tissue-specific state of lateral aggregation (Jelinski and Torchia 1979; Kramer et al. 1999). This allows modifiers of the fibril structure to affect interacting interfaces between neighboring collagen molecules. A tighter twist in the triple helix will result in different azimuthal surfaces exposed to neighboring molecules. This introduces a substantial amount of flexibility in collagen packing within the fibrils and can influence fibril structure.

3.2.3 Collagen Processing

The fibril-forming collagens as well as some of the other types, e.g., collagen VII, are synthesized and secreted as procollagens, which prevents molecular assembly into suprastructures. Removal of the propeptide regions occurs through an extracellular proteolytic process that is directed by collagen type-specific metalloproteinases. However, in some cases, it may begin during the transport of newly synthesized procollagens to the cell surface. Proteolytic shedding of membrane-bound collagen types also is a process regulated by different metalloproteinases, as is degradation of collagens during turnover of the extracellular matrix or in diseases. These same enzymes also are involved in the processing of other substrates, including developmental signaling molecules or noncollagenous extracellular matrix molecules. These subjects are covered in several recent reviews (Hopkins et al. 2007; Apte 2009) and will not be discussed further.

Collagens are further modified posttranslationally by reactions occurring during secretion and aggregate assembly. Extracellular lysyl oxidases can convert the amino groups on some of the hydroxylysine and lysine residues in the collagen polypeptide chain to aldehydes that form aldols or β-ketoamines by reacting with aldehydes or amino groups on lysines, respectively, in other chains to generate intra- and intermolecular covalent cross-links (see Chap. 9). In some cases, cross-linking of collagen molecules at early stages of aggregation can modulate the suprastructural outcome of fibrillogenesis or the formation of networks (see below).

3.3 Collagen-Containing Suprastructures

Collagens are assembled into polymeric structures, visualized in the electron microscope as suprastructures that constitute tissue scaffolds, such as fibrils, microfibrils, filaments, and networks. Only a few specialized tasks are reserved for isolated collagen molecules. These suprastructures further assemble into higher order tissue structures. For example, fibrils form fibers and lamellae. Beaded filaments combine to form broad-banded structures. Networks assemble into basement membranes, anchoring fibrils, and lattices. The diversity of extracellular matrix suprastructures is dependent on the collagen type and further diversity is achieved by copolymerization of several types of collagen and noncollagenous macromolecules. The components of the suprastructure can differ in their identities and relative abundance in different tissues or during development, growth, and aging in a single tissue; and during repair or in disease. Quantitatively minor components may dictate polymer properties although they represent only a minuscule mass fraction of the total aggregate (see below). Thus, suprastructures are biological composites that can have unique functional properties distinct from those of the individual molecules. Therefore, in different tissues, collagen suprastructures provide tissue-specific structure and functional properties. A classification of collagen types based on suprastructural organization is presented in Table 3.2.

3.3.1 Fibril-Forming Collagens: Fibrils

The fibril-forming collagen subfamily includes collagens I, II, III, V, XI, XXIV, and XXVII. These collagens have a long uninterrupted triple helical domain (ca. 300 nm). Fibril-forming collagen genes cluster into three distinct subclasses

Table 3.2 Suprastructural organization of collagens

Classification	Collagen types	Supramolecular structure
Fibril-forming collagens	I, II, III	Striated fibrils
Regulatory fibril-forming collagens	V, XI,	Striated fibrils, retain regulatory, noncollagenous N-terminal domains
	XXIV, XXVII	Unknown
FACIT[a] collagens	IX, XII, XIV	Associated with fibrils, other interactions
FACIT-like collagens	XVI, XIX, XXI, XXII	Interfacial regions, basement membrane zones
Basement membrane collagen	IV	Chicken-wire network with lateral association
Beaded filament-forming collagen	VI	Beaded filaments, networks
Anchoring fibrils	VII	Laterally associated anti-parallel dimers
Network-forming collagens	VIII, X	Hexagonal lattices
Transmembrane collagens	XIII, XVII, XXIII, XXV Gliomedins, ectodysplasin	Transmembrane and shed soluble ecto-domains
Multiplexin collagens (endostatin-XV and -XVIII)	XV, XVIII	Basement membrane proteoglycans, cleaved C-terminal domains influence angiogenesis
Other molecules with collagenous domains	XXVI, XXVIII C1q, collectins, acetylcholinesterase, adiponectin, surfactant protein, and others	Collagenous domains in primarily noncollagenous molecules

[a]Fibril-associated collagens with interupted triple helices

(Boot-Handford and Tuckwell 2003) and this carries over into functional subclasses. Collagens I, II, and III are the most abundant proteins in the vertebrate body and are the bulk components of all collagen fibrils. Collagens V and XI are quantitatively minor collagens found co-assembled with types I, II, and III in different tissues. This subclass retains portions of the N-terminal propeptide and is involved in the regulation of fibril assembly. Collagens XXIV and XXVII make up the third subclass and have differences relative to the other fibril-forming collagen types including shorter helical regions that are interrupted. Their structural organization and specific roles remain to be elucidated.

The fibril-forming collagens are synthesized and secreted as procollagens. Procollagens contain a noncollagenous C-terminal propeptide and an N-terminal propeptide. The N-propeptide is composed of several noncollagenous domains and a short collagenous domain. The presence of the propeptide prevents premature assembly of collagen molecules into fibrils. Processing of the propeptides, which requires a number of enzymes, regulates the initial assembly of collagen into fibrils. The C-propeptides are processed by BMP-1/tolloid proteinases or furin (Greenspan 2005). The processing of the N-propeptides involves ADAMTS 2, 3, and 14 as well as BMP-1 (Colige et al. 2005; Greenspan 2005). These processing enzymes have specificity for different collagen types (Kadler et al. 2007). Propeptide processing may be complete, i.e., collagens I and II, leaving a collagen molecule with one large central triple helical domain and terminal, short noncollagenous sequences termed the telopeptides. Processing also can be incomplete, i.e., in collagens III, V, and XI, leaving a C-telopeptide and a partially processed N-propeptide domain. Both the telopeptides and the N-terminal domain have been implicated in the regulation of fibrillogenesis.

After processing of the propeptides, collagen molecules self-assemble to form striated fibrils with a periodicity of 67 nm. Within the fibril, the collagen molecules are arranged in longitudinally staggered arrays. The length of the collagen molecule is a non-integer multiple of the lateral stagger between adjacent neighboring molecules. The stagger is equal to the D period and the molecular length is 4.4 D. Thus, a gap occurs between the ends of neighboring molecules. This generates a gap-overlap structure in all collagen fibrils with a D-periodic banding pattern. This is presented schematically in Fig. 3.1a.

Collagen fibrils are assembled from mixtures of two or more fibril-forming collagen types. Connective tissues consisting of collagens I, II, and/or III, the quantitatively major fibril-forming collagens, contain quantitatively minor amounts of collagens V and XI. These regulatory fibril-forming collagens (collagens V and XI) are characterized by a partial processing of the N-propeptide domain. The N-propeptides have a flexible or hinge domain (NC2) between the triple helical domain (COL1) and a short triple helical domain (COL2). The N-terminal domain (NC3) is composed of two domains, variable and PARP. Processing involves specific cleavage of the PARP domain with retention of the hinge, COL2, and variable domains (Linsenmayer et al. 1993; Gregory et al. 2000; Hoffman et al. 2010). These regulatory fibril-forming collagens co-assemble with the major fibril-forming collagens to form a heterotypic fibril. The N-terminal domain of the regulatory fibril-forming collagens cannot be integrated into the staggered packing of the helical domains. The hinge region (NC2) is flexible so that the rigid COL2 domain can project toward the fibril surface in the gap region and the variable domain is present in the gap and on the fibril surface (Fig. 3.1b). It is clear that interactions with fibrillar collagens, i.e., collagens I, II, III, and V/XI, act as key regulators of the collagen organization in the fibril, resulting in tissue-specific fibril differences.

3.3.2 Fibril-Associated Collagens with Interrupted Triple Helices

The supramolecular organization of fibril-associated collagens with interrupted triple helix (FACIT) collagens involves an interaction with fibril-forming collagens organized as fibrils and serves not only to modulate the surface properties of the fibril, but in some cases also is involved in packing of the fibril-forming collagens during fibril assembly (see collagen IX below). FACITs include collagens IX, XII,

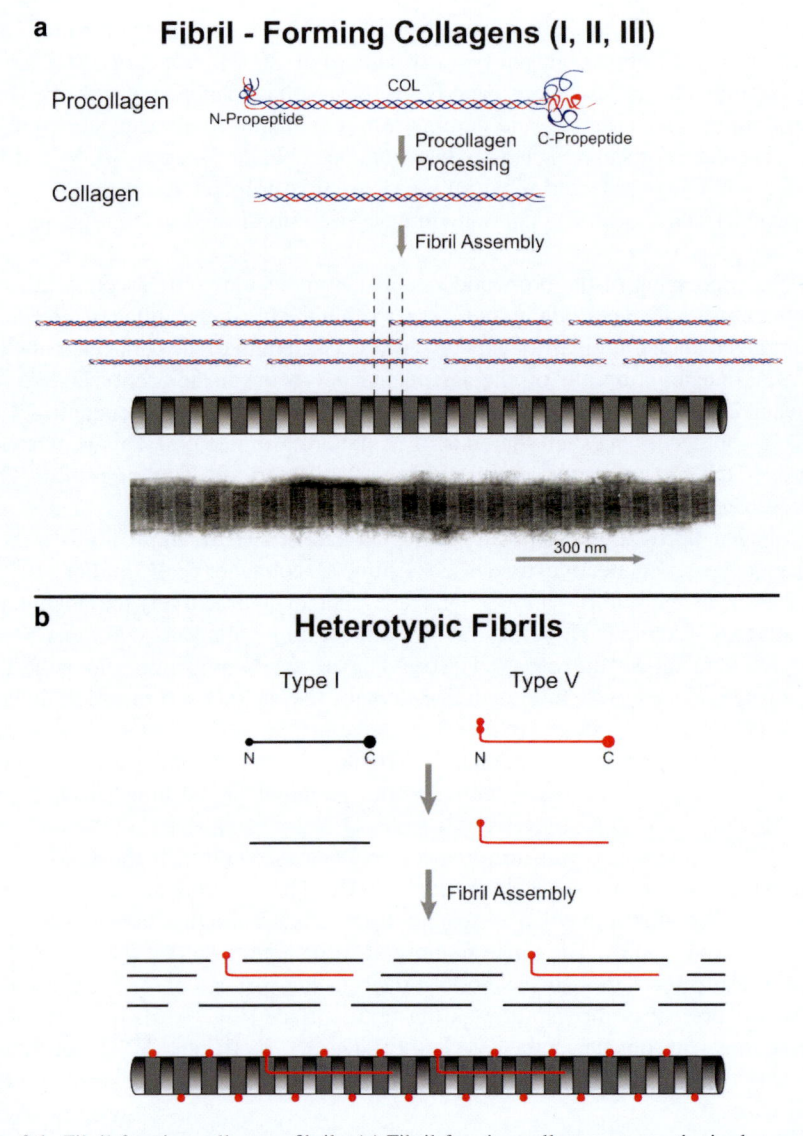

Fig. 3.1 Fibril-forming collagens: fibrils. (**a**) Fibril-forming collagens are synthesized as procollagen with a central COL domain and flanking N- and C-terminal NC domains, the propeptides. The propeptides are processed and the resulting collagen molecules assemble to form striated fibrils. The fibrillar collagen molecule is approximately 300 nm (4.4 D) in length and 1.5 nm in diameter. Within the fibril, the collagen molecules are staggered N to C and the staggered pattern of collagen molecules gives rise to the D-periodic repeat. A D-periodic collagen fibril from tendon is presented at the bottom of the panel. The negative stained fibril has a characteristic alternating light/dark pattern representing the gap (*dark*) and overlap (*light*) regions of the fibril. (**b**) Collagen fibrils are heterotypic, co-assembled from quantitatively major fibril-forming collagens, e.g., collagen I and regulatory fibril-forming collagens, e.g., collagen V. Regulatory fibril-forming collagens have a partially processed N-terminal propeptide, retaining a noncollagenous domain that must be in/on the gap region/fibril surface. The heterotypic interaction is involved in nucleation of fibril assembly

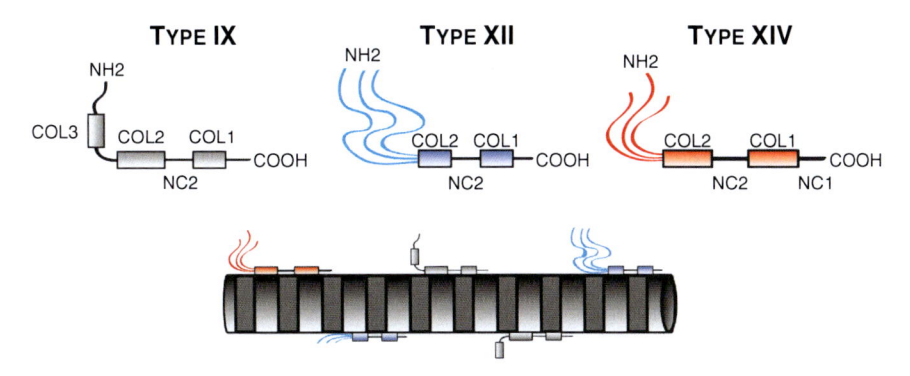

Fig. 3.2 FACIT collagens: associated with fibrils. The domain structures of FACIT collagens are illustrated. All FACITs have alternative spliced variants and collagen IX and XII can have glycosaminoglycan chains covalently attached. The FACIT collagens have 2–3 COL domains and 3–4 NC domains. Characteristic of this collagen type is a large N-terminal NC domain that projects into the interfibrillar space. The FACIT collagens all associate with the surface of collagen fibrils and this is illustrated, including N-truncated isoforms due to alternative splicing in collagens IX and XII. As described in the text, collagen IX can be integrated as a component of the cartilage fibril (not shown) and collagen XII is capable of other nonfibril interactions (not shown)

XIV, and XX collagen (collagen type XX is not present in humans). FACIT collagens have short COL domains interrupted by NC domains with an N-terminal NC domain that projects into the interfibrillar space (Fig. 3.2). Collagens IX and XII can be proteoglycans with covalently attached glycosaminoglycan chains.

A common feature of FACIT collagens is a FACIT domain, i.e., a relatively short C-terminal triple helical stretch flanked by a cysteine-containing motif, GXCXXXC, at the junction of the triple helix and the C-terminal non-helical region. The cysteines are essential for covalent bonding of the three constituent polypeptides. In collagen IX, the prototype FACIT collagen, the FACIT domain, may be incorporated into the gap between consecutive fibrillar collagen molecules in cartilage fibrils (Eyre et al. 2002). Thus, the selective expression of the FACITs would provide an elegant molecular mechanism for modulating the surface properties of collagen fibrils. It has been demonstrated that the N-terminal regions of collagens IX, XII, and XIV collagen protrude from the fibril surfaces in cartilage, skin, and tendon (Birk and Bruckner 2005). The biomechanical diversity of banded fibrils may be a direct consequence of distinct fibril surface properties afforded by FACITs.

At least in the case of the prototypic FACIT, collagen IX, triple helical domains other than the C-terminal FACIT domain are also incorporated into the fibril, possibly by an anti-parallel alignment with the fibrillar collagens of cartilage (see below). In doing so, FACITs become part of the fibrils, thereby modulating further and/or stabilizing the molecular organization within fibrils. However, this suprastructural association with fibrillar collagens is not a generalized feature of all FACITs.

3.3.3 FACIT-Like Collagens

The FACIT-like collagens have features in common with FACIT collagen, but are structurally and functionally unique. This FACIT-like group includes collagens XVI, XIX, XXI, and XXII (Pan et al. 1992; Yoshioka et al. 1992; Myers et al. 1994; Chou and Li 2002; Koch et al. 2004). These collagens are localized to basement membrane zones or interfacial regions separating different tissue types. One example is collagen XVI. In contrast to the FACIT collagens, this collagen has ten collagenous domains flanked by noncollagenous domains. It has different suprastructural organizations and can associate with fibrillin 1 at the dermal–epidermal basement membrane junction. In contrast, in cartilage it associates with fibrils, but only in the absence of collagen IX (Kassner et al. 2003). Multiple suprastructural forms are also observed in a true FACIT. Collagen XII can associate with collagen fibrils and also with basement membrane components (Gordon and Hahn 2010).

3.3.4 Collagen IV Networks: Basement Membranes

Basement membranes are extracellular matrices composed of several independent, but integrated, supramolecular networks (see Chap. 4). The focus here is the collagen IV network, whereas the others have laminin, or perlecan, as their major components. Additional macromolecules, including nidogen/entactin, mediate molecular contacts, thereby stabilizing the compound macromolecular networks in basement membranes.

Diverse basement membranes occur in different anatomical sites or may be coordinately formed during development. This also reflects the subtypes of collagen IV that are differentially expressed under different circumstances. There are six collagen IV-encoding genes, *COL4A1* through *COL4A6*, giving rise to the corresponding α chains, $\alpha 1$(IV) through $\alpha 6$(IV). However, these form a limited set of triple helical molecules with the stoichiometries $[\alpha 1(IV)]_2 \, \alpha 2(IV)$, $\alpha 3(IV) \, \alpha 4(IV) \, \alpha 5(IV)$, and $[\alpha 5(IV)]_2 \, \alpha 6(IV)$, respectively. Other chain combinations have not been described. Heterotypic interactions are possible involving the NC1 domains of different collagen IV isoforms. For instance, $[\alpha 1(IV)]_2 \, \alpha 2(IV)$ trimers can interact with $[\alpha 5(IV)]_2 \, \alpha 6(IV)$ trimers by interactions between $\alpha 1$- and $\alpha 5$-, as well as $\alpha 2$- and $\alpha 6$-NC1 domains. In contrast, $\alpha 3(IV) \, \alpha 4(IV) \, \alpha 5(IV)$ interact through their NC1 domains to yield pairs of two $\alpha 4(IV)$-NC1 domains or $\alpha 3(IV)$-$\alpha 5(IV)$-NC1 heterodimers. Thus, heterotypic networks arise with distinct supramolecular structures and functional properties (Khoshnoodi et al. 2006).

Collagen IV molecules aggregate into networks. This involves interactions between their N-terminal, triple helical domains, called 7S domains. This domain organizes four collagen IV molecules in an anti-parallel fashion. The long type IV collagen triple helices project out in all spatial directions from the 7S domains. The flexibility of the helical domains results from interruptions in the typical

Type IV Collagen

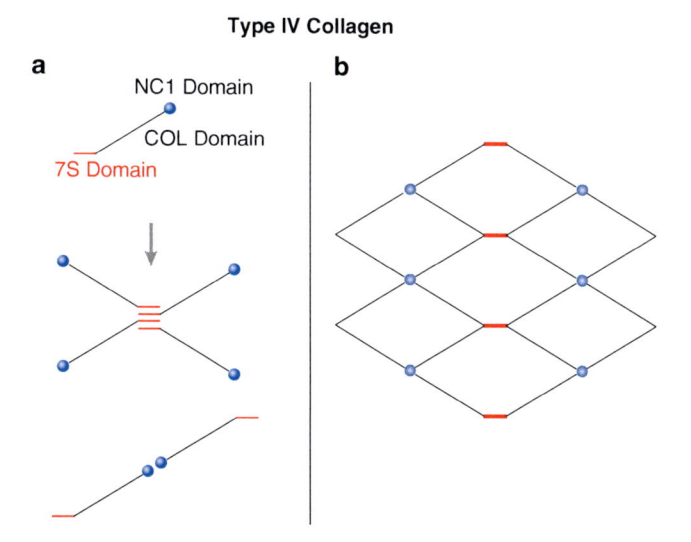

Fig. 3.3 Collagen IV: basement membrane networks. (**a**) A collagen IV monomer is illustrated with a central COL domain. This triple helical domain contains numerous interruptions of the Gly-X-Y sequence introducing flexibility into the COL domain. This is flanked by a C-terminal NC1 domain and an N-terminal domain, termed the 7S domain. Collagen IV molecules form dimers via interactions involving the NC1 domains, and tetramers via interactions involving the 7S domain. (**b**) These interactions produce supramolecular aggregates that generate an extended chicken-wire network of collagen IV molecules. In addition, there are lateral interactions involving the helical domains that generate a tighter, less regular network structure (not shown)

(Gly-X-Y)$_n$ sequences that, unlike in fibrillar collagens, occur abundantly, including at a site between the 7S and the triple helical domains. Such interruptions create points of flexibility in an otherwise stiff, rod-like molecule. At their C terminus, the noncollagenous NC1 domains interact head-to-head, creating in conjunction with the 7S interactions large supramacromolecular aggregates. These suprastructures resemble a chicken-wire-like network (Fig. 3.3). Superimposed on this basic network structure are lateral interactions that involve triple helical domains. This results in extended polygonal networks with variable mesh sizes.

3.3.5 Collagen VI: Beaded Filaments and Networks

Collagen VI is a ubiquitous component of connective tissues. It is found as an extensive filamentous network with collagen fibrils and is often enriched in pericellular regions. It is assembled into several different tissue forms, including beaded microfibrils, hexagonal networks, and broad-banded structures (Furthmayr et al. 1983; von der Mark et al. 1984; Bruns et al. 1986). Collagen VI interacts with a spectrum of extracellular molecules including collagens I, II, IV, XIV, microfibril-associated glycoprotein (MAGP-1), perlecan, decorin, biglycan, hyaluronan, heparin, and fibronectin as well as integrins and the cell-surface proteoglycan NG2. On the basis of tissue localization and large number of potential interactions, collagen VI has been proposed to integrate different components of the extracellular matrix, including cells (Kielty and Grant 2002). In addition, collagen VI may influence cell migration, differentiation, and apoptosis/proliferation. This indicates a roles in the development of tissue-specific extracellular matrices, repair processes, and in the maintenance of tissue homeostasis.

The best-characterized form of collagen VI is a heterotrimer composed of $\alpha1(VI)$, $\alpha2(VI)$, and $\alpha3(VI)$ chains (Chu et al. 1987; Kielty and Grant 2002).The monomer has a 105 nm triple helical domain with flanking N- and C-terminal globular domains. The molecular mass of the N-terminal domain is mainly derived from the $\alpha3(VI)$ chain and is approximately twice that of the C-terminal domain. The $\alpha3(VI)$ chain can be processed extracellularly. In addition, structural heterogeneity is introduced by alternative splicing of domains, primarily of the $\alpha3(VI)$ N–terminal domain. Three additional α chains of type VI collagen have been described, $\alpha4(VI)$, $\alpha5(VI)$, and $\alpha6(VI)$ (Gara et al. 2008; Fitzgerald et al. 2008). These chains have high homology with the $\alpha3(VI)$ chain and may form additional isoforms.

A property of collagen VI is that assembly of the supramolecular forms begins intracellularly (Fig. 3.4a) and involves a number of distinct steps. First, a dimer is formed via lateral, anti-parallel association of two monomers. The monomers are staggered by 30 nm with the C-terminal domains interacting with the helical domains. This overlap generates a central 75 nm helical domain flanked by a non-overlapped region with the N- and C-globular domains, each about 30 nm. Disulfide bonds near the ends of the overlapped region stabilize these interactions (Ball et al. 2003). A supercoil is formed in the overlapped helices of the two monomers in the central region (Knupp and Squire 2001). Second, tetramers form when two dimers align with the ends in register. Next, tetramers are secreted and are the building blocks used to assemble the tissue forms of collagen VI (Fig. 3.4b). In the extracellular environment, tetramers associate end-to-end forming beaded filaments. This is a noncovalent interaction that gives rise to thin, beaded filaments (3–10 nm) with a periodicity of approximately 100 nm. These beaded filaments laterally associate, forming beaded microfibrils (Bruns et al. 1986). In addition to beaded microfibrils, other collagen VI-containing supramolecular structures are found in the extracellular matrix including hexagonal lattices; and broad-banded fibrils with a 100 nm periodicity. The broad-banded fibrils represent continued lateral growth of beaded microfibrils and/or lateral association of preformed beaded microfibrils. In contrast, hexagonal lattices are formed via end-to-end interactions of tetramers in a nonlinear fashion (Wiberg et al. 2002).

Comparable to fibrillar collagen, supramolecular aggregates containing collagen VI are composite structures with other integrated molecules modulating the functional properties of the collagen VI-containing suprastructure. For example, biglycan interactions with the tetramer induced formation of hexagonal lattices rather than beaded microfibrils. This was dependent on the presence of the glycosaminoglycan chains. In contrast, decorin, which binds to the same site, was less effective

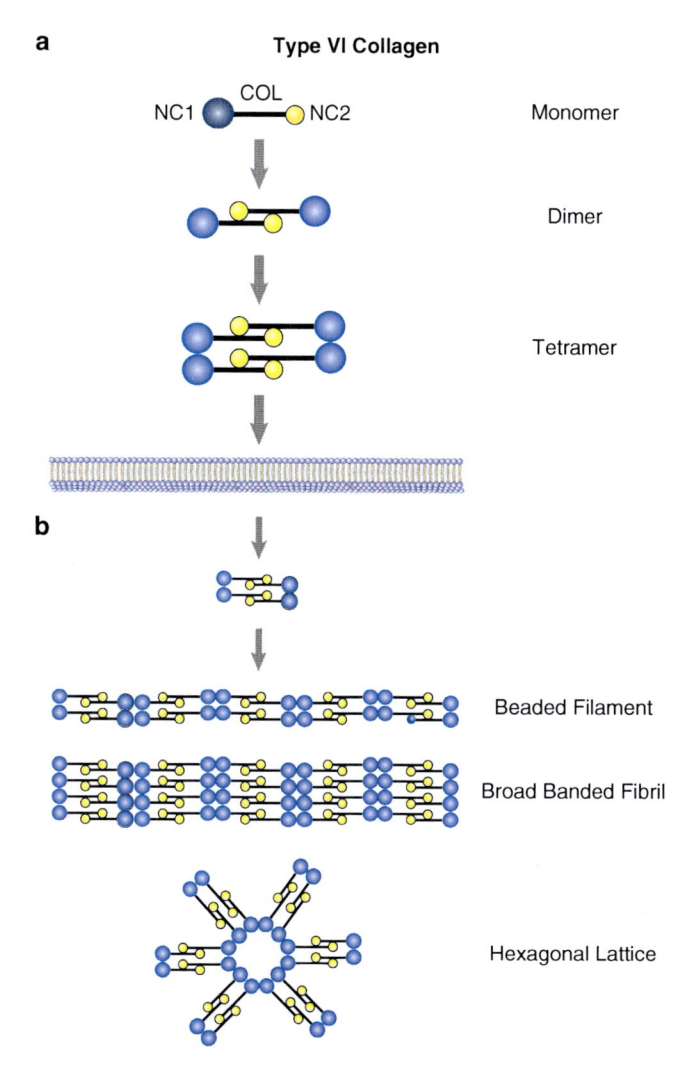

3 Collagens, Suprastructures, and Collagen Fibril Assembly

Fig. 3.4 Assembly of collagen VI suprastructures. (**a**) Collagen VI monomers have a C-terminal NC domain, a central triple helical domain, and an N-terminal NC domain. The monomers assemble N–C to form dimers. Tetramers are assembled from two dimers aligned in register. (**b**) The tetramers are secreted and form the building blocks of three different collagen VI suprastructures. Beaded filaments, broad-banded fibrils, and hexagonal lattices form via end-to-end interactions of tetramers and varying degrees of lateral association. (Note the change in scale from top to bottom indicated by change in arrow size)

in inducing hexagonal lattice formation (Wiberg et al. 2002). Analogous to fibril formation, the interaction of small leucine-rich proteoglycans with collagen VI influences the structure of the tissue aggregate and therefore its function. This provides a mechanism to assemble different suprastructures in adjacent regions or tissues with different functions. This illustrates how the composite structure of collagen suprastructures can contribute to the definition of structure/function associated with different tissues.

3.3.6 Collagen VII: Anchoring Fibrils

Collagen VII is a large collagen that is assembled into anchoring fibrils that tether the epidermal basement membrane to the underlying dermis (for review, see Burgeson and Christiano 1997). As a homotrimer, the large central COL domain of collagen VII is flanked by N- and C-terminal NC domains (Bruckner-Tuderman et al. 1999). The COL domain contains numerous interruptions that provide conformational flexibility to the COL domain.

Collagen VII is secreted into the extracellular matrix where it forms anti-parallel tail-to-tail dimers with a central C-terminal overlap and with the N-termini pointing outward. Intermolecular disulfide bonds stabilize the overlap. There is a proteolytic processing of a portion of the NC2 domain, which permits lateral association. Subsequently, the processed dimers aggregate laterally in a nonstaggered manner into the anchoring fibrils (Fig. 3.5). Mature anchoring fibrils are stabilized by transglutaminase cross-links.

Anchoring fibrils extend from the epidermal basement membrane to the upper papillary dermis, thus integrating the epidermis with the underlying dermis. It has been postulated that the NC1 domains of collagen VII at both ends of the anchoring

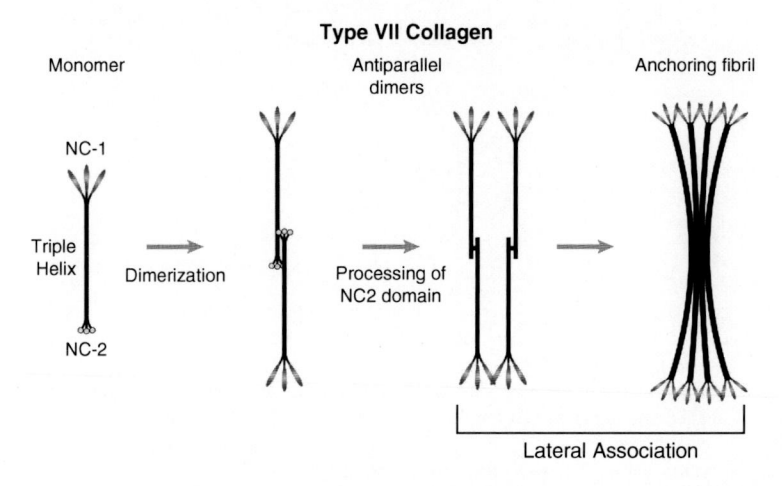

Fig. 3.5 Collagen VII: anchoring fibrils. Collagen VII molecules have a central triple helical COL domain with numerous interruptions conferring flexibility to the domain. The COL domain is flanked by noncollagenous N- (NC-1) and C-terminal (NC-2) domains. Two monomers interact to form an anti-parallel dimer with a central C-terminal overlap and the NC-1 domains pointing out. Processing occurs, with a cleavage of the NC-2 propeptide and covalent stabilization of the dimer by disulfide bonds. At this point, a nonstaggered lateral association of dimers occurs that generates the anchoring fibril

fibrils bind to the basement membrane macromolecules including collagen IV and laminins forming loops that entrap the collagen fibrils. Collagen VII has very poor affinity to most molecular collagens, including collagen I (Brittingham et al. 2006). However, anchoring fibrils bind tightly to cross-striated dermal collagen fibrils containing, among other types, collagen I. Therefore, there are binding determinants that do not exist in monomolecular matrix macromolecules, but only at the level of supramolecular aggregates. These binding sites are very important for the stabilization the dermo-epidermal cohesion (Villone et al. 2008) and are compromised in patients with dystrophic epidermolysis bullosa, a severe heritable skin blistering disease (Bruckner-Tuderman 2010).

3.3.7 Type VIII and X Collagen Networks

Collagens VIII and X are closely related short chain collagens, with comparable gene and protein structures (Yamaguchi et al. 1991; Kielty and Grant 2002). Collagen VIII is a major component located in the subendothelial region of blood vessels and in Descemet's membrane, separating the corneal endothelium from the stroma (Sawada et al. 1990; Shuttleworth 1997). Descemet's membrane is composed of layers of hexagonal lattices (Jakus 1956). These lattices are suprastructures containing collagen VIII (Sawada et al. 1990). Collagen VIII is a homo- or heterotrimer of $\alpha 1$(VIII) and $\alpha 2$(VIII) chains and evidence indicates that both homotrimers and the $\alpha 1$(VIII)$_2\alpha 2$(VIII) heterotrimer exist in tissues (Illidge et al. 1998, 2001). Collagen X has a restricted distribution, found only in hypertrophic cartilage. This collagen is a homotrimer composed of $\alpha 1$(X) chains and the supramolecular form is a hexagonal lattice similar to that formed by collagen VIII (Kwan et al. 1991).

The collagen VIII monomer has a short central COL domain and is flanked by N- and C-terminal NC domains. The monomers form lattices in vitro that are comparable to those seen in tissues (Stephan et al. 2004). On the basis of this work, it was proposed that collagen VIII monomers form a tetrahedron through the interaction of four molecules. The interaction is proposed to involve the C-terminal NC domains that have a conserved hydrophobic patch (Kvansakul et al. 2003). This structure serves as the building block that assembles into three-dimensional hexagonal lattices. The assembly of a layered hexagonal lattice could involve interaction of the N-terminal noncollagenous domains or anti-parallel interactions involving both helical and terminal domains. The anti-parallel interactions are consistent with thicker internodal struts observed and it is predicted that this is the primary mechanism for formation of hexagonal lattices both in vitro and in tissues (Fig. 3.6).

3.3.8 Transmembrane and Multiplexin Collagens

Transmembrane collagens include collagens XIII, XVII, XXIII, and XXV as well as at least seven related proteins including ectodysplasin and gliomedins

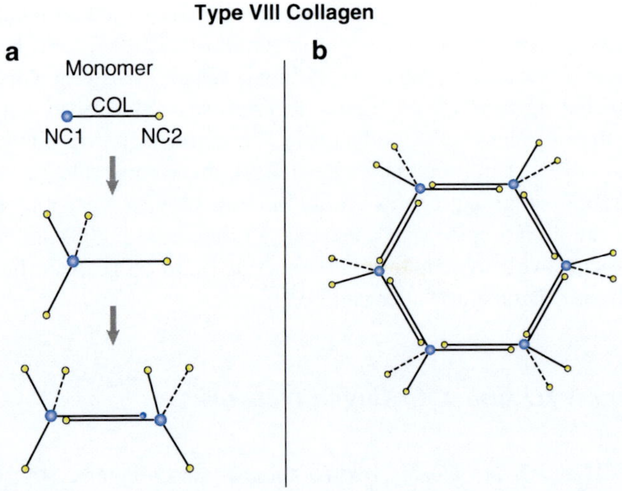

Fig. 3.6 Collagen VIII: assembly of hexagonal lattices. Collagen VIII is a short chain collagen with a central COL domain and flanking N- and C-terminal NC domains. The C-terminal (*blue*) NC domains of four collagen VIII molecules interact to form tetrahedrons. Tetrahedrons assemble further to form hexagonal lattices. A planar hexagonal lattice is diagramed. However, in tissues continued assembly, involving interactions of the N-terminal NC domains (*blue*) or anti-parallel interactions involving both helical and terminal domains (not shown) would generate a layered hexagonal lattice (not shown)

(for review see Franzke et al. 2003, 2005)). The transmembrane collagens are all homotrimers and contain an N-terminal cytoplasmic domain and a large C-terminal domain containing multiple COL domains with NC interruptions providing flexibility. There is a hydrophobic membrane spanning domain and adjacent extracellular linker domain between the membrane and the first COL domain involved in trimerization and also subject to proteolytic cleavage generating a shed extracellular domain.

Collagen XVII is a transmembrane collagen that is a component of the hemidesmosome anchoring complex. The cytoplasmic domain becomes organized as part of the hemidesmosomal plaque, while the extracellular domain localizes to the anchoring filaments that anchor the epithelial cell to the basement membrane (as opposed to anchoring fibrils that anchor the basement membrane to the stroma). Collagen XVII is the largest member of this group with 15 COL and 16 NC domains. Mutations in collagen XVII are associated with decreased epithelial adhesion, blistering, and alterations in hemidesmosome structure. This disorder is a junctional form of epidermolysis bullosa.

Multiplexin collagens include collagens XV and XVIII. The C-terminal domains can be cleaved generating endostatin-XV and endostatin-XVIII with anti-angiogenic properties. Both of these collagens have a central COL domain flanked by N- and C-terminal NC domains (Rehn et al. 1994). Collagens XV and XVIII are also proteoglycans with an attached chondroitin sulfate and heparan sulfate glycosaminoglycan

(Halfter et al. 1998; Li et al. 2000). In their unprocessed forms both collagens are found associated with basement membranes with distinctly different tissue distribution (Gordon and Hahn 2010).

Collagen XVIII is a component of basement membranes. The protein is colocalized in the dermo-epidermal basement membrane with perlecan-containing networks (Marneros et al. 2004). In the retinal pigment epithelium, this interaction appears to be important for the maintenance of an intact basement membrane zone. Mutations in the human *COL18A1* gene lead to Knobloch syndrome, a disease characterized by severe ocular alterations and occipital encephalocele and the absence of all collagen XVIII isoforms causes predisposition to epilepsy (Suzuki et al. 2002).

The transmembrane and multiplexin collagens have soluble proteolytic cleavage products with different functions in addition to their structural functions. The same is true for some other proteins usually not termed collagens, such as C1q, collectins, gliomedins, acetylcholinesterase, adiponectin, lung surfactant protein, and others.

3.4 Collagen Fibril Assembly

In the remainder of this chapter, we restrict our discussion to the mechanisms of fibril formation. Fibril formation will be used as an example of how the assembly of suprastructures is regulated. The chapter concludes with how these regulatory mechanisms can generate tissue-specific fibril structures and organization as well as tissue-specific function. Many of the principles discussed can be generalized to other suprastructural forms.

The assembly and deposition of collagen fibrils with tissue-specific structures and organizations involves a sequence of events that occur in both intracellular and extracellular compartments. Collagen molecules are synthesized, hydroxylated, glycosylated, assembled from three polypeptides, and folded in the rough endoplasmic reticulum. Packaging occurs in the Golgi, and transport is via specialized and elongated intracellular compartments with secretion at the cell surface. As described in the preceding section, suprastructures are composites of different matrix molecules. Control of heteropolymeric mixing and stoichiometry begins within the intracellular compartments. The secretion of different matrix molecules can also occur with different spatial and temporal patterns. Secretion of components such as procollagen processing enzymes; fibril-associated molecules, e.g., proteoglycans and FACITs; and adhesive glycoproteins, e.g., fibronectin, at unique sites or at different times would profoundly affect the character of the assembling matrix. This intracellular regulation of mixing during packaging/transport or at the sites of secretion provides a mechanism where a limited number of matrix molecules can be assembled in numerous ways to produce the diversity of structure and function observed in different tissues.

3.4.1 Extracellular Micro-domains

Extracellularly, the steps in fibril and matrix assembly also occur in compartments (Fig. 3.7). The relationship between extracellular domain structure and the assembly of collagen fibrils has been extensively studied in the developing tendon, but at least some principles can be extended to connective tissues in general (Birk and Linsenmayer 1994; Canty and Kadler 2002; Zhang et al. 2005). In developing tendon, fibril assembly begins in deep recesses or channels defined by the fibroblast surface (Trelstad and Hayashi 1979; Birk and Trelstad 1986; Canty et al. 2004; Canty and Kadler 2005). In these micro-domains, precursor suprastructures – the protofibrils – are assembled (Birk and Trelstad 1986; Birk et al. 1989). These immature fibrils have small and uniform diameters as well as short lengths compared with mature

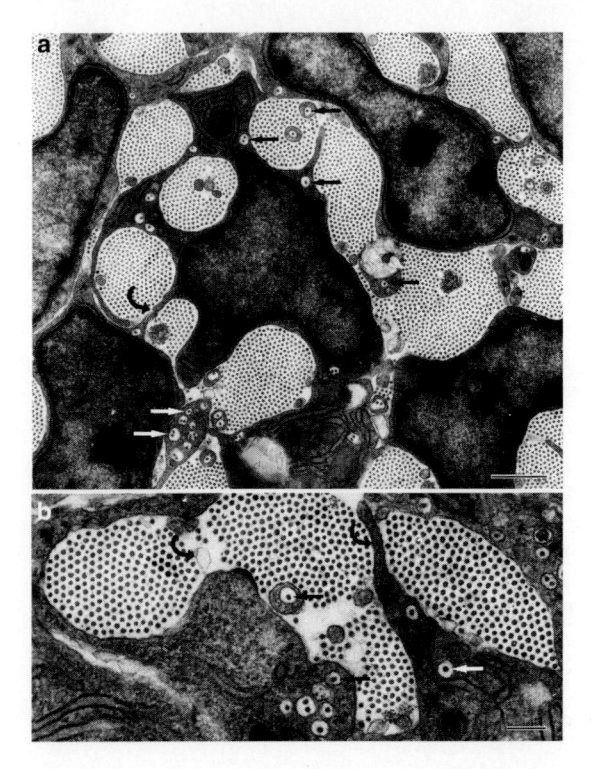

Fig. 3.7 Extracellular domain structure. Extracellular compartmentaliztion of the different levels of matrix assembly is seen in the developing chicken tendon. (**a, b**) Sections cut perpendicular to the tendon axis of a 14-day chicken embryo illustrate a series of micro-domains. Protofibrils are assembled in fibril-forming channels (*arrowheads*) and are deposited into the developing matrix in the fiber-forming spaces (F) where fibrils coalesce to form fibers. As development proceeds the fibers become larger as a result of aggregation of adjacent fibers in a third domain. These spaces form as the cytoplasmic processes that define the fiber-forming compartments retract (*curved arrow*) allowing fibers (*fibril bundles*) to coalesce into larger aggregates characteristic of the mature tissue. *Bar*, 1 μm (modified from Birk and Linsenmayer 1994)

fibrils. It was originally proposed that these extracellular channels form at the time of secretion as specialized post-Golgi secretory compartments, fuse with the fibroblast membrane, and are maintained due to slow membrane recycling associated with the presence of the assembled protofibril (Birk and Trelstad 1986; Birk et al. 1989). However, data also suggest that intracellular processing of procollagen within elongated Golgi-to-plasma membrane compartments (GPCs) may occur (Canty et al. 2004; Canty and Kadler 2005). In this case, extrusion from the fibroblast occurred through the formation of cellular protrusions, termed "fibripositors" that were formed by fusion of GPCs with the plasma membrane analogous to the formation of channels. It is possible that these morphologically distinct compartments are inter-related and represent different stages in the deposition of the protofibril.

Once the protofibrils are deposited into the extracellular matrix there is a second and third level of compartmentalization where fibrils form small fibers and then larger structures characteristic of the specific tissue, e.g., large fibers in tendon, layers or lamellae in bone and cornea, and interwoven network of fibers in dermis. This hierarchy of micro-domains within the extracellular space provides a mechanism for the fibroblast to exert control over the extracellular steps of matrix assembly; for instance, sequestering procollagen processing enzymes at the sites of initial fibril assembly and adding fibril-associated molecules during or after assembly of the protofibril.

3.4.2 Assembly and Growth of Mature Collagen Fibrils

In mature tissues, collagen fibrils are functionally continuous, i.e., are long and the lengths have not been measured, and have diameters in the 20–500 nm range depending on the tissue and developmental stage (Birk et al. 1995, 1997; Canty and Kadler 2002). However, during development, collagen fibrils are initially assembled as uniform and relatively short protofibrils (diameter ~20 nm, length 4–12 μm). The protofibrils are D-periodic with tapered ends (Birk et al. 1989, 1995; Kadler et al. 1996; Graham et al. 2000). The mature fibril is assembled by end-to-end and lateral association from protofibrils (Fig. 3.8). A model for the multistep assembly of mature fibrils from preformed intermediates, protofibrils, is presented in Fig. 3.9. Procollagen is processed to collagen that assembles into protofibrils closely associated with the cell surface. The protofibrils are deposited and incorporated into the developing extracellular matrix where they are stabilized via interactions with macromolecules such as FACITs and/or small leucine-rich proteoglycans. This stabilization can coincide with assembly; e.g., collagen IX or perhaps the NC domains of collagens V and XI (see below), or can occur with changing patterns at the time of and after assembly (see below for further discussion). Protofibril stabilization is not a single defined interaction, but rather a continuum that varies in a tissue- and development-specific manner.

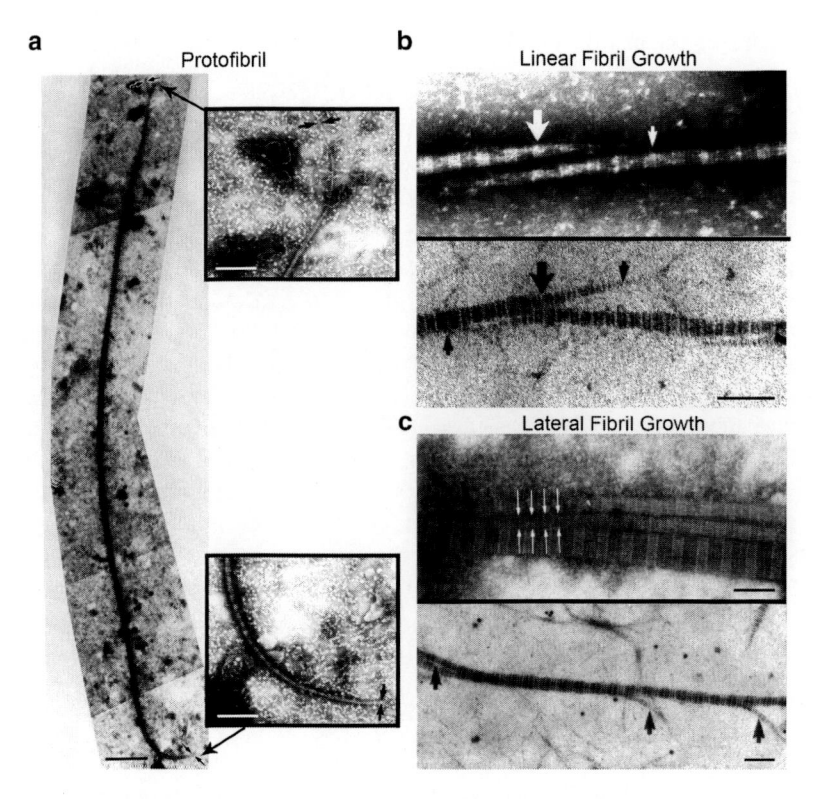

Fig. 3.8 Protofibrils and fibril growth. (**a**) Structure of protofibrils. This transmission electron micrograph illustrates an intact protofibril with short, measurable lengths. *Bar*, 1 μm. The ends of the protofibrils are asymmetric as is seen in situ, with α, long (*top*) and β, short (*bottom*) tapers (*insets*). *Bar,* 250 nm. Fibrils were extracted from chicken embryo tendons and negatively stained. Mature collagen fibrils result from linear and lateral associations of protofibrils. (**b**) Transmission electron microscopy of extracted (*top panel*) fibrils and from cryosectioned fibrils (*bottom panel*) from embryonic chicken tendons. These micrographs illustrate linear growth by end-to-end overlap and fusion of the tapered ends of protofibrils (*arrows*). This mechanism produces fibrils of increasing length without significantly altering fibril diameter. (**c**) Lateral fibril growth is illustrated in these transmission electron micrographs of extracted (*top panel*) and cryosectioned (*bottom panel*) fibrils. The extensive lateral association/fusion of growing protofibrils yields fibrils of increasing length and larger diameter. *Bars*, 100 nm (modified from Birk and Linsenmayer 1994)

As fibrillogenesis proceeds, there is linear fibril growth involving end overlap of the protofibrils. In most tissues, this is followed by lateral fibril growth where the fibrils merge laterally to generate large diameter fibrils seen in most mature tissues. These lateral associations involve molecular rearrangement necessary to regenerate the cylindrical fibril structure. In this process, some or all components stabilizing the protofibrils are lost or replaced during formation of the mature fibrils. The role and extent of collagen accretion during the linear and lateral growth steps is unclear. It does not appear to be a major feature in developing tissues, but a limited

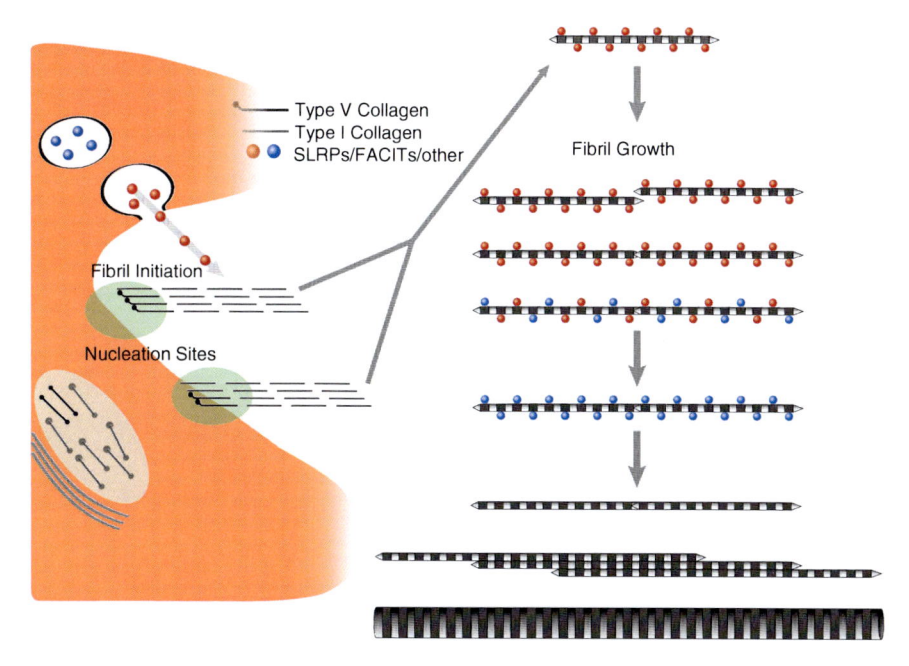

Fig. 3.9 Model of the regulation of fibril assembly. Fibril assembly involves a sequence of events. First, nucleators, e.g., collagen V and XI, initiate fibril assembly at the fibroblast cell surface. Immature, small diameter, short protofibrils are assembled. This nucleation process is cell directed involving interactions with organizers at the cell surface, e.g., integrins and syndecans (indicated by shaded *green* region). The protofibrils are deposited into the matrix and stabilized. This stabilization involves interactions with regulators, other matrix components such as SLRPs and FACITs. Changes in fibril stabilization resulting from processing, turnover, and/or displacement regulate linear and lateral growth to mature fibrils in a tissue-specific manner

role cannot be excluded. In addition, a role in normal turnover and repair is likely. Throughout the process, lysyl oxidase mediates intra- and intermolecular covalent cross-linking of collagen within the fibril. As the number of intermolecular cross-links increases with fibril maturation, molecular rearrangement is limited and mature fibril structure is stabilized. The growth in length and diameter as well as covalent cross-linking increases the mechanical strength of the connective tissue.

3.4.3 Regulation of Fibril Assembly

Regulation of collagen fibrillogenesis is tissue-specific (see sections below). One level of regulation involves control of the biosynthetic profile of the cells producing the extracellular matrix. This may involve genetic preprogramming and modulation via signaling events controlling the necessary sequential changes. The focus here is regulatory interactions during fibril assembly. Interactions with many different classes of molecules are involved, including processing enzymes, heterotypic interactions of fibril-forming collagens, FACITs, and small proteoglycans with leucine-rich motifs (SLRPs) as well as other glycoproteins. However, there are general mechanisms that are utilized. These include both temporal and spatial control of the nucleation of fibril formation, the recruitment of the initial sites of assembly to specialized cell-surface domains, and the regulation of linear and lateral fibril growth. It has been suggested that the regulation of these steps involves three general classes of interactions: organizers, nucleators, and regulators (Kadler et al. 2008). There may be disagreements as to which specific class a particular interaction belongs. In addition, it is likely that there will be tissue-specific differences in such interactions. Nevertheless, this division into organizers, nucleators, and regulators provides a foundation for discussion of the regulation of fibrillogenesis and tissue-specific modifications.

3.4.3.1 Nucleators

Collagens V and XI have been shown to nucleate collagen fibril formation in self-assembly assays in vitro, cell culture studies, and in mouse models (Marchant et al. 1996; Blaschke et al. 2000; Wenstrup et al. 2004a, b). In fact, collagens V and XI appear to represent variants of the same collagen type since their highly homologous α-chains seem to be interchangeable, at least to some extent. A mouse model with a targeted deletion in the *Col5a1* gene is embryonic lethal due to a virtual lack of fibril formation in the mesenchyme (Wenstrup et al. 2004a). This occurs in the presence of normal type I collagen synthesis and secretion and further demonstrated that type V collagen is essential for the assembly of type I-containing protofibrils in vivo. A similar situation also occurs in two mouse models in which the production of normal collagen XI is compromised. In a strain of mice (cho/cho) with naturally ablated *Col11a1*-alleles, collagen XI is absent. Homozygous animals develop a chondrodysplasia (cho) with cartilage essentially devoid of fibrils even though collagen II, the quantitatively major cartilage collagen, is produced normally (Seegmiller et al. 1971; Li et al. 1995b). In *Col2a1*-null mice, the collagen II deficiency leads to a massive upregulation of collagen I in cartilage. Although collagen I is not produced in normal hyaline cartilage it still ought to be capable of forming fibrils, but as in cho/cho-mice, the mutant cartilage lacks fibrils. Interestingly, the cartilage version of collagen XI contains an α3(XI) chain that is also derived from the *Col2a1* gene. As a consequence, *Col2a1*-null mice lack the normal cartilage fibril nucleator and, hence, cartilage fibrils (Li et al. 1995a; Aszodi et al. 1998).

Collagens I and II can self-assemble in vitro under physiological conditions after long lag phases. Aggregates with normal D-periodic cross-striation are generated, indicating that nucleation by collagens I and II alone is inefficient. However, other nucleators are not required when collagen concentrations are above relatively high critical levels. The nucleation of fibril formation by collagen V/XI provides a mechanism for the fibroblast to define the site of formation. By controlling the number of nucleation sites, the fibroblast controls the number of fibrils, and defining nucleation sites defines fibril organization. In addition, by regulating the number of type V/XI nucleators for a given collagen concentration, the fibroblast can regulate fibril diameter in a tissue-specific manner. For instance, collagen V makes up 10–20% of the total fibril-forming collagen in the corneal stroma and only 1–5% in dermis and tendon. The large numbers of nucleation sites in cornea contribute to the formation of small diameter fibrils necessary for transparency, while the lower number in tendon and dermis leads to large diameter fibrils required for mechanical strength. Classic Ehlers–Danlos syndrome is a generalized connective tissue disorder where a large percentage of the patients are heterozygous for mutations in collagen V, resulting in approximately 50% of the normal amount of collagen V. These patients have a dermal phenotype with large, structurally aberrant fibrils. The heterozygous ($Col5a1+/-$) mouse model had a reduction of 50% in collagen V and assembled fewer fibrils, indicating fewer nucleation events than the normal mice. In addition, there were two subpopulations of fibrils: the first slightly larger diameters and normal fibril structure and a second with very large diameters and aberrant fibril structures. This indicates that, with the 50% reduction, collagen V had become rate limiting. This suggests a regulated assembly involving interactions between collagens I and V and a dysfunctional assembly of collagen I in the absence of collagen I/V interactions. The regulation of these interactions is coordinated by the domain structure at the sites of assembly and by other molecules that organize and sequester these interactions at the cell surface. In this respect, it is noteworthy that collagen XI in cartilage, unlike collagen V in other tissues, appears to be sufficient as a nucleator. This may be related to the fact that the interval between fibrils and the nearest cell surface with nucleating/organizing potential in general is larger in cartilage than in other tissues. As a corollary, cells may control fibril suprastructure in tissues where cells form long projections reaching far into the extracellular matrix and modulate fibrillar precursors by new components in addition to collagen V.

3.4.3.2 Organizers

It is tempting to speculate that the nucleation of protofibrils is organized by the cell micro-domains and/or direct binding of nucleators such as collagen V to a cell-surface molecule such as an integrin or syndecan via the heparin-binding domain. While these direct interactions may have important roles, it is likely that the organizers could be more complex, involving multiple interactions or even cellular organelles, cytoskeletal components, or cytoplasmic membrane domains (rafts). The importance of cell-defined extracellular domains is emphasized by the observation that fibril assembly is very inefficient when the domain structure is disrupted. This is the case in typical monolayer cultures where much of the procollagen remains unprocessed, hence soluble components in the media. The elucidation of specific "organizers" is in the early stages and this remains a pristine field for future activity.

A discussion of cell-directed collagen fibril assembly must include a consideration of the potential roles of integrins and fibronectin. Fibronectin mediates interactions with cells and extracellular matrix molecules including collagen and assembles into fibronectin fibrils (Chap. 2). Assembly of fibronectin requires interactions with integrins inducing a conformational change that exposes a site required for polymerization (Zhong et al. 1998; Mao and Schwarzbauer 2005; Kadler et al. 2008). This generates a network of fibronectin fibrils with multiple matrix-binding sites that is organized by the fibroblast. A number of studies indicate roles for fibronectin and integrins in the assembly of collagen fibrils. In cell culture studies, blocking the collagen-binding site on fibronectin resulted in an inhibition of collagen fibril assembly (McDonald et al. 1982). Modulating fibronectin–integrin interactions have been shown to influence collagen fibril assembly (Li et al. 2003). This suggests functional integration of the cytoskeleton with collagen fibril assembly involving an intermediary integrin–fibronectin organizer complex. It is also possible that integrins and other cell-surface molecules such as syndecans can also interact directly with nucleators such as collagen V. A functional link to the cytoskeleton provides the basis for cell-directed collagen fibril and matrix assembly. It is probable that the organizers, like regulators and nucleators, are tissue-specific providing the basis for the diversity of structure and function observed.

It has been suggested that tenascin-X is involved in the regulation of fibril assembly. This is based on the work demonstrating that patients with mutations in tenascin-X have a classic Ehlers–Danlos phenotype (Schalkwijk et al. 2001; Bristow et al. 2005). A mouse model (Mao et al. 2002) demonstrated that collagen fibrils were of normal size and shape, but collagen content was reduced, as was fibril density. However, collagen I synthesis was normal. In addition, the tenascin-X-null fibroblasts did not assemble collagen I into a cell-associated matrix. It was suggested that tenascin-X is an essential regulator of collagen fibril assembly in the dermis. Many of the features described are comparable to those involving collagen V mutations, the exception being that collagen V mutations are associated with abnormal fibril structure. It is tempting to speculate that collagen V and tenascin-X are an organizer–nucleator complex. The absence of tenascin-X disrupts the cell-directed control, but the presence of collagen V results in regulated fibril assembly and therefore normal fibril structure. The decreased incorporation of collagen I into the matrix can be explained as dissociation of the nucleation from the regulated cell-surface domain, resulting in decreased efficiency of assembly.

Organizers provide a mechanism for tissue-specific coupling of fibril assembly to the cell surface. This allows cell-directed positioning of the deposited matrix. In addition, it provides a possible means for undocking of assembled protofibrils from the cell surface and incorporation into the assembling extracellular matrix, thus, freeing the surface for additional rounds of nucleation and protofibril assembly.

3.4.3.3 Regulators

Once the protofibrils are assembled and deposited into the extracellular matrix, further assembly to the mature fibrils involves linear and lateral growth of the preformed intermediates (Birk and Linsenmayer 1994; Birk 2001). Numerous molecules have been shown to be involved in the regulation of these steps in different tissues. Two classes of regulatory molecules are the SLRPs and the FACITs. Both classes are fibril-associated and have different tissue-specific as well as temporal and spatial expression patterns. The different expression patterns contribute to tissue-specific difference in structure and function.

SLRPs (see Chap. 6) are important regulators of linear and lateral fibril growth. Gene-targeting studies with SLRP-deficient mice demonstrate a cooperative relationship between SLRPs in this regulatory function (for reviews, see Ameye and Young 2002; Ameye et al. 2002; Chakravarti 2002; Kalamajski and Oldberg 2010). Members of each SLRP class, e.g., decorin and biglycan, have coordinate roles and members of different classes, e.g., decorin and fibromodulin, have synergistic roles in the regulation of fibrillogenesis and matrix assembly during development of mature, functional tissues. Targeted deletions of different SLRPs result in dysfunctional regulation of fibril growth in a tissue- and region-specific manner. For example, the decorin-deficient cornea has very large, structurally aberrant fibrils throughout the stroma compared with the homogeneous, cylindrical small diameter fibrils required for transparency (Zhang et al. 2009). In the lumican-deficient cornea, fibrils in the posterior stroma have increased diameters and abnormal structures associated with decreased transparency. In contrast, in the lumican-deficient tendon fibril structure is normal. However, the fibromodulin-deficient tendon has large diameter, structurally abnormal fibrils (Ezura et al. 2000). The phenotype is compounded when the tendons are deficient in both fibromodulin and biglycan (Ameye and Young 2002). Alterations in these regulatory interactions result in a tendon that is functionally deficient as well (Zhang et al. 2005). The interactions between SLRPs and fibrils differ between tissues or at different stages of development and regeneration.

As described above, FACIT collagens are fibril-associated molecules with large noncollagenous domains and may be modified with attached GAG chains (see above). This collagen class also demonstrates different tissue-specific and temporal expression patterns. Collagen IX is involved in regulation of fibril growth in cartilage (see below). Recently, collagen XIV has been implicated in regulation of tendon fibril growth (Ansorge et al. 2009). In the absence of collagen XIV, there is a premature increase in fibril diameter. This indicates that in some tissues FACITs may serve as "gate keepers" regulating the transition from protofibril assembly to fibril growth during development. The noncollagenous domains of the fibril-associated molecules have also been implicated in the regulation of fibril packing.

A functional overlap of nucleator/organizer/regulator needs to be considered. Nucleation and organization are likely to be tightly coupled at the cell surface and full function may involve complexes such as integrins, fibronectin and collagen, or tenascin-X and collagen V. Sometimes, the same molecules may serve more than one function. For instance, in assembly of cartilage fibrils, collagen IX is integrated into initial assembly and then requires processing prior to fibril growth and is therefore a regulator.

3.5 Tissue-Specific Regulation of Collagen Fibrillogenesis

The cornea, cartilage, and tendon are three tissues with distinctly different fibril structures and organizations. In the cornea, all fibrils have small (~30 nm) diameters and are regularly packed in orthogonal lamellae. This fibrillar architecture provides for the mechanical stability of the anterior eye and is the structural basis of corneal transparency. In cartilage, there are small diameter fibrils forming a network in the territorial matrix surrounding chondrocytes and larger diameter fibrils with an axial organization in the interterritorial matrix. In tendon, the fibrils have a heterogeneous distribution of large diameters and are arranged uniaxially. A comparison of these tissues demonstrates how similar regulatory mechanisms can be applied with markedly different outcomes.

3.5.1 Corneal Fibril Formation

The mature corneal stroma is composed of a single, homogeneous population of small diameter, regularly packed collagen fibrils that are macromolecular alloys co-assembled from collagens I and V. These heterotypic collagen I/V fibrils also incorporate FACITs on the fibril surface, collagens XII and/or XIV depending on developmental stage. In addition, the SLRPs decorin, biglycan, lumican, keratocan, and fibromodulin modulate the fibril surface properties making corneal fibrils a composite of an alloyed collagenous core adjoining another macromolecular alloy at the periphery. These heteropolymeric fibrils are responsible for the unique properties of the cornea.

In the cornea, nucleation and initial assembly of a protofibril is dependent on the collagen–collagen interactions that produce the alloy. The corneal keratocytes synthesize two fibril-forming collagens. Collagen I is the major collagen making up 80–90% of the total, while the $[\alpha1(V)]_2 \alpha2(V)$ isoform of collagen V is the minor regulatory fibril-forming collagen. Collagens I and V co-assemble so that the collagen V triple helix is internalized within the fibril, while the N-terminal domain projects through the gap region and is present on the fibril surface (Fig. 3.1b) (Linsenmayer et al. 1993; Birk 2001). The helical domain of collagen V is approximately 10% longer than the collagen I domain and does not perfectly fit a quarter-stagger arrangement with collagen I (Fessler et al. 1982; Silver and Birk 1984). Properties intrinsic to these collagen–collagen interactions regulate the nucleation and assembly of the protofibril. The corneal collagen V content is an order of magnitude greater than in other collagen I-containing tissues. This provides for a greater number of nucleation events that would generate a large number of small diameter protofibrils. The organization of this step along the keratocyte surface would provide for cornea-specific fibril organization within the developing matrix. While organization of the keratocyte micro-domains has been reported (Birk and Trelstad 1984), specific organizer molecules/complexes remain to be identified.

The first step in corneal fibrillogenesis involves nucleation of protofibril assembly. A number of fibril assembly assays modulating corneal collagen V in culture models and mouse models have demonstrated that reducing the percentage of collagen V relative to collagen I results in larger diameter fibrils. In cell-based analyses, this also results in decreased numbers of collagen protofibrils assembled (Marchant et al. 1996; Wenstrup et al. 2004a; Segev et al. 2006). This, coupled with the lack of fibril assembly in collagen V-null embryonic mice, indicates a major regulatory role in the nucleation of protofibril assembly (Wenstrup et al. 2004a). However, the large amount of type V collagen incorporated into corneal fibrils relative to other tissues suggests that the N-terminal domain may have an additional regulatory role on the corneal fibril surface. This domain contains sulfated tyrosines and there may be steric and/or electrostatic interactions modulating the fibril surface.

The second stage in corneal fibrillogenesis involves linear fibril growth needed to establish the tensile properties required to maintain the structure of the anterior eye and resist compressive forces due to hydration in the highly charged interfibrillar matrix. In addition, there is an inhibition of lateral fibril growth; large diameter fibrils are incompatible with transparency. It has been shown through targeted deletions that corneal SLRPs regulate the growth of protofibrils into mature long fibrils with no diameter increase. A major role for decorin in preventing lateral fibril growth throughout the stroma has been demonstrated (Zhang et al. 2009). In addition, lumican regulates fibril growth and lumican-null mice demonstrate abnormal lateral corneal fibril growth in the posterior stroma (Chakravarti et al. 2000; 2006). The keratocan-null cornea demonstrates corneal shape changes, but no obvious effect on fibril structure (Liu et al. 2003). Clearly, there is a redundancy in the block of lateral fibril growth in the cornea involving at least decorin and lumican. However, little evidence exists to address the regulation of linear growth. It is possible that in the noncompacted cornea prior to dehydration the protofibrils are separated and the frequency of interaction is low. It has been shown that end-to-end growth occurs ex vivo if the opportunities for interaction are increased (Graham et al. 2000). It is possible that the increasing charge density of the SLRP GAG chains with development begins to order the interfibrillar environment and optimized end-to-end interaction. Another possibility is that the tapered ends of corneal protofibrils alter binding properties providing stabilization to the ends and the normal turnover of bound molecules provides for increased opportunities for interaction in a controlled manner.

3.5.2 Cartilage Fibril Formation

Cartilage fibrils are the main tensile element containing the swelling pressure generated by binding of water to the highly polyanionic glycosaminoglycan chains of the interfibrillar matrix (Eikenberry and Bruckner 1999). Two major populations of fibrils exist in cartilage. In all hyaline cartilages, small uniform diameter fibrils

(~20 nm) are found throughout the extracellular matrix. These small diameter fibrils are enriched in the territorial matrix around the chondrocytes where they demonstrate a preferential orientation parallel to the chondrocyte surface. Thus, individual chondrocytes are embedded in and separated by fibrils that form basket-like structures (Poole 1992). The second fibril population is very restricted, distributed almost exclusively in specialized matrix compartments, termed interterritorial regions that are more remote from the chondrocytes. In growth plates and in articular cartilage, their preferential orientation is defined by the direction of forces generated by load bearing. The mechanism whereby the large diameter, interterritorial fibrils are formed has not been fully elucidated. However, it is probable that the thin territorial fibrils correspond to the cartilage protofibrils. Then, after appropriate processing, the protofibrils undergo growth in diameter by lateral association and fusion to form the larger interterritorial fibrils. Another possible lateral growth mechanism, i.e., the direct accretion of collagens and other macromolecules to pre-existing thin fibrils, is less likely to operate since the cells producing the macromolecular fibril constituents are separated by large distances from the thick and well-banded fibrils of the interterritorial matrix. This would necessitate extensive diffusion of fibril macromolecules through the dense network of cartilage matrix.

Cartilage fibrils exquisitely illustrate the concept of matrix suprastructures as macromolecular composites/alloys. Fibril-forming collagen II is the quantitatively major component in cartilage. This collagen is present in all hyaline cartilage, with a limited distribution in other tissues. By itself, collagen II is incapable of forming fibrils of the extensive lengths required in the tissues. Instead, collagen II forms tactoidal structures of limited lengths when the pure protein is subjected to aggregation in vitro. The tactoids essentially consist of two tapering ends joined together back-to-back. In addition, they are only formed at very high initial concentrations of collagen II and lack lateral growth control. In tissues, collagen II always occurs in macromolecular composites. In the case of the prototypic, territorial fibrils, the collagenous components include collagens II, IX, and XI (Mendler et al. 1989) or, more rarely, types II, XI, and XVI collagen (Kassner et al. 2003). In vitro fibrillogenesis studies demonstrated that mixtures of collagens II and XI form thin and uniform fibrils with a diameter of about 20 nm that closely resemble cartilage protofibrils. The tight regulation of fibril diameter occurs only when types II and XI collagen are present at molar fractions $f_{II/XI} = $ [collagen II]/[collagen XI] ≤ 8, similar to that occurring in cartilage prototypic fibrils.

The protofibrils formed in vitro by collagens II and XI alone are less stable in that the two collagens lose their aggregating capacity upon prolonged standing without demonstrable proteolytic alteration. Moreover, this loss of competence for fibrillogenesis is readily rescued by the addition of collagen IX. Thus, collagen IX is essential for the overall formation of cartilage fibrils by providing long-term stability. Therefore, collagen IX is a third collagenous component of the macromolecular alloy making up cartilage fibrils and can be important during assembly or after assembly in a tissue-specific manner (Blaschke et al. 2000). These heterotypic fibrils encompass parts of all three collagens and the N-terminal NC4 domain of collagen IX can project into the interfibrillar matrix with the collagenous region

COL3 serving as a spacer. In addition, collagen IX via the NC4 domain may interconnect individual cartilage fibrils in the tissue (Muller-Glauser et al. 1986; Eyre et al. 2002).

The macromolecular components of cartilage fibrils are not restricted to collagens. The SLRP decorin is known to occur preferentially in the interterritorial zones of cartilage matrix, i.e., regions containing the large diameter fibrils. In contrast to the territorial region with small diameter fibrils, the large diameter fibrils in the interterritorial zone lack collagen IX (Hagg et al. 1998). This suggests that interterritorial fibrils arise from small prototypic collagen IX-containing fibrils after proteolytic removal of, at least, the N-terminal COL3 and NC4 domains of collagen IX studding the fibril surface. These data indicate that FACIT collagens, i.e., collagen IX, function to maintain protofibrils and prevent the initiation of lateral growth (see collagen XIV and tendon below). After accommodation of such processed prototypic fibrils into the D-periodic stagger, fusion is thought to occur accompanied by a polyanionic conditioning of the fibril surface by incorporation of decorin. This mechanism of lateral growth provides for cell-directed fibril assembly with protofibril assembly occurring closely associated with chondrocytes. Further growth involves regulation of lateral growth from preformed protofibrils via changes in fibril-associated domains and/or molecules. Changes in fibril-associated molecules can occur via normal turnover or could involve selective cell-directed processing. The specific mechanisms regulating the spatially restricted lateral fibril growth in cartilage remain to be elucidated.

3.5.3 Tendon Fibril Formation

The collagen fibrils in tendon have significant increases in diameter during development and growth. The mature tendon contains uniaxial fibrils with a heterogeneous population of different size fibrils. The mechanical properties of the tendon are dependent on the increases in fibril diameter seen with development (Zhang et al. 2005). Tendon fibroblasts express collagen I as the quantitatively major fibril-forming collagen and minor amounts of collagen V. Tendon fibrils are heterotypic being alloys composed of collagens I and V. The nucleation of fibril assembly results in the formation of short, small diameter protofibrils. The protofibrils are deposited into the developing matrix and further growth is regulated by interactions with fibril-associated molecules. The nucleation would be comparable to that described above involving collagen V/XI. This is followed by the assembled protofibril being deposited into the developing matrix.

During tendon development there are changing expression patterns for FACITs and the SLRPs (Ezura et al. 2000; Young et al. 2000; Zhang et al. 2005). Following incorporation of protofibrils into the matrix, they are stabilized. Collagen XIV is expressed during early development when protofibrils predominate followed by little if any expression. A targeted deletion of collagen XIV demonstrated a premature entrance into the fibril growth stage in collagen XIV-deficient tendons.

This resulted in larger diameter fibrils in early developmental stages. This indicates that collagen XIV serves to temporarily stabilize protofibrils and to prevent the initiation of lateral fibril growth. While the specific mode of stabilization is unknown, the result is comparable to that in cartilage with collagen IX. Collagen XIV is expressed in a critical developmental period when tendon structure is being defined and protofibrils are being incorporated into developing fibers. A disruption in fiber development in the absence of collagen XIV has been observed. An additional role in fibril packing has been long suspected due to the large noncollagenous domain and its interfibrillar location. Control of fibril packing would also influence lateral associations necessary for growth.

Comparable to the cornea and cartilage, SLRPs are important in the regulation of tendon fibril growth. Two SLRP classes are expressed throughout tendon growth and maturation: decorin and biglycan (class I) as well as fibromodulin and lumican (class II). Studies of mouse models with single and compound mutations indicate regulatory roles for these SLRPs. Decorin and fibromodulin are dominant in this regulation and can be modulated by biglycan and lumican, respectively (Ezura et al. 2000; Zhang et al. 2005). An absence of decorin, biglycan, or fibromodulin leads to a disruption in fibril growth, generating larger diameter and structural abnormal fibrils as well as altered tendon function (Danielson et al. 1997; Ameye and Young 2002; Jepsen et al. 2002; Zhang et al. 2005). In addition, there is a synergistic effect between classes with compound biglycan and fibromodulin deficiencies have an additive effect (Ameye and Young 2002). These fibril-associated regulators turnover regularly, unlike the fibril-forming collagens, and provide a mechanism to regulate the sequential changes in fibrillogenesis.

3.6 Conclusion

There are 28 different collagen types. These collagens can be grouped into subfamilies based on their predominant suprastructural form including: fibril-forming, FACIT, network-forming, and transmembrane collagens. This provides considerable diversity in the possible functional suprastructures. These 28 collagen types can form isoforms with different alpha chain compositions, are subject to alternative splicing, and undergo different degrees of posttranslational modification such as glycosylation. This further compounds the diversity of the building blocks available for extracellular matrix assembly. It is, ultimately, the suprastructural organization of these collagen molecules that provide tissue-specific structure and function. Collagen suprastructures are composites of different collagens and other matrix molecules. The assembly of suprastructures is sequential within distinct micro-domains. The tissue-, region-, and development-specific expression of matrix molecules involved in suprastructure assembly provides for an almost unlimited degree of diversity. This diversity is necessary to assemble the vast array of extracellular matrices with tissue-specific structures and functions. While much is known about collagens and collagen-containing suprastructures, a major challenge is to elucidate the distinct composition and sequence of assembly required for a specific functional outcome. An understanding of the sequential changes during development will contribute to the understanding of tissue repair and regeneration. While general regulatory principles are evolving, a focus on tissue-specific assembly and regulation is required.

Acknowledgments Work from the author's laboratories was supported by grants from the National Institutes of Health, EY05129, AR44745, and AR55543 as well as by grants from Deutsche Forschungsgemeinschaft, Collaborative Research Centre 492, Projects A2, B9, and B18. Andre Holmes is gratefully acknowledged for his help in preparing the illustrations and Sheila Adams for the preparation of the electron micrographs.

References

Ameye L, Aria D, Jepsen K, Oldberg A, Xu T, Young MF (2002) Abnormal collagen fibrils in tendons of biglycan/fibromodulin-deficient mice lead to gait impairment, ectopic ossification, and osteoarthritis. FASEB J 16:673–680

Ameye L, Young MF (2002) Mice deficient in small leucine-rich proteoglycans: novel in vivo models for osteoporosis, osteoarthritis, Ehlers-Danlos syndrome, muscular dystrophy, and corneal diseases. Glycobiology 12:107R–116R

Ansorge HL, Meng X, Zhang G, Veit G, Sun M, Klement JF, Beason DP, Soslowsky LJ, Koch M, Birk DE (2009) Type XIV collagen regulates fibrillogenesis: premature collagen fibril growth and tissue dysfunction in null mice. J Biol Chem 284:8427–8438

Apte SS (2009) A disintegrin-like and metalloprotease (reprolysin-type) with thrombospondin type 1 motif (ADAMTS) superfamily: functions and mechanisms. J Biol Chem 284: 31493–31497

Aszodi A, Chan D, Hunziker E, Bateman JF, Fassler R (1998) Collagen II is essential for the removal of the notochord and the formation of intervertebral discs. J Cell Biol 143:1399–1412

Bachinger HP, Bruckner P, Timpl R, Prockop DJ, Engel J (1980) Folding mechanism of the triple helix in type-III collagen and type-III pN-collagen. Role of disulfide bridges and peptide bond isomerization. Eur J Biochem 106:619–632

Ball S, Bella J, Kielty C, Shuttleworth A (2003) Structural basis of type VI collagen dimer formation. J Biol Chem 278:15326–15332

Batge B, Winter C, Notbohm H, Acil Y, Brinckmann J, Muller PK (1997) Glycosylation of human bone collagen I in relation to lysylhydroxylation and fibril diameter. J Biochem (Tokyo) 122:109–115

Berg RA, Prockop DJ (1973) The thermal transition of a non-hydroxylated form of collagen. Evidence for a role for hydroxyproline in stabilizing the triple-helix of collagen. Biochem Biophys Res Commun 52:115–120

Birk DE (2001) Type V collagen: heterotypic type I/V collagen interactions in the regulation of fibril assembly. Micron 32:223–237

Birk DE, Bruckner P (2005) Collagen suprastructures. Top Curr Chem 247:185–205

Birk DE, Linsenmayer TF (1994) Collagen fibril assembly, deposition, and organization into tissue-specific matrices. In: Yurchenco PD, Birk DE, Mecham RP (eds) Extracellular matrix assembly and structure. Academic, NY, pp 91–128

Birk DE, Nurminskaya MV, Zycband EI (1995) Collagen fibrillogenesis in situ: fibril segments undergo post-depositional modifications resulting in linear and lateral growth during matrix development. Dev Dyn 202:229–243

Birk DE, Trelstad RL (1986) Extracellular compartments in tendon morphogenesis: collagen fibril, bundle, and macroaggregate formation. J Cell Biol 103:231–240

Birk DE, Trelstad RL (1984) Extracellular compartments in matrix morphogenesis: collagen fibril, bundle, and lamellar formation by corneal fibroblasts. J Cell Biol 99:2024–2033

Birk DE, Zycband EI, Winkelmann DA, Trelstad RL (1989) Collagen fibrillogenesis in situ: fibril segments are intermediates in matrix assembly. Proc Natl Acad Sci USA 86:4549–4553

Birk DE, Zycband EI, Woodruff S, Winkelmann DA, Trelstad RL (1997) Collagen fibrillogenesis in situ: fibril segments become long fibrils as the developing tendon matures. Dev Dyn 208: 291–298

Blaschke UK, Eikenberry EF, Hulmes DJ, Galla HJ, Bruckner P (2000) Collagen XI nucleates self-assembly and limits lateral growth of cartilage fibrils. J Biol Chem 275:10370–10378

Boot-Handford RP, Tuckwell DS (2003) Fibrillar collagen: the key to vertebrate evolution? A tale of molecular incest. Bioessays 25:142–151

Bristow J, Carey W, Egging D, Schalkwijk J (2005) Tenascin-X, collagen, elastin, and the Ehlers-Danlos syndrome. Am J Med Genet C Semin Med Genet 139C:24–30

Brittingham R, Uitto J, Fertala A (2006) High-affinity binding of the NC1 domain of collagen VII to laminin 5 and collagen IV. Biochem Biophys Res Commun 343:692–699

Brodsky B, Eikenberry EF, Belbruno KC, Sterling K (1982) Variations in collagen fibril structure in tendons. Biopolymers 21:935–951

Bruckner P, Eikenberry EF, Prockop DJ (1981) Formation of the triple helix of type I procollagen in cellulo. A kinetic model based on *cis-trans* isomerization of peptide bonds. Eur J Biochem 118:607–613

Bruckner-Tuderman L (2010) Dystrophic epidermolysis bullosa: pathogenesis and clinical features. Dermatol Clin 28:107–114

Bruckner-Tuderman L, Hopfner B, Hammami-Hauasli N (1999) Biology of anchoring fibrils: lessons from dystrophic epidermolysis bullosa. Matrix Biol 18:43–54

Bruns RR, Press W, Engvall E, Timpl R, Gross J (1986) Type VI collagen in extracellular, 100-nm periodic filaments and fibrils: identification by immunoelectron microscopy. J Cell Biol 103:393–404

Burgeson RE, Christiano AM (1997) The dermal-epidermal junction. Curr Opin Cell Biol 9:651–658

Canty EG, Kadler KE (2002) Collagen fibril biosynthesis in tendon: a review and recent insights. Comp Biochem Physiol A Mol Integr Physiol 133:979–985

Canty EG, Kadler KE (2005) Procollagen trafficking, processing and fibrillogenesis. J Cell Sci 118:1341–1353

Canty EG, Lu Y, Meadows RS, Shaw MK, Holmes DF, Kadler KE (2004) Coalignment of plasma membrane channels and protrusions (fibripositors) specifies the parallelism of tendon. J Cell Biol 165:553–563

Chakravarti S (2002) Functions of lumican and fibromodulin: lessons from knockout mice. Glycoconj J 19:287–293

Chakravarti S, Petroll WM, Hassell JR, Jester JV, Lass JH, Paul J, Birk DE (2000) Corneal opacity in lumican-null mice: defects in collagen fibril structure and packing in the posterior stroma. Invest Ophthalmol Vis Sci 41:3365–3373

Chakravarti S, Zhang G, Chervoneva I, Roberts L, Birk DE (2006) Collagen fibril assembly during postnatal development and dysfunctional regulation in the lumican-deficient murine cornea. Dev Dyn 235:2493–2506

Chou MY, Li HC (2002) Genomic organization and characterization of the human type XXI collagen (COL21A1) gene. Genomics 79:395–401

Chu ML, Mann K, Deutzmann R, Pribula-Conway D, Hsu-Chen CC, Bernard MP, Timpl R (1987) Characterization of three constituent chains of collagen type VI by peptide sequences and cDNA clones. Eur J Biochem 168:309–317

Colige A, Ruggiero F, Vandenberghe I, Dubail J, Kesteloot F, Van BJ, Beschin A, Brys L, Lapiere CM, Nusgens B (2005) Domains and maturation processes that regulate the activity of ADAMTS-2, a metalloproteinase cleaving the aminopropeptide of fibrillar procollagens types I-III and V. J Biol Chem 280:34397–34408

Danielson KG, Baribault H, Holmes DF, Graham H, Kadler KE, Iozzo RV (1997) Targeted disruption of decorin leads to abnormal collagen fibril morphology and skin fragility. J Cell Biol 136:729–743

Eikenberry EF, Bruckner P (1999) Supramolecular structure of cartilage matrix. In: Shaw MK, Rom E, Bilezekian JP (eds) Dynamics of bone and cartilage metabolism. Academic, San Diego, CA, pp 289–300

Engel J, Bachinger HP (2005) Structure, stability and folding of the collagen triple helix. Top Curr Chem 247:7–33

Eyre DR, Wu JJ, Fernandes RJ, Pietka TA, Weis MA (2002) Recent developments in cartilage research: matrix biology of the collagen II/IX/XI heterofibril network. Biochem Soc Trans 30:893–899

Ezura Y, Chakravarti S, Oldberg A, Chervoneva I, Birk DE (2000) Differential expression of lumican and fibromodulin regulate collagen fibrillogenesis in developing mouse tendons. J Cell Biol 151:779–788

Fessler JH, Bachinger HP, Lunstrum G, Fessler LI (1982) Biosynthesis and processing of some procollagens. In: Kuehn K, Schoene H, Timpl R (eds) New trends in basement membrane research. Raven, New York, pp 145–153

Fitzgerald J, Rich C, Zhou FH, Hansen U (2008) Three novel collagen VI chains, alpha4(VI), alpha5(VI), and alpha6(VI). J Biol Chem 283:20170–20180

Franzke CW, Bruckner P, Bruckner-Tuderman L (2005) Collagenous transmembrane proteins: recent insights into biology and pathology. J Biol Chem 280:4005–4008

Franzke CW, Tasanen K, Schumann H, Bruckner-Tuderman L (2003) Collagenous transmembrane proteins: collagen XVII as a prototype. Matrix Biol 22:299–309

Furthmayr H, Wiedemann H, Timpl R, Odermatt E, Engel J (1983) Electron-microscopical approach to a structural model of intima collagen. Biochem J 211:303–311

Gara SK, Grumati P, Urciuolo A, Bonaldo P, Kobbe B, Koch M, Paulsson M, Wagener R (2008) Three novel collagen VI chains with high homology to the alpha3 chain. J Biol Chem 283:10658–10670

Gordon MK, Hahn RA (2010) Collagens. Cell Tissue Res 339:247–257

Graham HK, Holmes DF, Watson RB, Kadler KE (2000) Identification of collagen fibril fusion during vertebrate tendon morphogenesis. The process relies on unipolar fibrils and is regulated by collagen-proteoglycan interaction. J Mol Biol 295:891–902

Greenspan DS (2005) Biosynthetic processing of collagen molecules. Top Curr Chem 247:149–183

Gregory KE, Oxford JT, Chen Y, Gambee JE, Gygi SP, Aebersold R, Neame PJ, Mechling DE, Bachinger HP, Morris NP (2000) Structural organization of distinct domains within the non-collagenous N-terminal region of collagen type XI. J Biol Chem 275:11498–11506

Hagg R, Bruckner P, Hedbom E (1998) Cartilage fibrils of mammals are biochemically heterogeneous: differential distribution of decorin and collagen IX. J Cell Biol 142:285–294

Halfter W, Dong S, Schurer B, Cole GJ (1998) Collagen XVIII is a basement membrane heparan sulfate proteoglycan. J Biol Chem 273:25404–25412

Hoffman GG, Branam AM, Huang G, Pelegri F, Cole WG, Wenstrup RM, Greenspan DS (2010) Characterization of the six zebrafish clade B fibrillar procollagen genes, with evidence for evolutionarily conserved alternative splicing within the pro-alpha1(V) C-propeptide. Matrix Biol 29:261–275

Hopkins DR, Keles S, Greenspan DS (2007) The bone morphogenetic protein 1/Tolloid-like metalloproteinases. Matrix Biol 26:508–523

Illidge C, Kielty C, Shuttleworth A (1998) The alpha 1(VIII) and alpha 2(VIII) chains of type VIII collagen can form stable homotrimeric molecules. J Biol Chem 273:22091–22095

Illidge C, Kielty C, Shuttleworth A (2001) Type VIII collagen: heterotrimeric chain association. Int J Biochem Cell Biol 33:521–529

Jakus MA (1956) Studies on the cornea. II. The fine structure of Descement's membrane. J Biophys Biochem Cytol 2:243–252

Jelinski LW, Torchia DA (1979) 13C/1H high power double magnetic resonance investigation of collagen backbone motion in fibrils and in solution. J Mol Biol 133:45–65

Jepsen KJ, Wu F, Peragallo JH, Paul J, Roberts L, Ezura Y, Oldberg A, Birk DE, Chakravarti S (2002) A syndrome of joint laxity and impaired tendon integrity in lumican- and fibromodulin-deficient mice. J Biol Chem 277:35532–35540

Kadler KE, Baldock C, Bella J, Boot-Handford RP (2007) Collagens at a glance. J Cell Sci 120:1955–1958

Kadler KE, Hill A, Canty-Laird EG (2008) Collagen fibrillogenesis: fibronectin, integrins, and minor collagens as organizers and nucleators. Curr Opin Cell Biol 20:495–501

Kadler KE, Holmes DF, Trotter JA, Chapman JA (1996) Collagen fibril formation. Biochem J 316 (Pt 1):1–11

Kalamajski S, Oldberg A (2010) The role of small leucine-rich proteoglycans in collagen fibrillogenesis. Matrix Biol 29:248–253

Kassner A, Hansen U, Miosge N, Reinhardt DP, Aigner T, Bruckner-Tuderman L, Bruckner P, Grassel S (2003) Discrete integration of collagen XVI into tissue-specific collagen fibrils or beaded microfibrils. Matrix Biol 22:131–143

Katz EP, Wachtel EJ, Maroudas A (1986) Extrafibrillar proteoglycans osmotically regulate the molecular packing of collagen in cartilage. Biochim Biophys Acta 882:136–139

Keller H, Eikenberry EF, Winterhalter KH, Bruckner P (1985) High post-translational modification levels in type II procollagen are not a consequence of slow triple-helix formation. Coll Relat Res 5:245–251

Khoshnoodi J, Cartailler JP, Alvares K, Veis A, Hudson BG (2006) Molecular recognition in the assembly of collagens: terminal noncollagenous domains are key recognition modules in the formation of triple helical protomers. J Biol Chem 281:38117–38121

Kielty CM, Grant ME (2002) The collagen family: structure, assembly, and organization in the extracellular matrix. In: Royce PM, Steinmann B (eds) Connective tissue and its heritable disorders. Wiley-Liss, New York, pp 159–222

Knupp C, Squire JM (2001) A new twist in the collagen story–the type VI segmented supercoil. EMBO J 20:372–376

Koch M, Schulze J, Hansen U, Ashwodt T, Keene DR, Brunken WJ, Burgeson RE, Bruckner P, Bruckner-Tuderman L (2004) A novel marker of tissue junctions, collagen XXII. J Biol Chem 279:22514–22521

Kramer RZ, Bella J, Mayville P, Brodsky B, Berman HM (1999) Sequence dependent conformational variations of collagen triple-helical structure. Nat Struct Biol 6:454–457

Kvansakul M, Bogin O, Hohenester E, Yayon A (2003) Crystal structure of the collagen alpha1 (VIII) NC1 trimer. Matrix Biol 22:145–152

Kwan AP, Cummings CE, Chapman JA, Grant ME (1991) Macromolecular organization of chicken type X collagen in vitro. J Cell Biol 114:597–604

Li D, Clark CC, Myers JC (2000) Basement membrane zone type XV collagen is a disulfide-bonded chondroitin sulfate proteoglycan in human tissues and cultured cells. J Biol Chem 275:22339–22347

Li S, Van Den DC, D'Souza SJ, Chan BM, Pickering JG (2003) Vascular smooth muscle cells orchestrate the assembly of type I collagen via alpha2beta1 integrin, RhoA, and fibronectin polymerization. Am J Pathol 163:1045–1056

Li SW, Prockop DJ, Helminen H, Fassler R, Lapvetelainen T, Kiraly K, Peltarri A, Arokoski J, Lui H, Arita M (1995a) Transgenic mice with targeted inactivation of the Col2 alpha 1 gene for collagen II develop a skeleton with membranous and periosteal bone but no endochondral bone. Genes Dev 9:2821–2830

Li Y, Lacerda DA, Warman ML, Beier DR, Yoshioka H, Ninomiya Y, Oxford JT, Morris NP, Andrikopoulos K, Ramirez F (1995b) A fibrillar collagen gene, Col1la1, is essential for skeletal morphogenesis. Cell 80:423–430

Linsenmayer T, Gibney E, Igoe F, Gordon M, Fitch J, Fessler L, Birk D (1993) Type V collagen: molecular structure and fibrillar organization of the chicken alpha 1(V) NH2-terminal domain, a putative regulator of corneal fibrillogenesis. J Cell Biol 121:1181–1189

Liu CY, Birk DE, Hassell JR, Kane B, Kao WW (2003) Keratocan-deficient mice display alterations in corneal structure. J Biol Chem 278:21672–21677

Malfait F, De PA (2009) Bleeding in the heritable connective tissue disorders: mechanisms, diagnosis and treatment. Blood Rev 23:191–197

Mao JR, Taylor G, Dean WB, Wagner DR, Afzal V, Lotz JC, Rubin EM, Bristow J (2002) Tenascin-X deficiency mimics Ehlers–Danlos syndrome in mice through alteration of collagen deposition. Nat Genet 30:421–425

Mao Y, Schwarzbauer JE (2005) Fibronectin fibrillogenesis, a cell-mediated matrix assembly process. Matrix Biol 24:389–399

Marchant JK, Hahn RA, Linsenmayer TF, Birk DE (1996) Reduction of type V collagen using a dominant-negative strategy alters the regulation of fibrillogenesis and results in the loss of corneal-specific fibril morphology. J Cell Biol 135:1415–1426

Marini JC, Cabral WA, Barnes AM (2010) Null mutations in LEPRE1 and CRTAP cause severe recessive osteogenesis imperfecta. Cell Tissue Res 339:59–70

Marneros AG, Keene DR, Hansen U, Fukai N, Moulton K, Goletz PL, Moiseyev G, Pawlyk BS, Halfter W, Dong S, Shibata M, Li T, Crouch RK, Bruckner P, Olsen BR (2004) Collagen XVIII/endostatin is essential for vision and retinal pigment epithelial function. EMBO J 23:89–99

McDonald JA, Kelley DG, Broekelmann TJ (1982) Role of fibronectin in collagen deposition: Fab' to the gelatin-binding domain of fibronectin inhibits both fibronectin and collagen organization in fibroblast extracellular matrix. J Cell Biol 92:485–492

Mendler M, Eich-Bender SG, Vaughan L, Winterhalter KH, Bruckner P (1989) Cartilage contains mixed fibrils of collagen types II, IX, and XI. J Cell Biol 108:191–197

Muller-Glauser W, Humbel B, Glatt M, Strauli P, Winterhalter KH, Bruckner P (1986) On the role of type IX collagen in the extracellular matrix of cartilage: type IX collagen is localized to intersections of collagen fibrils. J Cell Biol 102:1931–1939

Myers JC, Yang H, D'Ippolito JA, Presente A, Miller MK, Dion AS (1994) The triple-helical region of human type XIX collagen consists of multiple collagenous subdomains and exhibits limited sequence homology to alpha 1(XVI). J Biol Chem 269:18549–18557

Myllyharju J, Kivirikko KI (2004) Collagens, modifying enzymes and their mutations in humans, flies and worms. Trends Genet 20:33–43

Notbohm H, Nokelainen M, Myllyharju J, Fietzek PP, Muller PK, Kivirikko KI (1999) Recombinant human type II collagens with low and high levels of hydroxylysine and its glycosylated forms show marked differences in fibrillogenesis in vitro. J Biol Chem 274:8988–8992

Okuyama K, Hongo C, Wu G, Mizuno K, Noguchi K, Ebisuzaki S, Tanaka Y, Nishino N, Bachinger HP (2009) High-resolution structures of collagen-like peptides [(Pro-Pro-Gly)4-Xaa-Yaa-Gly-(Pro-Pro-Gly)4]: implications for triple-helix hydration and Hyp(X) puckering. Biopolymers 91:361–372

Pan TC, Zhang RZ, Mattei MG, Timpl R, Chu ML (1992) Cloning and chromosomal location of human alpha 1(XVI) collagen. Proc Natl Acad Sci USA 89:6565–6569

Poole CA (1992) Chondrons-the chondrocyte and its pericellular microenvironment. In: Kuettner KE, Schleyerbach R, Peyron JG, Hascall VC (eds) Articular cartilage and osteoarthritis. Raven, New York, pp 210–220

Rehn M, Hintikka E, Pihlajaniemi T (1994) Primary structure of the alpha 1 chain of mouse type XVIII collagen, partial structure of the corresponding gene, and comparison of the alpha 1 (XVIII) chain with its homologue, the alpha 1(XV) collagen chain. J Biol Chem 269: 13929–13935

Sawada H, Konomi H, Hirosawa K (1990) Characterization of the collagen in the hexagonal lattice of Descemet's membrane: its relation to type VIII collagen. J Cell Biol 110:219–227

Schalkwijk J, Zweers MC, Steijlen PM, Dean WB, Taylor G, van Vlijmen IM, van Haren B, Miller WL, Bristow J (2001) A recessive form of the Ehlers-Danlos syndrome caused by tenascin-X deficiency. N Engl J Med 345:1167–1175

Seegmiller R, Fraser FC, Sheldon H (1971) A new chondrodystrophic mutant in mice. Electron microscopy of normal and abnormal chondrogenesis. J Cell Biol 48:580–593

Segev F, Heon E, Cole WG, Wenstrup RJ, Young F, Slomovic AR, Rootman DS, Whitaker-Menezes D, Chervoneva I, Birk DE (2006) Structural abnormalities of the cornea and lid resulting from collagen V mutations. Invest Ophthalmol Vis Sci 47:565–573

Shoulders MD, Raines RT (2009) Collagen structure and stability. Annu Rev Biochem 78: 929–958

Shuttleworth CA (1997) Type VIII collagen. Int J Biochem Cell Biol 29:1145–1148

Silver FH, Birk DE (1984) Molecular structure of collagen in solution: comparison of types I, II, III and V. Int J Biol Macromol 6:125–132

Steinmann B, Bruckner P, Superti-Furga A (1991) Cyclosporin a slows collagen triple-helix formation in vivo: indirect evidence for a physiologic role of peptidyl-prolyl *cis-trans*-isomerase. J Biol Chem 266:1299–1303

Stephan S, Sherratt MJ, Hodson N, Shuttleworth CA, Kielty CM (2004) Expression and supramolecular assembly of recombinant alpha1(viii) and alpha2(viii) collagen homotrimers. J Biol Chem 279:21469–21477

Suzuki OT, Sertié AL, Der Kaloustian VM, Kok F, Carpenter M, Murray J, Czeizel AE, Kliemann SE, Rosemberg S, Monteiro M, Olsen BR, Passos-Bueno MR (2002) Molecular analysis of collagen XVIII reveals novel mutations, presence of a third isoform, and possible genetic heterogeneity in Knobloch syndrome. Am J Hum Genet 71:1320–1329

Torre-Blanco A, Adachi E, Hojima Y, Wootton JA, Minor RR, Prockop DJ (1992) Temperature-induced post-translational over-modification of type I procollagen. Effects of over-modification of the protein on the rate of cleavage by procollagen N-proteinase and on self-assembly of collagen into fibrils. J Biol Chem 267:2650–2655

Trelstad RL, Hayashi K (1979) Tendon collagen fibrillogenesis: intracellular subassemblies and cell surface changes associated with fibril growth. Dev Biol 71:228–242

Villone D, Fritsch A, Koch M, Bruckner-Tuderman L, Hansen U, Bruckner P (2008) Supramolecular interactions in the dermo-epidermal junction zone: anchoring fibril-collagen VII tightly binds to banded collagen fibrils. J Biol Chem 283:24506–24513

Vogel BE, Doelz R, Kadler KE, Hojima Y, Engel J, Prockop DJ (1988) A substitution of cysteine for glycine 748 of the alpha 1 chain produces a kink at this site in the procollagen I molecule and an altered N-proteinase cleavage site over 225 nm away. J Biol Chem 263:19249–19255

von der Mark H, Aumailley M, Wick G, Fleischmajer R, Timpl R (1984) Immunochemistry, genuine size and tissue localization of collagen VI. Eur J Biochem 142:493–502

Wenstrup RJ, Florer JB, Brunskill EW, Bell SM, Chervoneva I, Birk DE (2004a) Type V collagen controls the initiation of collagen fibril assembly. J Biol Chem 279:53331–53337

Wenstrup RJ, Florer JB, Cole WG, Willing MC, Birk DE (2004b) Reduced type I collagen utilization: a pathogenic mechanism in COL5A1 haplo-insufficient Ehlers-Danlos syndrome. J Cell Biochem 92:113–124

Wiberg C, Heinegard D, Wenglen C, Timpl R, Morgelin M (2002) Biglycan organizes collagen VI into hexagonal-like networks resembling tissue structures. J Biol Chem 277:49120–49126

Yamaguchi N, Mayne R, Ninomiya Y (1991) The alpha 1 (VIII) collagen gene is homologous to the alpha 1 (X) collagen gene and contains a large exon encoding the entire triple helical and carboxyl-terminal non-triple helical domains of the alpha 1 (VIII) polypeptide. J Biol Chem 266:4508–4513

Yoshioka H, Zhang H, Ramirez F, Mattei MG, Moradi-Ameli M, van der Rest M, Gordon MK (1992) Synteny between the loci for a novel FACIT-like collagen locus (D6S228E) and alpha 1 (IX) collagen (COL9A1) on 6q12-q14 in humans. Genomics 13:884–886

Young BB, Gordon MK, Birk DE (2000) Expression of type XIV collagen in developing chicken tendons: association with assembly and growth of collagen fibrils. Dev Dyn 217:430–439

Zhang G, Chen S, Goldoni S, Calder BW, Simpson HC, Owens RT, McQuillan DJ, Young MF, Iozzo RV, Birk DE (2009) Genetic evidence for the coordinated regulation of collagen fibrillogenesis in the cornea by decorin and biglycan. J Biol Chem 284:8888–8897

Zhang G, Young BB, Ezura Y, Favata M, Soslowsky LJ, Chakravarti S, Birk DE (2005) Development of tendon structure and function: regulation of collagen fibrillogenesis. J Musculoskelet Neuronal Interact 5:5–21

Zhong C, Chrzanowska-Wodnicka M, Brown J, Shaub A, Belkin AM, Burridge K (1998) Rho-mediated contractility exposes a cryptic site in fibronectin and induces fibronectin matrix assembly. J Cell Biol 141:539–551

Chapter 4
Basement Membranes

Jeffrey H. Miner

Abstract Basement membranes are thin sheets of specialized extracellular matrices. They are found at the basal surfaces of epithelial and endothelial cells, and they surround all muscle cells, fat cells, the central nervous system, and peripheral nerves. Basement membranes compartmentalize tissues, serve as macromolecular filters, provide sites for cell adhesion, and harbor signaling cues that mediate cell proliferation, migration, and differentiation. All basement membranes contain laminin, type IV collagen, nidogen, and sulfated proteoglycans such as perlecan, agrin, and collagen XVIII. Laminin and collagen IV can self-polymerize into networks that are linked by nidogen and proteoglycans. Cells recognize basement membranes and in some cases facilitate their assembly using both integrin and non-integrin receptors such as dystroglycan, DDR1, and Lutheran/BCAM. A number of human genetic diseases, including junctional epidermolysis bullosa, congenital muscular dystrophy, and Alport syndrome, are caused by mutations that affect basement membrane components. This chapter discusses these and other aspects of basement membranes.

4.1 Introduction

Basement membranes are specialized extracellular matrices (ECMs) found in most tissues and associated with diverse cell types. Basement membranes can be visualized by conventional light microscopy and are particularly well demarcated in paraffin sections by several histological stains, including periodic acid-Schiff (PAS) and Jones Methenamine Silver. Most notably, basement membranes are found between epithelial cell layers and the underlying connective tissue, stroma, or interstitium (Fig. 4.1). For most epithelial cells in the body, the basement membrane represents

---

J.H. Miner
Renal Division, Washington University School of Medicine, Box 8126, 660 S. Euclid Avenue, St. Louis, MO 63110, USA
e-mail: minerj@wustl.edu

Fig. 4.1 Histological identification of basement membranes. (**a**) Periodic acid-Schiff staining of an adult mouse kidney paraffin section reveals the basement membrane (*arrows*) around each tubule. (**b**) Anti-laminin-111 immunofluorescence staining of a frozen section of an adult mouse kidney reveals the same tubular basement membranes (*red*). Photo courtesy of Elizabeth Danka, Washington University School of Medicine

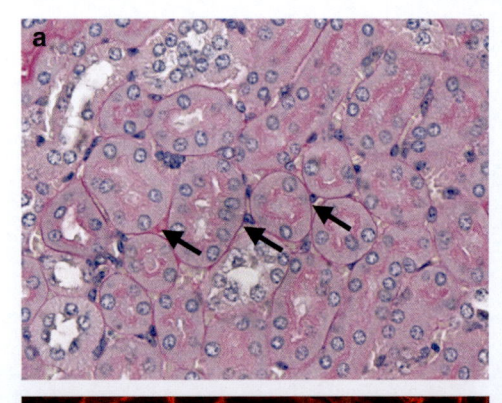

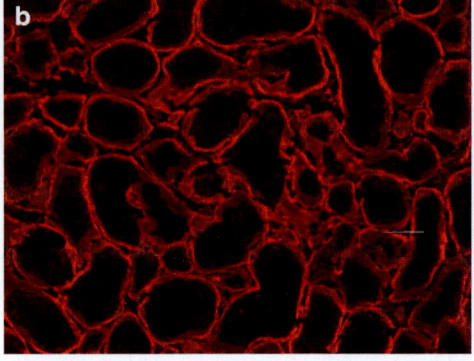

their only physical contact with the rest of the organism. This is true of tubular epithelial cells in the kidney, which are in contact with urine and basement membrane, alveolar epithelial cells in the lung, which are in contact with air and basement membrane, and the mucosal epithelial cells lining the luminal surface of the gastrointestinal tract, which are in contact with digesting food and basement membrane. In this regard, basement membranes are crucial for providing, either directly or indirectly, the appropriate signals to maintain epithelial cell homeostasis. They are also necessary for maintaining the integrity of the epithelium, among other important functions.

In addition to their association with epithelial cells, basement membranes also surround all muscle fibers, fat cells, blood vessels (endothelial cells), the central nervous system, and Schwann cell/axon units in peripheral nerves. The outer layer of most internal organs includes specialized mesothelial, capsular, or serosal cells that sit atop a basement membrane. Besides contributing to cell organization and polarity and influencing cell migration, proliferation, and differentiation, basement membranes also compartmentalize tissues and serve as filtration barriers for macromolecules.

With the advent of electron microscopy, it became clear that basement membranes are composed of more than one layer or type of ECM. Ultrastructurally, basement membranes contain a ribbon-like structure, referred to as the basal lamina, which is typically synthesized by the overlying epithelial cell or other cell type. When viewed in specimens that are fixed, dehydrated, and embedded in a conventional manner, the basal lamina is observed to consist of an electron dense component, the lamina densa, and flanking electron lucent components, the lamina lucidae, also known as the lamina rara interna and the lamina rara externa (Fig. 4.2a). However, the electron lucent components have been proposed to be artifacts of the dehydration procedure, as they are not visible in specimens treated in a different manner (Fig. 4.2b) (Chan and Inoue 1994; Chan et al. 1993). Interestingly, the features of both can be visualized by deep-etch electron microscopy of the same basal lamina (Fig. 4.2c). In addition to the basal lamina, the basement membrane includes the immediately adjacent amorphous or reticular ECM, which consists of collagen and other ECM proteins, that is secreted by underlying stromal or interstitial cells usually exhibiting a mesenchymal or fibroblastic phenotype.

In the literature, the distinction between "basement membrane" and "basal lamina" has become rather blurred. Many authors, this one included, have used the terms interchangeably. Although in some cases, such as the kidney glomerular basement membrane (GBM) and the synaptic basement membrane at the neuromuscular junction, the basement membrane does indeed consist of nothing but the basal lamina; there are many other cases where this clearly is not true. In any event, despite its title, this chapter focuses on the basal lamina portion of the basement membrane, in part because so much is known about the basal lamina's composition and function, as well as its malfunction in both genetic and acquired diseases. And because so much of what we understand about basement membrane function has been gleaned from analysis of knockout mice and human diseases in which genes encoding basement membrane protein genes have been mutated, much of this chapter deals with those aspects of basement membrane biology, in the context of mammalian physiology.

4.2 Basement Membrane Components

All basement membranes contain various isoforms of four major components: laminin, type IV collagen, nidogen, and sulfated proteoglycans (Fig. 4.3). The specific collection of isoforms of these and other components that are found in a given basement membrane, for example, the epidermal basement membrane or the kidney GBM, is usually stereotypical across species, as long as the evolutionary distance is not too great. Because of this conservation, it is presumed that a given basement membrane's composition is related, at least to some degree, to its specific function.

4.2.1 Laminin

Laminins are a family of large glycoproteins that are secreted into the ECM as α–β–γ heterotrimers. In mammals, there are five α ($\alpha1$–$\alpha5$), four β ($\beta1$–$\beta4$), and

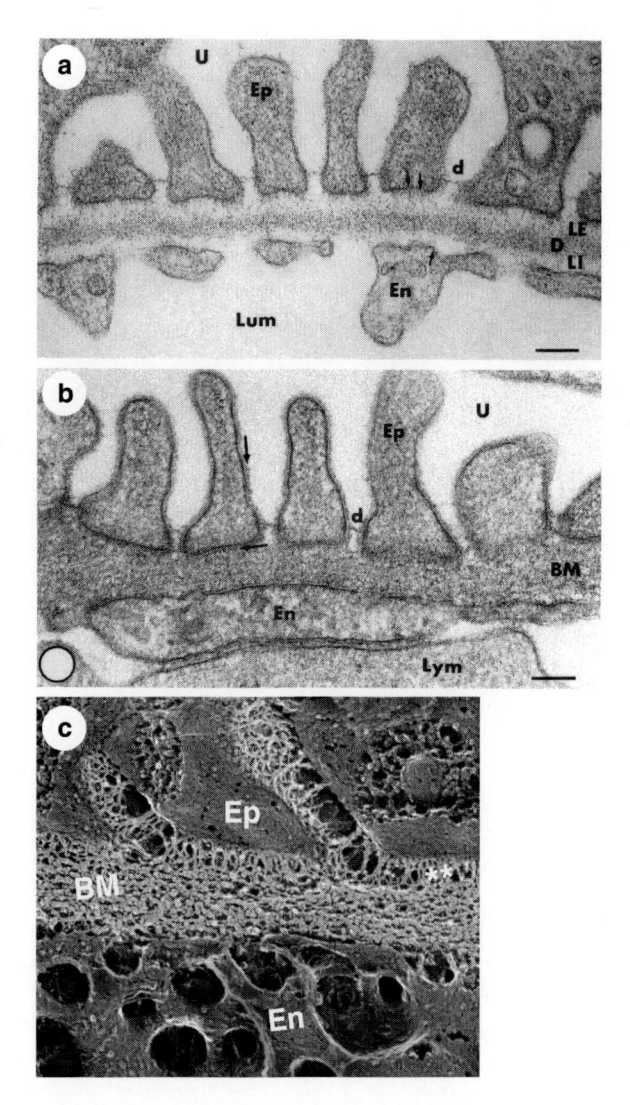

Fig. 4.2 The lamina lucida appears to be an artifact of conventional tissue processing. (**a**) Transmission electron micrograph showing the mouse kidney glomerular capillary wall in conventionally processed tissue. Note the lamina densa (D), lamina lucida externa (LE), and lamina lucida interna (LI). (**b**) Transmission electron micrograph showing the same anatomical structure after freeze substitution. The basement membrane (BM) lacks laminae lucidae, and the lamina densa is in direct contact with the cells. (**c**) Deep-etch electron microscopy of the same anatomical structure shows a less dense area beneath the podocyte foot processes but a tight association between the endothelium and the glomerular BM (GBM). Ep, podocyte foot process; U, urinary space; d, slit diaphragm; En, endothelial cell; Lym, lymphocyte; arrows, basal surface of podocytes. (**a**) and (**b**) Reprinted by permission from John Wiley and Sons: Chan and Inoue (1994), copyright 1994. (**c**) Courtesy of Dr. John Heuser, Washington University School of Medicine

4 Basement Membranes

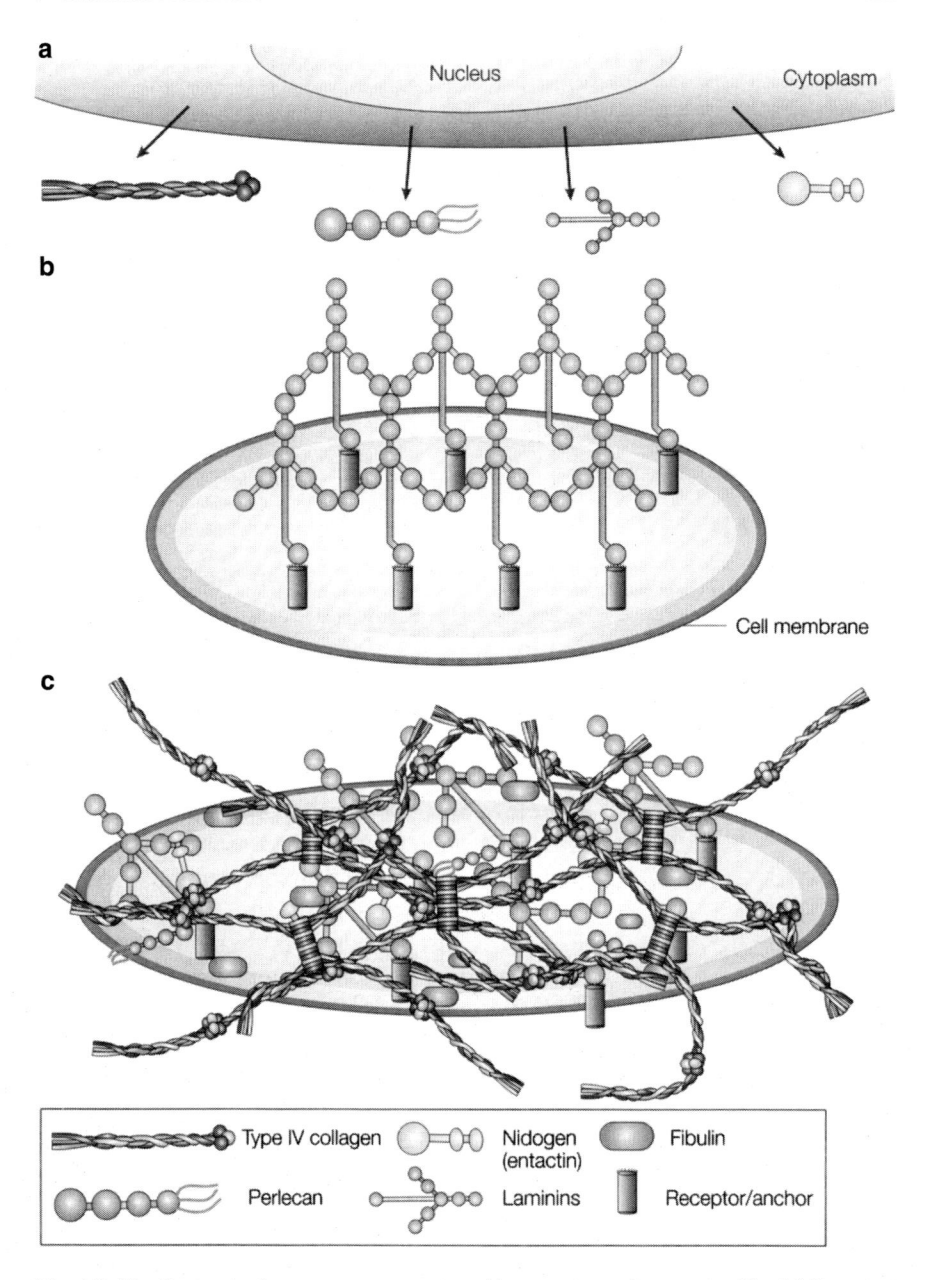

Fig. 4.3 The four major basement components and basement membrane assembly. (**a**) Basement membrane components (collagen IV heterotrimer, perlecan, laminin heterotrimer, and nidogen) are secreted by a cell into the extracellular matrix (ECM). (**b**) The laminin LG domain binds to receptors on the cell membrane (*cylinder*; dystroglycan or integrin), and laminin LN domains of α, β, and γ chains interact in a tripartite complex that promotes laminin polymerization. (**c**) The other components, including the collagen IV network, integrate with the laminin network to assemble the basement membrane. Reprinted by permission from Macmillan Publishers Ltd: Kalluri (2003), copyright 2003

three γ (γ1–γ3) chains (Fig. 4.4) that can assemble with each other in a nonrandom fashion to generate at least 15 different heterotrimers. Interestingly, mice and rats have only three β chains; the fourth is present in humans and horse, as well as in chicken. The evolutionary basis for this has not been investigated in detail, but the human *LAMB1* and *LAMB4* genes are only about 20 kb apart, and mouse *Lamb1* is located at the end of a conserved segment of orthologous human genes. What is clear is that the laminin chains are all evolutionarily related modular proteins (Fig. 4.4) containing globular, laminin type EGF-like (LE) repeat, and α-helical domains (known as laminin coiled-coil [LCC] domains). Laminin α chains are unique in that they bear a large, modular, COOH-terminal laminin globular (LG) domain that interacts with cellular receptors (see Sect. 8.3).

Laminins assemble into trimers in the endoplasmic reticulum via their LCC domains, and their association is further stabilized by limited interchain disulfide bonding (Timpl and Brown 1994). With the growth in the number of laminin chains and potential trimers, a new nomenclature was presented in 2005 (Aumailley et al. 2005) in which laminin trimers are named based solely on their constituent chains. For example, the laminin trimer containing the α2, β2, and γ1 chains is referred to as laminin-221 or abbreviated LM-221.

Depending on the constituent chains, laminin trimers are either cross-shaped, Y-shaped, or rod-shaped (Fig. 4.5); this determines in part how laminin trimers can interact with each other and with other basement membrane proteins in the ECM. For example, the cross-shaped trimers have three so-called short arms, and their

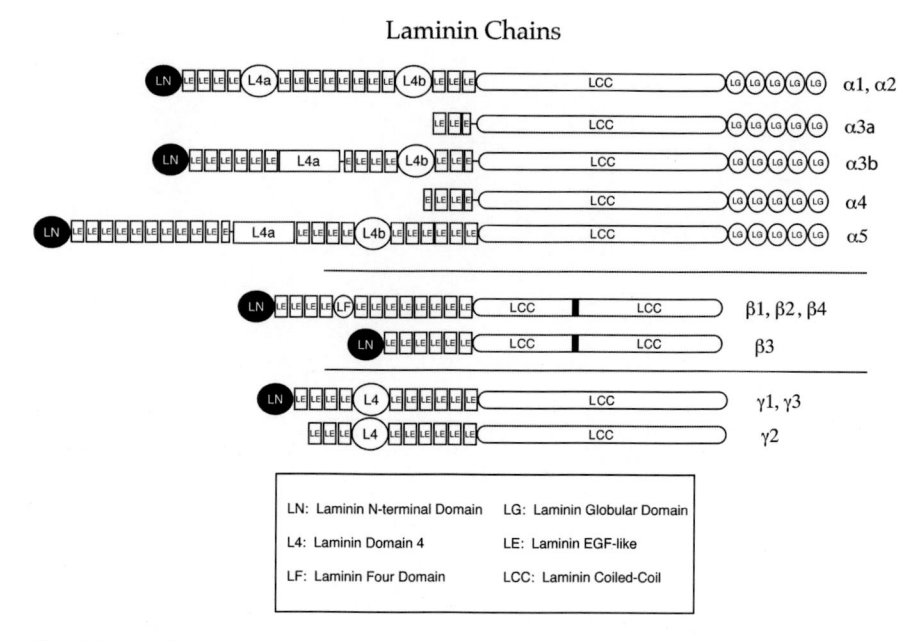

Fig. 4.4 Domain structure of the laminin chains. Laminins are evolutionarily related modular proteins. There are five α, four β, and three γ chains. Only α chains contain the COOH-terminal tandem of five laminin globular (LG) domains. Domains are identified in the key

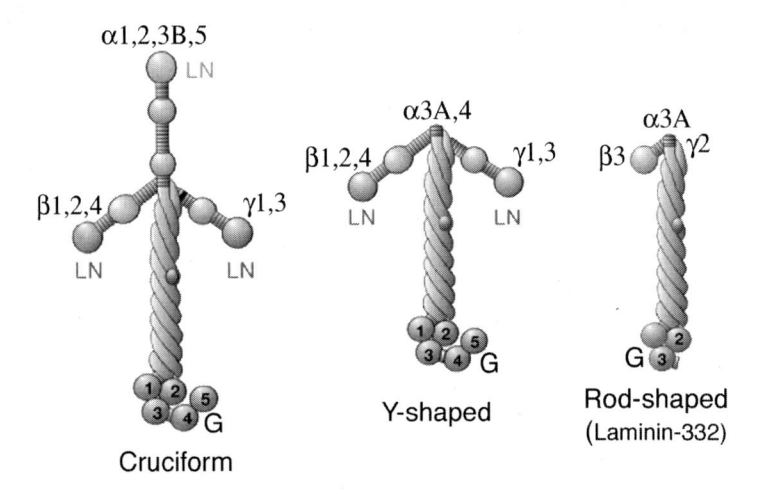

Fig. 4.5 Structure of laminin trimers. One α, one β, and one γ chain assemble to form trimers with one of the three shapes shown. Only cruciform trimers can self-polymerize into a network, as heteromeric interactions among α, β, and γ LN domains are required

laminin NH2-terminal (LN) domains (Figs. 4.3 and 4.5) mediate the inter-trimer α–β–γ associations that are absolutely required for polymerization of the laminin network (McKee et al. 2007). The Y-shaped trimers lack one short arm (Fig. 4.5), so they are unable to polymerize on their own, but can likely integrate into basement membranes via interactions with cross-shaped trimers. The rod-shaped LM-332 (Fig. 4.5) integrates into basement membrane via covalent linkage to laminin-311 and -321 (Champliaud et al. 1996).

There is some degree of alternative transcription and alternative splicing that generates diversity in laminin chain domain structure. For example, alternative promoters in *LAMA3* generate a short isoform lacking the short arm (α3A) as well as a full-length isoform (α3B) (Fig. 4.4) (Miner et al. 1997). More recently, it was shown that alternative splicing of both *LAMA3* and *LAMA5* RNAs generates truncated α3 and α5 short arms containing all or part of the LN domain and variable additional short arm segments (Hamill et al. 2009). These products are incapable of assembling into trimers because they lack the LCC domain, but the LN domain should allow them to incorporate into basement membranes and perhaps modulate laminin polymerization. Their functions remain to be determined, but some are similar in domain structure to netrins (Yurchenco and Wadsworth 2004).

A detailed discussion of laminin expression is beyond the scope of this chapter, but some generalizations are worth mentioning. Laminin γ1 is essentially ubiquitous in basement membranes. LM-111 is prominent in extraembryonic membranes of rodents (especially Reichert's membrane) and in kidney tubular basement membranes. LM-211 is found in muscle, pancreas, and the nervous system. LM-311 and -321 are prominent in skin and lung. LM-411 and -421 are found in muscle and in the vasculature. LM-511 is widely distributed in kidney, lung, vasculature, skin, salivary gland, and intestine. LM-521 is prominent in kidney glomeruli and at the neuromuscular junction, where LM-221 and LM-421 are also found (Miner and Patton 1999; Patton 2000). Laminin β2- and γ3-containing laminins are prominent in the eye (Libby et al. 2000; Pinzon-Duarte et al. 2010). Finally, LM-332, frequently called laminin-5 in the literature based on a previous nomenclature, is associated with a number of tumors and is prominent in the epidermal basement membrane, where it is required for hemidesmosome formation in the adjacent basal keratinocytes (Litjens et al. 2006).

Defining the function of individual laminin chains through mutagenesis in mice (shown in Table 4.1) has been ongoing for over 15 years, and much has been learned. The functions of those laminin chains that have been found to be mutated in human disease (α2, α3, β2, β3, and γ2), all of which have been studied in analogous mutant mice, are discussed in Sect. 8.5. Some of the results from mouse knockouts that have not yet been correlated with a human disease are discussed here.

The laminin β1 and γ1 chains are essentially ubiquitous in basement membranes. Therefore, it is not surprising that basement membranes cannot form without them, and the corresponding mutant mouse embryos are unable to gastrulate. In addition, mutations in *Lama1*, *Lamb1*, or *Lamc1* prevent the formation of Reichert's membrane and cause death of the embryo just after implantation (Miner et al. 2004; Smyth et al. 1999). But surprisingly, when a conditionally mutant *Lama1* allele was mutated specifically in the epiblast (which gives rise only to the embryo proper), thus sparing the extraembryonic cells, the resulting *Lama1* null embryos developed normally. Although *Lama1*−/− mice are viable and fertile, they do exhibit a retinal and inner limiting membrane defect that impacts vision (Edwards et al. 2010).

Laminin α5 is widely expressed during development and in adults. *Lama5*−/− null mice thus exhibit a corresponding wide range of defects, including defects in neural tube closure, digit septation, placentation, kidney function, and in kidney, lung, salivary gland, hair follicle, and tooth development (Miner and Yurchenco 2004; Rebustini et al. 2007). Exactly how laminin α5 is involved in all these processes is not known, but maintaining the integrity of basement membranes, signaling to the adjacent cells, and/or binding morphogens are likely to be involved.

4.2.2 Type IV Collagen

Like all collagens (the collagen family is discussed in detail in Chap. 3), type IV collagen is composed of trimerized ~180 kDa α chains that contain Gly-X-Y amino acid triplet repeats. There are six genetically distinct collagen IV chains that assemble into heterotrimeric protomers. These protomers are secreted by cells into the ECM, where they polymerize with other protomers to make a superstructure. However, unlike fibrillar collagen chains, the Gly-X-Y repeats in type IV collagen chains are interrupted multiple times. These interruptions are thought to impart flexibility to the collagen protomer, to the collagen network, and thus to the

4 Basement Membranes

Table 4.1 Major basement membrane protein knockout phenotypes

Protein	Phenotypes in brief
Laminin chains	
α1 (null)	Peri-implantation embryonic lethality; no Reichert's membrane
α1 (LG4-5 deletion)	Peri-implanatation embryonic lethality; failed epiblast polarization
α1 (cond)	Mild retinal defects in epiblast knockout
α2 (null)	Viable; congenital muscular dystrophy, peripheral neuropathy
α2 (LN deletion)	Viable; muscular dystrophy, peripheral neuropathy
α3 (null)	Neonatal lethality; severe skin blistering, tooth defects
α4 (null)	Viable; microvascular, cardiac, and motoneuron defects; kidney glomerular pathology
α5 (null)	Fetal lethality; syndactyly; neural tube, placental vascularization, kidney, lung, gut, and tooth defects
α5 (cond)	Lung, kidney, and neuromuscular defects
β1 (null)	Peri-implantation embryonic lethality; no Reichert's membrane
β2 (null)	Lethal at 3 weeks of age; neuromuscular junction, kidney glomerular, and retinal defects
β3 (null)	Neonatal lethality; severe skin blistering
γ1 (null)	Peri-implantation embryonic lethality; no Reichert's membrane
γ1 (cond)	Peripheral nervous system myelination
γ1 (LEb-deletion)	Embryonic lethality; kidney and lung developmental defects
γ2 (null)	Neonatal lethality; severe skin blistering
γ3 (null)	Viable; retinal defects
Collagen IV chains	
α1 (null)	Embryonic lethal
α1 (internal del)	Dominant; some neonatal lethality due to brain hemorrhaging and lung defects; retinal vasculature defects
α1 (Gly substitutions)	Dominant; eye defects, kidney glomerular defects
α2 (null)	Embryonic lethal
α3 (null)	Viable; kidney glomerular defects that lead to kidney failure in adults
α4 (null)	Viable; kidney glomerular defects that lead to kidney failure in adults
α5 (null)	Viable; kidney glomerular defects that lead to kidney failure in adults; synaptic maintenance defects
α6 (null)	Viable; no known defects
Perlecan (null)	Embryonic lethal; brain and heart basement membrane defects; cartilage defects
Perlecan (HS attachment sites deletion)	Viable; mild eye vessel defects; mild kidney filtration defect
Agrin (null)	Neonatal lethality; absence of neuromuscular junctions
Collagen XVIII (null)	Viable; retinal vessel defects
Nidogen-1 (null)	Viable; impaired wound healing
Nidogen-2 (null)	Viable; no known defects
Nidogen-1/2 (double null)	Perinatal lethality; lung and cardiac basement membrane defects

basement membrane. Additional important features include a short NH2-terminal domain called 7S and a larger COOH-terminal noncollagenous domain of ~20–25 kDa called NC1 (Khoshnoodi et al. 2008).

The six collagen IV chains are designated α1–α6. The protomers they can form are $(\alpha 1)_2 \alpha 2$, $\alpha 3 \alpha 4 \alpha 5$, and $(\alpha 5)_2 \alpha 6$. Assembly of these (and only these) protomers is governed by the NC1 domains, which bear a code that ensures proper chain recognition (Khoshnoodi et al. 2006). In addition to this role, the NC1 domains, once trimerized in the protomer, link to the trimerized NC1 domains of another protomer to form an NC1 hexamer. In some cases, inter-protomer covalent bonding occurs to further stabilize the interaction (Vanacore et al. 2009). Together with the interactions of trimerized 7S domains from four protomers, this leads to polymerization of collagen IV into a chicken wire-like network (Khoshnoodi et al. 2008).

Regarding collagen IV gene expression, the $(\alpha 1)_2 \alpha 2$ network is essentially ubiquitous. The $\alpha 3 \alpha 4 \alpha 5$ network is prominent in lung and kidney and at the neuromuscular junction. The $(\alpha 5)_2 \alpha 6$ network is found in smooth muscle, at the neuromuscular junction, and in kidney. Much of what is known about the function of collagen IV comes from studies of human disease and mouse knockouts and is discussed in Sect. 8.5. However, a notable finding from studies of *Drosophila*, which has only the $\alpha 1$ and $\alpha 2$ chains, is that collagen IV binds the decapentaplegic protein, which is homologous to mammalian bone morphogenetic proteins, and regulates its signaling during embryonic development by promoting gradient formation (Wang et al. 2008).

4.2.3 Nidogen

There are two different nidogen (also previously called entactin [Carlin et al., 1981)] genes, *Nid1* and *Nid2*, which encode related 150 kDa dumbbell-shaped proteins (Fig. 4.6) that bind to both laminin and type IV collagen (Fox et al. 1991). Its ability to bind to both laminin and type IV collagen led to the hypothesis that nidogen serves as a requisite bridge between the independent laminin and collagen IV networks, thereby promoting basement membrane formation and stability (Timpl 1996). However, studies of nidogen knockout mice are not fully consistent with this hypothesis. *Nid1* null and *Nid2* null mice are viable and exhibit no abnormal basement membrane phenotypes, suggesting that the nidogens might compensate for each other. Indeed, *Nid1/Nid2* double knockout mice exhibit perinatal lethality and localized basement membrane defects in lung, heart, and skin, but many other basement membranes and organs appear surprisingly normal (Bader et al. 2005). This and other data (e.g., Fox et al. 2008) indicate limited functions for nidogen in specific basement membranes rather than a global role in basement membrane formation and integrity. Similarly, in *Caenorhabditis elegans*, nidogen mutants show defects in neuronal migration and at neuromuscular synapses (Ackley et al. 2003; Kim and Wadsworth 2000).

4.2.4 Sulfated Proteoglycans

A number of sulfated proteoglycans are associated with basement membranes, including the heparan sulfate proteoglycans (HSPGs) perlecan (Fig. 4.6), agrin,

4 Basement Membranes

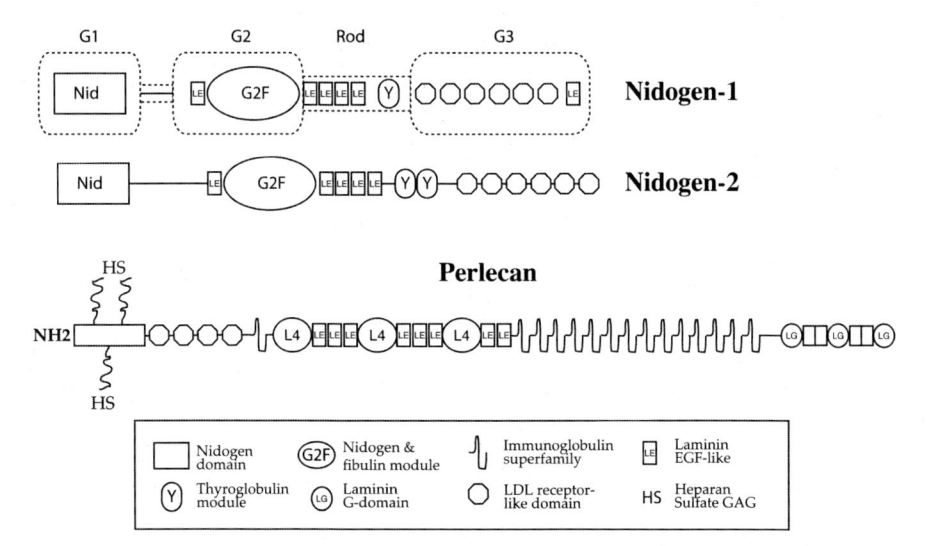

Fig. 4.6 Domain structure of nidogens and perlecan. Like most ECM proteins, nidogens and perlecan are modular proteins containing domains found in other extracellular proteins. Domain homologies are indicated in the key

and collagen XVIII, and the chondroitin sulfate proteoglycan bamacan. These and other proteoglycans are discussed in detail in Chaps. 5 and 6. Sulfated proteoglycans are thought to contribute much of the net negative charge of basement membranes due to their high degree of sulfation. All contain a modular core protein with glycosaminoglycan (GAG) side chains that dramatically increase overall molecular weight. Perlecan, agrin, and collagen XVIII are discussed further, and the mouse knockout phenotypes are shown in Table 4.1.

4.2.4.1 Perlecan

Perlecan is an abundant and almost ubiquitous basement membrane component whose protein core is about 400 kDa, but with the added GAG chains the molecular weight jumps to about 800 kDa. It is also found in cartilage ECM. Perlecan bears three GAG attachment sites encoded within a single exon. Perlecan has a distinctive modular structure (Fig. 4.6) that includes domains with homology to laminin LG and LE domains, to the SEA (sperm protein/enterokinase/agrin) domain, and to immunoglobulin-like repeats (Iozzo 2001). The COOH-terminal domain can be cleaved to generate a fragment of perlecan called endorepellin, which has antitumor and anti-angiogenic properties (Bix and Iozzo 2005).

Perlecan has affinity for both collagen IV and laminin and therefore might serve as a bridge between the two networks (Timpl 1989). The importance of perlecan in a subset of basement membranes is indicated by basement membrane discontinuities and embryonic lethality in perlecan mutant mice (Costell et al. 1999). However, a targeted mutation that removes the exon containing the heparan sulfate attachment sites results in surprisingly few abnormalities (Rossi et al. 2003), raising questions about the importance of perelcan's GAG chains and the associated charge. Mutations in human perlecan are discussed below in Sect. 8.5.7.

4.2.4.2 Agrin

Agrin is somewhat homologous to perlecan, and the genes are linked (though not tightly) in both human and mouse. The agrin core protein is about 210 kDa and contains a unique N-terminal agrin domain, follistatin-like repeats, a SEA domain, and laminin LE and LG domains (Iozzo 2001). The N-terminal agrin domain binds tightly to a site on the laminin γ1 coiled-coil domain and also binds to receptors on cells (integrins and dystroglycan, as discussed below in Sect. 8.3), thereby linking the basement membrane to the cell surface (Kammerer et al. 1999; Sanes et al. 1998). Although agrin is fairly widely expressed in basement membranes, its only well-characterized function is at the neuromuscular synapse. Here, a specific splice form of agrin called Z+-agrin is secreted into the myofiber basement membrane by a migrating motoneuron. The binding of Z+-agrin to its receptor on the muscle fiber, muscle-specific kinase (MUSK), initiates a signaling cascade that stabilizes and promotes maturation of the nascent postsynaptic apparatus (Kummer et al. 2006). In the absence of agrin, mice are born without neuromuscular junctions and die (Gautam et al. 1996). Furthermore, although agrin is highly concentrated in the kidney GBM, its removal, along with its concentrated negative charge, specifically from that basement membrane via Cre-lox technology did not result in glomerular filtration defects (Harvey et al. 2007).

4.2.4.3 Collagen XVIII

Collagen XVIII is the first identified collagen that is also an HSPG (Halfter et al. 1998). It consists of a protein core of about 180 kDa, plus GAG chains that add about 120 kDa to its mass. There is a single α chain that forms homotrimers. Collagen XVIII is found widely in basement membranes, most prominently in retina, epidermis, pia, heart and skeletal muscle, kidney, lung, and blood vessels (Halfter et al. 1998).

Collagen XVIII is a modular protein that contains a number of different domains, including a frizzled domain, a cysteine rich domain, a collagenous domain with multiple interruptions, a thrombospondin-type laminin G domain, and a COOH-terminal noncollagenous domain that is cleaved to release endostatin. Endostatin has received much attention for its anti-angiogenic and potential anti-tumor properties (Marneros and Olsen 2005). *Col18a1* null mice have eye defects and are a model for human Knobloch syndrome (Fukai et al. 2002).

4.2.5 Other Components

Although laminin, collagen IV, nidogen, and sulfated proteoglycans are found in all basement membranes, the remaining basement membrane proteome is likely substantial and highly variable from one basement membrane to another. Of the many additional ECM proteins that are known to be present in basement membranes to at least some extent, a few of particular interest are discussed in brief.

4.2.5.1 Fibronectin

Fibronectin (described in detail in Chap. 2) is best known as a sticky ECM protein secreted by fibroblasts; it harbors the Arg-Gly-Asp (RGD) motif (within an eponymous fibronectin type III repeat) that is an important ligand for several integrin receptors (Mao and Schwarzbauer 2005). Fibronectin thus plays an important role in mediating cell/ECM interactions. Although much of the fibronectin in organisms is probably not associated with basement membranes, but rather with stromal and interstitial matrices, many basement membranes contain fibronectin. Of note, fibronectin associated with developing salivary gland basement membrane has been shown to be critical for regulating the pattern of epithelial branching morphogenesis (Sakai et al. 2003).

4.2.5.2 Fibulin-1 and -2

Fibulins (discussed in more detail in Chap. 10) are a family of seven ECM proteins with a distinctive domain structure that includes tandem arrays of calcium-binding EGF-like motifs and a COOH-terminal fibulin type module. They are associated with elastic fibers, fibronectin fibrils, and basement membranes (Chu and Tsuda 2004). Of the seven fibulins, fibulin-1 and -2 are the ones present to a significant degree in basement membranes. Mice lacking fibulin-1 die neonatally with widespread vascular defects, primarily leakage of small vessels due to ruptures in their endothelial lining. In addition, kidney glomerular capillaries are dilated, indicative of a mesangial cell/matrix interaction defect, and lung development is delayed (Kostka et al. 2001). In contrast, mice lacking fibulin-2 show no apparent defects, perhaps due to compensation by fibulin-1 (Sicot et al. 2008).

4.3 Basement Membrane Receptors

Cells adhere to basement membranes via interactions between surface receptors and specific polypeptide sequences within basement membrane proteins that serve as their ligands. Although a complete discussion of basement membrane protein receptors would require its own chapter, the major relevant receptors are discussed in brief.

4.3.1 Integrins

Integrins are a large family of obligate αβ heterodimeric transmembrane receptors. They are involved in a multitude of biological processes, including platelet and immune cell activation, leukocyte extravasation, and cell adhesion to ECM. There are 22 different integrin heterodimers that result from the nonrandom association of 18 α and 8 β subunits, but only a small subset of these are known to be involved in binding to basement membrane proteins, primarily to collagen IV and laminin (Hynes 2002). Integrins do not have an intrinsic catalytic activity, in contrast to many growth factor receptors. Instead, integrin binding to ligand leads to clustering of the integrin and binding of numerous adapter and scaffolding proteins to the integrin's cytoplasmic tail (Miranti and Brugge 2002). Some of these proteins have an intrinsic kinase activity that becomes activated, resulting in signal transduction and changes in cell behavior. Others bind to the actin cytoskeleton and mediate cell adhesion or migration. And as discussed below in Sect. 8.4, integrins can facilitate the polymerization of laminin to initiate basement membrane formation.

The main collagen IV-binding integrins are α1β1 and α2β1; α1β1 exhibits higher affinity, and both of these integrins can bind other types of collagen as well. These integrins are fairly widely expressed, consistent with the widespread deposition of collagen in the ECM. Although the importance of integrin binding to collagen IV is assumed and is still under investigation, neither *Itga1* nor *Itga2* knockout mice show severe phenotypes. It is likely that there is some degree of cross-compensation, although studies have shown that they can exhibit very different activities (Abair et al. 2008). It will be interesting to determine the effect of deleting both integrins in mice, but the very tight linkage of *Itga1* and *Itga2* in the genome has hampered efforts to generate the double knockouts.

The main laminin-binding integrins are α3β1, α6β1, α6β4, and α7β1 (Hynes 1999) (see Chap. 2 for a discussion of laminin receptors). The binding of these to specific laminin chains has been very well characterized, and the identity of the α chain LG domain seems to be paramount for determining affinity. The knockout of these integrins in mice generates phenotypes that are mostly consistent with the knockout of the corresponding laminin ligands, which provides strong genetic evidence for some of the physical interactions that have been gleaned from biochemical studies (Nishiuchi et al. 2006). The highest affinity interactions between these integrins and the laminin trimers that have been purified and tested are shown in Table 4.2.

Further studies have revealed important additional insights into the "rules" of integrin binding to laminins. Although the α chain LG domain may be most important, it is clear that the identity of the β and γ chains also has an influence. For example, integrin α3β1 binds more avidly to LM-521 than to LM-511 (Taniguchi et al. 2009). In addition, a glutamic acid present near the COOH termini of the laminin γ1 and γ2 chains is required for integrin binding to the laminin trimers of which they are a part (Ido et al. 2007). Interestingly, the laminin γ3 chain lacks this

4 Basement Membranes

Table 4.2 Integrin binding to laminins (determined by α chain identity)

Integrin	Major ligands
α3β1	α3- and α5-containing laminins
α6β1	α1-, α3-, and α5-containing laminins
α6β4	α3- and α5-containing laminins
α7β1	α1- and α2-containing laminins

residue (Ido et al. 2008), suggesting that laminins containing γ3 might actually impair, rather than promote, integrin-mediated adhesion to the basement membranes in which it is found.

4.3.2 Dystroglycan

Dystroglycan is a highly glycosylated laminin-binding receptor. Proper glycosylation of dystroglycan is critical for its ability to bind to laminin with the necessary affinity (Michele et al. 2002), and recent data show that proper posttranslational modification of a phosphorylated O-linked mannosyl glycan in the mucin-like domain of dystroglycan is intimately involved in conferring its laminin-binding ability (Yoshida-Moriguchi et al. 2010). Dystroglycan has a transmembrane subunit (β-dystroglycan) and an extracellular subunit (α-dystroglycan) that derive from proteolytic cleavage of a single primary polypeptide encoded by the *Dag1* gene. Dystroglycan was first identified as part of the dystrophin–glycoprotein complex (DGC) (Fig. 4.7), which is normally found in skeletal muscle but is absent or defective in patients with some forms of muscular dystrophy. In skeletal muscle, α-dystroglycan binds to LM-211 in the basement membrane via the laminin α2 LG domain, while β-dystroglycan binds to dystrophin in the cytoplasm as well as to other transmembrane components of the DGC (Cohn and Campbell 2000). Dystrophin links the DGC to the actin cytoskeleton (Fig. 4.7), thus providing stability to the muscle fiber plasma membrane under the force of contractions.

Dystroglycan is also expressed in most epithelial cells, where it binds to laminin α chain LG domains and to agrin. Dystroglycan is thought to bind laminin α1 and α2 best and less well to α4 and α5, in all cases via the LG domain. The exact function of dystroglycan in nonmuscle cells is not known, but it is thought to be important for helping to organize laminin polymerization and basement membrane formation (Henry and Campbell 1999). In the absence of dystroglycan, Reichert's membrane, the major mouse extraembryonic basement membrane, does not form because laminin does not polymerize adjacent to the parietal epithelial cells that synthesize it (Williamson et al. 1997). This defect is phenocopied by null mutations in *Lama1*, *Lamb1*, and *Lamc1* (Miner et al. 2004; Smyth et al. 1999), which encode the chains of the major Reichert's membrane laminin, LM-111. However, the

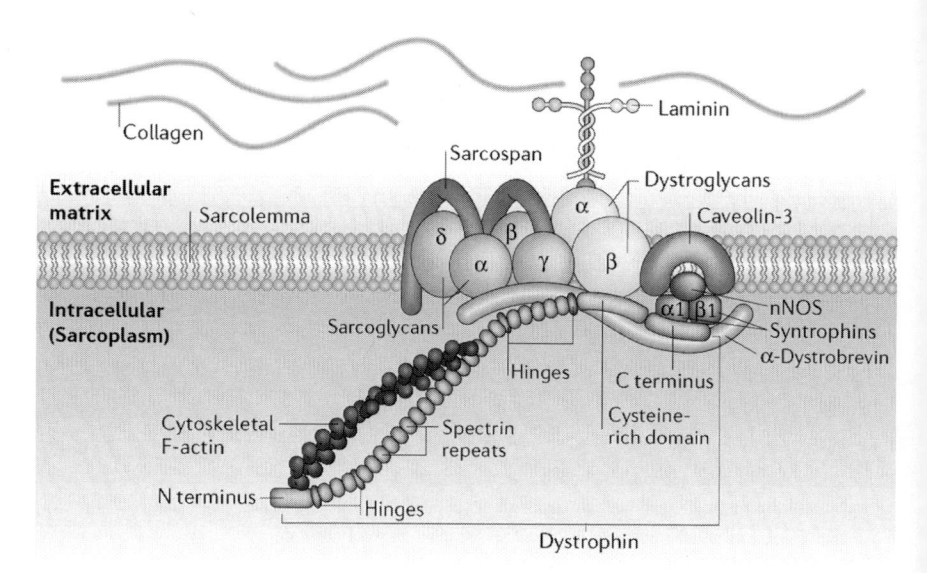

Fig. 4.7 The dystrophin–glycoprotein complex (DGC). The DGC of skeletal muscle is a link between the actin cytoskeleton and the myofiber basement membrane. α-Dystroglycan, which is associated with β-dystroglycan, binds to the laminin LG domain in the basement membrane. Sarcoglycans and sarcospan are also part of this transmembrane complex. Dystrophin is a cytoplasmic protein that links the transmembrane complex to filamentous actin with the help of the additional proteins shown. Reprinted by permission from Macmillan Publishers Ltd: Davies and Nowak (2006), copyright 2006

basement membranes of the embryo proper can form and function in the absence of dystroglycan (Cohn et al. 2002).

4.3.3 Lutheran/Basal Cell Adhesion Molecule

In humans, the Lutheran blood group glycoprotein/basal cell adhesion molecule (referred to collectively as BCAM) is found on the surface of red blood cells and is a meaningful blood group antigen. BCAM is an immunoglobulin superfamily transmembrane receptor that binds specifically to laminin α5 via α5's LG domain (Kikkawa and Miner 2005). The two different names for the protein refer to two different splice forms; the difference is that basal cell adhesion molecule has a shorter cytoplasmic tail. BCAM is not found on red blood cells in mice, but in both humans and mice it is found widely on endothelial, epithelial, and other cell types that are adjacent to a basement membrane, especially those basement membranes containing laminin α5. Mice lacking BCAM exhibit subtle defects in kidney and intestinal smooth muscle (Rahuel et al. 2008). There are humans who are null for the Lutheran blood group antigen and may not express it on non-red blood cells, but with no obvious clinical effects (Colin et al. 2008).

In addition to its importance as a blood group antigen in humans, BCAM has gained interest because it is overexpressed on the surface of sickled red blood cells in individuals with sickle cell disease (Colin et al. 2008). Because BCAM is able to bind laminin α5, and laminin α5 is found in many endothelial basement membranes, the BCAM/α5 interaction is thought to contribute to the vaso-occlusive crises that are the most painful aspect of sickle cell disease. Drugs or small molecule inhibitors that can prevent the association of BCAM with laminin α5 might reduce these episodes. The successful mapping of (1) the BCAM-binding site on laminin α5 and (2) the domain of BCAM that binds α5 (Kikkawa and Miner 2005; Mankelow et al. 2007) may facilitate development of such drugs.

4.3.4 Discoidin Domain Receptor 1

Discoidin domain receptor 1 (DDR1) is a transmembrane collagen receptor. Its extracellular domain shows homology to the *Dictyostelium* discoidin protein, which is involved in cell adhesion. DDR1 also contains a transmembrane domain and a cytoplasmic catalytically active tyrosine kinase domain. DDR1 can bind to both type I and type IV collagen, and this has been shown to mediate cellular adhesion (Fukunaga-Kalabis et al. 2006; Vogel 1999). The function of DDR1 is still under investigation, but roles in mammary gland function, kidney GBM architecture, vascular healthy, and melanocyte adhesion have been reported. *Ddr1*−/− mice are smaller than controls but nevertheless viable; mutant females exhibit fertility and lactation defects (Vogel et al. 2001).

4.4 Basement Membrane Assembly and Integrity

Existing data suggest that laminin polymerization at the cell surface (Fig. 4.3) initiates and is absolutely required for basement membrane assembly. Consistent with this, whereas basement membranes do not form in the absence of laminin, as in laminin β1 and γ1 knockout mice (Miner et al. 2004; Smyth et al. 1999), they form surprisingly well (but manifest impaired integrity) in *Col4a1/Col4a2*, *Nid1/Nid2*, and perlecan (*Hspg2*) knockouts (Arikawa-Hirasawa et al. 1999; Bader et al. 2005; Costell et al. 1999; Poschl et al. 2004). Investigations into the mechanisms of basement membrane formation have thus focused on laminin and its interactions with cell-surface components that concentrate laminin in the ECM over a critical threshold so that polymerization near the cell surface can occur (reviewed in Yurchenco and Patton 2009).

Aside from the cell-surface integrin receptors and dystroglycan that were discussed in the previous section, sulfatides (sulfated glycolipids such as galactosyl-3-sulfate ceramide) embedded in the plasma membrane are also important for laminin polymerization at cell surfaces (Li et al. 2005). Sulfatides, the relevant integrins, and dystroglycan all bind to sites in the LG domain of laminin α chains, leaving the laminin LN domains free to interact with each other to mediate laminin polymerization. The laminin-binding integrins (primarily α3β1, α6β1, α6β4, and α7β1) bind to sites in the LG1-3 segments, whereas dystroglycan and sulfatides bind to the LG4-5 segments (Yurchenco and Patton 2009).

Once laminin polymerizes at the cell surface, where it alone is capable of forming a lamina densa apparent by electron microscopy, it is thought, based on in vitro studies, that the self-polymerizing type IV collagen network integrates into the forming basement membrane primarily through the bridging activity of nidogen, which has affinity for both laminin and collagen IV. However, it is clear that collagen IV does have some intrinsic affinity for the laminin polymer (McKee et al. 2007); this may explain in part why basement membranes containing both laminin and collagen IV can form in mice lacking both nidogen-1 and -2, although they show impaired integrity in some tissues (Bader et al. 2005). Alternatively, perlecan likely has some bridging properties and may compensate for the missing nidogens in these mice.

In addition to initiating basement membrane formation, the polymerization of laminin and the subsequent reorganization of the bound integrins and dystroglycan have been shown to mediate intracellular signaling and restructuring of the cytoskeleton via outside-in signaling. This can have important functional consequences for cell behavior, including effects on proliferation, protection from apoptosis, and differentiation (Yurchenco et al. 2004). And as discussed below, the linkage between the cytoskeleton and the basement membrane is crucial for maintaining skeletal muscle fiber integrity.

4.5 Primary Basement Membrane Diseases

The importance of basement membranes to normal physiology is best exemplified by the many diseases that result from defects in basement membranes, either genetic or by acquisition. A few of the major diseases are discussed in this section.

4.5.1 Epidermolysis Bullosa

EB describes a group of inherited skin blistering diseases with severity that can vary from mild to lethal. Junctional EB (JEB) is a severe form that occurs when the epidermis becomes separated from the dermis in the plane of the basement membrane. JEB is caused by mutations that impair expression of laminin α3, laminin β3, or laminin γ2, the components of LM-332 (Pulkkinen and Uitto 1999). LM-332 is deposited by keratinocytes into the epidermal basement membrane (Fig. 4.8).

4 Basement Membranes

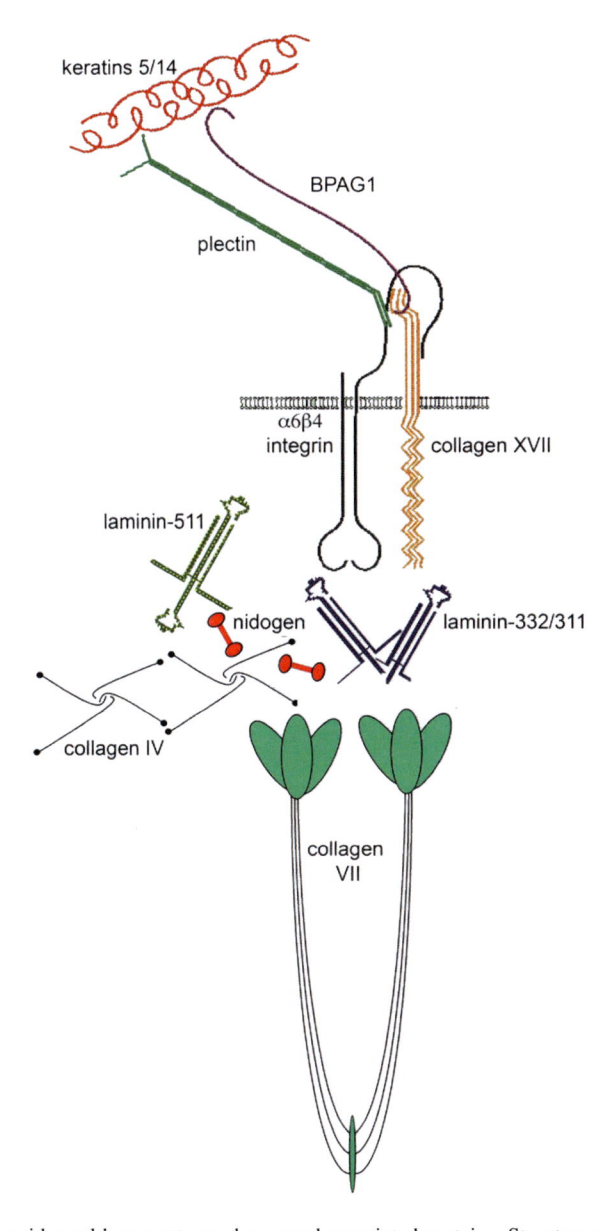

Fig. 4.8 The epidermal basement membrane and associated proteins. Structure and composition of dermal–epidermal basement membrane zone. Laminin-511 and collagen IV constitute the major polymeric networks of the basal lamina, connected by nidogens (shown) and heparan sulfate proteoglycans (not shown). Laminin-332 is a component of hemidesmosomes, formation of which requires laminin-332/integrin α6β4 binding and interactions with other proteins. In addition, laminin-332 is bound to the collagen VII anchoring filament complex, thus bridging the overlying keratinocyte cell surface with the dermis. Reprinted from Tzu and Marinkovich (2008), copyright 2008, with permission from Elsevier

There, its binding to integrin α6β4 on the basal surface of keratinocytes initiates formation of hemidesmosomes (Litjens et al. 2006); LM-332 also binds to anchoring fibrils in the dermis that are composed of type VII collagen (Tzu and Marinkovich 2008). Hemidesmosomes link the basement membrane to the overlying epithelial cell's intermediate filament cytoskeleton, which primarily comprises keratin filaments in keratinocytes (Fig. 4.8). In the skin, proper hemidesmosome formation and function are crucial for maintaining adhesion of the epidermis to the basement membrane. In JEB patients, hemidesmosomes do not form properly, and skin blistering results from even minor skin trauma. As the passage through the birth canal is a major trauma, JEB is readily apparent in newborns.

4.5.2 Muscular Dystrophies

The health of skeletal muscle fibers depends on stable and strong connections between the cell and the adjacent ECM. These connections, which can be followed molecularly from the nucleus to the ECM, must be maintained against the force of constant contractions. A large and diverse group of extracellular, transmembrane, cytoplasmic, cytoskeletal, and nuclear proteins have evolved to ensure stability of these cell/ECM interactions. Genetic defects in many of these proteins have been shown to cause various forms of muscular dystrophy in humans and/or in mice. Those proteins involved in direct connections between the myofiber and the basement membrane are discussed in the context of muscular dystrophy.

4.5.2.1 Congenital Muscular Dystrophy

Every skeletal muscle fiber is surrounded by a continuous basement membrane, the major laminin isoform of which is LM-211 (Patton 2000). Mutations that affect *LAMA2*, the gene encoding laminin α2, cause congenital muscular dystrophy, also known as merosin-deficient congenital muscular dystrophy type 1A (MDC1A). Features of this severe disease include congenital hypotonia and joint contractures, and progressive muscle weakness. In addition, because laminin α2 is also found in basement membranes of the central and peripheral nervous systems, neural defects, including aberrant myelination of peripheral nerves, also contribute significantly to the pathology of the disease (Jones et al. 2001).

The lack of LM-211 in the skeletal muscle basement membrane results in severe basement membrane defects and discontinuities. In addition, because the LG domain of laminin α2 harbors the major dystroglycan-binding site, its absence prevents tight binding of dystroglycan, and thus the DGC, to the basement membrane (Fig. 4.7). This causes damage to the muscle fiber plasma membrane during contractions and consequential muscle fiber injury and regeneration characteristic of muscular dystrophy.

The availability of several good mouse models for congenital muscular dystrophy, both spontaneous and engineered, has allowed for detailed investigation into the associated basement membrane defects and how the absence of laminin α2 affects the composition of the relevant basement membranes. For example, although laminin α1 is not usually found in skeletal muscle or peripheral nerve basement membranes either in normal or diseased muscle (Patton 2000), forced overexpression of α1 in muscle and nerve (via a transgene) on the *Lama2* mutant mouse background results in ectopic accumulation of LM-111 in muscle and nerve basement membranes. LM-111 compensates very well for the missing LM-211, and the mice exhibit only mild disease (Gawlik et al. 2004). One likely reason is that laminin α1, like laminin α2, is an excellent ligand for dystroglycan. These results suggest that upregulation of *LAMA1* in patients with *LAMA2* mutations should be beneficial.

A different approach for rescuing congenital muscular dystrophy in *Lama2* mutant mice took advantage of what was known from biochemical studies about the interactions among laminin, agrin, and dystroglycan. In the absence of laminin α2, the laminin α4 and α5 chains are upregulated and secreted as part of LM-411 and LM-511, but neither of these bind dystroglycan very well, and they are therefore poor substitutes for the missing LM-211. On the other hand, the COOH terminus of agrin does bind dystroglycan, and the NH2 terminus of agrin binds tightly to laminin γ1, which is present in both LM-411 and LM-511. Thus, agrin can theoretically serve as a bridge between these laminins and dystroglycan and perhaps improve laminin trimer organization (and therefore basement membrane organization). By removing much of the central portion of the agrin cDNA, a "mini-agrin" gene was created, and the encoded protein was expressed in skeletal muscle on the *Lama2*−/− background. The resulting mice had increased levels of laminin and dystroglycan in skeletal muscle, healthier muscles, and longer life spans (Bentzinger et al. 2005). Mini-agrin as a gene therapy "drug" may therefore be a plausible candidate for ameliorating MDC1A in patients (Meinen and Ruegg 2006).

4.5.2.2 Dystroglycanopathies

Most forms of muscular dystrophy target either cytoplasmic or cytoskeletal proteins, a discussion of which is beyond the scope of this chapter. However, a brief mention of the group of muscle diseases collectively called dystroglycanopathies, which all target dystroglycan's ability to bind to laminin, is warranted here.

No naturally occurring pathogenic mutations in the dystroglycan gene itself have yet been discovered. As a highly and complexly glycosylated protein, dystroglycan's maturation requires several steps, each of which is mediated by specific enzymes. Mutations in six genes that encode enzymes involved in posttranslational modification of dystroglycan, and which cause muscle, eye, and/or brain defects, have been found. These genes include *POMT1*, *POMT2*, *POMGnT1*, *fukutin*, *FKRP*, and *LARGE* (Hewitt 2009). The exact enzymatic activity of each of the encoded enzymes is still being investigated (Yoshida-Moriguchi et al. 2010), but biochemical studies have shown that dystroglycan isolated from patients or mice carrying mutations in these genes is impaired in binding to laminin (Hewitt 2009). This is believed to be the molecular basis for the various muscle, neural, and ocular defects observed.

4.5.3 Pierson Syndrome and Isolated Congenital Nephrotic Syndrome

Pierson syndrome is a recently identified basement membrane disease caused by inactivating mutations in *LAMB2*, which encodes the laminin β2 chain (Zenker et al. 2004). Pierson syndrome patients exhibit small pupils and neurological and motor deficits together with congenital nephrotic syndrome (high urinary protein content and associated systemic manifestations) leading to kidney failure. These characteristics are observed in patients carrying null or truncating *LAMB2* mutations. On the other hand, most patients carrying missense mutations show only the kidney disease aspect (isolated congenital nephrotic syndrome) or have both kidney and mild neurological problems (Hasselbacher et al. 2006). *Lamb2* null mice show comparable kidney and neuromuscular defects and serve as an excellent model for the human disease (Noakes et al. 1995a, b).

4.5.4 COL4A1 Diseases

A spectrum of human diseases, many of which affect the microvasculature, are caused by mutations in *COL4A1*. The discovery of the genetic basis for these diseases was facilitated in part by the similar phenotypes observed in mice carrying mutations in *Col4a1* (Gould et al. 2005, 2006; Van Agtmael et al. 2005). The various human conditions include brain small vessel disease with hemorrhage and/or stroke; retinal arteriolar tortuosity; porencephaly; leukoencephalopathy; and hereditary angiopathy with nephropathy, aneurysms, and muscle cramps (the HANAC syndrome) (Plaisier et al. 2007; Van Agtmael and Bruckner-Tuderman 2010). Most, if not all, of the pathogenic *COL4A1* mutations are heterozygous glycine substitutions in the collagenous domain. This should alter the structure of the triple helix because glycine is the only residue that can fit at its center (see Chap. 2). Because there are two α1 chains in each $(\alpha 1)_2 \alpha 2$ protomer, approximately three quarters of all protomers should theoretically be affected, and a third of the affected protomers should have two mutant α1 chains. Whether the severity of disease is related to this or not remains to be determined, but it is interesting that no pathogenic mutations in *COL4A2* have yet been discovered.

4.5.5 Alport Syndrome and Thin Basement Membrane Disease

Alport syndrome is a hereditary glomerulonephritis (inflammation of renal glomeruli) in combination with deafness. The kidney disease usually begins in children and first manifests as blood in the urine. Kidney failure usually occurs by adolescence or young adulthood (Kashtan and Michael 1993). This disease is caused by mutations that affect any one of the *COL4A3*, *COL4A4*, or *COL4A5* genes. These genes encode the collagen IV α chains that comprise the α3α4α5 heterotrimer, which is the major collagen IV component of the GBM. The absence of any one of the three chains prevents this trimer from forming and can cause disease. The *COL4A5* gene is on the X chromosome, so Alport syndrome is most frequently found in males (Hudson 2004).

In the absence of the collagen α3α4α5(IV) network there is compensation by the $(α1)_2α2$ network. However, although this network is sufficient for normal glomerular filtration early in life, over time the abnormal GBM thickens and splits, leading to a characteristic ultrastructural appearance. This somehow leads to infiltration of immune cells, glomerular injury and scarring, and eventually glomerular dropout that is responsible for reduced filtration.

A genetically related but much less severe condition is thin basement membrane disease, also called benign familial hematuria. In this disease the GBM is thinned, and patients show blood in the urine, but there is no deafness and usually no progression to overt kidney disease. This disease is caused by heterozygous mutations in *COL4A3* or *COL4A4* (Kashtan 2004; Thorner 2007). Patients with these mutations can be considered Alport "carriers," because if these mutations were homozygous they would likely cause the much more severe autosomal recessive Alport syndrome.

4.5.6 Goodpasture Syndrome

Goodpasture syndrome is an autoimmune disease in which pathogenic antibodies to a specific region of the COL4A3 NC1 domain are generated. Because the collagen α3α4α5(IV) network is prevalent in the kidney GBM and in the lung alveolar basement membrane, Goodpasture patients exhibit kidney disease (glomerulonephritis) and lung hemorrhage, both of which can lead to organ failure and death if not treated early during the disease course (Hudson 2004).

The specific epitope of COL4A3 that is targeted by autoantibodies is usually cryptic, because it lies sequestered within the cross-linked NC1 hexamer of the collagen IV network. It is thought that some injury to the basement membrane might expose the epitope, which would be viewed by the immune system as novel due its usually cryptic state (Vanacore et al. 2008). However, the rarity of Goodpasture syndrome suggests the existence of a complex mechanism, and perhaps also environmental contributions, to promote the onset of the disease.

4.5.7 Schwartz–Jampel Syndrome

Schwartz–Jampel syndrome is a recessive disease characterized by cartilage, skeletal, and neuromuscular defects (chondrodysplasia and neuromyotonia) due to missense mutations in *HSPG2*, the gene encoding perlecan. Chondrodysplasia results from perlecan deficiency in cartilage, where it is normally deposited at high levels. In contrast, analysis of perlecan knockout mice and of mice expressing perlecan point mutants that cause Schwartz–Jampel syndrome revealed that the neuromyotonia results from acetylcholinesterase deficiency in the synaptic basement membrane (Arikawa-Hirasawa et al. 2002; Stum et al. 2008). Perlecan is required for anchoring the collagen-tailed form of acetylcholinesterase to the synaptic basement membrane (Arikawa-Hirasawa et al. 2002); the proper localization of the enzyme is required to prevent myotonia via degradation of acetycholine, which allows muscle relaxation.

References

Abair TD, Sundaramoorthy M, Chen D, Heino J, Ivaska J, Hudson BG, Sanders CR, Pozzi A, Zent R (2008) Cross-talk between integrins alpha1beta1 and alpha2beta1 in renal epithelial cells. Exp Cell Res 314:3593–3604

Ackley BD, Kang SH, Crew JR, Suh C, Jin Y, Kramer JM (2003) The basement membrane components nidogen and type XVIII collagen regulate organization of neuromuscular junctions in *Caenorhabditis elegans*. J Neurosci 23:3577–3587

Arikawa-Hirasawa E, Watanabe H, Takami H, Hassell JR, Yamada Y (1999) Perlecan is essential for cartilage and cephalic development. Nat Genet 23:354–358

Arikawa-Hirasawa E, Rossi SG, Rotundo RL, Yamada Y (2002) Absence of acetylcholinesterase at the neuromuscular junctions of perlecan-null mice. Nat Neurosci 5:119–123

Aumailley M, Bruckner-Tuderman L, Carter WG, Deutzmann R, Edgar D, Ekblom P, Engel J, Engvall E, Hohenester E, Jones JC, Kleinman HK, Marinkovich MP, Martin GR, Mayer U, Meneguzzi G, Miner JH, Miyazaki K, Patarroyo M, Paulsson M, Quaranta V, Sanes JR, Sasaki T, Sekiguchi K, Sorokin LM, Talts JF, Tryggvason K, Uitto J, Virtanen I, von der Mark K, Wewer UM, Yamada Y, Yurchenco PD (2005) A simplified laminin nomenclature. Matrix Biol 24:326–332

Bader BL, Smyth N, Nedbal S, Miosge N, Baranowsky A, Mokkapati S, Murshed M, Nischt R (2005) Compound genetic ablation of nidogen 1 and 2 causes basement membrane defects and perinatal lethality in mice. Mol Cell Biol 25:6846–6856

Bentzinger CF, Barzaghi P, Lin S, Ruegg MA (2005) Overexpression of mini-agrin in skeletal muscle increases muscle integrity and regenerative capacity in laminin-alpha2-deficient mice. FASEB J 19:934–942

Bix G, Iozzo RV (2005) Matrix revolutions: "tails" of basement-membrane components with angiostatic functions. Trends Cell Biol 15:52–60

Carlin B, Jaffe R, Bender B, Chung AE (1981) Entactin, a novel basal lamina-associated sulfated glycoprotein. J Biol Chem 256:5209–5214

Champliaud MF, Lunstrum GP, Rousselle P, Nishiyama T, Keene DR, Burgeson RE (1996) Human amnion contains a novel laminin variant, laminin 7, which like laminin 6, covalently associates with laminin 5 to promote stable epithelial-stromal attachment. J Cell Biol 132:1189–1198

Chan FL, Inoue S (1994) Lamina lucida of basement membrane: an artefact. Microsc Res Tech 28:48–59

Chan FL, Inoue S, Leblond CP (1993) The basement membranes of cryofixed or aldehyde-fixed, freeze-substituted tissues are composed of a lamina densa and do not contain a lamina lucida. Cell Tissue Res 273:41–52

Chu ML, Tsuda T (2004) Fibulins in development and heritable disease. Birth Defects Res C Embryo Today 72:25–36

Cohn RD, Campbell KP (2000) Molecular basis of muscular dystrophies. Muscle Nerve 23:1456–1471

Cohn RD, Henry MD, Michele DE, Barresi R, Saito F, Moore SA, Flanagan JD, Skwarchuk MW, Robbins ME, Mendell JR, Williamson RA, Campbell KP (2002) Disruption of *Dag1* in differentiated skeletal muscle reveals a role for dystroglycan in muscle regeneration. Cell 110:639–648

Colin Y, Rahuel C, Wautier MP, El Nemer W, Filipe A, Cartron JP, Le Van KC, Wautier JL (2008) Red cell and endothelial Lu/BCAM beyond sickle cell disease. Transfus Clin Biol 15:402–405

Costell M, Gustafsson E, Aszodi A, Morgelin M, Bloch W, Hunziker E, Addicks K, Timpl R, Fassler R (1999) Perlecan maintains the integrity of cartilage and some basement membranes. J Cell Biol 147:1109–1122

Davies KE, Nowak KJ (2006) Molecular mechanisms of muscular dystrophies: old and new players. Nat Rev Mol Cell Biol 7:762–773

Edwards MM, Mammadova-Bach E, Alpy F, Klein A, Hicks WL, Roux MJ, Simon-Assmann P, Smith RS, Orend G, Wu J, Peachey NS, Naggert JK, Lefebvre O, Nishina PM (2010) Mutations in Lama1 disrupt retinal vascular development and inner limiting membrane formation. J Biol Chem 285:7697–7711

Fox JW, Mayer U, Nischt R, Aumailley M, Reinhardt D, Wiedemann H, Mann K, Timpl R, Krieg T, Engel J et al (1991) Recombinant nidogen consists of three globular domains and mediates binding of laminin to collagen type IV. EMBO J 10:3137–3146

Fox MA, Ho MS, Smyth N, Sanes JR (2008) A synaptic nidogen: developmental regulation and role of nidogen-2 at the neuromuscular junction. Neural Dev 3:24

Fukai N, Eklund L, Marneros AG, Oh SP, Keene DR, Tamarkin L, Niemela M, Ilves M, Li E, Pihlajaniemi T, Olsen BR (2002) Lack of collagen XVIII/endostatin results in eye abnormalities. EMBO J 21:1535–1544

Fukunaga-Kalabis M, Martinez G, Liu ZJ, Kalabis J, Mrass P, Weninger W, Firth SM, Planque N, Perbal B, Herlyn M (2006) CCN3 controls 3D spatial localization of melanocytes in the human skin through DDR1. J Cell Biol 175:563–569

Gautam M, Noakes PG, Moscoso L, Rupp F, Scheller RH, Merlie JP, Sanes JR (1996) Defective neuromuscular synaptogenesis in agrin-deficient mutant mice. Cell 85:525–535

Gawlik K, Miyagoe-Suzuki Y, Ekblom P, Takeda S, Durbeej M (2004) Laminin alpha1 chain reduces muscular dystrophy in laminin alpha2 chain deficient mice. Hum Mol Genet 13:1775–1784

Gould DB, Phalan FC, Breedveld GJ, van Mil SE, Smith RS, Schimenti JC, Aguglia U, van der Knaap MS, Heutink P, John SW (2005) Mutations in Col4a1 cause perinatal cerebral hemorrhage and porencephaly. Science 308:1167–1171

Gould DB, Phalan FC, van Mil SE, Sundberg JP, Vahedi K, Massin P, Bousser MG, Heutink P, Miner JH, Tournier-Lasserve E, John SW (2006) Role of COL4A1 in small-vessel disease and hemorrhagic stroke. N Engl J Med 354:1489–1496

Halfter W, Dong S, Schurer B, Cole GJ (1998) Collagen XVIII is a basement membrane heparan sulfate proteoglycan. J Biol Chem 273:25404–25412

Hamill KJ, Langbein L, Jones JC, McLean WH (2009) Identification of a novel family of laminin N-terminal alternate splice isoforms: structural and functional characterization. J Biol Chem 284:35588–35596

Harvey SJ, Jarad G, Cunningham J, Rops AL, van der Vlag J, Berden JH, Moeller MJ, Holzman LB, Burgess RW, Miner JH (2007) Disruption of glomerular basement membrane charge through podocyte-specific mutation of agrin does not alter glomerular permselectivity. Am J Pathol 171:139–152

Hasselbacher K, Wiggins RC, Matejas V, Hinkes BG, Mucha B, Hoskins BE, Ozaltin F, Nurnberg G, Becker C, Hangan D, Pohl M, Kuwertz-Broking E, Griebel M, Schumacher V, Royer-Pokora B, Bakkaloglu A, Nurnberg P, Zenker M, Hildebrandt F (2006) Recessive missense mutations in LAMB2 expand the clinical spectrum of LAMB2-associated disorders. Kidney Int 70:1008–1012

Henry MD, Campbell KP (1999) Dystroglycan inside and out. Curr Opin Cell Biol 11:602–607

Hewitt JE (2009) Abnormal glycosylation of dystroglycan in human genetic disease. Biochim Biophys Acta 1792:853–861

Hudson BG (2004) The molecular basis of Goodpasture and Alport syndromes: beacons for the discovery of the collagen IV family. J Am Soc Nephrol 15:2514–2527

Hynes R (1999) Fibronectins. In: Kreis T, Vale R (eds) Guidebook to the extracellular matrix, anchor, and adhesion Proteins. Oxford University Press, New York, pp 422–425

Hynes RO (2002) Integrins: bidirectional, allosteric signaling machines. Cell 110:673–687

Ido H, Nakamura A, Kobayashi R, Ito S, Li S, Futaki S, Sekiguchi K (2007) The requirement of the glutamic acid residue at the third position from the carboxyl termini of the laminin gamma chains in integrin binding by laminins. J Biol Chem 282:11144–11154

Ido H, Ito S, Taniguchi Y, Hayashi M, Sato-Nishiuchi R, Sanzen N, Hayashi Y, Futaki S, Sekiguchi K (2008) Laminin isoforms containing the gamma3 chain are unable to bind to integrins due to the absence of the glutamic acid residue conserved in the C-terminal regions of the gamma1 and gamma2 chains. J Biol Chem 283:28149–28157

Iozzo RV (2001) Heparan sulfate proteoglycans: intricate molecules with intriguing functions. J Clin Investig 108:165–167

Jones KJ, Morgan G, Johnston H, Tobias V, Ouvrier RA, Wilkinson I, North KN (2001) The expanding phenotype of laminin alpha2 chain (merosin) abnormalities: case series and review. J Med Genet 38:649–657

Kalluri R (2003) Basement membranes: structure, assembly and role in tumour angiogenesis. Nat Rev Cancer 3:422–433

Kammerer RA, Schulthess T, Landwehr R, Schumacher B, Lustig A, Yurchenco PD, Ruegg MA, Engel J, Denzer AJ (1999) Interaction of agrin with laminin requires a coiled-coil conformation of the agrin-binding site within the laminin gamma1 chain. EMBO J 18:6762–6770

Kashtan CE (2004) Familial hematuria due to type IV collagen mutations: Alport syndrome and thin basement membrane nephropathy. Curr Opin Pediatr 16:177–181

Kashtan CE, Michael AF (1993) Alport syndrome: from bedside to genome to bedside. Am J Kidney Dis 22:627–640

Khoshnoodi J, Cartailler JP, Alvares K, Veis A, Hudson BG (2006) Molecular recognition in the assembly of collagens: terminal noncollagenous domains are key recognition modules in the formation of triple helical protomers. J Biol Chem 281:38117–38121

Khoshnoodi J, Pedchenko V, Hudson BG (2008) Mammalian collagen IV. Microsc Res Tech 71:357–370

Kikkawa Y, Miner JH (2005) Review: Lutheran/B-CAM: a laminin receptor on red blood cells and in various tissues. Connect Tissue Res 46:193–199

Kim S, Wadsworth WG (2000) Positioning of longitudinal nerves in C. elegans by nidogen. Science 288:150–154

Kostka G, Giltay R, Bloch W, Addicks K, Timpl R, Fassler R, Chu ML (2001) Perinatal lethality and endothelial cell abnormalities in several vessel compartments of fibulin-1-deficient mice. Mol Cell Biol 21:7025–7034

Kummer TT, Misgeld T, Sanes JR (2006) Assembly of the postsynaptic membrane at the neuromuscular junction: paradigm lost. Curr Opin Neurobiol 16:74–82

Li S, Liquari P, McKee KK, Harrison D, Patel R, Lee S, Yurchenco PD (2005) Laminin-sulfatide binding initiates basement membrane assembly and enables receptor signaling in Schwann cells and fibroblasts. J Cell Biol 169:179–189

Libby RT, Champliaud MF, Claudepierre T, Xu Y, Gibbons EP, Koch M, Burgeson RE, Hunter DD, Brunken WJ (2000) Laminin expression in adult and developing retinae: evidence of two novel CNS laminins. J Neurosci 20:6517–6528

Litjens SH, de Pereda JM, Sonnenberg A (2006) Current insights into the formation and breakdown of hemidesmosomes. Trends Cell Biol 16:376–383

Mankelow TJ, Burton N, Stefansdottir FO, Spring FA, Parsons SF, Pedersen JS, Oliveira CL, Lammie D, Wess T, Mohandas N, Chasis JA, Brady RL, Anstee DJ (2007) The Laminin 511/521-binding site on the Lutheran blood group glycoprotein is located at the flexible junction of Ig domains 2 and 3. Blood 110:3398–3406

Mao Y, Schwarzbauer JE (2005) Fibronectin fibrillogenesis, a cell-mediated matrix assembly process. Matrix Biol 24:389–399

Marneros AG, Olsen BR (2005) Physiological role of collagen XVIII and endostatin. FASEB J 19:716–728

McKee KK, Harrison D, Capizzi S, Yurchenco PD (2007) Role of laminin terminal globular domains in basement membrane assembly. J Biol Chem 282:21437–21447

Meinen S, Ruegg MA (2006) Congenital muscular dystrophy: mini-agrin delivers in mice. Gene Ther 13:869–870

Michele DE, Barresi R, Kanagawa M, Saito F, Cohn RD, Satz JS, Dollar J, Nishino I, Kelley RI, Somer H, Straub V, Mathews KD, Moore SA, Campbell KP (2002) Post-translational disruption of dystroglycan–ligand interactions in congenital muscular dystrophies. Nature 418: 417–422

Miner JH, Patton BL (1999) Laminin-11. Int J Biochem Cell Biol 31:811–816

Miner JH, Yurchenco PD (2004) Laminin functions in tissue morphogenesis. Annu Rev Cell Dev Biol 20:255–284

Miner JH, Patton BL, Lentz SI, Gilbert DJ, Snider WD, Jenkins NA, Copeland NG, Sanes JR (1997) The laminin alpha chains: expression, developmental transitions, and chromosomal locations of alpha1–5, identification of heterotrimeric laminins 8–11, and cloning of a novel alpha3 isoform. J Cell Biol 137:685–701

Miner JH, Li C, Mudd JL, Go G, Sutherland AE (2004) Compositional and structural requirements for laminin and basement membranes during mouse embryo implantation and gastrulation. Development 131:2247–2256

Miranti CK, Brugge JS (2002) Sensing the environment: a historical perspective on integrin signal transduction. Nat Cell Biol 4:E83–E90

Nishiuchi R, Takagi J, Hayashi M, Ido H, Yagi Y, Sanzen N, Tsuji T, Yamada M, Sekiguchi K (2006) Ligand-binding specificities of laminin-binding integrins: a comprehensive survey of laminin–integrin interactions using recombinant alpha3beta1, alpha6beta1, alpha7beta1 and alpha6beta4 integrins. Matrix Biol 25:189–197

Noakes PG, Gautam M, Mudd J, Sanes JR, Merlie JP (1995a) Aberrant differentiation of neuromuscular junctions in mice lacking s-laminin/laminin β2. Nature 374:258–262

Noakes PG, Miner JH, Gautam M, Cunningham JM, Sanes JR, Merlie JP (1995b) The renal glomerulus of mice lacking s-laminin/laminin β2: nephrosis despite molecular compensation by laminin β1. Nat Genet 10:400–406

Patton BL (2000) Laminins of the neuromuscular system. Microsc Res Tech 51:247–261

Pinzon-Duarte GA, Daly GH, Li YN, Koch M, Brunken WJ (2010) Defective formation of the inner limiting membrane in the laminin {beta}2 and {gamma}3 null retina results in retinal dysplasia. Invest Ophthalmol Vis Sci. 51:1773–1782

Plaisier E, Gribouval O, Alamowitch S, Mougenot B, Prost C, Verpont MC, Marro B, Desmettre T, Cohen SY, Roullet E, Dracon M, Fardeau M, Van Agtmael T, Kerjaschki D, Antignac C, Ronco P (2007) COL4A1 mutations and hereditary angiopathy, nephropathy, aneurysms, and muscle cramps. N Engl J Med 357:2687–2695

Poschl E, Schlotzer-Schrehardt U, Brachvogel B, Saito K, Ninomiya Y, Mayer U (2004) Collagen IV is essential for basement membrane stability but dispensable for initiation of its assembly during early development. Development 131:1619–1628

Pulkkinen L, Uitto J (1999) Mutation analysis and molecular genetics of epidermolysis bullosa. Matrix Biol 18:29–42

Rahuel C, Filipe A, Ritie L, El Nemer W, Patey-Mariaud N, Eladari D, Cartron JP, Simon-Assmann P, Le Van KC, Colin Y (2008) Genetic inactivation of the laminin {alpha}5 chain receptor Lu/BCAM leads to kidney and intestinal abnormalities in the mouse. Am J Physiol Renal Physiol 294:F393–406

Rebustini IT, Patel VN, Stewart JS, Layvey A, Georges-Labouesse E, Miner JH, Hoffman MP (2007) Laminin alpha5 is necessary for submandibular gland epithelial morphogenesis and influences FGFR expression through beta1 integrin signaling. Dev Biol 308:15–29

Rossi M, Morita H, Sormunen R, Airenne S, Kreivi M, Wang L, Fukai N, Olsen BR, Tryggvason K, Soininen R (2003) Heparan sulfate chains of perlecan are indispensable in the lens capsule but not in the kidney. EMBO J 22:236–245

Sakai T, Larsen M, Yamada KM (2003) Fibronectin requirement in branching morphogenesis. Nature 423:876–881

Sanes JR, Apel ED, Gautam M, Glass D, Grady RM, Martin PT, Nichol MC, Yancopoulos GD (1998) Agrin receptors at the skeletal neuromuscular junction. Ann NY Acad Sci 841:1–13

Sicot FX, Tsuda T, Markova D, Klement JF, Arita M, Zhang RZ, Pan TC, Mecham RP, Birk DE, Chu ML (2008) Fibulin-2 is dispensable for mouse development and elastic fiber formation. Mol Cell Biol 28:1061–1067

Smyth N, Vatansever HS, Murray P, Meyer M, Frie C, Paulsson M, Edgar D (1999) Absence of basement membranes after targeting the LAMC1 gene results in embryonic lethality due to failure of endoderm differentiation. J Cell Biol 144:151–160

Stum M, Girard E, Bangratz M, Bernard V, Herbin M, Vignaud A, Ferry A, Davoine CS, Echaniz-Laguna A, Rene F, Marcel C, Molgo J, Fontaine B, Krejci E, Nicole S (2008) Evidence of a dosage effect and a physiological endplate acetylcholinesterase deficiency in the first mouse models mimicking Schwartz–Jampel syndrome neuromyotonia. Hum Mol Genet 17: 3166–3179

Taniguchi Y, Ido H, Sanzen N, Hayashi M, Sato-Nishiuchi R, Futaki S, Sekiguchi K (2009) The C-terminal region of laminin beta chains modulates the integrin binding affinities of laminins. J Biol Chem 284:7820–7831

Thorner PS (2007) Alport syndrome and thin basement membrane nephropathy. Nephron Clin Pract 106:c82–88

Timpl R (1989) Structure and biological activity of basement membrane proteins. Eur J Biochem 180:487–502

Timpl R (1996) Macromolecular organization of basement membranes. Curr Opin Cell Biol 8:618–624

Timpl R, Brown JC (1994) The laminins. Matrix Biol 14:275–281

Tzu J, Marinkovich MP (2008) Bridging structure with function: structural, regulatory, and developmental role of laminins. Int J Biochem Cell Biol 40:199–214

Van Agtmael T, Bruckner-Tuderman L (2010) Basement membranes and human disease. Cell Tissue Res 339:167–188

Van Agtmael T, Schlotzer-Schrehardt U, McKie L, Brownstein DG, Lee AW, Cross SH, Sado Y, Mullins JJ, Poschl E, Jackson IJ (2005) Dominant mutations of Col4a1 result in basement membrane defects which lead to anterior segment dysgenesis and glomerulopathy. Hum Mol Genet 14:3161–3168

Vanacore RM, Ham AJ, Cartailler JP, Sundaramoorthy M, Todd P, Pedchenko V, Sado Y, Borza DB, Hudson BG (2008) A role for collagen IV cross-links in conferring immune privilege to the Goodpasture autoantigen: structural basis for the crypticity of B cell epitopes. J Biol Chem 283:22737–22748

Vanacore R, Ham AJ, Voehler M, Sanders CR, Conrads TP, Veenstra TD, Sharpless KB, Dawson PE, Hudson BG (2009) A sulfilimine bond identified in collagen IV. Science 325:1230–1234

Vogel W (1999) Discoidin domain receptors: structural relations and functional implications. FASEB J 13(Suppl):S77–S82

Vogel WF, Aszodi A, Alves F, Pawson T (2001) Discoidin domain receptor 1 tyrosine kinase has an essential role in mammary gland development. Mol Cell Biol 21:2906–2917

Wang X, Harris RE, Bayston LJ, Ashe HL (2008) Type IV collagens regulate BMP signalling in Drosophila. Nature 455:72–77

Williamson RA, Henry MD, Daniels KJ, Hrstka RF, Lee JC, Sunada Y, Ibraghimov-Beskrovnaya O, Campbell KP (1997) Dystroglycan is essential for early embryonic development: disruption of Reichert's membrane in Dag1-null mice. Hum Mol Genet 6:831–841

Yoshida-Moriguchi T, Yu L, Stalnaker SH, Davis S, Kunz S, Madson M, Oldstone MB, Schachter H, Wells L, Campbell KP (2010) O-Mannosyl phosphorylation of alpha-dystroglycan is required for laminin binding. Science 327:88–92

Yurchenco PD, Patton BL (2009) Developmental and pathogenic mechanisms of basement membrane assembly. Curr Pharm Des 15:1277–1294

Yurchenco PD, Wadsworth WG (2004) Assembly and tissue functions of early embryonic laminins and netrins. Curr Opin Cell Biol 16:572–579

Yurchenco PD, Amenta PS, Patton BL (2004) Basement membrane assembly, stability and activities observed through a developmental lens. Matrix Biol 22:521–538

Zenker M, Aigner T, Wendler O, Tralau T, Muntefering H, Fenski R, Pitz S, Schumacher V, Royer-Pokora B, Wuhl E, Cochat P, Bouvier R, Kraus C, Mark K, Madlon H, Dotsch J, Rascher W, Maruniak-Chudek I, Lennert T, Neumann LM, Reis A (2004) Human laminin beta2 deficiency causes congenital nephrosis with mesangial sclerosis and distinct eye abnormalities. Hum Mol Genet 13:2625–2632

Chapter 5
Hyaluronan and the Aggregating Proteoglycans

Thomas N. Wight, Bryan P. Toole, and Vincent C. Hascall

Abstract Proteoglycans that interact specifically with hyaluronan are known as the "hyalectins". This family includes aggrecan, versican, neurocan, and brevican. These proteoglycans form macromolecular complexes with hyaluronan and contribute to the structural and mechanical stability of different tissues. The synthesis and turnover of the individual components of these complexes are highly regulated. In addition, different parts of these complexes interact with cells and influence cellular phenotype. Specific qualitative and quantitative changes take place in these macromolecules during development and disease and, in part, regulate key events that determine normal and pathological tissue phenotype. This chapter reviews both past and present evidence for the critical role that these ECM components play in the biology and pathology of human tissues.

5.1 Introduction

The family of hyaluronan-binding proteoglycans includes aggrecan, versican, neurocan, and brevican and constitutes a gene family collectively termed the "hyalectins" and/or "lecticans" (Iozzo 1998; Margolis and Margolis 1994; Yamaguchi 1996; Zimmermann 2000) (Fig. 5.1). The core proteins of proteoglycans of the hyalectin family share extensive structural similarity within N- and C-terminal globular domains and have central domains of variable length with multiple sites for the addition of chondroitin sulfate or dermatan sulfate chains and O-linked

---

T.N. Wight (✉)
The Hope Heart Matrix Biology Program, Benaroya Research Institute at Virginia Mason, 1201 Ninth Avenue, Seattle, WA, USA
e-mail: twight@benaroyaresearch.org

B.P. Toole
Department of Regenerative Medicine and Cell Biology, Medical University of South Carolina, Charleston, SC, USA

V.C. Hascall
Department of Biomedical Engineering, Cleveland Clinic Foundation, Cleveland, OH, USA

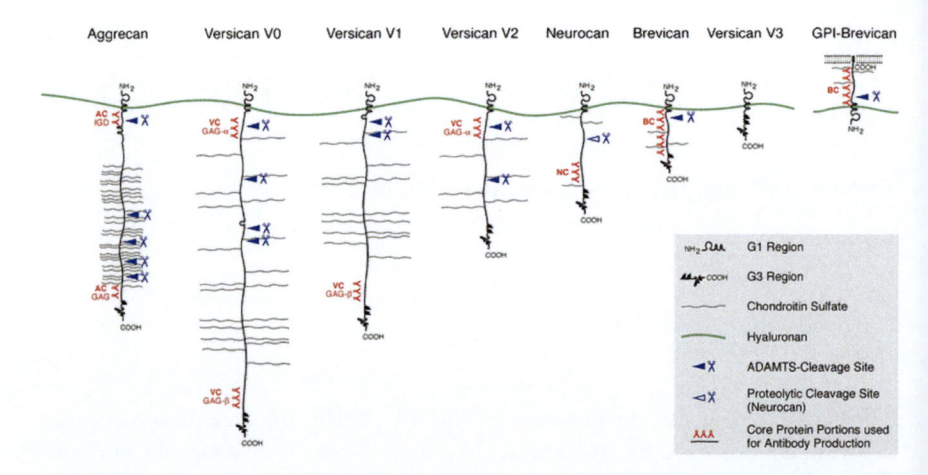

Fig. 5.1 Structured models of the proteoglycans of the hyalectin family. Proteolytic cleavage sites (ADAMTS) and core protein portions used in antibody production are indicated. From Zimmermann (2008). *Reproduced with permission*

oligosaccharides. Considerable evidence suggests that the large hydrodynamic domains occupied by hyalectins influence tissue turgidity and viscoelasticity. These proteoglycans are present in the extracellular matrix (ECM) of many tissues with aggrecan prominent in cartilage, brevican, and neurocan enriched in the central nervous system, and versican present in most soft tissues of the body.

The functions of these proteoglycans depend to a great extent on their ability to form aggregates that contribute to many of the biomechanical properties of tissues and to the protective and signaling functions of the pericellular microenvironment. Crucial components required for aggregate formation are the link proteins and the glycosaminoglycan, hyaluronan. The so-called *link module* domains of link proteins and the hyalectins bind specifically to hyaluronan. However, proteoglycan aggregate formation is only one of the numerous binding functions of hyaluronan. Unlike heparan sulfate, where its many functions derive from heterogeneity in the polysaccharide sequence, the multifunctionality of hyaluronan derives from multivalent interactions with a wide range of hyaluronan-binding proteins [termed "hyaladherins" (Toole 1990)], which include the four proteoglycans of the hyalectin family, the four link proteins and several cell surface signaling receptors, and a variety of other pericellular proteins.

5.2 Hyaluronan

5.2.1 Structure and Biosynthesis of Hyaluronan

The glycosaminoglycan hyaluronan has a simple disaccharide structure, glucuronate-*N*-acetylglucosamine. Its biosynthesis is distinctly different from the mechanisms of the other glycosaminoglycans. First, it does not require a core protein and hence is not normally a proteoglycan. Second, it requires only a single enzyme – one of the three hyaluronan synthases, *Has1, 2, or 3* (Weigel and DeAngelis 2007). Third, Has enzymes are normally transported to the plasma membrane before they are activated. Fourth, Has enzymes utilize cytoplasmic substrates. Fifth, the alternate UDP-sugar substrate is added to the reducing end of the elongating chain with the release of the anchoring UDP moiety. Sixth, the elongating chain is extruded into the extracellular space, which allows the final hyaluronan chain to be tens of thousands of disaccharides long without any modifications such as sulfation. Seventh, biosynthesis of hyaluronan is energetically efficient as it bypasses all the machinery required for the synthesis of other glycosaminoglycans.

Catabolism – Hyaluronan is actively synthesized and catabolized by many cells, maintaining a steady-state metabolism. For example, synovial lining cells synthesize and secrete hyaluronan into the synovial fluid of articulating joints where it functions to distribute load and reduce surface friction during joint movement. Hyaluronan is removed from synovial joints into the lymphatics (half-life 2–3 days), and lining cells in the lymphatics remove a large proportion (70–80%) of the hyaluronan before the remainder drains into the vascular system (Laurent and Fraser 1986; Prevo et al. 2001). The liver sinusoidal cells efficiently remove hyaluronan from circulation and degrade it in lysosomes to recover the sugar residues. At the cellular level, the human genome has six related enzymes in the hyaluronidase family Hyal1–5 and PH-20 (Stern et al. 2007). Hyal1 and Hyal2 appear to be the critical enzymes for the catabolic pathway that many cells utilize to maintain a steady-state metabolism of hyaluronan. Hyal2 is a GPI-anchored protein that is located on cell surfaces and appears to be critical for degrading the macromolecular hyaluronan to fragments in the 30–50 kDa range (Duterme et al. 2009). These fragments are then internalized by the cell surface hyaluronan receptor, CD44, which contains a link module and is present on most, if not all, cells. The hyaluronan fragments are deposited in an intracellular membrane compartment that is distinct from clathrin-coated pits or pinocytotic vesicles, and CD44 re-cycles to the cell surface (Tammi et al. 2001). This cycling has a half-life of ~15 min. The internalized hyaluronan fragments are then transported to lysosomes for further degradation by Hyal1, an endolytic hexosaminidase, and by exoglycosidases to the sugar units. The half-life for transport to the golgi is 2–3 h.

5.2.2 Evolution of Hyaluronan and Hyaladherins

The original Has enzyme evolved late in evolution, quite likely from a chitin synthase in an animal with a primitive notochord, and the original link module protein subsequently evolved in due course, possibly from a heparan sulfate binding protein. Neither protein family is represented in the genomes of early organisms (insects, arthropods, arachnids, and crustaceans). A major change in biology was the emerging ability of precursor cells to migrate long distances through hyaluronan-rich matrices before differentiating. This is particularly well represented in the role of hyaluronan in the development of vertebrate tissues and organs by the formation of stem cell "niches" and in the emergence of numerous extracellular, cell surface, and intracellular hyaladherins. Examination of vertebrate genomes reveals the likely presence of at least 13 hyaluronan-binding proteins with one or two link module domains, plus an unknown number with other hyaluronan-binding motifs such $B(X_7)B$ (see below).

5.2.3 Hyaladherins

Hyaluronan interacts with several cell surface receptors such as CD44, RHAMM, LYVE-1, Hare/stabilin-2, layilin, the link proteins, and TNFα-stimulated gene 6 (TSG-6) [reviewed in Jiang et al. (2007), Ponta et al. (2003), Turley et al. (2002)]. Among the best characterized hyaladherins are those in which the hyaluronan-binding domain is a "link module" motif, so-called since these domains are homologous to the hyaluronan-binding domains of link proteins. Members of this family of hyaladherins include the link proteins; the four hyalectins; the cell surface hyaluronan receptors CD44, HARE, and LYVE-1; and the extracellular protein TSG-6. The link modules contain characteristic disulfide-bonded loops and sequence homologies of 30–40%. Two link modules form the hyaluronan-binding regions of link proteins and the hyalectins, whereas a single link module is found in CD44 and TSG-6. Some hyaladherins, notably RHAMM, do not have link modules. Molecular studies of RHAMM have revealed a hyaluronan-binding motif which is present, not only within RHAMM, but also within or adjacent to the link modules of several of the proteins mentioned above, e.g., link proteins and CD44. This motif is $B(X_7)B$, where B is arginine or lysine and X is any non-acidic amino acid (Yang et al. 1994). Several other hyaladherins, especially intracellular hyaladherins (Deb and Datta 1996; Grammatikakis et al. 1995; Huang et al. 2000), lack domains with homology to link modules but contain $B(X_7)B$ and related sequences. Although clusters of basic amino acids contribute to hyaluronan binding in most hyaladherins, other structural features, e.g., glycosylation and conformational effects, are also involved.

Hyaluronan-receptor interactions mediate at least three important physiological processes, i.e., receptor-mediated internalization, assembly of pericellular matrices, and signal transduction [reviewed in Evanko et al. (2007), Knudson et al. (2002), Toole (2001, 2009)] (Fig. 5.2). Receptor-mediated internalization usually leads to degradation of the hyaluronan ligand. HARE/stabilin-2 is a scavenging receptor that clears hyaluronan and other glycosaminoglycans from the circulation (Pandey et al. 2008). The involvement of CD44 in catabolism of hyaluronan has been shown by the failure of CD44-null tissues to clear excess hyaluronan, e.g., in skin (Kaya et al. 1997) and lung (Teder et al. 2002). Inability to clear hyaluronan produced in

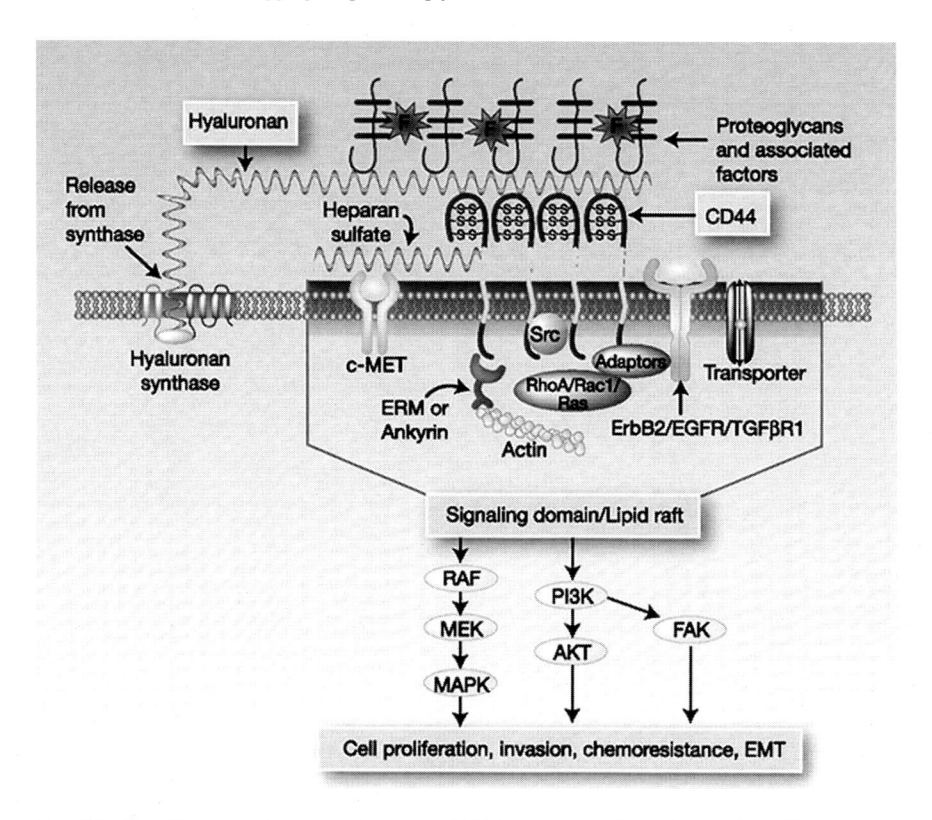

5 Hyaluronan and the Aggregating Proteoglycans

Fig. 5.2 Signaling cascades by hyaluronan–CD44 interactions. Upon interaction of hyaluronan with CD44, signaling domains within the plasma membrane may contain receptor tyrosine kinases (e.g., ErbB2 and EGFR), other signaling receptors (e.g., TGFβR1), and non-receptor kinases (e.g., Src family) that drive several signaling pathways (e.g., the MAP kinase and PI3 kinase/Akt cell proliferation and survival pathways), as well as various transporters that influence a variety of cell properties. Various adaptor proteins, such as Vav2, Grb2, and Gab-1, mediate interaction of CD44 with upstream effectors (e.g., RhoA, Rac1, and Ras), which drive these pathways. In other cases, carbohydrate side groups on variant regions of CD44 (e.g., heparan sulfate chains) bind regulatory factors, such as FGF, and co-activate receptor tyrosine kinases, such as the c-Met receptor. Hyaluronan–CD44 interactions also induce cytoskeletal changes that promote cell motility and invasion. In this case, actin filaments are joined to the cytoplasmic tail of CD44 via members of the ezrin–radixin–moiesin (ERM) family or ankyrin. Proteoglycans and associated factors attached to pericellular hyaluronan may also influence these activities. From Toole (2009), with permission

lungs of CD44-null mice after a bleomycin inflammatory challenge results in death of the animals (Teder et al. 2002).

Several cell types exhibit highly hydrated, hyaluronan-dependent, pericellular matrices or "coats" that can be visualized indirectly by their ability to exclude particles. They are usually 5–10 μm in thickness, and they are removed by treatment with hyaluronan-specific hyaluronidases [reviewed in Evanko et al. (2007)] (Fig. 5.3). These pericellular matrices provide the milieu in which numerous cellular activities take place and influence the behavior of cells. During tissue

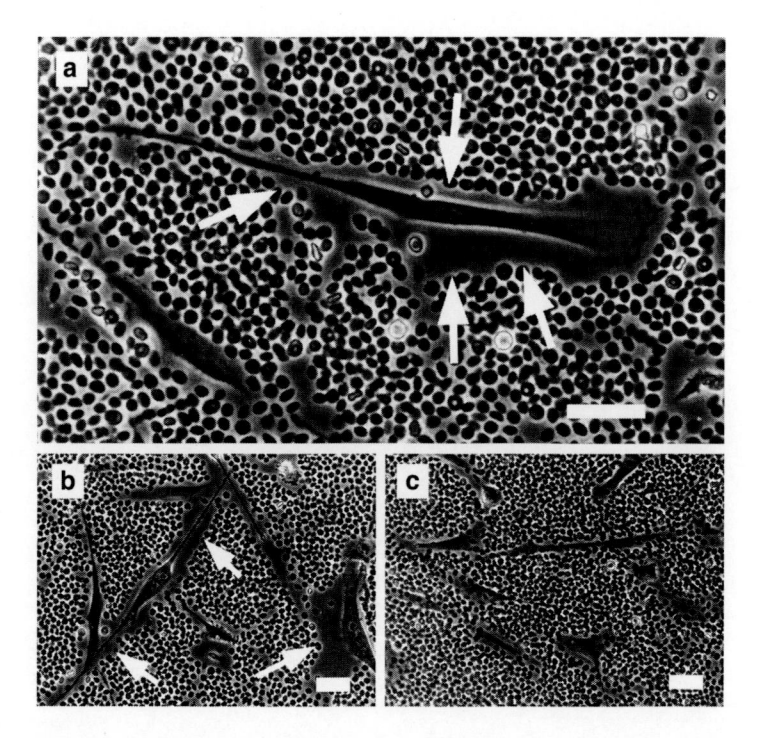

Fig. 5.3 Hyaluronan-dependent pericellular matrix in human smooth muscle cells visualized using the particle exclusion assay. The cell coat excludes the fixed erythrocytes and is seen as a clear zone surrounding the cell (*arrows*). (**a**) A typical locomoting cell with a small amount of pericellular matrix at the lammellipodium in front and more abundant matrix along the cell flanks and trailing uropod. (**b** and **c**) Pericellular matrices were visualized before (**b**) or after (**c**) digestion with *Streptomyces* hyaluronidase. *Bars* equal 50 µm. From Evanko (2007), used with permission

formation or remodeling, such matrices provide a hydrated, fluid pericellular environment in which assembly of other matrix components and presentation of growth and differentiation factors can readily occur without interference from the highly structured fibrous matrix usually found in fully differentiated tissues. In some cases, such as in cartilage, the pericellular matrix is a unique structural component that protects cells and contributes to the characteristic properties of the differentiated tissue. The assembly and function of these pericellular matrices are dependent on three features. First, hyaluronan is crucial to their integrity. Second, the assembly and density of pericellular matrices require specific interaction of hyaluronan with a hyalectin, usually aggrecan or versican. Third, hyaluronan must be tethered to the cell surface. Tethering of hyaluronan to different cell types occurs by at least two mechanisms (1) via binding to the hyaluronan receptor, CD44, and (2) sustained attachment to hyaluronan synthase or associated proteins on the cytoplasmic face of the plasma membrane. Another hyaladherin, TSG-6, also contributes to the properties of some extracellular matrices.

In several physiological and pathological processes, multivalent binding of hyaluronan to its cell surface receptors leads to multiple signaling pathways and numerous downstream cellular phenomena (Fig. 5.2). For example, hyaluronan–CD44 binding can result in direct or indirect interactions of CD44 with signaling receptors, such as ErbB2, EGFR, and TGF-β receptor type I, that influence the activity of these receptors. This can, in turn, activate non-receptor kinases of the Src family or Ras family GTPases, thus influencing the activity of a variety of downstream signaling pathways such as the MAP kinase and PI3 kinase/Akt pathways. In addition to its action as a co-receptor or co-activator of membrane-associated signaling molecules, CD44 can influence cellular events such as proliferation and motility through cross-linking to the actin cytoskeleton via ankyrin or members of the ezrin–radixin–moiesin family. It is likely that CD44 is recruited into lipid microdomains in response to ligand interactions and associates indirectly or directly therein with signaling proteins, transporters, and cytoskeletal elements [reviewed in Toole (2009)]. Moreover, endocytosis of hyaluronan and CD44 occurs from these domains (Thankamony and Knudson 2006).

RHAMM can be present either in the cytoplasm or on the cell surface, and is also an important factor in cell motility and proliferation in a variety of systems (Maxwell et al. 2008). CD44 and RHAMM can exhibit both cooperative and interchangeable signaling functions. For example, interactions at the plasma membrane between CD44 and RHAMM have been shown to activate CD44 signaling through ERK1/2 and promote cell motility. In some cases, e.g., in animal models of autoimmune diseases, RHAMM can compensate for CD44, a very important consideration when interpreting experiments in CD44-null mice (Naor et al. 2007). LYVE-1 is a close relative of CD44 that is mainly restricted to lymphatic vessel and lymph node endothelia, but its function is not well established (Jackson 2009).

5.2.4 Hyaluronan in Development

The pericellular matrices surrounding cells during morphogenesis of embryonic organs are enriched in hyaluronan and resemble the pericellular matrices described above. Striking examples of such hyaluronan-rich matrices are seen around migrating and proliferating cells during gastrulation and during formation of the cornea, peripheral ganglia, vertebrae, heart valves, brain, and limb, as well as during salamander limb regeneration and vertebrate wound repair (Toole 2001). In many cases hyaluronan is removed during final differentiation of cells subsequent to these morphogenetic events. These hyaluronan-rich matrices most likely contribute to cell behavior by creating hydrated pathways that separate barriers to cell invasion and by promoting signaling cascades necessary for epithelial–mesenchymal transitions (EMTs), cell invasion, and cell proliferation. Two systems serve well to illustrate the crucial role of hyaluronan in developmental processes, i.e., development of embryonic heart valves and limbs.

Of the three vertebrate hyaluronan synthase genes that have been characterized (*Has1*, *Has2*, and *Has3*), *Has2* appears to be the most important in early embryonic development (Tien and Spicer 2005). $Has2^{-/-}$ mice exhibit serious cardiovascular defects, which lead to their death during mid-gestation (Camenisch et al. 2000). At this stage of development, $Has2^{-/-}$ mouse embryos contain virtually no hyaluronan, and they exhibit multiple structural abnormalities in yolk sac, vasculature and heart morphogenesis, notably defective heart valve formation. At the tissue level, extracellular matrices are more compact than normal and the organization of other matrix components, especially versican, is altered. Strikingly, an insertional transgene mutation in the versican gene, which leads to loss of versican expression, causes similar defects in heart valve formation to those seen in the $Has2^{-/-}$ mice, suggesting that interaction of versican with hyaluronan is critical (Mjaatvedt et al. 1998). In addition to the alterations in tissue structure in the $Has2^{-/-}$ mice, changes also occur in cell behavior. Cardiac cushion morphogenesis involves transformation of endothelial to mesenchymal cells that then migrate into the hyaluronan-rich cardiac jelly and eventually form valves and other structures. To investigate the role of hyaluronan in this process, an explant system was used in which the developing cushion tissues were cultured on collagen gels. Endothelial–mesenchymal transformation and migration occurred when wild-type cushion tissue was cultured in this way but not when $Has2^{-/-}$ tissue was used. Transformation and migration were rescued when the $Has2^{-/-}$ tissue was transfected with *Has2* cDNA. Hyaluronan-mediated rescue of transformation, but not migration, was mimicked by transfection with constitutively active Ras and inhibited by dominant-negative Ras, thus implicating the Ras signaling pathway in this effect of hyaluronan (Camenisch et al. 2000). Thus, this model reveals the major, overlapping, molecular functions of hyaluronan, its biophysical properties, interactions with structural extracellular macromolecules, and instructive effects on cell signaling and behavior.

A second illustrative system is limb development, in which hyaluronan has been implicated in various aspects of morphogenesis. The outgrowth and patterning of the developing limb is regulated by reciprocal interactions between the apical ectodermal ridge, a thickened cap of ectoderm along the distal periphery of the limb bud, and the underlying distal subridge mesenchymal cells, which undergo proliferation, directed migration, and patterning in response to the apical ectodermal ridge and other signaling centers. These mesenchymal cells express *Has2* (Li et al. 2007), produce high amounts of hyaluronan, form an expansive hydrated extracellular matrix between the cells, and express voluminous, hyaluronan-dependent, pericellular matrices in culture (Knudson and Toole 1985; Kosher et al. 1981; Singley and Solursh 1981; Toole 1972). *Has2* is also expressed, and hyaluronan secreted by the apical ridge cells themselves (Li et al. 2007). Thus the cell and tissue interactions controlling the outgrowth and patterning of the limb occur in an environment rich in extracellular and pericellular hyaluronan. Subsequent to this early stage of limb development, the mesoderm condenses, i.e., the intercellular matrix decreases in volume, at sites of future cartilage and muscle differentiation. This is paralleled by decreased *Has2* expression and hyaluronan production during condensation of the mesoderm and loss of ability of the mesodermal cells to form hydrated pericellular matrices in culture (Knudson and Toole 1985; Kosher et al. 1981), most likely due to Hyal2 cleavage of hyaluronan and CD44 receptor-mediated endocytosis (Knudson et al. 2002) and degradation of hyaluronan by the hyaluronidase, Hyal1 (Kulyk and Kosher 1987; Nicoll et al. 2002). Overexpression of *Has2* in the mesoderm of the chick limb bud in vivo results in severely malformed limbs, which lack one or more skeletal elements and/or possess skeletal elements that exhibit abnormal morphology and are positioned inappropriately (Li et al. 2007). Thus, sustained production of hyaluronan in vivo perturbs limb growth, patterning, and cartilage differentiation. Furthermore, sustained hyaluronan production in micromass cultures of limb mesenchymal cells inhibits formation of precartilage condensations and subsequent chondrogenesis, indicating that downregulation of hyaluronan is necessary for the formation of the precartilage condensations that trigger cartilage differentiation (Li et al. 2007). In support of this conclusion, conditional inactivation of *Has2* in the developing limb mesoderm causes major defects in tissue patterning, chondrocyte maturation, skeletal growth, and synovial joint formation in the developing limb (Matsumoto et al. 2009).

Other systems in which hyaluronan clearly plays a critical role include, but are not restricted to, gastrulation (Bakkers et al. 2004), neural crest migration, and formation of somites (Ori et al. 2006), branching morphogenesis during early development of the prostate (Gakunga et al. 1997) and kidney (Pohl et al. 2000), maturation of the ductus arteriosus (Yokoyama et al. 2006), and expansion of the cumulus oophorus during ovulation (Salustri et al. 1999). Importantly, growing evidence indicates an important role for hyaluronan-cell interactions in migration, homing and differentiation of precursor cells of various types (Avigdor et al. 2004; Choudhary et al. 2007; Matrosova et al. 2004; Nilsson et al. 2003; Ori et al. 2006; Shukla et al. 2010; Smith et al. 2008).

5.2.5 Hyaluronan in Cancer

Extensive experimental evidence implicating hyaluronan in tumor growth and metastasis has been obtained in animal models of several tumor types. The approaches used include manipulation of levels of hyaluronan and perturbation of endogenous hyaluronan–receptor interactions by a number of methods. However, it has become evident that turnover of hyaluronan by hyaluronidases is an essential aspect of the promotion of tumor progression by hyaluronan and that the balance of synthesis and degradation is critical (Simpson and Lokeshwar 2008). Recent work in which hyaluronan synthesis was upregulated conditionally in mammary tumors that arise spontaneously in MMTV-Neu mice highlights the importance of hyaluronan in tumor promotion, especially via recruitment of stromal cells and angiogenesis (Koyama et al. 2007). Numerous studies have demonstrated an important role for hyaluronan–CD44 interactions in recruitment or homing of various cell types, including circulating immune cells and precursor cells. The MMTV-Neu studies also confirmed the importance of hyaluronan in EMT. As discussed above, a major defect in the *Has2*-null mouse is failure to undergo EMT during early cardiac development. Moreover, upregulation of *Has2* in phenotypically normal epithelium induces the characteristics of EMT, including anchorage-independent growth and invasiveness, two of the major properties of malignant cells (Zoltan-Jones et al. 2003).

A large body of evidence indicates that activation of hyaluronan-mediated signaling via interaction with CD44 and RHAMM promotes tumor progression. Many studies have implicated variants of CD44 rather than standard CD44 in tumor progression, but this depends on the stage of progression and type of tumor (Sherman et al. 1994). A striking development in recent years is the emergence of CD44 as a marker for subpopulations of several types of human carcinomas, often termed cancer stem cells (CSCs), that exhibit highly malignant and chemoresistant properties (Polyak and Weinberg 2009; Visvader and Lindeman 2008). Interestingly, the characteristics of EMT have recently been linked to the properties of these cell subpopulations. A $CD44^+/CD24^-$ subpopulation exhibiting CSC properties is induced by upregulation of EMT-associated transcription factors in primary human breast epithelium, and a similar subpopulation with both EMT and CSC properties can be isolated from transformed epithelial cells (Hollier et al. 2009; Polyak and Weinberg 2009). Notably, these cells exhibit anchorage-independent growth of colonies in soft agar, a property that usually reflects resistance to apoptosis, which in turn is linked to chemoresistance. Numerous studies have shown that the CSC subpopulation of carcinomas and other tumor types is resistant to chemotherapeutic agents, most likely due to increased antiapoptotic pathway activity and enrichment of multidrug transporters (Hollier et al. 2009; Polyak and Weinberg 2009; Toole and Slomiany 2008). Another important outcome of EMT is invasiveness (Kalluri and Weinberg 2009; Turley et al. 2008), and accordingly, CSCs have been linked to invasiveness and metastasis (Polyak and Weinberg 2009; Sleeman and Cremers 2007; Visvader and Lindeman 2008). As noted above, hyaluronan is closely associated with EMT, and these same properties of anchorage-independent growth, resistance to apoptosis, drug resistance, and invasiveness are induced or increased by upregulation of hyaluronan synthesis and reversed by antagonists of hyaluronan–CD44 interactions. (Toole 2004; Toole et al. 2008). In particular, strong evidence has been published showing hyaluronan-dependent association of CD44 with receptor kinases (Bourguignon 2009; Ponta et al. 2003; Toole 2004) and transporters (Bourguignon et al. 2004; Colone et al. 2008; Miletti-Gonzalez et al. 2005; Slomiany et al. 2009a, b; Toole et al. 2008) that are important in drug resistance and malignancy. Recently, hyaluronan–CD44 interactions were examined in a CSC-like subpopulation of cells isolated from human patient ovarian carcinoma ascites. It was found that the CSCs are enriched in receptor tyrosine kinases and multidrug transporters, that these proteins are present in close association with CD44 in the plasma membrane of the CSCs, and that this association depends on constitutive hyaluronan interactions (Slomiany et al. 2009a, b).

5.2.6 Hyaluronan Matrices in Inflammation

The hyaluronan matrix formed by cumulus cells in follicles prior to ovulation is an example of an inflammatory process. In the preovulatory follicle, the cumulus cells around the oocyte upregulate *Has2*, which initiates hyaluronan, and TSG-6, which contains a link module that binds to the hyaluronan (Salustri et al. 1999). The follicle becomes permeable to serum, which brings the serum macromolecule, IαI, into the follicle. IαI is a chondroitin sulfate proteoglycan with bikunin, a trypsin inhibitor, as the core protein. Two heavy chains (HCs) are bound to the chondroitin sulfate by an aspartate ester bond to 6-hydroxyls on galNAc residues in the chain. TSG-6 transesterifies HCs from IαI to 6-hydroxyls on glcNAc residues in the hyaluronan (Mukhopadhyay et al. 2004). This reaction is required for successful synthesis of the hyaluronan matrix that surrounds the ovulating oocyte. Mice lacking either bikunin (unable to synthesize IαI) (Zhuo et al. 2001) or TSG-6 (Fulop et al. 2003) do not organize the matrix and are female infertile. The HC modification of hyaluronan occurs in other inflammatory processes, which generally involve upregulation of hyaluronan synthesis and formation of monocyte-adhesive hyaluronan matrices. This was initially demonstrated by stressing colon smooth muscle cells with poly I:C, which initiates cell responses similar to viral infection (de La Motte et al. 1999; de La Motte et al. 2003). Large cable structures of coalesced hyaluronan were formed within 18 h. U937 monocytes (or normal circulating monocytes) bind to these structures at 4°C and degrade them rapidly when warmed to 37°C. In contrast, the monocytes do not bind to the normal pericellular hyaluronan coats. Cells undergoing various stress responses, such as ER stress (Majors et al. 2003), idiopathic pulmonary hypertension (Aytekin et al. 2008), and wound healing (Pienimaki et al. 2001), also produce these monocyte-adhesive matrices. Removal of these matrices by inflammatory cells is essential and requires CD44 on the inflammatory cells (monocytes/macrophages). Mice null in CD44 do not survive an inflammation in the lung (Teder et al. 2002). After bleomycin inhalation, the hyaluronan content of the lung increases during the first week in both wild-type and CD44 null mice with an increasing influx of monocytes/macrophages. During the second week in the wild type, the hyaluronan matrix is removed, and the number of monocytes/macrophages decreases to a normal level. In contrast, in the CD44 null mice, both the hyaluronan content and the monocytes/macrophages continue to increase until the animals die. Thus, the removal of the abnormal hyaluronan matrix by the monocytes/macrophages is CD44 dependent and is essential to restore normal oxygen exchange and lung function.

Another way that hyaluronan may influence inflammation is by its direct interaction with immune cells such as T-lymphocytes (Bollyky et al. 2007; Bollyky et al. 2009; Firan et al. 2006). For example, intact high molecular weight hyaluronan interacts with CD44 on the surface of T-regulatory cells and promotes their functional suppression of T-cell responder cell proliferation, whereas low molecular weight hyaluronan does not exhibit this activity (Bollyky et al. 2007; Bollyky et al. 2009). On the other hand, fragments of hyaluronan may induce inflammatory cytokine release from the immune cells through CD44 and exhibit proinflammatory properties (Noble 2002; Powell and Horton 2005). Such results provide an excellent example of how a single ECM component can have opposite activities depending upon whether it is intact or degraded.

5.2.7 Hyaluronan Matrices in Diabetes

Smooth muscle cells and mesangial cells that divide in hyperglycemic medium initiate synthesis of hyaluronan inside the cell after activation of hyaluronan synthases in various intracellular compartments – endoplasmic reticulum/golgi/ transport vesicles. The deposition of the high MW, polyanionic hyaluronan in these compartments creates an ER stress response that drives autophagy with subsequent upregulation of cyclin D3, which mediates extrusion and formation of a monocyte-adhesive hyaluronan matrix after the completion of cell division (Wang and Hascall 2009). Some mesangial cells in glomeruli of hyperglycemic rats undergo cell division and the autophagic response that creates the hyaluronan matrix within the first week, and macrophages are recruited into the glomeruli and localize within the abnormal matrix. This pathological process then leads to nephropathy and proteinuria. Hyaluronan is also associated with increased inflammation observed in adipose tissue associated with diabetes and obesity (Han et al. 2007). For example, high glucose stimulates the production of hyaluronan in 3T3-L1 adipocytes which in turn bind and trap monocytes in in vitro *assays*. Furthermore, hyaluronan is enriched in adipose tissue in a diabetic mouse model coincident with macrophage accumulation (Han et al. 2007). Such results indicate that hyperglycemia is an important effector in promoting hyaluronan enriched ECM accumulation and influencing the pathogenesis of diabetic disease in several tissues.

5.3 The Cartilage Proteoglycan Aggrecan

A large proportion of the early work on proteoglycans focused on cartilage, which was known to be a rich source of the glycosaminoglycan chondroitin sulfate (CS). Karl Meyer's experiments in the 1930s led him to propose that the extracellular matrix of cartilage was an ionic interaction between the polyanionic CS and collagen (Meyer et al. 1937). However, subsequent work in the laboratories of Helen Muir (Muir 1958) and Maxwell Schubert (Gerber et al. 1960; Shatton and Schubert 1954) in the 1950s and 1960s showed that the CS is covalently bound to a noncollagen protein, and Schubert introduced the term Protein Polysaccharide, later replaced by Proteoglycan, to describe this new class of macromolecules. During this time period, Meyer also discovered a second class of glycosaminoglycans, keratan sulfate (KS), in cornea (Meyer et al. 1953) and showed that KS was also present in the cartilage proteoglycans (Seno et al. 1965). The development of the dissociative extraction procedure and CsCl density gradient methods (Sajdera and Hascall 1969) succeeded in purifying the cartilage proteoglycan, which is now named aggrecan.

5.3.1 Biosynthesis of Aggrecan

Biosynthesis of aggrecan, like all proteoglycans, is very complex. The core protein, which is synthesized in the endoplasmic reticulum, contains sequence motifs with serine residues that serve as sites for initiating the CS chains and serine and threonine residues that serve as sites for initiating most of the KS chains. The core protein is then packaged into transport vesicles for delivery to the golgi. During transport, a xylose is added to those serines that will eventually have CS chains (Fig. 5.4). A resident xylosyl transferase enzyme transfers xylose from the substrate, UDP-xylose, onto the serine hydroxyl with release of UDP that is then

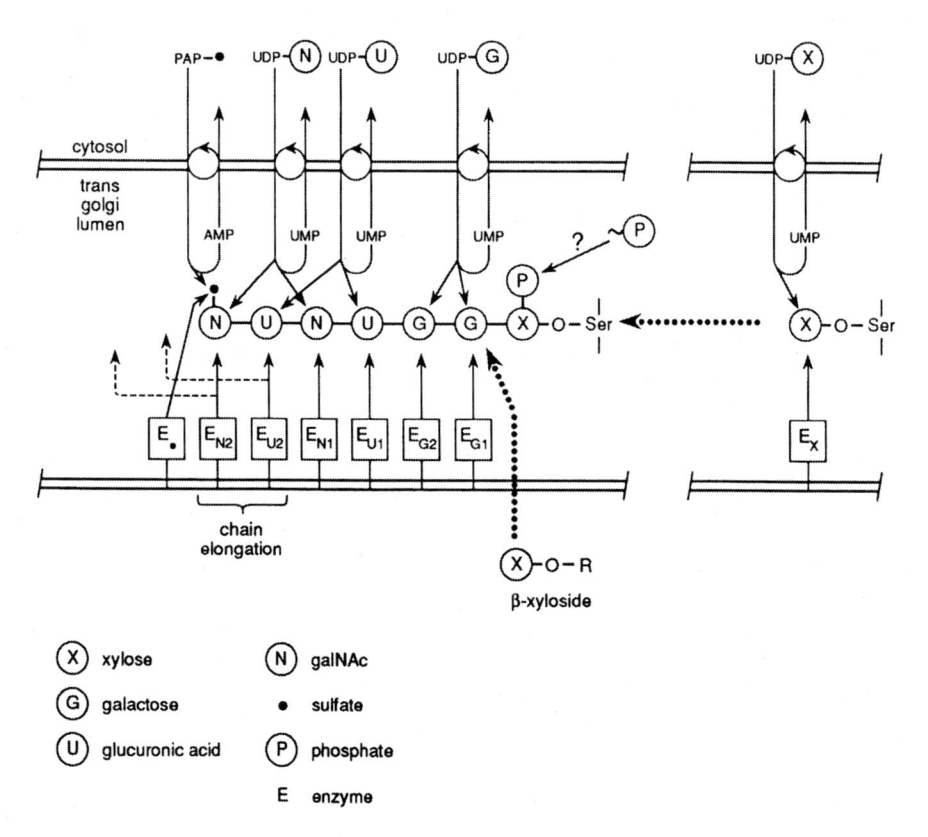

Fig. 5.4 Schematic diagram illustrating the synthesis of CS chains that are part of the hyalectins. From Wight (1991). *Reproduced with permission*

converted to UMP by a phosphatase. The UDP-xylose substrate enters the vesicle from the cytoplasm by a specific antiporter that simultaneously transports UMP into the cytoplasm. Once in the golgi, an oligosaccharide linkage region is synthesized by adding sequentially a galactose to the xylose, a second galactose to the first galactose and a glucuronate to the second galactose. This involves three specific glycosyl transferases and the UDP-galactose and UDP-glucuronate substrates as well as their antiporters. This linkage structure is the same for CS and heparan sulfate/heparin biosynthesis. The backbone CS structure, *N*-acetylgalactosamine and glucuronate, can now be added by an enzyme that alternately utilizes the UPD-*N*-acetylgalactosamine and UDP-glucuronate substrates, thereby forming chains that can reach a few hundred disaccharides in length. An additional UDP-*N*-acetylgalactosamine antiporter is also needed. This antiport mechanism is a very efficient way to control the concentrations of the different sugar substrates inside the vesicles and golgi to meet the rates that they are utilized by the different glycosyl transferases. The final modification of CS involves the addition of sulfate moieties by site specific sulfotransferases (to the four and six hydroxyls of the *N*-acetylgalactosamine residues for aggrecan CS) utilizing the phosphoadenosine phosphosulfate (PAPS) substrate that is also likely added to the golgi through an antiporter mechanism. This mechanism elongates the glycosaminoglycan by adding the sugar residues onto the nonreducing end of the elongating chain, which is a major reason that a core protein is needed as a primer.

Most of the KS chains are initiated by the addition of *N*-acetylgalactosamines to serine and threonine residues followed by elongation of the galactose-*N*-acetylglucosamine backbone structure. Some KS chains also are elongated on N-linked oligosaccharides on asparagine residues. The elongation and sulfation mechanisms for KS biosynthesis are less well known, but most likely involve the same mechanisms involving a cohort of glycosyl transferases, sulfotransferases, and sialic acid transferases and antiporters that would be needed. Upon completion, the aggrecan molecules will contain ~100 CS and ~50 KS chains.

5.3.2 Cartilage Proteoglycan Aggregates

The G1 protein domain of aggrecan and the link protein each have two link modules that interact specifically with five disaccharides of HA (HA10). The G1 and the link protein also have separate domains that interact with each other, which locks aggrecan onto the HA. This is critical for survival. Aggrecan in the transgenic mouse that lacks the link protein diffuses from the cartilage with resulting malformation of limbs and death shortly after birth (Czipri et al. 2003). It is likely that aggrecan and link protein interact in a 1:1 ratio inside the cell sometime during their progression from the endoplasmic reticulum through the golgi and the transport vesicles that carry them to the cell surface. Once they are secreted, they will interact with the extracellular HA molecules to form the stable aggregate structures (Fig. 5.5) (Kimura et al. 1979).

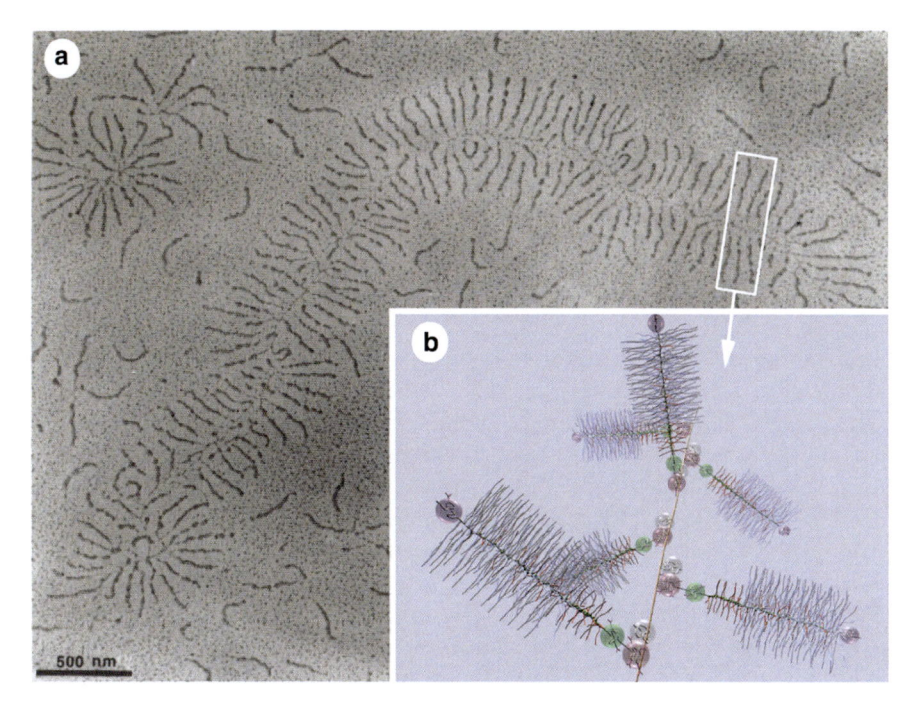

Fig. 5.5 Structure of cartilage proteoglycan aggregates. (a) shows an electron micrograph of proteoglycan aggregate isolated and purified from cartilage. The long thin core of this structure is hyaluronan, and the strands emanating from both sides of this core are aggrecan monomers. (b) shows a model of what exists in (a), indicating the involvement of link protein in stabilizing the interaction of aggrecan with hyaluronan. From Roseman (2001). © The American Society for Biochemistry and Molecular Biology. *Reproduced with permission*

5.3.3 Function of Aggrecan in Cartilage

The extracellular matrix of cartilage is a composite structure with a type II collagen network that contains interspersed, compressed aggrecan aggregates. Typically, the aggrecan macromolecules are compressed to 20–30% of the volume they would occupy in physiological solvents when fully expanded. This increases the charge density of the anionic sulfoesters and carboxyl groups in the glycosaminoglycan side chains. This increased charge density as well as the numerous partially constrained glycosaminoglycan chains creates a swelling pressure on the collagen network that resists compressive loads, a necessary physical property of cartilages.

An example of this property is observed in the brachymorphic mouse. These mice have a partial defect in the synthesis of PAPS, which becomes rate limiting in chondrocytes in growth plates that must synthesize large amounts of aggrecan to form the columnar matrix necessary for endochondral bone formation. Therefore, the aggrecans synthesized by the brachymorphic chondrocytes contain ~15% fewer sulfoesters, and hence fewer negative charges, than aggrecans synthesized by normal chondrocytes. To achieve the same charge density in the tissue, the undersulfated aggrecans occupy a smaller domain, which provides a narrower growth plate. Thus endochondral bone formation results in shorter bones.

5.4 Versican

Versican like aggrecan is a large proteoglycan with a core protein of similar molecular weight but with markedly fewer chondroitin sulfate glycosaminoglycan chains attached to the core protein. Versican was first isolated from the medium of cultured fibroblasts (Coster et al. 1979) and subsequently cloned from placental fibroblasts and named *versican* in recognition of its domain structure and versatility as a highly interactive molecule (Zimmermann and Ruoslahti 1989). Versican is also known as PG-M (Coster et al. 1979; Kimata et al. 1986) and CSPG-2 (Naso et al. 1994). In humans, versican is encoded by a single gene locus on chromosome 5q14.3 (Iozzo et al. 1992) and is 86% identical between mouse and human (Naso et al. 1994), indicating the importance and highly conserved nature of this proteoglycan.

Versican is encoded by 15 exons that are arrayed over 90 kb of continuous genomic DNA (Zimmermann 2000). The central, GAG-bearing domain of versican core protein is coded by two large exons, α-GAG and β-GAG, which can be alternately spliced (Fig. 5.6). Exon 7 codes for the α-GAG region and exon 8 codes for the β-GAG region. When both the entire exons 7 and 8 are present and no splicing occurs, versican V0 is formed. When exon 7 is spliced out, versican V1 is formed. When exon 8 is spliced out, versican V2 is formed. When both exons 7

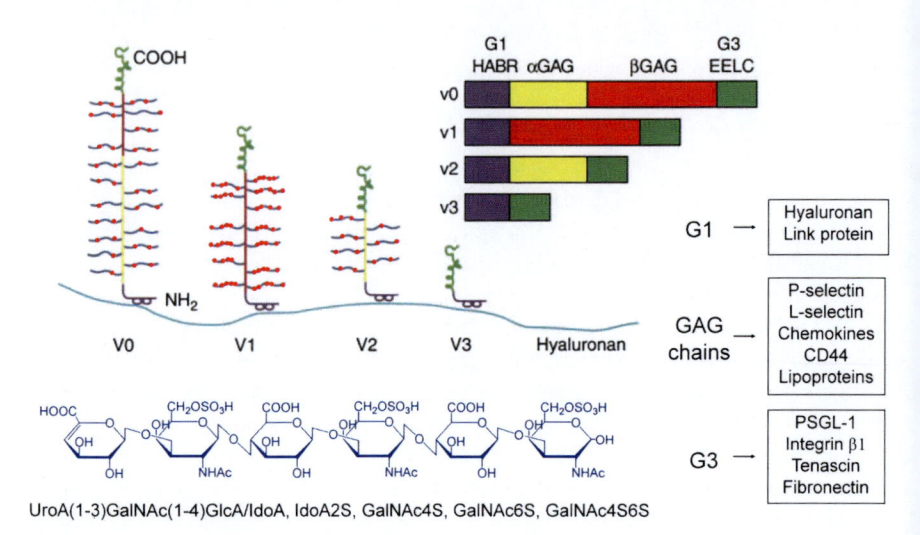

Fig. 5.6 Schematic model of the four variants of versican, indicating the domains within the variants that exhibit selective binding activity. Shown at the *bottom* of the figure is a model of the CS chain attached to versican (details in text). From Wight (2002). *Reproduced with permission*

and 8 are spliced out, versican V3 is formed. This form of alternate splicing gives rise to versican variants that differ in the number of CS chains attached to the consensus sequence attachment sites in the core proteins. Since V3 does not contain any CS chains, it cannot be considered a proteoglycan, but it is frequently grouped with proteoglycans and characterized as such (Zako et al. 1995). It is of interest that while V0, V1, and V3 are found in most tissues, V2 is mostly restricted to the central nervous system (Zimmermann and Dours-Zimmermann 2008). Additional isoform variants have been identified in chick versican (Zako et al. 1997) involving a short sequence near the N-terminus that may specify a keratan sulfate addition sequence (Zimmermann 2000). Recently a new versican isoform, V4 – consisting of the G1 domain, the first 398 aa of the β-GAG region and the G3 domain – has been found to be upregulated in human breast cancer lesions (Kischel et al. 2010). Other isoforms potentially may exist such as a V5 isoform, consisting of essentially only the G1 domain, found by new gene discovery techniques and listed as a reference sequence for mouse versican in Entrez Gene.

The CS GAG chains attached to the different isoforms of versican may differ in size and composition, depending upon the species, the tissue of origin, or the culture conditions that promote versican synthesis. For example, CS chains isolated from versican synthesized by ascending aorta smooth muscle cells have a 6S:4S ratio of 2, which increases to approximately 4 upon PDGF stimulation of the cells (Cardoso et al. 2010; Schönherr et al. 1991). Such stimulation also increases the length of the CS chains attached to versican from 45 to 70k leading to an overall increase in the hydrodynamic size of the proteoglycan. These changes in the structure of CS attached to versican influence the ability of versican to interact with other molecules such as has been shown for low density lipoproteins (Little et al. 2002, 2008). Such results indicate that the machinery that controls the addition of sulfate to the growing CS on versican as well as the machinery that controls CS chain elongation is highly regulated. It is of interest that some aspects of the signaling pathways that control the posttranslational processing of CS chains attached to versican differ from those that control the transcription of versican core protein synthesis (Cardoso et al. 2010).

5.4.1 Regulated Synthesis and Turnover of Versican in Development

The synthesis of versican is highly regulated in different tissues during development [reviewed in Kinsella et al. (2004)]. For example, changes in the expression of versican isoforms characterize the developing brain such that versican V2 replaces versican V0 and V1 during prenatal development (Milev et al. 1998). In vitro studies show that V2 has potent inhibitory properties on axonal growth (Schmalfeldt et al. 2000). However, knockout of V2 in the adult mouse CNS produced no obvious phenotype in the nervous system, and postnatal brain development appears unaffected (Dours-Zimmermann et al. 2009). Such results highlight the possibility that other hyaluronan-binding proteoglycans such as brevican and neurocan (discussed below) may substitute V2 in developmental events. Expression of versican and accumulation occur in the dermal papillae and associated mesenchyme in the skin in a distinct temporal and spatial pattern during hair follicle development implicating versican in hair follicle maturation (du Cros et al. 1995; Kishimoto et al. 1999). Versican also increases with hyaluronan during preovulatory follicular development period and expansion of the cumulus cell oocyte complex during ovulation (Russell and Salustri 2006; Russell et al. 2003a, b). The uterine cervix undergoes changes during pregnancy and labor that transforms it from a closed rigid structure to a dilated distensible structure to allow birth. This involves significant remodeling of the ECM with increases in hyaluronan and versican (Ruscheinsky et al. 2008). Versican expression is also high in the developing mesenchyme during limb development but downregulated during mesenchymal condensation as aggrecan synthesis is upregulated in the prechondrogenic core of developing cartilage (Shinomura et al. 1990, 1993). However, in vitro studies of chondrogenesis reveal a role for versican in mesenchymal condensation and indicate possible involvement of the different splice variants of versican for controlling different aspects of the differentiation process (Kamiya et al. 2006). For example, forced expression of V3 in a chondrogenic cell line disrupts the deposition and organization of V0/V1 and inhibits mesenchymal condensation and chondrogenesis. Whether these effects are seen in vivo awaits future investigations (Kamiya et al. 2006). Transient expression of versican also occurs in migratory pathways of melanoblasts (Perris et al. 1990; Stigson et al. 1997) and in neural crest migratory pathways in the mutant mouse, *Splotch*, which is characterized by a mutation in Pax 3 transcription factor and defective neural crest cell migration (Henderson et al. 1997). These studies demonstrate that versican is a downstream target of Pax 3 and that versican V2 is upregulated while versican V3 is downregulated suggesting that different isoforms of versican can have different effects on cellular behavior (Mayanil et al. 2001). Expression and processing of versican appears important during embryonic stem cell differentiation since the different splice variants of versican are upregulated and deposited along with ADAMTS generated versican fragments during embryoid body formation with localization in the developing mesenchyme consistent with a role in EMT (Choudhary et al. 2007; Nairn et al. 2007; Shukla et al. 2010). While a distinct mechanistic role for these ECM components in stem cell differentiation is not known, it is of interest that changes in the expression of hyaluronan and versican accompany cardiomyocyte differentiation from undifferentiated human embryonic stem cells (Chan et al. 2010), and it will be important how critical these changes can be in controlling terminal cardiomyocyte differentiation.

Versican is also critical to the migration of endocardial cushion tissue cells and in other regions of the developing heart (Mjaatvedt et al. 1998; Yamamura et al. 1997). The expression of versican is seen in areas of the developing myocardium involved in differentiation rather than in zones associated with active and rapid proliferation (Henderson and Copp 1998). It may be that versican plays a role in cardiomyocyte differentiation, perhaps by directing a switch from rapid proliferation to differentiation. Cardiomyocytes undergo terminal differentiation in the ventricle during the neonatal period when versican is completely switched off. This suggests that versican may need to be downregulated before terminal differentiation can take place (Henderson and Copp 1998). The different isoforms of versican may control different aspects of cardiomyocyte differentiation and heart development. For example, Kern et al (2007) recently transduced mouse embryonic cardiomyocytes with V3 and noticed a marked reduction in proliferation of the cardiomyocytes and a significant increase in myocardial cell–cell association. Furthermore, injection of an adenovirus that contained V3 into the heart field of a developing mouse heart led to an increase in the outflow track myocardium and at least a twofold increase in the compact layer of the ventricular myocardium. Such findings indicate that the noncleavable, V3 form of versican, may lead to increase in myocardial cell survival and stabilization of the myocardial cell layer. It is of interest that recent studies show that microRNA-138 represses versican expression in the heart ventricles but allows versican expression in the arterioventricular canal. Knockdown of this microRNA leads to ventricular expansion and abnormal ventricular cardiomyocyte development (Morton et al. 2008). In addition, microRNAs appear to be involved in regulating versican expression and controlling the differentiation of arterial smooth muscle cells (Wang et al. 2010). For example, myocardin is a transcription factor that regulates smooth muscle cell specific gene expression and has been found to stimulate miRNA-143 expression (Wang et al. 2010). This miRNA binds to the 3′ untranslated region of the versican gene, which suppresses versican expression. These findings indicate that part of the smooth muscle cell differentiation program involves decreased versican production. Collectively, these studies highlight the need to fine tune the expression of versican during heart and blood vessel development.

Regulation of versican expression by its promoter sequence has only been partially described. Mutational analysis of the versican promoter coupled to reporter constructs reveal that approximately 650 bases upstream of the ATG start codon are important to its expression (Naso et al. 1994; Rahmani et al. 2005, 2006). Sequence analysis indicates that several transcription factor binding site consensus sequences exist in addition to a classic TATA box sequence; specifically XRE, C/EBP, AP2, SP1, CTF/CBF, TCF/LEF, and CREB regulatory elements. The canonical Wnt/β-catenin pathway, which is critical in early embryogenesis, cell differentiation, and neoplasms (Huang and He 2008; Korswagen and Clevers 1999; Taipale and Beachy 2001), upregulates versican expression (Rahmani et al. 2006). The accumulation of β-catenin and the subsequent formation of a complex with T cell factors (TCF) or lymphoid enhancing factors (LEF) on the versican promoter lead to an increased versican gene expression (Rahmani et al. 2005, 2006).

A number of different growth factors and cytokines regulate versican synthesis and accumulation in a number of different cell types. The growth factors, platelet-derived growth factor (PDGF) and transforming growth factor-beta (TGF-β), up-regulate versican mRNA and core protein synthesis and cause elongation of the CS chains attached to the core protein of versican in ASMCs (Little et al. 2002; Schönherr et al. 1991; Schönherr et al. 1997). Although PDGF stimulates the proliferation and migration of ASMCs, TGF-β1 inhibits ASMC proliferation in vitro suggesting that versican synthesis is not directly causatively linked to proliferative and migratory stimulation. However, interference with versican synthesis in ASMCs, in fibroblasts, and in some cancer cells inhibits the proliferation of these cells suggesting that versican synthesis and accumulation is necessary but not sufficient to cause these changes in cell behavior (Huang et al. 2006; Merrilees et al. 2002). The effect of PDGF on versican transcription appears to be mediated by tyrosine kinases as it is abolished by genestein (Schönherr et al. 1997) and signaling pathways involving PKC and ERK (Cardoso et al. 2010). Positive regulation of versican synthesis by both PDGF and TGF β1 has been reported for a variety of different cell types (Kähäri et al. 1991; Kaji et al. 2000), while some effectors such as angiotensin II affect versican synthesis by ASMCs through accessory growth factor receptors such as EGF by a mechanism of transactivation (Shimizu-Hirota et al. 2001). Mechanical stretch also influences versican expression. For example, biaxial strain applied to ASMCs upregulates versican expression while decreasing decorin expression (Lee et al. 2001). In contrast, fetal lung fibroblasts decrease versican expression when exposed to elevated oxygen levels (Caniggia et al. 1996), while hypoxia induces versican synthesis in monocyte/macrophages (Asplund et al. 2009). While most growth factors and cytokines positively influence versican synthesis, IL 1B decreases versican synthesis in gingival and skin fibroblasts (Ostberg et al. 1995; Qwarnström et al. 1993) and ASMCs (Lemire et al. 2007) while this same growth factor increases versican expression in human lung fibroblasts (Tufvesson and Westergren-Thorsson 2000).

5.4.2 Turnover of Versican

The degradation of versican is associated with several tissue remodeling events including organogenesis, cancer, inflammation, and ovulation (Brown et al. 2006; Carpizo and Iruela-Arispe 2000; Kenagy et al. 2006). Numerous matrix metalloproteinases (MMPs), including MMP-1, -2, -3, -7, and -9, degrade versican (Halpert et al. 1996; Passi et al. 1999; Perides et al. 1995; Sandy 2006).The serine protease plasmin (Kenagy et al. 2002) and at least five ADAMTS metalloproteinases (*a d*isintegrin *a*nd *m*etalloproteinase with *t*hrombo*s*pondin motifs), specifically ADAMTS-1, -4, -5, -9, and -20, also mediate versican proteolysis (Apte 2004, 2009; Koo et al. 2006, 2007; Longpre et al. 2009; Sandy et al. 2001; Silver et al. 2008; Sommerville et al. 2008). ADAMTSs are cell secreted enzymes which belong to the superfamily of zinc-dependent metalloproteinases which degrade ECM molecules (Apte 2004, 2009; Rocks et al. 2008). Originally identified for their aggrecanse activity, ADAMTSs were later shown to degrade versican and gelatin (Jonsson-Rylander et al. 2005; Kuno et al. 2000; Sandy et al. 2001). Of the ADAMTS enzymes, ADAMTS-4 has the highest versican digestive activity – five to tenfold greater when compared to ADAMTS-1 per microgram enzyme (Sandy et al. 2001). All ADAMTS proteases contain consensus sites for cleavage by proprotein convertases (such as furin) at the junction of the propeptides with the catalytic domains (Longpre et al. 2009). These enzymes differ, however, in how and where they are activated. The removal of the propeptide is necessary for the activation of ADAMTS-1 and -4 and occurs intracellularly, while with ADAMTS-5 and -9, propeptide processing occurs extracellularly (Koo et al. 2006; Longpre et al. 2009). For most ADAMTS pro-forms, cleavage by convertase enzymes increases substrate activity; ADAMTS-9 is the exception in that its activity against versican decreases with pro ADAMTS-9 processing (Koo et al. 2007). Within the βGAG domain of versican, there exists an ADAMTS consensus sequence by which versican is cleaved. Cleavage of versican by ADAMTS-1, -4, -5, and -9 leads to production of amino-terminal fragments that are detected using an antibody recognizing the neoepitope sequence DPEAAE (DPE) created by the cleavage (Sandy et al. 2001; Somerville et al. 2003). Notably, areas of high ADAMTS-1 and -4 in the vascular wall neointima correlate with greater versican degradation and production of the amino terminal DPE-containing versican fragment (Sandy et al. 2001). In fact, experiments altering blood flow after neointima formation demonstrated that in the high flow situation, the versican cleavage fragment was increased compared to normal flow conditions (Kenagy et al. 2005). The increase in cleaved versican also correlated with regression of neointimal thickenings and loss of versican (Kenagy et al. 2005). Interesting, these changes in versican integrity also correlated with cell death in these regressed lesions possibly partly regulated by the versicanase ADAMTS4 (Kenagy et al. 2009). These findings are of interest because of recent studies that suggest that cleaved versican may regulate apoptosis during mammalian interdigital web regression (McCulloch et al. 2009). Thus, cleaved fragments of versican can elicit changes to cellular phenotypes, and a balance between intact and cleaved versican may be critical to maintain the appropriate ECM structure for optimal functional activity. Tipping that balance may determine the nature of the pathological changes associated with versican in developing atherosclerotic lesions (Wight and Merrilees 2004).

Recent work has shown that the ADAMTS-9 heterozygote mouse develops anomalies in the aortic wall, valvulosinus, and valve leaflets coincident with versican accumulation at these sites together with lack of versican degradation (Kern et al. 2010). Such observations suggest that failure of versican processing due to a deficiency of ADAMTS 9 may be responsible for these cardiac anomalies. Loss of myocardium from the distal cardiac outlet during development of the heart correlates with reduced levels of versican and increased production of the N-terminal cleavage fragment of versican (Kern et al. 2007). ADAMTS-1 mRNA transcript is also shown to be abundant in human aorta and increases as arterial SMCs migrate and proliferate in vitro (Jonsson-Rylander et al. 2005). Furthermore, high levels of ADAMTS-1 in brain tissues are associated with neurodegenerative diseases such as Down syndrome, Alzheimer's, and Pick's disease (Miguel et al. 2005). These authors suggest that proteoglycan degradation mediated by ADAMTS-1 has a major role in the disease process. Of interest is that ADAMTS-1 is also localized to these versican-enriched regions and appears to cleave versican immediately preceding ovulation suggesting that versican processing may be critical for ovulation to proceed (Russell et al. 2003a, b). While the functional consequences of this cleavage is not known, it may be that cleavage of versican is important to allow the matrix to expand by removing the versican constraints on the expanding hyaluronan matrix during ovulation or on the other hand to rapidly disassemble the COC matrix following ovulation which could aid in sperm penetration of the COC (Russell et al. 2003a, b).

Versican cleavage occurs throughout cardiac development (Kern et al. 2006), during atrioventricular remodeling, and in the growth and compaction of the trabeculae in the ventricular myocardium (Cooley et al. 2008; Stankunas et al. 2008). Recent studies show reduced cleavage of versican in the ADAMTS9 haploinsufficient mouse and cardiac and aortic anomalies suggesting a critical role for versican processing in heart development (Kern et al. 2010). However, the importance of ADAMTS activities on ECM substrates such as versican, and the potential roles of the resulting degradation products in the pathology of diseases, requires further study (Wight 2005). Furthermore, the significance of processing or turnover of versican and other matrix molecules remains to be clarified for natural cellular processes of angiogenesis, cell proliferation, and apoptosis (Gustavsson et al. 2008). Nevertheless, it is clear that degradation of versican is important in development and disease.

5.4.3 Versican Effects on Cell Phenotype

5.4.3.1 Cell Adhesion

A number of in vitro cell biological studies have shown that versican has an effect on cell phenotype. Early studies showed that versican is antiadhesive (Ang et al. 1999; Yamagata et al. 1989, 1993; Yamagata and Kimata 1994), and this activity appears to reside in the G1 domain of versican (Ang et al. 1999; Yang et al. 1999) However, the carboxy-terminal domain of versican interacts with the β1 integrin of glioma cells, activating focal adhesion kinase (FAK), promoting cell adhesion, and preventing apoptosis in this cell type (Ang et al. 1999; Wu et al. 2002; Yang et al. 1999). The proadhesive property of the G3 domain of versican raises the possibility that different breakdown products of versican might differentially affect cell adhesion in different ways. In fact, a number of studies have indicated that isolated subdomains of versican can have profound influence on cell behavior (LaPierre et al. 2007; Sheng et al. 2005; Wu et al. 2004; Yang et al. 2003; Zhang et al. 1999; Zheng et al. 2006). Interestingly, overexpression of V3 which lacks the CS chains, in ASMCs, leads to extreme cell spreading and increased adhesion (Lemire et al. 2002).

5.4.3.2 Cell Proliferation

Versican is also involved in cell proliferation. For example, mitogens such as platelet-derived growth factor (PDGF) upregulate versican expression in arterial

5 Hyaluronan and the Aggregating Proteoglycans 169

smooth muscle cells (Evanko et al. 1999, 2001; Schönherr et al. 1991, 1997) and together with hyaluronan contribute to the expansion of the pericellular ECM that is required for the proliferation and migration of these cells (Evanko et al. 1999, 2001, 2007). These complexes increase the viscoelastic nature of the pericellular matrix, creating a highly malleable extracellular environment that supports a cell-shape change necessary for cell proliferation and migration (Evanko et al. 2007). Furthermore, these versican-enriched macromolecular complexes may have a dramatic effect on the tension exerted on the cells themselves and the traction forces generated by the cell. Such mechanical changes could impact mechanically coupled signaling (Chicurel et al. 1998a, b; Wang et al. 2009a, b). Thus, the versican–hyaluronan complex that surrounds cells serves as an important, but infrequently considered, mechanism for controlling cell shape and cell division. In fact, inhibiting the formation of this pericellular coat blocks the proliferation of arterial smooth muscle cells in response to PDGF (Evanko et al. 1999, 2007).

Another mechanism by which versican could influence proliferation is by acting as a mitogen itself, through the EGF sequences in the G3 domain of the molecule. For example, expression of G3 minigenes in NIH3T3 cells enhances cell proliferation, and the effect can be blocked by deletion of the EGF domains in the G3 construct (Zhang et al. 1999). This same construct exerts a dominant negative effect on cell proliferation through inhibiting the binding of G3 to the cell surface, via the lectin domain in G3 (Wu et al. 2001; Wu et al. 2005) The concentration of versican associated with the cell surface appears to be a critical factor, and loss of versican from the cell surface is associated with decreased cell proliferation. Maximal growth-promoting activity is achieved in NIH3T3 cells and chondrocytes with both G1 and G3 minigene constructs, supporting the concept that versican regulates proliferation by binding directly to a growth factor receptor and by interfering with cell adhesion (Yang et al. 1999; Zhang et al. 1999). Thus, versican expression is associated with a proliferative cell phenotype and is often found in tissues exhibiting elevated proliferation, such as in development and in a variety of tumors (see reviews) (Evanko et al. 2007; Ricciardelli et al. 2009).

5.4.3.3 Cell Migration

Versican is expressed along neural crest pathways and influences neural cell migration (Perissinotto et al. 2000). A number of other studies suggest that versican blocks neural crest migration because cells do not enter tissues that express versican (Henderson and Copp 1997; Landolt et al. 1995). Pax3 is a transcription factor associated with defective neural cell migration. Splotch mice are characterized by mutations in the Pax3 gene and exhibit neural-crest-related abnormalities, including the failure of neural crest cells to colonize target tissues. However, neural crest cells derived from these mutant mice migrate as controls in vitro, so it has been suggested that the defect may not reside in the neural crest cells themselves, but rather in the ECM environment through which they migrate. Indeed, earlier studies (Henderson and Copp 1997) demonstrated that versican was markedly overexpressed in Splotch mutants in neural crest cell migration pathways, suggesting that versican may be responsible for defective cell migration in this species. Recent studies show that overexpression of Pax3 in a medulloblastoma cell line causes upregulation of the V2 splice variant of versican and a downregulation of the V3 variant (Mayanil et al. 2001). Such differential regulation of the versican isoforms may explain, in part, the migratory defect in the Splotch mouse. It is of interest that the V3 isoform lacks chondroitin sulfate chains, which should reduce the exclusionary properties of the ECM.

Versican also appears to plays a role in the migration of embryonic cells in the development of the heart. Versican gene expression occurs at high levels during the development of the heart (Henderson and Copp 1998). Versican is expressed in a chamber-specific manner, with high levels in trabeculations of the right ventricle. In addition, versican is expressed in the endocardial cushion of the atrioventricular and outflow tract regions and in the atrioventricular, semilunar and venous valves. That versican plays an essential role in the development of the heart has been demonstrated by the identification of an insertional transgene mutation in the versican gene in the heart-defect (hdf) mouse (Mjaatvedt et al. 1998). The loss of versican expression in the homozygous hdf is associated with the failure of the endocardial cushion cells to migrate. It is of interest that this endocardial cushion phenotype also resembles the phenotype in the hyaluronan synthase 2 knockout mouse, suggesting that the interaction of versican with hyaluronan is critical to cell movement in this tissue (Camenisch et al. 2000).

Versican influences the migration of a variety of other cell types, and this activity appears to be mostly associated with the antiadhesive activities involving the G1 domain of the molecule. In the nervous system and in the axonal growth, the V2 splice variant inhibits axonal outgrowth and migration (Fidler et al. 1999; Niederost et al. 1999; Schmalfeldt et al. 2000). This inhibiting activity of versican can be reduced, but not eliminated, by removing chondroitin sulfate chains, indicating that multiple domains of versican are involved in controlling axon regeneration. Although the V2 isoform is widely present in the CNS (Yamaguchi 2000), it is predominately localized to the myelinated fiber tracts. Oligodendrocytes are the likely source of V2 (Asher et al. 2002; Milev et al. 1998). How versican inhibits axonal growth remains open. The finding that both the GAGs and core protein domains of the molecule are involved in the inhibitory activity suggests a direct interaction with the cells or modification of the surrounding matrix to form exclusionary boundaries. The fact that versican plays a fundamental role in axonal migration is highlighted by studies that show upregulation of versican along with other hyalectins such as neurocan (see below) following CNS injury (Asher et al. 2002). These changes have been associated with the failure of nerves to regenerate. The importance of the hyalectins in preventing nerve regeneration is highlighted by studies that show that degradation of chondroitin sulfate chains by chondroitinase ABC lyase treatment following spinal cord injury in experimental animals promotes regeneration of both ascending and descending corticospinal-tract axons (Bradbury et al. 2002) (also see below). Such results suggest that manipulating versican synthesis in spinal cord injury may be a useful intervention for therapeutic treatment of this condition. Failure of axons to regenerate is also characteristic of multiple sclerosis, and versican appears to increase in plaques present in the white matter of the brain from patients with multiple sclerosis (Sobel and Ahmed 2001).

5.4.4 Extracellular Matrix Assembly

Versican interacts with several different ECM molecules and, in part, plays a central role in ECM assembly. The domain structure of versican lends itself to multiple types of interactions through either protein–protein or protein–carbohydrate interactions. Perhaps the best known of these interactions involves a specific interaction between the amino-terminal domain of versican (G1) and hyaluronan. The binding of versican to hyaluronan involves a tandem double-loop sequence (two link modules) in the G1 domain of versican and a stretch of five repeat disaccharides in hyaluronan. This interaction is stabilized by another protein – link protein – which exhibits selective binding specificity for both hyaluronan and versican [reviewed in Wight et al. (1991)].

In addition to hyaluronan, versican interacts with other ECM molecules such as tenascin R (Aspberg et al. 1995, 1997; Lundell et al. 2004). Versican interacts with tenascin R through the lectin-binding domain of versican and involves protein–carbohydrate interactions. The lectin-binding domain participates in other ligand interactions as well. For example, versican interacts with fibulin-1 and fibulin-2 (Aspberg et al. 1999; Olin et al. 2001) (also see Chap. 10) a growing family of ECM proteins that are expressed in particularly high levels in the developing heart valve. In adults, however, fibulin-1 and -2 are found associated with microfibrils that are part of elastic fibers. Versican also can interact with proteins associated with elastin in elastic fibers. For example, versican interacts with the elastic-fiber-associated protein fibrillin (Aspberg et al. 1999; Isogai et al. 2002; Olin et al. 2001), and versican has been shown to colocalize with elastic fibers in skin (Isogai et al. 2002). Furthermore, fibrillins bind fibulin 2, and fibulin is preferentially localized to the elastin/microfibril interface in some tissues but not in others (Reinhardt et al. 1996). It may be that fibulin serves as a bridge between versican and fibrillin, forming high-ordered multimolecular structures important in the assembly of elastic fibers. The relationship of versican to elastic fiber assembly is interesting and unusual. For example, rat pup arterial smooth muscle cells have high levels of tropoelastin expression, but no detectable levels of versican synthesis, while the opposite is true for adult arterial smooth muscle cells (Lemire et al. 1994, 1996). Furthermore, it is known that chondroitin sulfate inhibits the formation of elastic fibers, through a mechanism involving interference with the binding of the elastin receptor to the surface of arterial smooth muscle cells (Hinek et al. 1991). Recent studies have shed light on the relationship of versican and elastic fiber assembly. For example, overexpressing the versican splice variant that lacks chondroitin sulfate chains (V3) or inhibiting versican synthesis by antisense RNA dramatically alters arterial smooth muscle cell and fibroblast phenotype by enhancing cell adhesion, decreasing growth and migration, and upregulating tropoelastin expression (Hinek et al. 2004; Huang et al. 2006; Lemire et al. 2002; Merrilees et al. 2002). Placement of V3-transduced arterial smooth muscle cells into injured blood vessels results in the formation of multiple elastic laminae (Merrilees et al. 2002) during injury repair. Thus, it may be that overexpressing the form of versican that lacks chondroitin sulfate chains competes for binding sites with versican molecules that contain chondroitin sulfate chains associated with hyaluronan on the cell surface. This would allow the elastin-binding protein to associate with the cell surface and promote elastic fiber assembly (Hinek et al. 1991; Hinek and Wilson 2000; Merrilees et al. 2002).

5.4.5 Versican in Inflammation and Disease

Inflammatory responses require the emigration of leukocytes from the vasculature into damaged underlying tissue areas as part of the innate immune response. Upon extravasation into the subendothelial compartment, leukocytes encounter the ECM, which functions as a scaffold for cell adhesion, migration, activation, and retention (Vaday and Lider 2000; Vaday et al. 2001). For example, specific components of the ECM such as versican can interact with chemokines, growth factors, proteases, and receptors on the surface of the immune cells to provide intrinsic signals and influence immune cell phenotype (Hirose et al. 2001; Taylor and Gallo 2006). A number of studies have shown that fragments of ECM (Adair-Kirk and Senior 2008), such as versican, exhibit proinflammatory properties. For example, the G3 domain of versican can interact with P selectin glycoprotein-1 (PSGL-1) on the surface of macrophages and cause macrophage aggregation (Zheng et al. 2004). Exciting recent studies demonstrate that versican can interact with macrophage TLR2 to induce secretion of inflammatory cytokines, such as tumor necrosis factor-α (TNFα) and IL-1-6 (Kim et al. 2009). It is of interest that highly sulfated CS GAG chains on versican (Kawashima et al. 2002) may be critical to promote inflammatory cytokine release and activity (Kawashima et al. 2002; Li et al. 2008). Furthermore, once bound to the versican-containing ECM, leukocytes may degrade the ECM to generate proinflammatory fragments that further drive the inflammatory response (Schor et al. 2000; Vaday and Lider 2000) through effects on a variety of inflammatory and immune cell regulatory processes (Adair-Kirk and Senior 2008; Arroyo and Iruela-Arispe 2010). Extravasation of monocytes depends on the nature of the ECM, which in turn influences macrophage phenotype. Recent studies show that inflammatory cells such as monocytes interact with specific components of the ECM, such as hyaluronan (see above) and versican (Potter-Perigo et al. 2009). In fact, high resolution confocal microscopy show that these two components are organized into discrete ECM filaments that emanate from the cell surface and form a matrix that binds myeloid cells (Evanko et al. 2009). Furthermore, binding of myeloid cells to this matrix depends both on the presence of hyaluronan and versican (Potter-Perigo et al. 2009) Thus, versican may be part of proinflammatory ECM (Gill et al. 2010) and a useful and novel therapeutic target to control the immune response associated with inflammation in a variety of diseases (Jarvelainen et al. 2009).

5.4.5.1 Versican in Diseases of the Eye

Versican is a normal ECM component of the trabecular meshwork and ciliary muscle of the human eye (Miyamoto et al. 2005; Zhao and Russell 2005). Versican is believed to maintain the structure of the vitreous body in the human eye by keeping the collagen molecules apart (Bishop 2000). Wagner syndrome is a hereditary vitroretinopathy that maps to chromosome 5q13–q14 and is associated with mutations in *CSPG2* encoding versican (Kloeckener-Gruissem et al. 2006; Miyamoto et al. 2005; Mukhopadhyay et al. 2006; Ronan et al. 2009; Zhao and Russell 2005). Mutations have been found in introns 7 and 8 leading to reduced expression of the V0 and V1 forms of versican and increase in the V2 and V3 forms (Mukhopadhyay et al. 2006). This imbalance in the variants of versican is believed to lead to the pathology, but the actual mechanism by which altered versican causes this retinopathy is not known.

5.4.5.2 Versican in Cancer

Tumor cells also express versican such as that seen in malignant melanoma (Touab et al. 2002), and upregulation of versican expression in highly metastatic and invasive human tumor cells is regulated by several transcription factors such as AP-1, Sp-1, AP-2, and two TCF-4 sites. Promoter activation requires ERK/MAPK and JNK signaling pathways acting on the AP-1 site (Domenzain-Reyna et al. 2009). Such results may indicate a link between the superactivation of ERK that has been tied to malignant melanoma (Gorden et al. 2003; Maldonado et al. 2003) and the production of a versican-rich ECM by the tumor cells to promote their own ability to metastasize. This study demonstrated also that there was cross talk shown between ERK and β-catenin involvement in versican upregulation in the highly metastatic melanoma cells, indicating once again for a central role of the Wnt/β-catenin pathway in regulating versican expression in this epithelial cancer (Domenzain-Reyna et al. 2009). In addition, a recent study shows that versican can activate tumor infiltrating myeloid cells through toll 2 and its coreceptors TLR6 and CD14, which can elicit the production of proinflammatory cytokines including TNF alpha that enhance tumor metastasis in an animal model of Lewis lung carcinoma (Kim et al. 2009; Wang et al. 2009a, b). This study combined with an earlier study shows that CS isolated from highly metastatic Lewis Lung carcinoma cells have a higher proportion of their CS as highly sulfated CSE compared to low metastatic lung carcinoma cells (Li et al. 2008). This suggests that versican may carry different proportions of highly sulfated CS chains capable of interacting with Toll 2 on myeloid cells promoting inflammation and driving metastasis. This hypothesis awaits further testing. Versican expression by prostate stromal cells is regulated by binding the androgen receptor to the proximal promoter of the versican gene (Read et al. 2007), and β-catenin is required for androgen receptor driven transcription of versican in these cells. This study identifies a novel role for β-catenin in nuclear hormone receptor-mediated transcription in prostate stromal cells and may be a central axis as to why versican accumulates in prostate cancer (Ricciardelli et al. 1998, 2009).

5.4.5.3 Versican in Cardiovascular Disease

Versican is a major ECM component that accumulates throughout early and late human coronary atherosclerotic lesions in defined locations such as in early pathologic intimal thickenings, associated with lipid and macrophage accumulation and at the plaque thrombus interface (Chung et al. 2002; Farb et al. 2004; Geary et al. 1998; Gutierrez et al. 1997; Kolodgie et al. 2002; Lin et al. 1996; Wight et al. 1997; Wight and Merrilees 2004). Versican also accumulates in human lesions that develop in carotid (Formato et al. 2004) and cerebral arteries (Hara et al. 2009). In addition, proteolytic cleavage products of versican are present in human plaques from endarterectomy segments consistent with their generation in a proinflammatory microenvironment (Formato et al. 2004). In fact versican content of the artery has been linked to propensity of the artery to occlude when used in grafting procedures (Merrilees et al. 2001). In addition to the human studies, versican has been identified as major ECM component in atherosclerotic lesions in experimental animals including atherosclerotic lesions in the mouse (Jonsson-Rylander et al. 2005; Karra et al. 2005; Seidelmann et al. 2008; Strom et al. 2006). Mechanistically, versican may play multiple roles in promoting atherogenesis in that it influences lipid retention within the blood vessel wall (Williams and Tabas 1995), regulates arterial smooth muscle cell proliferation and migration (Wight and Merrilees 2004), interacts with proinflammatory leukocytes such as macrophages (Evanko et al. 2009; Potter-Perigo et al. 2009), and influences coagulation and thrombosis (Mazzucato et al. 2002; McGee and Wagner 2003; Zheng et al. 2006).

5.5 Neurocan and Brevican

Neurocan is a major component of brain ECM that forms link stabilized aggregates with hyaluronan (Margolis et al. 1996). It has a calculated molecular size of approximately 300 kDa, and after chondroitinase ABC digestion, it runs at 245 kDa on SDS-PAGE. The cDNA-deduced sequence gives a core of 133 kDa (Rauch et al. 1992). The difference between actual and calculated core protein size is no doubt due to the presence of N-linked oligosaccharides not removed by the chondroitinase digestion and the slow electrophoretic migration of glycosylated proteins due to decreased binding to SDS-PAGE. There are seven potential glyosaminoglycan binding sites on the core protein, but calculations indicate that only three of those sites are occupied by chondroitin sulfate chains (Rauch et al. 1992). Neurocan is expressed by neurons under normal physiological conditions, but it is also expressed by astrocytes after brain injury (Haas et al. 1999). Surprisingly, inactivation of the neurocan gene in mice does not lead to an altered phenotype.

Brevican is restricted to the central nervous system and is not found in peripheral nervous tissue (Hartmann and Maurer 2001). Different forms of brevican exist due to alternative splicing that generates a soluble form and a GPI-linked form (Seidenbecher et al. 1995a, b). The full-length cDNA sequence predicts a molecular mass of 99,510 Da with 3–4 glycosaminoglycan attachment sites (Yamada et al. 1994). A lower molecular weight 80 kDa form exists as an N-terminally truncated cleavage of the 145-kDa form (Yamada et al. 1994). In adult rat brain, the majority of the brevican core protein is of the 80-kDa cleaved form, while no cleaved form is found in the brain of newborn rats (Yamaguchi 1996). Brevican also exists as a C-terminally truncated splice variant which lacks the entire C-terminal domain that is replaced by an attachment sequence for a GPI anchor. The core proteins can be synthesized and secreted without CS chains (Yamada et al. 1994; Yamaguchi 1996). Brevican structure is further complicated by a number of variants that differ in the amount of glycosylation, which may determine its capacity to interact with membranes and/or other proteins (Viapiano et al. 2003). The roles for brevican in normal developing brain include regulation of cell adhesion and neurite outgrowth and involvement in synaptic plasticity (Miura et al. 2001; Yamada et al. 1997). Brevican is markedly upregulated during ventricular brain development coincident with gliogenesis (Jaworski et al. 1995).

The brain has an unusual ECM, which contains a large number of different proteoglycans and hyaluronan while lacking most of the common ECM proteins, such as collagens and fibronectin. In fact, hyaluronan can be organized into fiber-like structures in the brain and serve as a backbone forming aggregates with CSPGs, such as neurocan (Baier et al. 2007). Such structures may guide the migration of neuronal precursors and other cells during development. Also, such an organization of hyaluronan may serve to guide the diffusion of hyalectins and aid in establishing concentration gradients of hyalectins to influence development events such as axon guidance by virtue of their ability to bind guidance cues such a growth factors and cytokines (Kappler et al. 2009). Interestingly, in the vitreous humor of the eye, where hyaluronan is abundant, a similar fiber-like organization of hyaluronan was observed (Zhang et al. 2004).

Both neurocan and brevican are expressed in the nervous system in highly specific manners. For example, neurocan is expressed early in the developing brain, whereas brevican tends to be expressed later and during the postnatal period (Yamaguchi 2000; Zimmermann and Dours-Zimmermann 2008). Thus, during the development of the brain, the matrix takes on two different characteristics, an early "juvenile matrix" type in which neurocan is expressed by neurons, and other hyalectins such as aggrecan and versican V0 and V1 are expressed by astroglial lineage cells along with a high content of hyaluronan. Following birth, "this juvenile matrix" is replaced by a more mature condensed matrix which includes appearance of different CSPGs, including the hyalectins brevican and versican V2 (Rauch 2004; Zimmermann and Dours-Zimmermann 2008). The juvenile more open matrix is thought to promote events such as neuronal migration, whereas the mature more condensed matrix is the stable matrix that persists through adulthood. The appearance of increased amounts of brevican as the matrix matures has its origin in the astrocytes.

5.5.1 Perineuronal Net

Neurocan and brevican are present in the perineruonal net (PNN) which is a lattice-like ECM that surrounds nerve cell bodies, proximal dendrites, and a specific subset of axons. This net-like matrix is a collection of a number of different components, including hyaluronan, and is an excellent example of a condensed matrix. Most of the components of the PNN seem to be expressed by the neurons themselves, but surrounding astrocytes may contribute neurocan and brevican (Carulli et al. 2007). PNNs are formed late in development, and they selectively enwrap large neurons to create a suitable polyanionic microenvironment. They are thought to function as first local buffers for strong variations in the extracellular cationic concentrations (Hartig et al. 1999) and/or as a protective shield for the neuron against oxidative stress (Morawski et al. 2004). Other studies suggest that they may contribute stabilization and electrical insulation, as well as supply necessary key regulatory molecules to maintain synapses and neurotransmission properties of the CNS (Bruckner et al. 1993; Celio et al. 1998). Electrophysiological studies of brevican-deficient mice reveal a phenotype of significantly impaired synaptic plasticity (Brakebusch et al. 2002).

5.5.2 Regulated Synthesis and Turnover

The dynamic changes in the brain ECM that take place from embryonic and early postnatal stages to their mature form in the adult must be due to a combination of differential expression (Friedlander et al. 1994; Katoh-Semba et al. 1998; Tuttle et al. 1998) and turnover. Neurocan and brevican exhibit important differences in the spatiotemporal expression in the CNS. Neurocan expression peaks during early development along with versican, whereas brevican and aggrecan are expressed at low levels during early development, but increase markedly in the adult CNS [reviewed in Viapiano and Matthews (2006)]. While differences do exist with the expression of these lecticans in the CNS, there appears to be functional redundancy as demonstrated by the lack of a major neuroglial alteration in animals that are deficient in one or more of these hyalectins (Rauch et al. 2005). During development, amounts of both the soluble and phosphatidylinositol-linked brevican

increase suggesting a role in the differentiating nervous system (Seidenbecher et al. 1995a, b, 1998).
Like aggrecan and versican, the ADAMTS family of proteins degrade brevican (Matthews et al. 2000; Nakada et al. 2005; Nakamura et al. 2000) at a site located next to the globular NH_2 terminal domain, which allows for the selective release of the large GAG carrying protein region of the proteoglycan (Apte 2009; Sandy 2006). ADAMTS 4 and 5 degrade brevican (Nakada et al. 2005; Nakamura et al. 2000). Neurocan, on the other hand, has not been shown to be cleaved by the ADAMTS enzymes, but can be cleaved by MMP2 (Zimmermann and Dours-Zimmermann 2008).

5.5.3 Brain Injury

The brain ECM is quite stable with low turnover. However, when the adult CNS is injured, this ECM changes dramatically. Several components of this ECM are upregulated including neurocan and brevican (Beggah et al. 2005; Yamaguchi 2000; Zimmermann and Dours-Zimmermann 2008) which eventually contributes to a glial scar. These components are produced by a variety of cells such as reactive astrocytes, oligodendrocyte precursors, microglia/macrophages, and eventually by mesangial cells. The matrix that forms in these lesions resembles the juvenile matrix type. The formation of the glial scar results from an astrocytic response to CNS injury. This reactive gliosis is rich in CSPGs, and these molecules act as a barrier and inhibit re-extension of axons from undamaged areas (Fawcett and Asher 1999).The hyalectins deposit in a dense fashion around the lesion site (Silver and Miller 2004) and are identified as inhibiting growing axons into and through the scar (Morgenstern et al. 2002; Properzi and Fawcett 2004). Neurocan is strongly upregulated after injury (Asher et al. 2000; Matsui et al. 2002; McKeon et al. 1999) and remains elevated for sometime following injury, and this is thought to contribute to the inhibitory ECM (Viapiano and Matthews 2006). Brevican, on the other hand, decreases during the early phases of injury but increases strongly during the later astroglial infiltration stages.

Since neurocan is a key molecule involved in axonal guidance (Friedlander et al. 1994; Katoh-Semba et al. 1998; Tuttle et al. 1998), removing the chondroitin sulfate chains from neurocan and other chondroitin sulfate proteoglycans in this scar appears to have beneficial effects by promoting axonal regeneration (Bradbury et al. 2002; McKeon et al. 1995; Snow et al. 1990). Interestingly, other attempts to inhibit chondroitin sulfate containing neurocan and other CSPGs in response to injury have been used such as interference with chondroitin sulfate elongation by targeting chondroitin sulfate polymerizing factor (Laabs et al. 2007) using siRNA. There is some evidence that there are differences in the nature of the chondroitin sulfate chains produced following brain injury. Whereas, chondroitin sulfate A and C predominate in normal and uninjured CNS (Properzi et al. 2005), oversulfated types, such as chondroitin sulfate D and -E, have been found following injury (Dobbertin et al. 2003; Gilbert et al. 2005). The functional significance of these changes is not clear but could play a role in binding specific cytokines and growth factors and influencing the necessary signaling pathways to allow cell migration and extension.

5.5.4 Role in Cancer

Malignant gliomas are the most common and deadly form of primary brain tumors due to their invasion of normal neural tissue that makes them impossible to completely eliminate. Upregulation and cleavage of brevican is a necessary step in mediating the promotion of glioma invasion. Upregulation of brevican cleavage products has been observed in human gliomas, and overexpression of these fragments increases glioma cell mobility in vitro as well as tumor progression in vivo (Nutt et al. 2001; Viapiano et al. 2008; Zhang et al. 1998). Proteolytic digestion of matrix components and subsequent invasion of the tissue are tightly linked to angiogenesis, a critical step in the progression of brain tumors. Current evidence suggests ADAMTS 4 and 5 are likely candidates. They are upregulated in gliomas (Held-Feindt et al. 2006; Matthews et al. 2000; Nakada et al. 2005). Interestingly, brevican cleavage products bind other ECM proteins such as fibronectin which in turn promotes cell adhesion and mobility of the glioma cells (Hu et al. 2008). Thus, although brevican is highly upregulated in glioma and possibly contributing to the production of an ECM that resists invasion, brevican is also broken down quickly and fragments of brevican interact with other ECM proteins such as fibronectin to stabilize and organize a matrix that becomes proadhesive and promigratory.

5.6 Concluding Remarks

The aggregating proteoglycans together with hyaluronan are critical components in maintaining tissue integrity and homeostasis. As evident from the studies reviewed in this chapter, they play enormous and varied roles in all tissues. Not only are these components critical for maintaining tissue structure by their capacity to form higher ordered molecular structures, but also their capacity to interact with cells and influence cell phenotype makes them ideal candidates for targets when considering new pharmacotherapeutic drugs in the future. Outstanding progress has been made in the understanding of these complex ECM components over the years but we need to continue to probe the direct and/or indirect molecular mechanisms involved in their capacity to determine cell fate and cell behavior. It has become increasing clear that not only do these ECM components possess unique biological activity as intact molecules and complexes but perhaps even more interesting fragments or breakdown products of these components have other unique biological activities as well. Such discoveries open up new avenues for future research. Indeed, the stage has been set for exciting new developments in the future.

Acknowledgements This chapter was prepared with grant support from the National Heart, Lung and Blood Institute of the National Institutes of Health (#18645, 5RO1HL064387 to TNW) and from the National Cancer Institute (R01 CA073839 and R01 CA082867 to BPT). The authors wish to thank Dr. Virginia M. Green for careful editing and Dr. Michael G. Kinsella for helpful discussions in the preparation of this manuscript.

References

Adair-Kirk TL, Senior RM (2008) Fragments of extracellular matrix as mediators of inflammation. Int J Biochem Cell Biol 40:1101–1110

Ang LC, Zhang Y, Cao L, Yang BL, Young B, Kiani C, Lee V, Allan K, Yang BB (1999) Versican enhances locomotion of astrocytoma cells and reduces cell adhesion through its G1 domain. J Neuropathol Exp Neurol 58:597–605

Apte SS (2004) A disintegrin-like and metalloprotease (reprolysin type) with thrombospondin type 1 motifs: the ADAMTS family. Int J Biochem Cell Biol 36:981–985

Apte SS (2009) A disintegrin-like and metalloprotease (reprolysin-type) with thrombospondin type 1 motif (ADAMTS) superfamily: functions and mechanisms. J Biol Chem 284:31493–31497

Arroyo AG, Iruela-Arispe ML (2010) Extracellular matrix, inflammation, and the angiogenic response. Cardiovasc Res 86:226–235

Asher RA, Morgenstern DA, Fidler PS, Adcock KH, Oohira A, Braistead JE, Levine JM, Margolis RU, Rogers JH, Fawcett JW (2000) Neurocan is upregulated in injured brain and in cytokine-treated astrocytes. J Neurosci 20:2427–2438

Asher RA, Morgenstern DA, Shearer MC, Adcock KH, Pesheva P, Fawcett JW (2002) Versican is upregulated in CNS injury and is a product of oligodendrocyte lineage cells. J Neurosci 22:2225–2236

Aspberg A, Binkert C, Ruoslahti E (1995) The versican C-type lectin domain recognizes the adhesion protein tenascin-R. Proc Natl Acad Sci USA 92:10590–10594

Aspberg A, Miura R, Bourdoulous S, Shimonaka M, Heinegård D, Schachner M, Ruoslahti E, Yamaguchi Y (1997) The C-type lectin domains of lecticans, a family of aggregating chondroitin sulfate proteoglycans, bind tenascin-R by protein-protein interactions independent of carbohydrate moiety. Proc Natl Acad Sci USA 94:10116–10121

Aspberg A, Adam S, Kostka G, Timpl R, Heinegard D (1999) Fibulin-1 is a ligand for the C-type lectin domains of aggrecan and versican. J Biol Chem 274:20444–20449

Asplund A, Stillemark-Billton P, Larsson E, Rydberg EK, Moses J, Hulten LM, Fagerberg B, Camejo G, Bondjers G (2009) Hypoxic regulation of secreted proteoglycans in macrophages. Glycobiology 20:33–40

Avigdor A, Goichberg P, Shivtiel S, Dar A, Peled A, Samira S, Kollet O, Hershkoviz R, Alon R, Hardan I, Ben-Hur H, Naor D, Nagler A, Lapidot T (2004) CD44 and hyaluronic acid cooperate with SDF-1 in the trafficking of human CD34+ stem/progenitor cells to bone marrow. Blood 103:2981–2989

Aytekin M, Comhair SA, de la Motte C, Bandyopadhyay SK, Farver CF, Hascall VC, Erzurum SC, Dweik RA (2008) High levels of hyaluronan in idiopathic pulmonary arterial hypertension. Am J Physiol Lung Cell Mol Physiol 295:L789–799

Baier C, Baader SL, Jankowski J, Gieselmann V, Schilling K, Rauch U, Kappler J (2007) Hyaluronan is organized into fiber-like structures along migratory pathways in the developing mouse cerebellum. Matrix Biol 26:348–358

Bakkers J, Kramer C, Pothof J, Quaedvlieg NE, Spaink HP, Hammerschmidt M (2004) Has2 is required upstream of Rac1 to govern dorsal migration of lateral cells during zebrafish gastrulation. Development 131:525–537

Beggah AT, Dours-Zimmermann MT, Barras FM, Brosius A, Zimmermann DR, Zurn AD (2005) Lesion-induced differential expression and cell association of Neurocan, Brevican, Versican V1 and V2 in the mouse dorsal root entry zone. Neuroscience 133:749–762

Bishop PN (2000) Structural macromolecules and supramolecular organisation of the vitreous gel. Prog Retin Eye Res 19:323–344

Bollyky PL, Lord JD, Masewicz SA, Evanko SP, Buckner JH, Wight TN, Nepom GT (2007) Cutting edge: high molecular weight hyaluronan promotes the suppressive effects of CD4+CD25+ regulatory T cells. J Immunol 179:744–747

Bollyky PL, Falk BA, Wu RP, Buckner Thomas NWJH, Nepom GT (2009) Intact extracellular matrix and the maintenance of immune tolerance: high molecular weight hyaluronan promotes persistence of induced CD4+CD25+ regulatory T cells. J Leukoc Biol 86:567–572

Bourguignon LY (2009) Hyaluronan-mediated CD44 interaction iwth receptor and non-receptor kinases promotes oncogenic signaling, cytoskeleton activation and tumor progression. In: Stern R (ed) Hyaluronan in cancer biology. Academic, San Diego, pp 89–107

Bourguignon LY, Singleton PA, Diedrich F, Stern R, Gilad E (2004) CD44 interaction with Na+-H+ exchanger (NHE1) creates acidic microenvironments leading to hyaluronidase-2 and cathepsin B activation and breast tumor cell invasion. J Biol Chem 279:26991–27007

Bradbury EJ, Moon LD, Popat RJ, King VR, Bennett GS, Patel PN, Fawcett JW, McMahon SB (2002) Chondroitinase ABC promotes functional recovery after spinal cord injury. Nature 416:636–640

Brakebusch C, Seidenbecher CI, Asztely F, Rauch U, Matthies H, Meyer H, Krug M, Bockers TM, Zhou X, Kreutz MR, Montag D, Gundelfinger ED, Fassler R (2002) Brevican-deficient mice display impaired hippocampal CA1 long-term potentiation but show no obvious deficits in learning and memory. Mol Cell Biol 22:7417–7427

Brown HM, Dunning KR, Robker RL, Pritchard M, Russell DL (2006) Requirement for ADAMTS-1 in extracellular matrix remodeling during ovarian folliculogenesis and lymphangiogenesis. Dev Biol 300:699–709

Bruckner G, Brauer K, Hartig W, Wolff JR, Rickmann MJ, Derouiche A, Delpech B, Girard N, Oertel WH, Reichenbach A (1993) Perineuronal nets provide a polyanionic, glia-associated form of microenvironment around certain neurons in many parts of the rat brain. Glia 8:183–200

Camenisch TD, Spicer AP, Brehm-Gibson T, Biesterfeldt J, Augustine ML, Calabro A Jr, Kubalak S, Klewer SE, McDonald JA (2000) Disruption of hyaluronan synthase-2 abrogates normal cardiac morphogenesis and hyaluronan-mediated transformation of epithelium to mesenchyme. J Clin Invest 106:349–360

Caniggia I, Liu J, Kuliszewski M, Tanswell AK, Post M (1996) Fetal lung fibroblasts selectively down-regulate proteoglycan synthesis in response to elevated oxygen. J Biol Chem 271: 6625–6630

Cardoso LE, Little PJ, Ballinger ML, Chan CK, Braun KR, Potter-Perigo S, Bornfeldt KE, Kinsella MG, Wight TN (2010) Platelet-derived growth factor differentially regulates the expression and post-translational modification of versican by arterial smooth muscle cells through distinct protein kinase C and extracellular signal-regulated kinase pathways. J Biol Chem 285:6987–6995

Carpizo D, Iruela-Arispe ML (2000) Endogenous regulators of angiogenesis–emphasis on proteins with thrombospondin–type I motifs. Cancer Metastasis Rev 19:159–165

Carulli D, Rhodes KE, Fawcett JW (2007) Upregulation of aggrecan, link protein 1, and hyaluronan synthases during formation of perineuronal nets in the rat cerebellum. J Comp Neurol 501:83–94

Celio MR, Spreafico R, De Biasi S, Vitellaro-Zuccarello L (1998) Perineuronal nets: past and present. Trends Neurosci 21:510–515

Chan CK, Rolle MW, Potter-Perigo S, Braun KR, Van Biber BP, Laflamme MA, Murry CE, Wight TN (2010) Differentiation of cardiomyocytes from human embryonic stem cells is accompanied by changes in the extracellular matrix production of versican and hyaluronan. J Cell Biochem 111(3):585–596

Chicurel ME, Chen CS, Ingber DE (1998a) Cellular control lies in the balance of forces. Curr Opin Cell Biol 10:232–239

Chicurel ME, Singer RH, Meyer CJ, Ingber DE (1998b) Integrin binding and mechanical tension induce movement of mRNA and ribosomes to focal adhesions. Nature 392:730–733

Choudhary M, Zhang X, Stojkovic P, Hyslop L, Anyfantis G, Herbert M, Murdoch AP, Stojkovic M, Lako M (2007) Putative role of hyaluronan and its related genes, HAS2 and RHAMM, in human early preimplantation embryogenesis and embryonic stem cell characterization. Stem Cells 25: 3045–3057

Chung IM, Gold HK, Schwartz SM, Ikari Y, Reidy MA, Wight TN (2002) Enhanced extracellular matrix accumulation in restenosis of coronary arteries after stent deployment. J Am Coll Cardiol 40:2072–2081

Colone M, Calcabrini A, Toccacieli L, Bozzuto G, Stringaro A, Gentile M, Cianfriglia M, Ciervo A, Caraglia M, Budillon A, Meo G, Arancia G, Molinari A (2008) The multidrug transporter P-glycoprotein: a mediator of melanoma invasion? J Invest Dermatol 128:957–971

Cooley MA, Kern CB, Fresco VM, Wessels A, Thompson RP, McQuinn TC, Twal WO, Mjaatvedt CH, Drake CJ, Argraves WS (2008) Fibulin-1 is required for morphogenesis of neural crest-derived structures. Dev Biol 319:336–345

Coster L, Carlstedt I, Malmstrom A (1979) Isolation of 35S- and 3H-labelled proteoglycans from cultures of human embryonic skin fibroblasts. Biochem J 183:669–681

Czipri M, Otto JM, Cs-Szabo G, Kamath RV, Vermes C, Firneisz G, Kolman KJ, Watanabe H, Li Y, Roughley PJ, Yamada Y, Olsen BR, Glant TT (2003) Genetic rescue of chondrodysplasia and the perinatal lethal effect of cartilage link protein deficiency. J Biol Chem 278:39214–39223

de La Motte CA, Hascall VC, Calabro A, Yen-Lieberman B, Strong SA (1999) Mononuclear leukocytes preferentially bind via CD44 to hyaluronan on human intestinal mucosal smooth muscle cells after virus infection or treatment with poly(I.C). J Biol Chem 274:30747–30755

de La Motte CA, Hascall VC, Drazba J, Bandyopadhyay SK, Strong SA (2003) Mononuclear leukocytes bind to specific hyaluronan structures on colon mucosal smooth muscle cells treated with polyinosinic acid:polycytidylic acid: inter-α-trypsin inhibitor is crucial to structure and function. Am J Pathol 163:121–133

Deb TB, Datta K (1996) Molecular cloning of human fibroblast hyaluronic acid-binding protein confirms its identity with P-32, a protein co-purified with splicing factor SF2. Hyaluronic acid-binding protein as P-32 protein, co-purified with splicing factor SF2. J Biol Chem 271:2206–2212

Dobbertin A, Rhodes KE, Garwood J, Properzi F, Heck N, Rogers JH, Fawcett JW, Faissner A (2003) Regulation of RPTPbeta/phosphacan expression and glycosaminoglycan epitopes in injured brain and cytokine-treated glia. Mol Cell Neurosci 24:951–971

Domenzain-Reyna C, Hernandez D, Miquel-Serra L, Docampo MJ, Badenas C, Fabra A, Bassols A (2009) Structure and regulation of the versican promoter: the versican promoter is regulated by AP-1 and TCF transcription factors in invasive human melanoma cells. J Biol Chem 284:12306–12317

Dours-Zimmermann MT, Maurer K, Rauch U, Stoffel W, Fassler R, Zimmermann DR (2009) Versican V2 assembles the extracellular matrix surrounding the nodes of ranvier in the CNS. J Neurosci 29:7731–7742

du Cros DL, LeBaron RG, Couchman JR (1995) Association of versican with dermal matrices and its potential role in hair follicle development and cycling. J Invest Dermatol 105:426–431

Duterme C, Mertens-Strijthagen J, Tammi M, Flamion B (2009) Two novel functions of hyaluronidase-2 (Hyal2) are formation of the glycocalyx and control of CD44-ERM interactions. J Biol Chem 284:33495–33508

Evanko SP, Angello JC, Wight TN (1999) Formation of hyaluronan- and versican-rich pericellular matrix is required for proliferation and migration of vascular smooth muscle cells. Arterioscler Thromb Vasc Biol 19:1004–1013

Evanko SP, Johnson PY, Braun KR, Underhill CB, Dudhia J, Wight TN (2001) Platelet-derived growth factor stimulates the formation of versican-hyaluronan aggregates and pericellular matrix expansion in arterial smooth muscle cells. Arch Biochem Biophys 394:29–38

Evanko SP, Tammi MI, Tammi RH, Wight TN (2007) Hyaluronan-dependent pericellular matrix. Adv Drug Deliv Rev 59:1351–1365

Evanko SP, Potter-Perigo S, Johnson PY, Wight TN (2009) Organization of hyaluronan and versican in the extracellular matrix of human fibroblasts treated with the viral mimetic poly I:C. J Histochem Cytochem 57:1041–1060

Farb A, Kolodgie FD, Hwang JY, Burke AP, Tefera K, Weber DK, Wight TN, Virmani R (2004) Extracellular matrix changes in stented human coronary arteries. Circulation 110:940–947

Fawcett JW, Asher RA (1999) The glial scar and central nervous system repair. Brain Res Bull 49:377–391

Fidler PS, Schuette K, Asher RA, Dobbertin A, Thornton SR, Calle-Patino Y, Muir E, Levine JM, Geller HM, Rogers JH, Faissner A, Fawcett JW (1999) Comparing astrocytic cell lines that are inhibitory or permissive for axon growth: the major axon-inhibitory proteoglycan is NG2. J Neurosci 19:8778–8788

Firan M, Dhillon S, Estess P, Siegelman MH (2006) Suppressor activity and potency among regulatory T cells is discriminated by functionally active CD44. Blood 107:619–627

Formato M, Farina M, Spirito R, Maggioni M, Guarino A, Cherchi GM, Biglioli P, Edelstein C, Scanu AM (2004) Evidence for a proinflammatory and proteolytic environment in plaques from endarterectomy segments of human carotid arteries. Arterioscler Thromb Vasc Biol 24:129–135

Friedlander DR, Milev P, Karthikeyan L, Margolis RK, Margolis RU, Grumet M (1994) The neuronal chondroitin sulfate proteoglycan neurocan binds to the neural cell adhesion molecules Ng-CAM/L1/NILE and N-CAM, and inhibits neuronal adhesion and neurite outgrowth. J Cell Biol 125:669–680

Fulop C, Szanto S, Mukhopadhyay D, Bardos T, Kamath RV, Rugg MS, Day AJ, Salustri A, Hascall VC, Glant TT, Mikecz K (2003) Impaired cumulus mucification and female sterility in tumor necrosis factor-induced protein-6 deficient mice. Development 130:2253–2261

Gakunga P, Frost G, Shuster S, Cunha G, Formby B, Stern R (1997) Hyaluronan is a prerequisite for ductal branching morphogenesis. Development 124:3987–3997

Geary RL, Nikkari ST, Wagner WD, Williams JK, Adams MR, Dean RH (1998) Wound healing: a paradigm for lumen narrowing after arterial reconstruction. J Vasc Surg 27:96–106

Gerber BR, Franklin EC, Schubert M (1960) Ultracentrifugal fractionation of bovine nasal chondromucoprotein. J Biol Chem 235:2870–2875

Gilbert RJ, McKeon RJ, Darr A, Calabro A, Hascall VC, Bellamkonda RV (2005) CS-4, 6 is differentially upregulated in glial scar and is a potent inhibitor of neurite extension. Mol Cell Neurosci 29:545–558

Gill S, Wight TN, Frevert CW (2010) Proteoglycans: key regulators of pulmonary inflammation and the innate immune response to lung infection. Anat Rec (Hoboken) 293:968–981

Gorden A, Osman I, Gai W, He D, Huang W, Davidson A, Houghton AN, Busam K, Polsky D (2003) Analysis of BRAF and N-RAS mutations in metastatic melanoma tissues. Cancer Res 63:3955–3957

Grammatikakis N, Grammatikakis A, Yoneda M, Yu Q, Banerjee SD, Toole BP (1995) A novel glycosaminoglycan-binding protein is the vertebrate homologue of the cell cycle control protein, cdc37. J Biol Chem 270:16198–16205

Gustavsson H, Jennbacken K, Welen K, Damber JE (2008) Altered expression of genes regulating angiogenesis in experimental androgen-independent prostate cancer. Prostate 68:161–170

Gutierrez P, O'Brien KD, Ferguson M, Nikkari ST, Alpers CE, Wight TN (1997) Differences in the distribution of versican, decorin, and biglycan in atherosclerotic human coronary arteries. Cardiovasc Pathol 6:271–278

Haas CA, Rauch U, Thon N, Merten T, Deller T (1999) Entorhinal cortex lesion in adult rats induces the expression of the neuronal chondroitin sulfate proteoglycan neurocan in reactive astrocytes. J Neurosci 19:9953–9963

Halpert I, Sires U, Potter-Perigo S, Wight TN, Shapiro DS, Welgus HG, Wickline SA, Parks WC (1996) Matrilysin is expressed by lipid-laden macrophages at sites of potential rupture in atherosclerotic lesions and localized to areas of versican deposits. Proc Natl Acad Sci USA 93:9748–9753

Han CY, Subramanian S, Chan CK, Omer M, Chiba T, Wight TN, Chait A (2007) Adipocyte-derived serum amyloid A3 and hyaluronan play a role in monocyte recruitment and adhesion. Diabetes 56:2260–2273

Hara K, Shiga A, Fukutake T, Nozaki H, Miyashita A, Yokoseki A, Kawata H, Koyama A, Arima K, Takahashi T, Ikeda M, Shiota H, Tamura M, Shimoe Y, Hirayama M, Arisato T, Yanagawa S, Tanaka A, Nakano I, Ikeda S, Yoshida Y, Yamamoto T, Ikeuchi T, Kuwano R, Nishizawa M, Tsuji S, Onodera O (2009) Association of HTRA1 mutations and familial ischemic cerebral small-vessel disease. N Engl J Med 360:1729–1739

Hartig W, Derouiche A, Welt K, Brauer K, Grosche J, Mader M, Reichenbach A, Bruckner G (1999) Cortical neurons immunoreactive for the potassium channel Kv3.1b subunit are predominantly surrounded by perineuronal nets presumed as a buffering system for cations. Brain Res 842:15–29

Hartmann U, Maurer P (2001) Proteoglycans in the nervous system–the quest for functional roles in vivo. Matrix Biol 20:23–35

Held-Feindt J, Paredes EB, Blomer U, Seidenbecher C, Stark AM, Mehdorn HM, Mentlein R (2006) Matrix-degrading proteases ADAMTS4 and ADAMTS5 (disintegrins and metalloproteinases with thrombospondin motifs 4 and 5) are expressed in human glioblastomas. Int J Cancer 118:55–61

Henderson DJ, Copp AJ (1997) Role of the extracellular matrix in neural crest cell migration. J Anat 191:507–515

Henderson DJ, Copp AJ (1998) Versican expression is associated with chamber specification, septation, and valvulogenesis in the developing mouse heart. Circ Res 83:523–532

Henderson DJ, Ybot-Gonzalez P, Copp AJ (1997) Over-expression of the chondroitin sulphate proteoglycan versican is associated with defective neural crest migration in the Pax3 mutant mouse (splotch). Mech Dev 69:39–51

Hinek A, Wilson SE (2000) Impaired elastogenesis in Hurler disease: dermatan sulfate accumulation linked to deficiency in elastin-binding protein and elastic fiber assembly. Am J Pathol 156:925–938

Hinek A, Mecham RP, Keeley F, Rabinovitch M (1991) Impaired elastin fiber assembly related to reduced 67-kD elastin-binding protein in fetal lamb ductus arteriosus and in cultured aortic smooth muscle cells treated with chondroitin sulfate. J Clin Invest 88:2083–2094

Hinek A, Braun KR, Liu K, Wang Y, Wight TN (2004) Retrovirally mediated overexpression of versican v3 reverses impaired elastogenesis and heightened proliferation exhibited by fibroblasts from Costello syndrome and Hurler disease patients. Am J Pathol 164:119–131

Hirose J, Kawashima H, Yoshie O, Tashiro K, Miyasaka M (2001) Versican interacts with chemokines and modulates cellular responses. J Biol Chem 276:5228–5234

Hollier BG, Evans K, Mani SA (2009) The epithelial-to-mesenchymal transition and cancer stem cells: a coalition against cancer therapies. J Mammary Gland Biol Neoplasia 14:29–43

Hu B, Kong LL, Matthews RT, Viapiano MS (2008) The proteoglycan brevican binds to fibronectin after proteolytic cleavage and promotes glioma cell motility. J Biol Chem 283:24848–24859

Huang H, He X (2008) Wnt/beta-catenin signaling: new (and old) players and new insights. Curr Opin Cell Biol 20:119–125

Huang L, Grammatikakis N, Yoneda M, Banerjee SD, Toole BP (2000) Molecular characterization of a novel intracellular hyaluronan-binding protein. J Biol Chem 275:29829–29839

Huang R, Merrilees MJ, Braun K, Beaumont B, Lemire J, Clowes AW, Hinek A, Wight TN (2006) Inhibition of versican synthesis by antisense alters smooth muscle cell phenotype and induces elastic fiber formation in vitro and in neointima after vessel injury. Circ Res 98:370–377

Iozzo RV (1998) Matrix proteoglycans: from molecular design to cellular function. Annu Rev Biochem 67:609–652

Iozzo RV, Naso MF, Cannizzaro LA, Wasmuth JJ, McPherson JD (1992) Mapping of the versican proteoglycan gene (CSPG2) to the long arm of human chromosome 5 (5q12–5q14). Genomics 14:845–851

Isogai Z, Aspberg A, Keene DR, Ono RN, Reinhardt DP, Sakai LY (2002) Versican interacts with fibrillin-1 and links extracellular microfibrils to other connective tissue networks. J Biol Chem 277:4565–4572

Jackson DG (2009) Immunological functions of hyaluronan and its receptors in the lymphatics. Immunol Rev 230:216–231

Jarvelainen H, Sainio A, Koulu M, Wight TN, Penttinen R (2009) Extracellular matrix molecules: potential targets in pharmacotherapy. Pharmacol Rev 61:198–223

Jaworski DM, Kelly GM, Hockfield S (1995) The CNS-specific hyaluronan-binding protein BEHAB is expressed in ventricular zones coincident with gliogenesis. J Neurosci 15:1352–1362

Jiang D, Liang J, Noble PW (2007) Hyaluronan in tissue injury and repair. Annu Rev Cell Dev Biol 23:435–461

Jonsson-Rylander AC, Nilsson T, Fritsche-Danielson R, Hammarstrom A, Behrendt M, Andersson JO, Lindgren K, Andersson AK, Wallbrandt P, Rosengren B, Brodin P, Thelin A, Westin A, Hurt-Camejo E, Lee-Sogaard CH (2005) Role of ADAMTS-1 in atherosclerosis: remodeling of carotid artery, immunohistochemistry, and proteolysis of versican. Arterioscler Thromb Vasc Biol 25:180–185

Kähäri V-M, Larjava H, Uitto J (1991) Differential regulation of extracellular matrix proteoglycan (PG) gene expression. J Biol Chem 266:10609–10615

Kaji T, Yamada A, Miyajima S, Yamamoto C, Fujiwara Y, Wight TN, Kinsella MG (2000) Cell density-dependent regulation of proteoglycan synthesis by transforming growth factor-beta(1) in cultured bovine aortic endothelial cells. J Biol Chem 275:1463–1470

Kalluri R, Weinberg RA (2009) The basics of epithelial-mesenchymal transition. J Clin Invest 119:1420–1428

Kamiya N, Watanabe H, Habuchi H, Takagi H, Shinomura T, Shimizu K, Kimata K (2006) Versican/PG-M regulates chondrogenesis as an extracellular matrix molecule crucial for mesenchymal condensation. J Biol Chem 281:2390–2400

Kappler J, Hegener O, Baader SL, Franken S, Gieselmann V, Haberlein H, Rauch U (2009) Transport of a hyaluronan-binding protein in brain tissue. Matrix Biol 28:396–405

Karra R, Vemullapalli S, Dong C, Herderick EE, Song X, Slosek K, Nevins JR, West M, Goldschmidt-Clermont PJ, Seo D (2005) Molecular evidence for arterial repair in atherosclerosis. Proc Natl Acad Sci USA 102:16789–16794

Katoh-Semba R, Matsuda M, Watanabe E, Maeda N, Oohira A (1998) Two types of brain chondroitin sulfate proteoglycan: their distribution and possible functions in the rat embryo. Neurosci Res 31:273–282

Kawashima H, Atarashi K, Hirose M, Hirose J, Yamada S, Sugahara K, Miyasaka M (2002) Oversulfated chondroitin/dermatan sulfates containing GlcAbeta1/IdoAalpha1–3GalNAc(4, 6-O-disulfate) interact with L- and P-selectin and chemokines. J Biol Chem 277:12921–12930

Kaya G, Rodriguez I, Jorcano JL, Vassalli P, Stamenkovic I (1997) Selective suppression of CD44 in keratinocytes of mice bearing an antisense CD44 transgene driven by a tissue-specific promoter disrupts hyaluronate metabolism in the skin and impairs keratinocyte proliferation. Genes Dev 11:996–1007

Kenagy RD, Fischer JW, Davies MG, Berceli SA, Hawkins SM, Wight TN, Clowes AW (2002) Increased plasmin and serine proteinase activity during flow-induced intimal atrophy in baboon PTFE grafts. Arterioscler Thromb Vasc Biol 22:400–404

Kenagy RD, Fischer JW, Lara S, Sandy JD, Clowes AW, Wight TN (2005) Accumulation and loss of extracellular matrix during shear stress-mediated intimal growth and regression in baboon vascular grafts. J Histochem Cytochem 53:131–140

Kenagy RD, Plaas AH, Wight TN (2006) Versican degradation and vascular disease. Trends Cardiovasc Med 16:209–215

Kenagy RD, Min SK, Clowes AW, Sandy JD (2009) Cell death-associated ADAMTS4 and versican degradation in vascular tissue. J Histochem Cytochem 57:889–897

Kern CB, Twal WO, Mjaatvedt CH, Fairey SE, Toole BP, Iruela-Arispe ML, Argraves WS (2006) Proteolytic cleavage of versican during cardiac cushion morphogenesis. Dev Dyn 235:2238–2247

Kern CB, Norris RA, Thompson RP, Argraves WS, Fairey SE, Reyes L, Hoffman S, Markwald RR, Mjaatvedt CH (2007) Versican proteolysis mediates myocardial regression during outflow tract development. Dev Dyn 236:671–683

Kern CB, Wessels A, McGarity J, Dixon LJ, Alston E, Argraves WS, Geeting D, Nelson CM, Menick DR, Apte SS (2010) Reduced versican cleavage due to Adamts9 haploinsufficiency is associated with cardiac and aortic anomalies. Matrix Biol 29:304–316

Kim S, Takahashi H, Lin WW, Descargues P, Grivennikov S, Kim Y, Luo JL, Karin M (2009) Carcinoma-produced factors activate myeloid cells through TLR2 to stimulate metastasis. Nature 457:102–106

Kimata K, Oike Y, Tani K, Shinomura T, Yamagata M, Uritani M, Suzuki S (1986) A large chondroitin sulfate proteoglycan (PG-M) synthesized before chondrogenesis in the limb bud of chick embryo. J Biol Chem 261:13517–13525

Kimura JH, Hardingham TE, Hascall VC, Solursh M (1979) Biosynthesis of proteoglycans and their assembly into aggregates in cultures of chondrocytes from the Swarm rat chondrosarcoma. J Biol Chem 254:2600–2609

Kinsella M, Bressler S, Wight T (2004) The regulated synthesis of versican, decorin and biglycan: extracellular proteoglycans that influence cell phenotype. Crit Rev Eukaryot Gene Expr 14:203–234

Kischel P, Waltregny D, Dumont B, Turtoi A, Greffe Y, Kirsch S, De Pauw E, Castronovo V (2010) Versican overexpression in human breast cancer lesions: known and new isoforms for stromal tumor targeting. Int J Cancer 126:640–650

Kishimoto J, Ehama R, Wu L, Jiang S, Jiang N, Burgeson RE (1999) Selective activation of the versican promoter by epithelial- mesenchymal interactions during hair follicle development. Proc Natl Acad Sci USA 96:7336–7341

Kloeckener-Gruissem B, Bartholdi D, Abdou MT, Zimmermann DR, Berger W (2006) Identification of the genetic defect in the original Wagner syndrome family. Mol Vis 12:350–355

Knudson CB, Toole BP (1985) Changes in the pericellular matrix during differentiation of limb bud mesoderm. Dev Biol 112:308–318

Knudson W, Chow G, Knudson CB (2002) CD44-mediated uptake and degradation of hyaluronan. Matrix Biol 21:15–23

Kolodgie FD, Burke AP, Farb A, Weber DK, Kutys R, Wight TN, Virmani R (2002) Differential accumulation of proteoglycans and hyaluronan in culprit lesions: insights into plaque erosion. Arterioscler Thromb Vasc Biol 22:1642–1648

Koo BH, Longpre JM, Somerville RP, Alexander JP, Leduc R, Apte SS (2006) Cell-surface processing of pro-ADAMTS9 by furin. J Biol Chem 281:12485–12494

Koo BH, Longpre JM, Somerville RP, Alexander JP, Leduc R, Apte SS (2007) Regulation of ADAMTS9 secretion and enzymatic activity by its propeptide. J Biol Chem 282:16146–16154

Korswagen HC, Clevers HC (1999) Activation and repression of wingless/Wnt target genes by the TCF/LEF-1 family of transcription factors. Cold Spring Harb Symp Quant Biol 64:141–147

Kosher RA, Savage MP, Walker KH (1981) A gradation of hyaluronate accumulation along the proximodistal axis of the embryonic chick limb bud. J Embryol Exp Morphol 63:85–98

Koyama H, Hibi T, Isogai Z, Yoneda M, Fujimori M, Amano J, Kawakubo M, Kannagi R, Kimata K, Taniguchi S, Itano N (2007) Hyperproduction of hyaluronan in neu-induced mammary tumor accelerates angiogenesis through stromal cell recruitment: possible involvement of versican/PG-M. Am J Pathol 170:1086–1099

Kulyk WM, Kosher RA (1987) Temporal and spatial analysis of hyaluronidase activity during development of the embryonic chick limb bud. Dev Biol 120:535–541

Kuno K, Okada Y, Kawashima H, Nakamura H, Miyasaka M, Ohno H, Matsushima K (2000) ADAMTS-1 cleaves a cartilage proteoglycan, aggrecan. FEBS Lett 478:241–245

Laabs TL, Wang H, Katagiri Y, McCann T, Fawcett JW, Geller HM (2007) Inhibiting glycosaminoglycan chain polymerization decreases the inhibitory activity of astrocyte-derived chondroitin sulfate proteoglycans. J Neurosci 27:14494–14501

Landolt RM, Vaughan L, Winterhalter KH, Zimmermann DR (1995) Versican is selectively expressed in embryonic tissues that act as barriers to neural crest cell migration and axon outgrowth. Development 121:2303–2312

LaPierre DP, Lee DY, Li SZ, Xie YZ, Zhong L, Sheng W, Deng Z, Yang BB (2007) The ability of versican to simultaneously cause apoptotic resistance and sensitivity. Cancer Res 67:4742–4750

Laurent TC, Fraser JR (1986) The properties and turnover of hyaluronan. Ciba Found Symp 124:9–29

Lee RT, Yamamoto C, Feng Y, Potter-Perigo S, Briggs WH, Landschulz KT, Turi TG, Thompson JF, Libby P, Wight TN (2001) Mechanical strain induces specific changes in the synthesis and organization of proteoglycans by vascular smooth muscle cells. J Biol Chem 276:13847–13851

Lemire JM, Covin CW, White S, Giachelli CM, Schwartz SM (1994) Characterization of cloned aortic smooth muscle cells from young rats. Am J Pathol 144:1068–1081

Lemire JM, Potter-Perigo S, Hall KL, Wight TN, Schwartz SM (1996) Distinct rat aortic smooth muscle cells differ in versican/PG-M expression. Arterioscler Thromb Vasc Biol 16:821–829

Lemire JM, Merrilees MJ, Braun KR, Wight TN (2002) Overexpression of the V3 variant of versican alters arterial smooth muscle cell adhesion, migration, and proliferation in vitro. J Cell Physiol 190:38–45

Lemire JM, Chan CK, Bressler S, Miller J, LeBaron RG, Wight TN (2007) Interleukin-1beta selectively decreases the synthesis of versican by arterial smooth muscle cells. J Cell Biochem 101:753–766

Li Y, Toole BP, Dealy CN, Kosher RA (2007) Hyaluronan in limb morphogenesis. Dev Biol 305:411–420

Li F, Ten Dam GB, Murugan S, Yamada S, Hashiguchi T, Mizumoto S, Oguri K, Okayama M, van Kuppevelt TH, Sugahara K (2008) Involvement of highly sulfated chondroitin sulfate in the metastasis of the lewis lung carcinoma cells. J Biol Chem 283:34294–34304

Lin H, Ignatescu M, Wilson JE, Roberts CR, Horley KJ, Winters GL, Costanzo MR, McManus BM (1996) Prominence of apolipoproteins B, (a), and E in the intimae of coronary arteries in transplanted human hearts: geographic relationship to vessel wall proteoglycans. J Heart Lung Transplant 15:1223–1232

Little PJ, Tannock L, Olin KL, Chait A, Wight TN (2002) Proteoglycans synthesized by arterial smooth muscle cells in the presence of transforming growth factor-beta1 exhibit increased binding to LDLs. Arterioscler Thromb Vasc Biol 22:55–60

Little PJ, Ballinger ML, Burch ML, Osman N (2008) Biosynthesis of natural and hyperelongated chondroitin sulfate glycosaminoglycans: new insights into an elusive process. Open Biochem J 2:135–142

Longpre JM, McCulloch DR, Koo BH, Alexander JP, Apte SS, Leduc R (2009) Characterization of proADAMTS5 processing by proprotein convertases. Int J Biochem Cell Biol 41:1116–1126

Lundell A, Olin AI, Morgelin M, al-Karadaghi S, Aspberg A, Logan DT (2004) Structural basis for interactions between tenascins and lectican C-type lectin domains: evidence for a crosslinking role for tenascins. Structure 12:1495–1506

Majors AK, Austin RC, de la Motte CA, Pyeritz RE, Hascall VC, Kessler SP, Sen G, Strong SA (2003) Endoplasmic reticulum stress induces hyaluronan deposition and leukocyte adhesion. J Biol Chem 278:47223–47231

Maldonado JL, Fridlyand J, Patel H, Jain AN, Busam K, Kageshita T, Ono T, Albertson DG, Pinkel D, Bastian BC (2003) Determinants of BRAF mutations in primary melanomas. J Natl Cancer Inst 95:1878–1890

Margolis RU, Margolis RK (1994) Aggrecan-versican-neurocan family proteoglycans. Methods Enzymol 245:105–126

Margolis RK, Rauch U, Maurel P, Margolis RU (1996) Neurocan and phosphacan: two major nervous tissue-specific chondroitin sulfate proteoglycans. Perspect Dev Neurobiol 3: 273–290

Matrosova VY, Orlovskaya IA, Serobyan N, Khaldoyanidi SK (2004) Hyaluronic acid facilitates the recovery of hematopoiesis following 5-fluorouracil administration. Stem Cells 22:544–555

Matsui F, Kawashima S, Shuo T, Yamauchi S, Tokita Y, Aono S, Keino H, Oohira A (2002) Transient expression of juvenile-type neurocan by reactive astrocytes in adult rat brains injured by kainate-induced seizures as well as surgical incision. Neuroscience 112:773–781

Matsumoto K, Li Y, Jakuba C, Sugiyama Y, Sayo T, Okuno M, Dealy CN, Toole BP, Takeda J, Yamaguchi Y, Kosher RA (2009) Conditional inactivation of Has2 reveals a crucial role for hyaluronan in skeletal growth, patterning, chondrocyte maturation and joint formation in the developing limb. Development 136:2825–2835

Matthews RT, Gary SC, Zerillo C, Pratta M, Solomon K, Arner EC, Hockfield S (2000) Brain-enriched hyaluronan binding (BEHAB)/brevican cleavage in a glioma cell line is mediated by a disintegrin and metalloproteinase with thrombospondin motifs (ADAMTS) family member. J Biol Chem 275:22695–22703

Maxwell CA, McCarthy J, Turley E (2008) Cell-surface and mitotic-spindle RHAMM: moonlighting or dual oncogenic functions? J Cell Sci 121:925–932

Mayanil CS, George D, Freilich L, Miljan EJ, Mania-Farnell B, McLone DG, Bremer EG (2001) Microarray analysis detects novel Pax3 downstream target genes. J Biol Chem 276: 49299–49309

Mazzucato M, Cozzi MR, Pradella P, Perissinotto D, Malmstrom A, Morgelin M, Spessotto P, Colombatti A, De Marco L, Perris R (2002) Vascular PG-M/versican variants promote platelet adhesion at low shear rates and cooperate with collagens to induce aggregation. FASEB J 16:1903–1916

McCulloch DR, Nelson CM, Dixon LJ, Silver DL, Wylie JD, Lindner V, Sasaki T, Cooley MA, Argraves WS, Apte SS (2009) ADAMTS metalloproteases generate active versican fragments that regulate interdigital web regression. Dev Cell 17:687–696

McGee M, Wagner WD (2003) Chondroitin sulfate anticoagulant activity Is linked to water transfer: relevance to proteoglycan structure in atherosclerosis. Arterioscler Thromb Vasc Biol 23:1921–1927

McKeon RJ, Hoke A, Silver J (1995) Injury-induced proteoglycans inhibit the potential for laminin-mediated axon growth on astrocytic scars. Exp Neurol 136:32–43

McKeon RJ, Jurynec MJ, Buck CR (1999) The chondroitin sulfate proteoglycans neurocan and phosphacan are expressed by reactive astrocytes in the chronic CNS glial scar. J Neurosci 19:10778–10788

Merrilees MJ, Beaumont B, Scott LJ (2001) Comparison of deposits of versican, biglycan and decorin in saphenous vein and internal thoracic, radial and coronary arteries: correlation to patency. Coron Artery Dis 12:7–16

Merrilees MJ, Lemire JM, Fischer JW, Kinsella MG, Braun KR, Clowes AW, Wight TN (2002) Retrovirally mediated overexpression of versican v3 by arterial smooth muscle cells induces tropoelastin synthesis and elastic fiber formation in vitro and in neointima after vascular injury. Circ Res 90:481–487

Meyer K, Palmer JW, Smyth EM (1937) On glycoproteins: V. Protein complexes of chrondroitinsulfuric acid. J Biol Chem 119:501–506

Meyer K, Linker A, Davidson EA, Weissmann B (1953) The mucopolysaccharides of bovine cornea. J Biol Chem 205:611–616

Miguel RF, Pollak A, Lubec G (2005) Metalloproteinase ADAMTS-1 but not ADAMTS-5 is manifold overexpressed in neurodegenerative disorders as Down syndrome, Alzheimer's and Pick's disease. Brain Res Mol Brain Res 133:1–5

Miletti-Gonzalez KE, Chen S, Muthukumaran N, Saglimbeni GN, Wu X, Yang J, Apolito K, Shih WJ, Hait WN, Rodriguez-Rodriguez L (2005) The CD44 receptor interacts with P-glycoprotein to promote cell migration and invasion in cancer. Cancer Res 65:6660–6667

Milev P, Maurel P, Chiba A, Mevissen M, Popp S, Yamaguchi Y, Margolis RK, Margolis RU (1998) Differential regulation of expression of hyaluronan-binding proteoglycans in

developing brain: aggrecan, versican, neurocan, and brevican. Biochem Biophys Res Commun 247:207–212
Miura R, Ethell IM, Yamaguchi Y (2001) Carbohydrate-protein interactions between HNK-1-reactive sulfoglucuronyl glycolipids and the proteoglycan lectin domain mediate neuronal cell adhesion and neurite outgrowth. J Neurochem 76:413–424
Miyamoto T, Inoue H, Sakamoto Y, Kudo E, Naito T, Mikawa T, Mikawa Y, Isashiki Y, Osabe D, Shinohara S, Shiota H, Itakura M (2005) Identification of a novel splice site mutation of the CSPG2 gene in a Japanese family with Wagner syndrome. Invest Ophthalmol Vis Sci 46:2726–2735
Mjaatvedt CH, Yamamura H, Capehart AA, Turner D, Markwald RR (1998) The Cspg2 gene, disrupted in the hdf mutant, is required for right cardiac chamber and endocardial cushion formation. Dev Biol 202:56–66
Morawski M, Bruckner MK, Riederer P, Bruckner G, Arendt T (2004) Perineuronal nets potentially protect against oxidative stress. Exp Neurol 188:309–315
Morgenstern DA, Asher RA, Fawcett JW (2002) Chondroitin sulphate proteoglycans in the CNS injury response. Prog Brain Res 137:313–332
Morton SU, Scherz PJ, Cordes KR, Ivey KN, Stainier DY, Srivastava D (2008) microRNA-138 modulates cardiac patterning during embryonic development. Proc Natl Acad Sci USA 105:17830–17835
Muir H (1958) The nature of the link between protein and carbohydrate of a chondroitin sulphate complex from hyaline cartilage. Biochem J 69:195–204
Mukhopadhyay D, Asari A, Rugg MS, Day AJ, Fulop C (2004) Specificity of the tumor necrosis factor-induced protein 6-mediated heavy chain transfer from inter-alpha-trypsin inhibitor to hyaluronan: implications for the assembly of the cumulus extracellular matrix. J Biol Chem 279:11119–11128
Mukhopadhyay A, Nikopoulos K, Maugeri A, de Brouwer AP, van Nouhuys CE, Boon CJ, Perveen R, Zegers HA, Wittebol-Post D, van den Biesen PR, van der Velde-Visser SD, Brunner HG, Black GC, Hoyng CB, Cremers FP (2006) Erosive vitreoretinopathy and wagner disease are caused by intronic mutations in CSPG2/Versican that result in an imbalance of splice variants. Invest Ophthalmol Vis Sci 47:3565–3572
Nairn AV, Kinoshita-Toyoda A, Toyoda H, Xie J, Harris K, Dalton S, Kulik M, Pierce JM, Toida T, Moremen KW, Linhardt RJ (2007) Glycomics of proteoglycan biosynthesis in murine embryonic stem cell differentiation. J Proteome Res 6:4374–4387
Nakada M, Miyamori H, Kita D, Takahashi T, Yamashita J, Sato H, Miura R, Yamaguchi Y, Okada Y (2005) Human glioblastomas overexpress ADAMTS-5 that degrades brevican. Acta Neuropathol 110:239–246
Nakamura H, Fujii Y, Inoki I, Sugimoto K, Tanzawa K, Matsuki H, Miura R, Yamaguchi Y, Okada Y (2000) Brevican is degraded by matrix metalloproteinases and aggrecanase-1 (ADAMTS4) at different sites. J Biol Chem 275:38885–38890
Naor D, Nedvetzki S, Walmsley M, Yayon A, Turley EA, Golan I, Caspi D, Sebban LE, Zick Y, Garin T, Karussis D, Assayag-Asherie N, Raz I, Weiss L, Slavin S (2007) CD44 involvement in autoimmune inflammations: the lesson to be learned from CD44-targeting by antibody or from knockout mice. Ann NY Acad Sci 1110:233–247
Naso MF, Zimmermann DR, Iozzo RV (1994) Characterization of the complete genomic structure of the human versican gene and functional analysis of its promoter. J Biol Chem 269:32999–33008
Nicoll SB, Barak O, Csoka AB, Bhatnagar RS, Stern R (2002) Hyaluronidases and CD44 undergo differential modulation during chondrogenesis. Biochem Biophys Res Commun 292:819–825
Niederost BP, Zimmermann DR, Schwab ME, Bandtlow CE (1999) Bovine CNS myelin contains neurite growth-inhibitory activity associated with chondroitin sulfate proteoglycans. J Neurosci 19:8979–8989

Nilsson SK, Haylock DN, Johnston HM, Occhiodoro T, Brown TJ, Simmons PJ (2003) Hyaluronan is synthesized by primitive hemopoietic cells, participates in their lodgment at the endosteum following transplantation, and is involved in the regulation of their proliferation and differentiation in vitro. Blood 101:856–862

Noble PW (2002) Hyaluronan and its catabolic products in tissue injury and repair. Matrix Biol 21:25–29

Nutt CL, Zerillo CA, Kelly GM, Hockfield S (2001) Brain enriched hyaluronan binding (BEHAB)/ brevican increases aggressiveness of CNS-1 gliomas in Lewis rats. Cancer Res 61:7056–7059

Olin AI, Morgelin M, Sasaki T, Timpl R, Heinegard D, Aspberg A (2001) The proteoglycans aggrecan and Versican form networks with fibulin-2 through their lectin domain binding. J Biol Chem 276:1253–1261

Ori M, Nardini M, Casini P, Perris R, Nardi I (2006) XHas2 activity is required during somitogenesis and precursor cell migration in Xenopus development. Development 133:631–640

Ostberg CO, Zhu P, Wight TN, Qwarnstrom EE (1995) Fibronectin attachment is permissive for IL-1 mediated gene regulation. FEBS Lett 367:93–97

Pandey MS, Harris EN, Weigel JA, Weigel PH (2008) The cytoplasmic domain of the hyaluronan receptor for endocytosis (HARE) contains multiple endocytic motifs targeting coated pit-mediated internalization. J Biol Chem 283:21453–21461

Passi A, Negrini D, Albertini R, Miserocchi G, De Luca G (1999) The sensitivity of versican from rabbit lung to gelatinase A (MMP-2) and B (MMP-9) and its involvement in the development of hydraulic lung edema. FEBS Lett 456:93–96

Perides G, Asher RA, Lark MW, Lane WS, Robinson RA, Bignami A (1995) Glial hyaluronate-binding protein: a product of metalloproteinase digestion of versican? Biochem J 312:377–384

Perissinotto D, Iacopetti P, Bellina I, Doliana R, Colombatti A, Pettway Z, Bronner-Fraser M, Shinomura T, Kimata K, Morgelin M, Lofberg J, Perris R (2000) Avian neural crest cell migration is diversely regulated by the two major hyaluronan-binding proteoglycans PG-M/versican and aggrecan. Development 127:2823–2842

Perris R, Lofberg J, Fallstrom C, von Boxberg Y, Olsson L, Newgreen DF (1990) Structural and compositional divergencies in the extracellular matrix encountered by neural crest cells in the white mutant axolotl embryo. Development 109:533–551

Pienimaki JP, Rilla K, Fulop C, Sironen RK, Karvinen S, Pasonen S, Lammi MJ, Tammi R, Hascall VC, Tammi MI (2001) Epidermal growth factor activates hyaluronan synthase 2 in epidermal keratinocytes and increases pericellular and intracellular hyaluronan. J Biol Chem 276:20428–20435

Pohl M, Sakurai H, Stuart RO, Nigam SK (2000) Role of hyaluronan and CD44 in in vitro branching morphogenesis of ureteric bud cells. Dev Biol 224:312–325

Polyak K, Weinberg RA (2009) Transitions between epithelial and mesenchymal states: acquisition of malignant and stem cell traits. Nat Rev Cancer 9:265–273

Ponta H, Sherman L, Herrlich PA (2003) CD44: from adhesion molecules to signalling regulators. Nat Rev Mol Cell Biol 4:33–45

Potter-Perigo S, Johnson PY, Evanko SP, Chan CK, Braun KR, Wilkinson TS, Altman LC, Wight TN (2009) Poly I:C stimulates versican accumulation in the extracellular matrix promoting monocyte adhesion. Am J Respir Cell Mol Biol 43:109–120

Powell JD, Horton MR (2005) Threat matrix: low-molecular-weight hyaluronan (HA) as a danger signal. Immunol Res 31:207–218

Prevo R, Banerji S, Ferguson DJ, Clasper S, Jackson DG (2001) Mouse LYVE-1 is an endocytic receptor for hyaluronan in lymphatic endothelium. J Biol Chem 276:19420–19430

Properzi F, Fawcett JW (2004) Proteoglycans and brain repair. News Physiol Sci 19:33–38

Properzi F, Carulli D, Asher RA, Muir E, Camargo LM, van Kuppevelt TH, ten Dam GB, Furukawa Y, Mikami T, Sugahara K, Toida T, Geller HM, Fawcett JW (2005) Chondroitin 6-sulphate synthesis is up-regulated in injured CNS, induced by injury-related cytokines and enhanced in axon-growth inhibitory glia. Eur J Neurosci 21:378–390

Qwarnström EE, Järveläinen HT, Kinsella MK, Ostberg CO, Sandell LJ, Page RC, Wight TN (1993) Interleukin-1β regulation of fibroblast proteoglycan synthesis involves a decrease in versican steady-state mRNA levels. Biochem J 294:613–620

Rahmani M, Read JT, Carthy JM, McDonald PC, Wong BW, Esfandiarei M, Si X, Luo Z, Luo H, Rennie PS, McManus BM (2005) Regulation of the versican promoter by the beta-catenin-T-cell factor complex in vascular smooth muscle cells. J Biol Chem 280:13019–13028

Rahmani M, Wong BW, Ang L, Cheung CC, Carthy JM, Walinski H, McManus BM (2006) Versican: signaling to transcriptional control pathways. Can J Physiol Pharmacol 84:77–92

Rauch U (2004) Extracellular matrix components associated with remodeling processes in brain. Cell Mol Life Sci 61:2031–2045

Rauch U, Karthikeyan L, Maurel P, Margolis RU, Margolis RK (1992) Cloning and primary structure of neurocan, a developmentally regulated, aggregating chondroitin sulfate proteoglycan of brain. J Biol Chem 267:19536–19547

Rauch U, Zhou XH, Roos G (2005) Extracellular matrix alterations in brains lacking four of its components. Biochem Biophys Res Commun 328:608–617

Read JT, Rahmani M, Boroomand S, Allahverdian S, McManus BM, Rennie PS (2007) Androgen receptor regulation of the versican gene through an androgen response element in the proximal promoter. J Biol Chem 282:31954–31963

Reinhardt DP, Sasaki T, Dzamba BJ, Keene DR, Chu ML, Gohring W, Timpl R, Sakai LY (1996) Fibrillin-1 and fibulin-2 interact and are colocalized in some tissues. J Biol Chem 271:19489–19496

Ricciardelli C, Mayne K, Sykes PJ, Raymond WA, McCaul K, Marshall VR, Horsfall DJ (1998) Elevated levels of versican but not decorin predict disease progression in early-stage prostate cancer. Clin Cancer Res 4:963–971

Ricciardelli C, Sakko AJ, Ween MP, Russell DL, Horsfall DJ (2009) The biological role and regulation of versican levels in cancer. Cancer Metastasis Rev 28:233–245

Rocks N, Paulissen G, El Hour M, Quesada F, Crahay C, Gueders M, Foidart JM, Noel A, Cataldo D (2008) Emerging roles of ADAM and ADAMTS metalloproteinases in cancer. Biochimie 90:369–379

Ronan SM, Tran-Viet KN, Burner EL, Metlapally R, Toth CA, Young TL (2009) Mutational hot spot potential of a novel base pair mutation of the CSPG2 gene in a family with Wagner syndrome. Arch Ophthalmol 127:1511–1519

Roseman S (2001) Reflections on glycobiology. J Biol Chem 276:41527–41542

Ruscheinsky M, De la Motte C, Mahendroo M (2008) Hyaluronan and its binding proteins during cervical ripening and parturition: dynamic changes in size, distribution and temporal sequence. Matrix Biol 27:487–497

Russell D, Salustri A (2006) Extracellular matrix of the cumulus-oocyte complex. Semin Reprod Med 24:217–227

Russell DL, Doyle KM, Ochsner SA, Sandy JD, Richards JS (2003a) Processing and localization of ADAMTS-1 and proteolytic cleavage of versican during cumulus matrix expansion and ovulation. J Biol Chem 278:42330–42339

Russell DL, Ochsner SA, Hsieh M, Mulders S, Richards JS (2003b) Hormone-regulated expression and localization of versican in the rodent ovary. Endocrinology 144:1020–1031

Sajdera SW, Hascall VC (1969) Proteinpolysaccharide complex from bovine nasal cartilage. A comparison of low and high shear extraction procedures. J Biol Chem 244:77–87

Salustri A, Camaioni A, Di Giacomo M, Fulop C, Hascall VC (1999) Hyaluronan and proteoglycans in ovarian follicles. Hum Reprod Update 5:293–301

Sandy JD (2006) A contentious issue finds some clarity: on the independent and complementary roles of aggrecanase activity and MMP activity in human joint aggrecanolysis. Osteoarthritis Cartilage 14:95–100

Sandy JD, Westling J, Kenagy RD, Iruela-Arispe ML, Verscharen C, Rodriguez-Mazaneque JC, Zimmermann DR, Lemire JM, Fischer JW, Wight TN, Clowes AW (2001) Versican V1 proteolysis in human aorta in vivo occurs at the Glu441-Ala442 bond, a site that is cleaved by recombinant ADAMTS-1 and ADAMTS-4. J Biol Chem 276:13372–13378

Schmalfeldt M, Bandtlow CE, Dours-Zimmermann MT, Winterhalter KH, Zimmermann DR (2000) Brain derived versican V2 is a potent inhibitor of axonal growth. J Cell Sci 113:807–816

Schönherr E, Järveläinen HT, Sandell LJ, Wight TN (1991) Effects of platelet-derived growth factor and transforming growth factor-β 1 on the synthesis of a large versican-like chondroitin sulfate proteoglycan by arterial smooth muscle cells. J Biol Chem 266:17640–17647

Schönherr E, Kinsella MG, Wight TN (1997) Genistein selectively inhibits platelet-derived growth factor stimulated versican biosynthesis in monkey arterial smooth muscle cells. Arch Biochem Biophys 339:353–361

Schor H, Vaday GG, Lider O (2000) Modulation of leukocyte behavior by an inflamed extracellular matrix. Dev Immunol 7:227–238

Seidelmann SB, Kuo C, Pleskac N, Molina J, Sayers S, Li R, Zhou J, Johnson P, Braun K, Chan C, Teupser D, Breslow JL, Wight TN, Tall AR, Welch CL (2008) Athsq1 is an atherosclerosis modifier locus with dramatic effects on lesion area and prominent accumulation of versican. Arterioscler Thromb Vasc Biol 28:2180–2186

Seidenbecher CI, Richter K, Rauch U, Fassler R, Garner CC, Gundelfinger ED (1995a) Brevican, a chondroitin sulfate proteoglycan of rat brain, occurs as secreted and cell surface glycosylphosphatidylinositol-anchored isoforms. J Biol Chem 270:27206–27212

Seidenbecher CI, Richter K, Rauch U, Fässler R, Garner CC, Gundelfinger ED (1995b) Brevican, a chondroitin sulfate proteoglycan of rat brain, occurs as secreted and cell surface glycosylphosphatidylinositol-anchored isoforms. J Biol Chem 270:27206–27212

Seidenbecher CI, Gundelfinger ED, Bockers TM, Trotter J, Kreutz MR (1998) Transcripts for secreted and GPI-anchored brevican are differentially distributed in rat brain. Eur J Neurosci 10:1621–1630

Seno N, Meyer K, Anderson B, Hoffman P (1965) Variations in Keratosulfates. J Biol Chem 240:1005–1010

Shatton J, Schubert M (1954) Isolation of a mucoprotein from cartilage. J Biol Chem 211:565–573

Sheng W, Wang G, Wang Y, Liang J, Wen J, Zheng PS, Wu Y, Lee V, Slingerland J, Dumont D, Yang BB (2005) The roles of versican V1 and V2 isoforms in cell proliferation and apoptosis. Mol Biol Cell 16:1330–1340

Sherman L, Sleeman J, Herrlich P, Ponta H (1994) Hyaluronate receptors: key players in growth, differentiation, migration and tumor progression. Curr Opin Cell Biol 6:726–733

Shimizu-Hirota R, Sasamura H, Mifune M, Nakaya H, Kuroda M, Hayashi M, Saruta T (2001) Regulation of vascular proteoglycan synthesis by angiotensin II type 1 and type 2 receptors. J Am Soc Nephrol 12:2609–2615

Shinomura T, Jensen KL, Yamagata M, Kimata K, Solursh M (1990) The distribution of mesenchyme proteoglycan (PG-M) during wing bud outgrowth. Anat Embryol (Berl) 181:227–233

Shinomura T, Nishida Y, Ito K, Kimata K (1993) DNA cloning of PG-M, a large chondroitin sulfate proteoglycan expressed during chondrogenesis in chick limb buds. Alternative spliced multiforms of PG-M and their relationship to versican. J Biol Chem 268:14461–14469

Shukla S, Nair R, Rolle MW, Braun KR, Chan CK, Johnson PY, Wight TN, McDevitt TC (2010) Synthesis and organization of hyaluronan and versican by embryonic stem cells undergoing embryoid body differentiation. J Histochem Cytochem 58:345–358

Silver J, Miller JH (2004) Regeneration beyond the glial scar. Nat Rev Neurosci 5:146–156

Silver DL, Hou L, Somerville R, Young ME, Apte SS, Pavan WJ (2008) The secreted metalloprotease ADAMTS20 is required for melanoblast survival. PLoS Genet 4:e1000003

Simpson MA, Lokeshwar VB (2008) Hyaluronan and hyaluronidase in genitourinary tumors. Front Biosci 13:5664–5680

Singley CT, Solursh M (1981) The spatial distribution of hyaluronic acid and mesenchymal condensation in the embryonic chick wing. Dev Biol 84:102–120

Sleeman JP, Cremers N (2007) New concepts in breast cancer metastasis: tumor initiating cells and the microenvironment. Clin Exp Metastasis 24:707–715

Slomiany MG, Dai L, Tolliver LB, Grass GD, Zeng Y, Toole BP (2009a) Inhibition of functional hyaluronan-CD44 interactions in CD133-positive primary human ovarian carcinoma cells by small hyaluronan oligosaccharides. Clin Cancer Res 15:7593–7601

Slomiany MG, Grass GD, Robertson AD, Yang XY, Maria BL, Beeson C, Toole BP (2009b) Hyaluronan, CD44, and emmprin regulate lactate efflux and membrane localization of monocarboxylate transporters in human breast carcinoma cells. Cancer Res 69:1293–1301

Smith LS, Kajikawa O, Elson G, Wick M, Mongovin S, Kosco-Vilbois M, Martin TR, Frevert CW (2008) Effect of Toll-like receptor 4 blockade on pulmonary inflammation caused by mechanical ventilation and bacterial endotoxin. Exp Lung Res 34:225–243

Snow DM, Lemmon V, Carrino DA, Caplan AI, Silver J (1990) Sulfated proteoglycans in astroglial barriers inhibit neurite outgrowth in vitro. Exp Neurol 109:111–130

Sobel RA, Ahmed AS (2001) White matter extracellular matrix chondroitin sulfate/dermatan sulfate proteoglycans in multiple sclerosis. J Neuropathol Exp Neurol 60:1198–1207

Somerville RP, Longpre JM, Jungers KA, Engle JM, Ross M, Evanko S, Wight TN, Leduc R, Apte SS (2003) Characterization of ADAMTS-9 and ADAMTS-20 as a distinct ADAMTS subfamily related to Caenorhabditis elegans GON-1. J Biol Chem 278:9503–9513

Sommerville LJ, Kelemen SE, Autieri MV (2008) Increased smooth muscle cell activation and neointima formation in response to injury in AIF-1 transgenic mice. Arterioscler Thromb Vasc Biol 28:47–53

Stankunas K, Hang CT, Tsun ZY, Chen H, Lee NV, Wu JI, Shang C, Bayle JH, Shou W, Iruela-Arispe ML, Chang CP (2008) Endocardial Brg1 represses ADAMTS1 to maintain the microenvironment for myocardial morphogenesis. Dev Cell 14:298–311

Stern R, Kogan G, Jedrzejas MJ, Soltes L (2007) The many ways to cleave hyaluronan. Biotechnol Adv 25:537–557

Stigson M, Lofberg J, Kjellen L (1997) PG-M/versican-like proteoglycans are components of large disulfide-stabilized complexes in the axolotl embryo. J Biol Chem 272:3246–3253

Strom A, Olin AI, Aspberg A, Hultgardh-Nilsson A (2006) Fibulin-2 is present in murine vascular lesions and is important for smooth muscle cell migration. Cardiovasc Res 69:755–763

Taipale J, Beachy PA (2001) The Hedgehog and Wnt signalling pathways in cancer. Nature 411:349–354

Tammi R, Rilla K, Pienimaki JP, MacCallum DK, Hogg M, Luukkonen M, Hascall VC, Tammi M (2001) Hyaluronan enters keratinocytes by a novel endocytic route for catabolism. J Biol Chem 276:35111–35122

Taylor KR, Gallo RL (2006) Glycosaminoglycans and their proteoglycans: host-associated molecular patterns for initiation and modulation of inflammation. FASEB J 20:9–22

Teder P, Vandivier RW, Jiang D, Liang J, Cohn L, Pure E, Henson PM, Noble PW (2002) Resolution of lung inflammation by CD44. Science 296:155–158

Thankamony SP, Knudson W (2006) Acylation of CD44 and its association with lipid rafts are required for receptor and hyaluronan endocytosis. J Biol Chem 281:34601–34609

Tien JY, Spicer AP (2005) Three vertebrate hyaluronan synthases are expressed during mouse development in distinct spatial and temporal patterns. Dev Dyn 233:130–141

Toole BP (1972) Hyaluronate turnover during chondrogenesis in the developing chick limb and axial skeleton. Dev Biol 29:321–329

Toole BP (1990) Hyaluronan and its binding proteins, the hyaladherins. Curr Opin Cell Biol 2:839–844

Toole BP (2001) Hyaluronan in morphogenesis. Semin Cell Dev Biol 12:79–87

Toole BP (2004) Hyaluronan: from extracellular glue to pericellular cue. Nat Rev Cancer 4:528–539

Toole BP (2009) Hyaluronan-CD44 Interactions in cancer: paradoxes and possibilities. Clin Cancer Res 15:7462–7468

Toole BP, Slomiany MG (2008) Hyaluronan, CD44 and Emmprin: partners in cancer cell chemoresistance. Drug Resist Updat 11:110–121

Toole BP, Ghatak S, Misra S (2008) Hyaluronan oligosaccharides as a potential anticancer therapeutic. Curr Pharm Biotechnol 9:249–252

Touab M, Villena J, Barranco C, Arumi-Uria M, Bassols A (2002) Versican is differentially expressed in human melanoma and may play a role in tumor development. Am J Pathol 160:549–557

Tufvesson E, Westergren-Thorsson G (2000) Alteration of proteoglycan synthesis in human lung fibroblasts induced by interleukin-1beta and tumor necrosis factor-alpha. J Cell Biochem 77:298–309

Turley EA, Noble PW, Bourguignon LY (2002) Signaling properties of hyaluronan receptors. J Biol Chem 277:4589–4592

Turley EA, Veiseh M, Radisky DC, Bissell MJ (2008) Mechanisms of disease: epithelial-mesenchymal transition–does cellular plasticity fuel neoplastic progression? Nat Clin Pract Oncol 5:280–290

Tuttle R, Braisted JE, Richards LJ, O'Leary DD (1998) Retinal axon guidance by region-specific cues in diencephalon. Development 125:791–801

Vaday GG, Lider O (2000) Extracellular matrix moieties, cytokines, and enzymes: dynamic effects on immune cell behavior and inflammation. J Leukoc Biol 67:149–159

Vaday GG, Franitza S, Schor H, Hecht I, Brill A, Cahalon L, Hershkoviz R, Lider O (2001) Combinatorial signals by inflammatory cytokines and chemokines mediate leukocyte interactions with extracellular matrix. J Leukoc Biol 69:885–892

Viapiano MS, Matthews RT (2006) From barriers to bridges: chondroitin sulfate proteoglycans in neuropathology. Trends Mol Med 12:488–496

Viapiano MS, Matthews RT, Hockfield S (2003) A novel membrane-associated glycovariant of BEHAB/brevican is up-regulated during rat brain development and in a rat model of invasive glioma. J Biol Chem 278:33239–33247

Viapiano MS, Hockfield S, Matthews RT (2008) BEHAB/brevican requires ADAMTS-mediated proteolytic cleavage to promote glioma invasion. J Neurooncol 88:261–272

Visvader JE, Lindeman GJ (2008) Cancer stem cells in solid tumours: accumulating evidence and unresolved questions. Nat Rev Cancer 8:755–768

Wang A, Hascall VC (2009) Hyperglycemia, intracellular hyaluronan synthesis, cyclin D3 and autophagy. Autophagy 5:864–865

Wang N, Tytell JD, Ingber DE (2009a) Mechanotransduction at a distance: mechanically coupling the extracellular matrix with the nucleus. Nat Rev Mol Cell Biol 10:75–82

Wang W, Xu GL, Jia WD, Ma JL, Li JS, Ge YS, Ren WH, Yu JH, Liu WB (2009b) Ligation of TLR2 by versican: a link between inflammation and metastasis. Arch Med Res 40:321–323

Wang X, Hu G, Zhou J (2010) Repression of versican expression by microRNA-143. J Biol Chem 285:23241–23250

Weigel PH, DeAngelis PL (2007) Hyaluronan synthases: a decade-plus of novel glycosyltransferases. J Biol Chem 282:36777–36781

Wight TN (2002) Versican: a versatile extracellular matrix proteoglycan in cell biology. Curr Opin Cell Biol 14:617–623

Wight TN (2005) The ADAMTS proteases, extracellular matrix, and vascular disease: waking the sleeping giant(s)! Arterioscler Thromb Vasc Biol 25:12–14

Wight TN, Merrilees MJ (2004) Proteoglycans in atherosclerosis and restenosis: key roles for versican. Circ Res 94:1158–1167

Wight TN, Heinegård DK, Hascall VC (1991) Proteoglycans: structure and function. In: Hay ED (ed) Cell biology of extracellular matrix. Plenum, New York, pp 45–78

Wight TN, Lara S, Reissen R, LeBaron R, Isner J (1997) Selective deposits of versican in the extracellular matrix of restenotic lesions from human peripheral arteries. Am J Pathol 151: 963–973

Williams KJ, Tabas I (1995) The response-to-retention hypothesis of early atherogenesis. Arterioscler Thromb Vasc Biol 15:551–561

Wu Y, Zhang Y, Cao L, Chen L, Lee V, Zheng PS, Kiani C, Adams ME, Ang LC, Paiwand F, Yang BB (2001) Identification of the motif in versican G3 domain that plays a dominant-negative effect on astrocytoma cell proliferation through inhibiting versican secretion and binding. J Biol Chem 276:14178–14186

Wu Y, Chen L, Zheng PS, Yang BB (2002) beta 1-Integrin-mediated glioma cell adhesion and free radical-induced apoptosis are regulated by binding to a C-terminal domain of PG-M/versican. J Biol Chem 277:12294–12301

Wu Y, Chen L, Cao L, Sheng W, Yang BB (2004) Overexpression of the C-terminal PG-M/versican domain impairs growth of tumor cells by intervening in the interaction between epidermal growth factor receptor and β_1-integrin. J Cell Sci 117:2227–2237

Wu YJ, La Pierre DP, Wu J, Yee AJ, Yang BB (2005) The interaction of versican with its binding partners. Cell Res 15:483–494

Yamada H, Watanabe K, Shimonaka M, Yamaguchi Y (1994) Molecular cloning of brevican, a novel brain proteoglycan of the aggrecan/versican family. J Biol Chem 13:10119–10126

Yamada H, Fredette B, Shitara K, Hagihara K, Miura R, Ranscht B, Stallcup WB, Yamaguchi Y (1997) The brain chondroitin sulfate proteoglycan brevican associates with astrocytes ensheathing cerebellar glomeruli and inhibits neurite outgrowth from granule neurons. J Neurosci 17:7784–7795

Yamagata M, Kimata K (1994) Repression of a malignant cell-substratum adhesion phenotype by inhibiting the production of the anti-adhesive proteoglycan, PG-M/versican. J Cell Sci 107:2581–2590

Yamagata M, Suzuki S, Akiyama SK, Yamada KM, Kimata K (1989) Regulation of cell-substrate adhesion by proteoglycans immobilized on extracellular substrates. J Biol Chem 264: 8012–8018

Yamagata M, Saga S, Kato M, Bernfield M, Kimata K (1993) Selective distributions of proteoglycans and their ligands in pericellular matrix of cultured fibroblasts. Implications for their roles in cell-substratum adhesion. J Cell Sci 106:55–65

Yamaguchi Y (1996) Brevican: a major proteoglycan in adult brain. Perspect Dev Neurobiol 3:307–317

Yamaguchi Y (2000) Chondroitin sulfate proteoglycans in the nervous system. In: Iozzo R (ed) Proteoglycans-structure, functions and interactions. Marcel Dekker, New York, pp 379–402

Yamamura H, Zhang M, Markwald RR, Mjaatvedt CH (1997) A heart segmental defect in the anterior-posterior axis of a transgenic mutant mouse. Dev Biol 186:58–72

Yang B, Yang BL, Savani RC, Turley EA (1994) Identification of a common hyaluronan binding motif in the hyaluronan binding proteins RHAMM, CD44 and link protein. EMBO J 13:286–296

Yang BL, Zhang Y, Cao L, Yang BB (1999) Cell adhesion and proliferation mediated through the G1 domain of versican. J Cell Biochem 72:210–220

Yang BL, Yang BB, Erwin M, Ang LC, Finkelstein J, Yee AJ (2003) Versican G3 domain enhances cellular adhesion and proliferation of bovine intervertebral disc cells cultured in vitro. Life Sci 73:3399–3413

Yokoyama U, Minamisawa S, Quan H, Ghatak S, Akaike T, Segi-Nishida E, Iwasaki S, Iwamoto M, Misra S, Tamura K, Hori H, Yokota S, Toole BP, Sugimoto Y, Ishikawa Y (2006) Chronic activation of the prostaglandin receptor EP4 promotes hyaluronan-mediated neointimal formation in the ductus arteriosus. J Clin Invest 116:3026–3034

Zako M, Shinomura T, Ujita M, Ito K, Kimata K (1995) Expression of PG-M (V3), an alternatively spliced form of PG-M without a chondroitin sulfate attachment region in mouse and human tissues. J Biol Chem 270:3914–3918

Zako M, Shinomura T, Kimata K (1997) Alternative splicing of the unique "PLUS" domain of chicken PG- M/versican is developmentally regulated. J Biol Chem 272:9325–9331

5 Hyaluronan and the Aggregating Proteoglycans 195

Zhang H, Kelly G, Zerillo C, Jaworski DM, Hockfield S (1998) Expression of a cleaved brain-specific extracellular matrix protein mediates glioma cell invasion In vivo. J Neurosci 18:2370–2376

Zhang Y, Cao L, Kiani C, Yang BL, Hu W, Yang BB (1999) Promotion of chondrocyte proliferation by versican mediated by G1 domain and EGF-like motifs. J Cell Biochem 73:445–457

Zhang H, Baader SL, Sixt M, Kappler J, Rauch U (2004) Neurocan-GFP fusion protein: a new approach to detect hyaluronan on tissue sections and living cells. J Histochem Cytochem 52:915–922

Zhao X, Russell P (2005) Versican splice variants in human trabecular meshwork and ciliary muscle. Mol Vis 11:603–608

Zheng PS, Vais D, Lapierre D, Liang YY, Lee V, Yang BL, Yang BB (2004) PG-M/versican binds to P-selectin glycoprotein ligand-1 and mediates leukocyte aggregation. J Cell Sci 117:5887–5895

Zheng PS, Reis M, Sparling C, Lee DY, La Pierre DP, Wong CK, Deng Z, Kahai S, Wen J, Yang BB (2006) Versican G3 domain promotes blood coagulation through suppressing the activity of tissue factor pathway inhibitor-1. J Biol Chem 281:8175–8182

Zhuo L, Yoneda M, Zhao M, Yingsung W, Yoshida N, Kitagawa Y, Kawamura K, Suzuki T, Kimata K (2001) Defect in SHAP-hyaluronan complex causes severe female infertility. A study by inactivation of the bikunin gene in mice. J Biol Chem 276:7693–7696

Zimmermann D (2000) Versican. In: Iozzo R (ed) Proteoglycans: structure, biology and molecular interactions. Marcel Dekker, New York, pp 327–341

Zimmermann R, Dours-Zimmermann MT (2008) Extracellular matrix of the central nervous system: from neglect to challenge. Histochem Cell Biol 130:635–653

Zimmermann R, Ruoslahti E (1989) Multiple domains of the large fibroblast proteoglycan, versican. EMBO J 8:2975–2981

Zoltan-Jones A, Huang L, Ghatak S, Toole BP (2003) Elevated hyaluronan production induces mesenchymal and transformed properties in epithelial cells. J Biol Chem 278:45801–45810

Chapter 6
Small Leucine-Rich Proteoglycans

Renato V. Iozzo, Silvia Goldoni, Agnes D. Berendsen, and Marian F. Young

Abstract The small leucine-rich proteoglycans (SLRPs) comprise an expanding family of proteoglycans and glycoproteins that now encompass five distinct groups including three canonical and two noncanonical classes based on shared structural and functional parameters. SLRPs are tissue organizers by orienting and ordering various collagenous matrices during ontogeny, wound repair, and cancer and interact with a number of surface receptors and growth factors, thereby regulating cell behavior. The focus of this chapter is on novel conceptual and functional advances in our understanding of SLRP biology with special emphasis on genetic diseases, cancer growth, fibrosis, osteoporosis, and other biological processes where these proteoglycans play a central role.

6.1 Introduction

Small leucine-rich proteoglycans (SLRPs) (Iozzo 1999) are present within the extracellular matrix of all tissues and within the thin membranes that envelop all the major parenchymal organs such as pericardium, pleura, periosteum, perimesium, and adventitia of blood vessels. This strategic location suggests that SLRPs are involved in the control of organ shape and size (Iozzo 1998). The characteristic hallmark of all the SLRPs is their intrinsic ability to interact with other proteins. Foremost among these interactions are those with collagens, growth factors, and various plasma membrane receptors. Various SLRPs interact with fibrils of collagen

---

R.V. Iozzo (✉) and S. Goldoni
Department of Pathology, Anatomy and Cell Biology and the Cancer Cell Biology and Signaling Program, Kimmel Cancer Center, Thomas Jefferson University, Room 249 JAH, 1020 Locust Street, Philadelphia, PA 19107, USA
e-mail: iozzo@mail.jci.tju.edu

A.D. Berendsen and M.F. Young
Craniofacial and Skeletal Diseases Branch, Molecular Biology of Bones and Teeth Section, NIDCR, National Institutes of Health, Bethesda, MD 20892, USA

R.P. Mecham (ed.), *The Extracellular Matrix: an Overview*, Biology of Extracellular Matrix, 197
DOI 10.1007/978-3-642-16555-9_6, © Springer-Verlag Berlin Heidelberg 2011

type I, II, III, V, and XI, forming a "surface coat". Indeed, the eponym "decorin" is based on its ability to decorate fibrillar (banded) collagen in a periodic fashion. The surface coat formed by various SLRPs is a sort of biological processor that regulates the physiology of collagenous matrices in a tissue-specific manner. This coat plays two fundamental roles: (1) it regulates proper fibril assembly, which occurs through lateral association of collagen molecules, and (2) it protects collagen fibrils from cleavage by collagenases by acting as a steric barrier limiting the access of the collagenases to their cleavage sites. This biological activity is governed by SLRP dual activities evoked by the glycosaminoglycan or protein core moieties. Some of the SLRPs contain stretches of amino acids that can be sulfated such as the polytyrosine sulfate in fibromodulin or the polyaspartate region of asporin. The region containing either the sulfated glycosaminoglycan(s) or the charged amino acid residues is consistently located at the N terminus, outside the leucine-rich repeats (LRRs). First, we will review recent advances in the biology of SLRPs with special emphasis on the molecular interactions and mechanisms of action of SLRP signaling and as causative agents of genetic diseases. Then, we will critically assess the involvement of SLRPs in various pathologies, including inflammation, fibrosis, bone diseases, cancer, and angiogenesis.

6.2 Structure, Evolutionary Conservation, and Specificity of Function

SLRPs are grouped into five classes based mainly on evolutionary conservation, homology at both the protein and genomic level, and chromosomal organization (Fig.6.1a) (Schaefer and Iozzo 2008). In total, there are 18 genes that encode SLRPs spread over seven chromosomes. Until recently, SLRPs were grouped into only three canonical classes (Iozzo and Murdoch 1996; Iozzo 1997). However, two new noncanonical classes have been recently introduced (Schaefer and Iozzo 2008): class IV, which includes chondroadherin, nyctalopin, and tsukushi, and class V with podocan and podocan-like protein 1. Regardless of the classification used, SLRPs share common functionality. For example, decorin, biglycan, asporin, and podocan bind to type I collagen, while decorin, biglycan, and lumican inhibit cell growth and various SLRPs interact with TGF-β and bone morphogenetic protein (BMP).

The typical, easily recognizable, structural features of SLRPs include a variable number of LRRs in the central portion of the protein (Fig. 6.1b). LRRs are units of ~24 amino acids characterized by a conserved pattern of hydrophobic residues. Each LRR folds into a secondary structure comprising a short parallel β-sheet, a turn, and a more variable region. Essentially, the LRRs form a curved, solenoid structure where specific protein interactions are mediated through the side chains of variable residues protruding from the short parallel β-strands that form the inner (concave) surface of the solenoid, a sort of 3D coil. The LRRs are preceded by a cysteine-rich region at the N terminus, comprising four cysteine residues with a variable number of intervening amino acids, which defines the various classes (Fig. 6.1b). A C-terminal

6 Small Leucine-Rich Proteoglycans

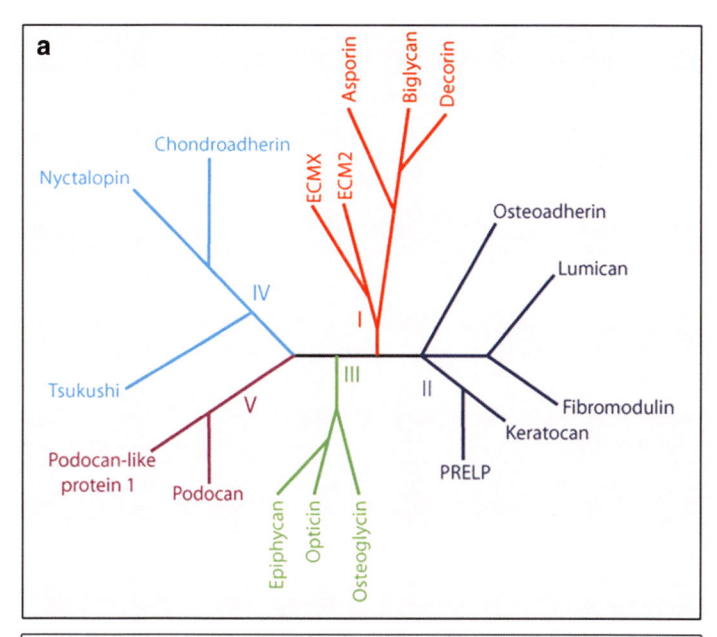

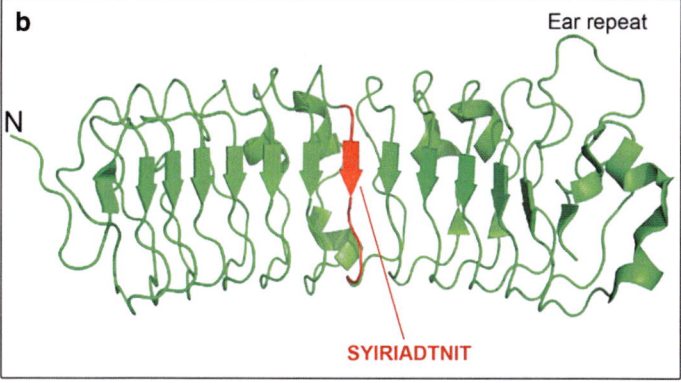

Fig. 6.1 Phylogenetic tree of small leucine-rich proteoglycans (SLRPs) and structure of the prototype decorin. (**a**) Dendogram of the five SLRP classes, *color-coded*. Various human sequences were first aligned with CLUSTALW and then an unrooted dendogram was generated using Biology Workbench. (**b**) *Ribbon diagram* of the crystal structure of monomeric bovine decorin rendered with Pymol2 (PDB accession number 1XKU). *Arrows* indicate β-strands. The area highlighted in *red* corresponds to the sequence (SYIRIADTNIT) involved in the binding to collagen type I. The ear repeat in the terminal leucine-rich repeat cysteine capping motif is also indicated (Park et al. 2008)

capping motif encompasses two terminal LRRs and includes the so-called ear repeat (Fig. 6.1b), which is present in the canonical SLRPs (classes I–III) but absent in the other two noncanonical classes (Park et al. 2008).

Decorin protein core, a Zn^{2+} metalloprotein (Yang et al. 1999; Dugan et al. 2003), is biologically active as a monomer in solution (Goldoni et al. 2004) and binds noncovalently to an intraperiod site on the surface of collagen fibrils every D period, approximately every 67 nm (Scott 1988). Specifically, decorin protein core binds near the C terminus of collagen $\alpha 1$(I) in isolated procollagen molecules close to an intermolecular cross-linking site (Keene et al. 2000). The glycosaminoglycan chains can also be involved in collagen interaction (Rühland et al. 2007; Henninger et al. 2006; Raspanti et al. 2008). Indeed, an interesting feature of decorin (and biglycan) is that the glycosaminoglycan-binding region is located near the N terminus. This feature provides a degree of mobility for the dermatan sulfate chain, which can align orthogonally or parallel to the major axis of the collagen fibril. This leads to two major properties: (1) it maintains interfibrillar space in the corneal collagen, thereby providing transparency (Scott 1988), and (2) in tendon and skin, and perhaps in other connective tissues, it guarantees the mechanical coupling of fibrils and could distribute the mechanical stress throughout the whole tissue (Vesentini et al. 2005; Reed and Iozzo 2002). The estimated binding force of ~12 × 10^3 nN of decorin core to collagen fibrils is greater than that exerted by the binding of the dermatan sulfate chain for the collagen fibrils. This suggests that overloads or other forms of mechanical stress are likely to damage the collagen mechanical integrity by disrupting the glycosaminoglycan/collagen interaction rather than decorin/collagen interaction (Vesentini et al. 2005). This is interesting insofar as genetic disruption of dermatan sulfate epimerase 1, the enzyme required for the modification of glucuronic acid to iduronic acid (responsible for the generation of dermatan from chondroitin sulfate chains), causes a mild skin fragility phenotype reminiscent of the decorin-null mice (Maccarana et al. 2009). Thus, the absence of dermatan sulfate could disrupt the proper interaction of decorin with collagen fibrils during development as also suggested by studies involving 3D collagenous matrices, which support a negative role for the glycosaminoglycan chain of decorin on collagen fibril diameter at early stages of fibril assembly (Rühland et al. 2007).

Asporin binds to collagen with an affinity in the low nanomolar range as decorin does (Kalamajski et al. 2009). Notably, asporin and decorin bind on the same region in fibrillar collagen insofar as they can effectively compete with each other at equimolar concentrations, whereas biglycan does not. However, the collagen-binding domains in these two class I SLRPs differ, being LRR_7 and LRR_{10-12} in decorin and asporin, respectively (Kalamajski et al. 2009). Another example of diversified functional activity is provided by two members of class II SLRPs, fibromodulin and lumican, both of which utilize LRR_{5-7} to bind collagen (Kalamajski and Oldberg 2009, 2010). However, during development of tendons, both lumican and fibromodulin regulate the initial assembly of collagen protomers, but only fibromodulin facilitates growth steps leading to mature fibrils (Ezura et al. 2000). Similarly, there is strong genetic evidence for the coordinated control of collagen fibrillogenesis by decorin and biglycan during development (Zhang et al. 2009), and for regulating acquisition of biomechanical properties during tendon development (Robinson et al. 2005; Zhang et al. 2006).

Another level of intricacy is provided by the potential SLRP substitution with glycosaminoglycan side chains of various types. For example, canonical class I members contain chondroitin or dermatan sulfate side chains with the exception of asporin, ECM2, and ECMX. All class II members contain polylactosamine or keratan sulfate chains in their LRRs and sulfated tyrosine residues in the N-terminal ends. Class III members contain chondroitin/dermatan sulfate (epiphycan), keratan sulfate (osteoglycin), or no glycosaminoglycan (opticin) chain. Noncanonical class IV and V members do not contain any glycosaminoglycan chain with the exception of chondroadherin, which is substituted with keratan sulfate. The attachment of the chondroitin and dermatan sulfate chains is tissue specific. For instance, in bone, the chains on both decorin and biglycan are primarily chondroitin sulfate, while in skin they are primarily dermatan sulfate. Thus, the presence or absence of specific glycosaminoglycans, together with changes in degree of sulfation or epimerization (chondroitin versus dermatan sulfate, for example), endows this class of proteoglycans with an additional layer of structural complexity.

Overall, there is specificity of binding that presumably dictates specificity of "function" among various SLRPs, in spite of their highly conserved structure. The differential binding of various combinations of SLRPs, together with differential temporal expression of SLRPs binding to collagen via the same LRRs, may indeed shape collagenous matrices into "stromal compartments" (Kalamajski and Oldberd 2010) that characterize specialization of tissues and organs.

6.3 Lessons from Gene Targeting Studies

Key information has been gathered regarding the function and tissue expression pattern of SLRPs from the available knockout mice and it has become clear that these mice can represent valuable in vivo models for various diseases such as skin fragility, osteoporosis, and muscular dystrophy (Ameye and Young 2002). The first SLRP-encoding gene to be targeted was decorin (Danielson et al. 1997), which shows a complex genomic organization and transcriptional control (Santra et al. 1994; Iozzo and Danielson 1999) as well as a widespread tissue distribution (Danielson et al. 1993; Scholzen et al. 1994). The phenotype of the decorin-deficient mice provides strong genetic evidence, in a defined animal model, for the essential role of SLRPs in regulating collagen fibrillogenesis, which was until then mostly based on cell-free experimental systems (Vogel et al. 1984). These mice present with abnormal collagen fibril morphology in the skin and tail tendon (Table 6.1). Presumably, collagen fibrils lacking decorin might be less stable due to abnormal posttranslational modifications such as cross-linking (Keene et al. 2000) or enhanced susceptibility to collagenases (Geng et al. 2006). The most obvious phenotype of the decorin-null mice, explainable with the high decorin expression in the dermis, is skin fragility resulting from thinner dermis and reduced tensile strength, a mechanical impairment directly linked to the abnormal collagen network. This phenotype mimics some of the cutaneous defects observed in the human Ehlers–Danlos syndrome, also known as *Cutis hyperelastica*, characterized by skin hyperextensibility and tissue fragility. Ultrastructural analysis of dermal collagen

Table 6.1 Pathological consequences of targeted ablation of various SLRP genes in mice

Targeted gene	Molecular pathology	Phenotype	References
Decorin	Abnormal collagen fibril structure in dermis and tendon	Skin fragility	Danielson et al. (1997)
	Disruption of enteric cell maturation	Intestinal tumor formation	Bi et al. (2008)
Biglycan	Reduced growth rate and decreased bone mass	Progressive osteoporosis	Xu et al. (1998)
	Structural abnormalities of collagen fibrils in aortic media	Spontaneous aortic dissection and rupture	Heegaard et al. (2007)
Lumican	Abnormal collagen fibril architecture in cornea and dermis	Skin fragility and corneal opacity	Chakravarti et al. (1998)
Fibromodulin	Abnormal collagen fibril structure in tendon	No overt phenotype	Svensson et al. (1999)
Keratocan	Abnormal collagen fibril structure in the corneal stroma	Altered cornea shape and reduced visual acuity	Liu et al. (2003)
Biglycan/Decorin	Abnormal collagen fibril formation in bone, tendon and dermis	Mimics the progeroid variant of human Ehlers–Danlos syndrome	Corsi et al. (2002)
	Hypomineralization of frontal and parietal craniofacial bones	Impaired posterior frontal sutural fusion	Wadhwa et al. (2007)
	Impaired amelogenesis: massive deposition of enamel and hypomineralization of dentin	Abnormal tooth development	Goldberg et al. (2005)
Biglycan/fibromodulin	Structural and mechanical abnormalities in collagen fibrils in tendon affected by exercise	Progressive gait impairment, ectopic tendon ossification and premature osteoarthritis	Ameye et al. (2002); Kilts et al. (2009)
		Impaired tendon function	
Decorin/p53	Cooperative action of germ-line mutation permissive for tumorigenesis	Rapid development of T cell lymphomas	Iozzo et al. (1999a)
Decorin/dentin sialophosphoprotein	Lack of decorin rescues the abnormal dentin mineralization caused by deficiency of dentin sialophosphoprotein	Restoration of predentin structure	Haruyama et al. (2009)
Lumican/fibromodulin	Abnormal collagen maturation and architecture in tendons	Joint laxity and impaired tendon function	Jepsen et al. (2002)
Epiphycan/biglycan	Abnormal collagen fibrils in sclera, increased ocular axial length, thin sclera and retinal detachment	Mimics high myopia	Chakravarti et al. (2003); Nuka et al. (2010)
	Damage and erosion of articular cartilage		
	Increased osteophyte formation within joint and ossification of tendons	Premature onset of osteoarthritis	

fibrils in decorin-null mice displays irregular outlines and size variability with uncontrolled lateral fusion (Fig. 6.2a, b). The periodicity of collagen fibers is maintained in the decorin-null mice, likely because of compensatory occupation of the *d* band by other SLRPs (Fig. 6.2c, d), suggesting that the concerted action of multiple SLRPs might determine the final structure and function of collagenous matrices. Accordingly, the decorin-null mouse has a mild phenotype and has become one of the most utilized animal models to investigate the role of this SLRP under various experimental challenges (Brown et al. 2001; Schaefer et al. 2002;

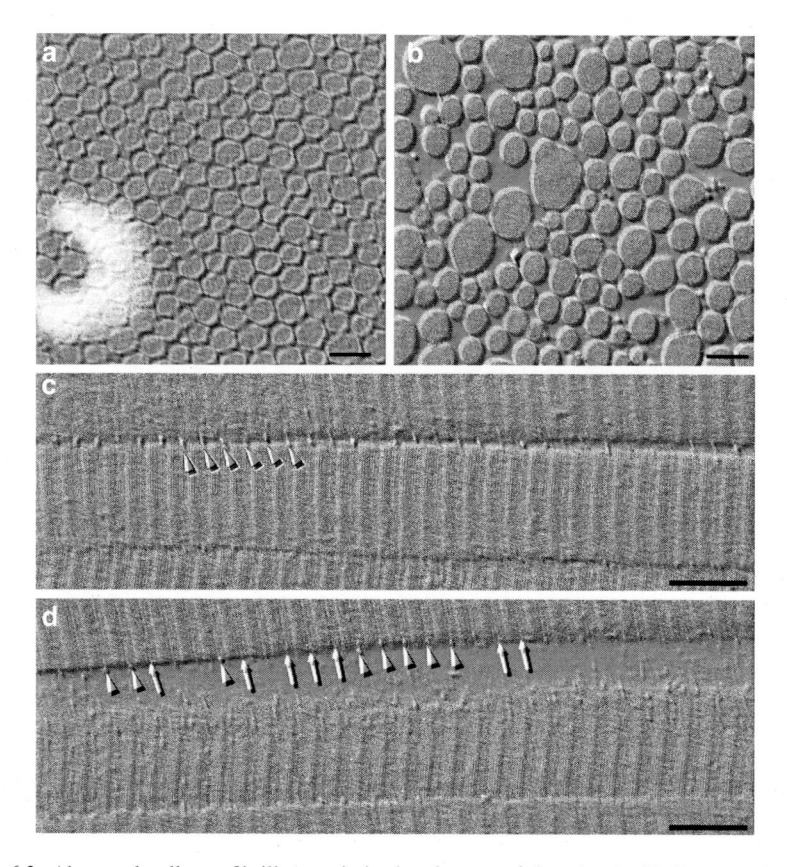

Fig. 6.2 Abnormal collagen fibrillogenesis in the absence of decorin. (**a, b**) Cross sections of wild-type (**a**) and decorin-deficient (**b**) dermal collagen fibers. Note the presence of larger and irregular fibrils in the decorin-deficient animals. *Scale bars* ~120 nm. (**c, d**) Longitudinal sections of tendon collagens from wild-type (**c**) and decorin-deficient animals (**d**) following staining with *cuprolinic blue*. Notice the presence of proteoglycan granules (*arrowheads*) in nearly all the *d* bands of banded collagen in wild-type tendons (**d**). In contrast, decorin-deficient animals show areas lacking proteoglycan granules (*arrows*), albeit the cross-banding of collagen is relatively well maintained (**d**). All the electron micrographs were modified (*embossed*) using Adobe Photoshop CS2 to enhance visualization of fibril architecture. *Scale bars* ~200 nm

Häkkinen et al. 2000; Weis et al. 2005; Elliott et al. 2003; Liang et al. 2004; Fust et al. 2005; Williams et al. 2007; Merline et al. 2009) (Table 6.1).

Notably, knockdown of zebrafish decorin causes a severe phenotype, presumably because of lack of compensation by other SLRPs, which is characterized by abnormal convergent extension, craniofacial abnormalities, and cyclopia (Zoeller et al. 2009). These features are similar to several zebrafish mutants affecting the noncanonical Wnt signaling pathway, suggesting that decorin might play a role in this pathway.

The targeted deletion of the biglycan gene, which codes for a proteoglycan with a widespread tissue distribution and a pronounced expression in bone (Bianco et al. 1990; Wegrowski et al. 1995), reveals a central role for this SLRP in regulating postnatal skeletal growth (Xu et al. 1998). Bones grow more slowly and are ultimately shorter and the bone mass is reduced compared with wild-type mice due to a significant decline in osteoblast number and progressive depletion of the bone marrow stromal cells (Xu et al. 1998). For this reason biglycan-null mice represent a good model to study osteoporosis (Table 6.1). These mutant mice also display broader metadentin and altered dentin mineralization leading to enamel structural defects. Although decorin deficiency results in changes in collagen fibril size and organization in the bone, it does not affect bone mass and growth like in the case of biglycan deficiency, pointing at non-overlapping and specific functions that have evolved for these two highly homologous class I SLRPs in vivo.

The complete lack of decorin in mouse cornea can be compensated by biglycan and the lack of both decorin and biglycan results in a more severe phenotype, supporting the idea that these two SLRPs have some overlapping functions (Zhang et al. 2009). Indeed, decorin, biglycan, and lumican play an interactive role in regulating collagen fibrillogenesis in the mouse endometrium, a biological process linked to the stage of pregnancy (Sanches et al. 2010).

Fibromodulin-deficient mice develop structural and mechanical collagen alterations in their tendons (Svensson et al. 1999), which could explain the observed joint laxity (Jepsen et al. 2002) and increased incidence of osteoarthritis (Gill et al. 2002) in these mutant animals. Fibromodulin-deficient mice also have impaired collagen fibrillogenesis in predentin, which could be the basis for altered dentin mineralization directly and indirectly for defects in enamel formation arising from abnormal epithelial (enamel)–mesenchymal (dentin) interaction (Goldberg et al. 2006).

Lumican is highly expressed in the cornea and lumican-null mice develop progressive corneal opacification, indicating that lumican is not essential for a correct embryonic corneal development but plays an important role during postnatal life (Chakravarti 2003). Collagen fibrils in the posterior area of the cornea are thicker and loosely packed and consequently the light is poorly reflected. In addition to the corneal phenotype, these mice display a skin phenotype similar to the decorin-null mice. Notably, transplant of human stem cells isolated from corneal stroma into the corneas of lumican-null mice is capable of restoring corneal transparency (Du et al. 2009). This is an exciting translational study and supports the idea of the "immune privilege" of adult stem cells and the ability to regenerate tissue in a fashion analogous to organogenesis and noticeably different from normal wound healing. Stem cell-based therapy might become an effective treatment of human corneal diseases in the near future.

Keratocan is another major component of the cornea and the keratocan-deficient mouse displays a cornea-specific phenotype where the corneal stroma is thinner, due to minor collagen fibrillogenesis alterations, and the cornea–iris angle is narrower (Liu et al. 2003). Interestingly, in the keratocan-null cornea, the expression of the other SLRPs, decorin, lumican, and fibromodulin is not affected. Due to the altered corneal shape, vision acuity is reduced, a feature evident in the human mutation of the keratocan gene (see below).

6.4 Human Genetic Diseases Caused by Mutations in SLRP Genes

There are very few human genetic diseases linked to specific mutations of SLRP genes. With the exception of asporin, in which a D14 allele (14 aspartate residues) has been linked to an increased susceptibility to osteoarthritis predominantly in Asian patients (Kizawa et al. 2005), all the SLRP-linked genetic defects cause ocular defects (Table 6.2). Mutations in the lumican and keratocan genes lead to high myopia and cornea plana, respectively (Wang et al. 2006; Majava et al. 2007). The lack of keratocan results in a flattened curvature of the cornea that leads to hypermetropia, astigmatism, and poor acuity. The case of decorin is particularly interesting because the decorin-null mice do not display any corneal abnormalities, whereas mutations in the human decorin gene cause a rare form of congenital stromal dystrophy of the cornea (Bredrup et al. 2005b; Rødahl et al. 2006). Specifically, a single base pair deletion in exon 10 leads to a loss of the terminal 33 amino acid residues, including the "ear repeat." This truncated decorin would act in a dominant-negative fashion and disrupt the collagen-regulating activity of the intact decorin (Bredrup et al. 2005a).

Mutations of human nyctalopin, a GPI-anchored SLRP expressed in the retina (O'Connor et al. 2005), cause X-linked congenital stationary blindness (Bech-Hansen et al. 2000; Pusch et al. 2000). This suggests that nyctalopin might be directly involved in establishing or maintaining functional contacts between rod photoreceptor cells and postsynaptic neurons, involved in the transmission of visual information.

It has been suggested that the expression of decorin and biglycan is altered in various forms of muscular dystrophies (Brandan et al. 2008). Specifically, both SLRPs are upregulated in skeletal muscle biopsies of Duchenne muscular dystrophy patients and the source was identified in the muscle fibroblasts (Fadic et al. 2006). The increased synthesis suggests a response of the muscle to the dystrophic damage and the fibrotic process. Interestingly, the biglycan-null mice develop a

Table 6.2 Human diseases linked to mutations in SLRP-encoding genes

Gene	Type of mutation	Type of inheritance and affected chromosome (s)	Molecular pathology	Clinical phenotype and references
Decorin	Frameshift mutation generating a C-terminal truncated decorin protein core	Autosomal dominant chromosome 12	Corneal opacities caused by deposition of white fluffy material in the corneal stroma	Congenital stromal dystrophy of the cornea (Bredrup et al. 2005a)
Lumican, fibromodulin, PRELP, and opticin	Intronic variations, nonsynonymous and synonymous changes, SNPs in promoter	Autosomal dominant chromosomes 1 and 2	Corneal detachment and choroidal neovascularization	High myopia (Majava et al. 2007; Wang et al. 2006; Chen et al. 2009).
Keratocan	Missense and frameshift mutations generating a single amino acid substitution or a C-terminal truncated keratocan	Autosomal recessive chromosome 12	Corneal radius of curvature larger than normal producing high hypermetropia with astigmatism and poor acuity	Cornea plana (CNA2) (Pellegata et al. 2000)
Nyctalopin	Intragenic deletions, missense mutations, nonsense mutations, and in-frame insertions	X-linked X chromosome	Disruption of developing retinal interconnections between rod photoreceptors and postsynaptic neurons	Congenital stationary night blindness, with associated myopia, hyperopia, nystagmus, and reduced visual acuity (Bech-Hansen et al. 2000; Pusch et al. 2000)

mild muscular dystrophy explained by lack of the complex between biglycan and α-dystroglycan, suggesting a key role for biglycan in maintaining the structure of the muscle extracellular matrix (Rafii et al. 2006). The regulation and sarcolemmal localization of other critical muscle components including dystrobrevin, syntrophin, and nNOS are also altered in biglycan-null mice and the mild dystrophic phenotype could be "rescued" by injecting biglycan into skeletal muscles (Mercado et al. 2006).

The involvement of SLRPs in muscular dystrophy certainly needs more investigation also considering the small number of patients tested in the available studies. This field of research, once expanded, certainly deepens our knowledge regarding the role of decorin and other SLRPs in inflammation, diabetes, fibrosis, and metabolic pathways.

6.5 Interaction with Growth Factors

It has become clear that SLRPs are signaling molecules in addition to playing structural functions in the extracellular matrix. Through binding to growth factors and receptors on the cell surface, they can regulate the complex intracellular signaling cascade and determine cell fate (Iozzo and Schaefer 2010). The protein core and specifically the LRR motifs have been demonstrated to retain the biological function, but certainly more needs to be investigated regarding the possible role of the glycosaminoglycan side chains during signaling. The high-affinity interaction between decorin and various TGF-β isoforms was discovered two decades ago and explains the antifibrotic effects of decorin in damaged tissues (Yamaguchi et al. 1990). The association is disrupted by matrix metalloproteinases that cleave decorin and cause the release of TGF-β. Decorin/TGF-β interactions are quite complex and lead to a variety of outcomes such as controlling growth and survival of normal and neoplastic cells (Ständer et al. 1998, 1999), regulating matrix organization and mechanical characteristics of 3D matrices (Ferdous et al. 2007, 2008, 2010; Seidler et al. 2005), blocking fibrosis in various animal models (Iozzo 1999), and preventing intimal thickening (Fischer et al. 2001). In addition, decorin and biglycan interact with tumor necrosis factor-α (TNF-α) (Tufvesson and Westergren-Thorsson 2002).

In the muscle, decorin and the TGF-β signaling pathways cooperate in regulating myoblast proliferation and differentiation. Specifically, decorin inhibits the expression of myogenin, a muscle-specific transcription factor that promotes myoblasts differentiation. Decorin has also been reported to modulate myoblasts proliferation in vitro through binding of myostatin (Miura et al. 2006), a member of the TGF-β family of growth factors (Kishioka et al. 2008). Decorin sequesters myostatin in the extracellular matrix and, as a consequence, favors myogenic cell proliferation and differentiation, as proved by increased expression of $p21^{WAF1}$, a cyclin-dependent kinase inhibitor, negative regulator of cell cycle progression, MyoD, and myogenin. Notably, decorin interacts with the insulin-like growth factor I (IGFI) as well as its receptor (Schönherr et al. 2005; Schaefer et al. 2007; Merline et al. 2009) and can bind to platelet-derived growth factor (PDGF) BB via its protein core (Nili et al. 2003) or its dermatan sulfate side chain (Kozma et al. 2009). Decorin overexpression can indeed block PDGF-evoked activation of PDGF receptor and smooth muscle cell growth, thus providing a potential mechanism for the decorin-mediated inhibition of intimal hyperplasia following balloon angioplasty (Nili et al. 2003).

Decorin and biglycan might have different roles during skeletal muscle formation and repair (Brandan et al. 2008). Biglycan expression levels decrease during development and are normally very low in the adult muscle unless muscle damage has occurred. They both sequester TGF-β but, in addition to this, decorin binds to the cell surface receptor low-density lipoprotein receptor-related protein 1 affecting muscle signaling through activation of phosphoinositide 3-kinase and indirect enhancement of the Smad pathway downstream of the TGF-β receptor (Brandan et al. 2006). Overall, decorin and biglycan favor bone formation by sequestering TGF-β.

BMPs are growth factors involved in bone and cartilage formation, also part of the TGF-β superfamily. Both decorin and biglycan have been shown to interact with some members of this family. Specifically, decorin regulates BMP2 signaling during the conversion of myoblasts to osteoblasts (Gutierrez et al. 2006). In *Xenopus*, biglycan binds BMP4 and regulates BMP4 signaling through modulation of the antagonist Chordin (Moreno et al. 2005). Another SLRP, tsukushi, inhibits BMP activity (Ohta et al. 2004, 2006; Kuriyama et al. 2005). In the mouse, lack of biglycan leads to reduced BMP4 binding to osteoblasts, indicating that this SLRP modulates BMP4-evoked signaling to control osteoblast differentiation (Chen et al. 2004). Asporin binds BMP2 and negatively regulates BMP2-induced cytodifferentiation of periodontal ligament cells by preventing binding of BMP2 to its receptor (Yamada et al. 2007). Notably, the binding for BMP2 was mapped to asporin LRR$_5$, the LRR that binds collagen in the homologous decorin, and some of the BMP2 regulatory activity of asporin could be blocked by a peptide encompassing asporin LRR$_5$ (Tomoeda et al. 2008).

Additional studies need to be performed to resolve the multiplicity of activities, some specific and some overlapping, of SLRP/growth factor interactions and to help rationalize SLRP complexity.

6.6 Signaling Through Multiple Receptors

6.6.1 EGFR and Met

An emerging body of data indicates that decorin and perhaps other SLRPs play a physiological role in negatively regulating cell proliferation primarily by attenuating receptor tyrosine kinase (RTK) such as members of the ErbB family of RTKs. Decorin binds to a region on the EGFR extracellular domain overlapping with the EGF-binding domain (Santra et al. 2002). Upon binding, the receptor dimerizes and

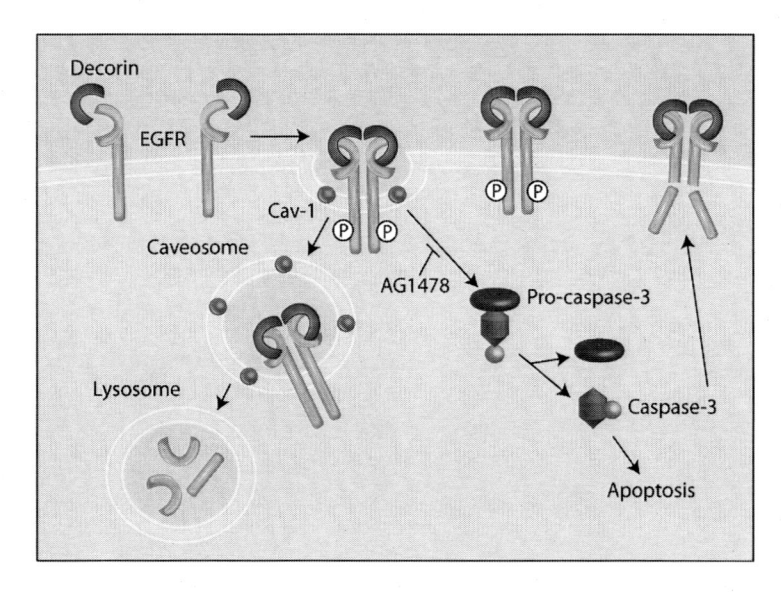

Fig. 6.3 Diagram depicting the mechanism of action of decorin in inhibiting the EGFR signaling pathway. Decorin binds to the receptor and induces caveolin-mediated (Cav-1) internalization and degradation in the lysosomes. An additional mechanism to shut down this pathway is activation of the caspase-3 cascade following transient EGFR phosphorylation, and consequent cleavage of the EGFR kinase domain. AG1478 is a tyrphostin that specifically blocks the activity of EGFR tyrosine kinase and its downstream signaling

is removed from the cell surface through caveolin-mediated endocytosis (Zhu et al. 2005) (Fig. 6.3). Following internalization, EGFR is downregulated by degradation in the lysosome. In contrast, EGF triggers EGFR internalization via clathrin-coated pits, an event that has been associated with signaling and recycling of the receptor to the cell surface. It is very intriguing that EGF, the natural ligand of EGFR, induces the same changes, specifically, dimerization and degradation, but leads to the opposite outcome in terms of signaling and biological effects. Decorin also transiently activates the EGFR and mobilizes intracellular Ca^{2+} stores (Patel et al. 1998), but induces cell growth suppression by evoking the expression of $p21^{WAF1}$ (De Luca et al. 1996). By affecting the EGFR, decorin can inhibit other members of the ErbB family of receptor tyrosine kinases, such as ErbB2 (Santra et al. 2000), which heterodimerizes with EGFR. The consequence of decorin interaction is a prolonged suppression of cellular signaling required for cell survival and proliferation, making decorin a natural "pan-RTK" inhibitor. This is in agreement with the fact that decorin inhibits the proliferation and migration of human trophoblasts via different RTKs (Iacob et al. 2008). In addition, decorin induces apoptosis in carcinoma cells via activation of caspase-3, an event downstream of EGFR signaling (Seidler et al. 2006; Goldoni and Iozzo 2008).

The Met receptor has been found to be directly affected by decorin (Goldoni et al. 2009). Decorin binds to the Met receptor and triggers specific signaling that

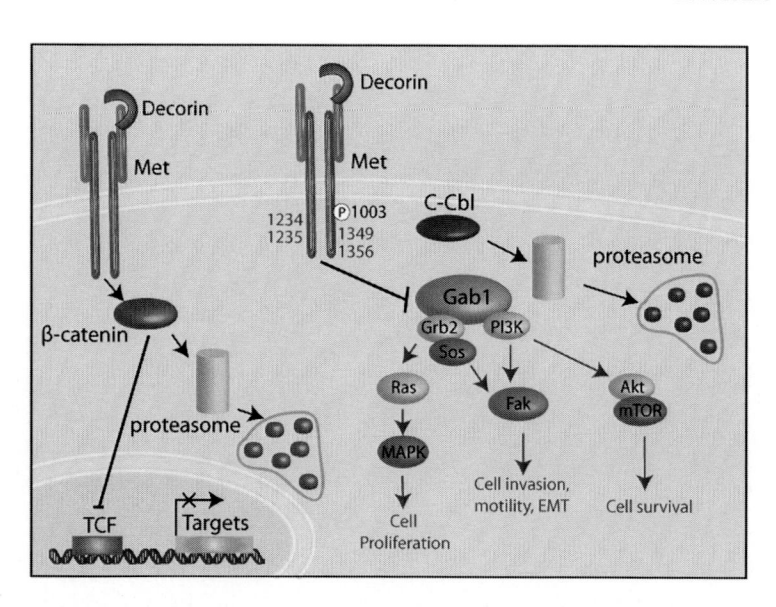

Fig. 6.4 Diagram depicting the mechanism of action of decorin in inhibiting the Met receptor signaling pathway. Following decorin binding, Met-Tyr^{1003} is activated, the ubiquitin ligase c-Cbl is recruited and the receptor is sent for degradation into the proteasome. Note that, at the same time, decorin inhibits Met-Tyr^{1349} impeding the major signaling events downstream of Met that would lead to cell survival, proliferation, and invasion. β-Catenin is also degraded along with the Met receptor, depriving cells of a transcription factor essential for cell cycle progression

leads to its downregulation by both shedding of the extracellular domain and internalization/degradation (Fig. 6.4). In the same study, downregulation of Met following decorin treatment has been linked to degradation of β-catenin, a transcription factor essential for cell cycle progression. Considering that both EGFR and Met are often deregulated in various forms of cancer and that coexpression (and/or co-amplification) of these two receptors drives the tumorigenesis process, a deep understanding of decorin's mechanism of action could lead to novel therapeutic approaches against malignancies.

Considering the opposite biological effects achieved by decorin vis-à-vis EGF/TGF-α and HGF following binding to the same receptors, we envision a scenario where stromal cells, the producers of decorin, a natural EGFR and Met antagonist, could potentially counteract the growth-promoting and prosurvival activities of RTKs within a growing neoplasm. The importance of this biological interplay in a pathological condition such as cancer has been reasonably explored, whereas the role of decorin in physiological tissue homeostasis has not. The most common route by which cell fate is regulated is by signaling through different receptors and ligands and by positive or negative feedbacks originating from inside the cell following the extracellular stimuli. Decorin represents a novel example of cell cycle regulator which, by a unique mode of binding to the EGF and Met receptors, triggers specific downstream signaling that differs from the one evoked by EGF and HGF. This type of regulation adds an additional layer of complexity to the known canonical pathways by which cells respond to extracellular cues.

6.6.2 Type 1 Insulin-Like Growth Factor Receptor and Toll-Like Receptors

Decorin is also involved in the insulin-like growth factor receptor (IGF-IR) pathway. This interaction has been studied in endothelial cells (Schönherr et al. 2005) where it could represent a major player in the regulation of physiological and pathological angiogenesis (Schönherr et al. 2004, 2005). In the kidney, decorin regulates the deposition of fibrillin-1 by triggering specific signaling in renal fibroblasts through the IGF-IR (Schaefer et al. 2007).

Biglycan is an endogenous ligand of Toll-like receptor 2 (TLR2) and TLR4 in macrophages and stimulates the expression of TNF-α and macrophage inflammatory protein-2 via activation of p38 and NF-κB (Schaefer et al. 2005). This activity is highly proinflammatory (Schaefer 2010) explaining why biglycan-null mice present an advantage in LPS-induced sepsis. The molecular mechanisms linking biglycan action to TLRs require the formation of a receptor cluster with P2x (Babelova et al. 2009). This cooperative receptor clustering triggers the NALP3 inflammasome expression which, in turn, activates caspase-1 and IL-1β release. Biglycan, along with decorin, can also function as an anti-inflammatory protein by binding to and blocking the complement protein C1q, thereby inhibiting activation of the complement cascade and proinflammatory cytokine production at the tissue level (Groeneveld et al. 2005).

Lumican is also involved in innate immune response by affecting TLR4 signaling pathway. Lumican-deficient macrophages show impaired response to LPS resulting in lower production of TNF-α and IL-6 (Wu et al. 2007). A possible mechanism of action involves CD14 and the presentation of LPS to TLR4 through this cell surface molecule. Lumican produced by vascular endothelial cells binds to the surface of extravasating leukocytes via β2-containing integrins and promotes leukocyte migration during inflammation (Lee et al. 2009). In addition to its role in inflammation, lumican has been shown to inhibit proliferation of stromal keratocytes in the cornea through activation of p21^{WAF1} and p53 and to induce apoptosis through enhancing Fas–Fas ligand signaling (Vij et al. 2004).

Biglycan may work through other receptors for immune responses including selectin/CD44 where it can selectively recruit peripheral blood CD16(−) natural killer cells into human endometrium (Kitaya and Yasuo 2009). Other factors may also modulate the LPS-induced inflammation. Both keratocan and lumican regulate neutrophil infiltration and corneal clarity in LPS-induced keratitis by direct interaction with CXCL1 (Carlson et al. 2007). Biglycan, decorin, fibromodulin, and lumican can all bind to C1q and differentially activate the classical complement pathway, thereby having implications in chronic inflammatory processes.

6.7 Skeletal Connective Tissues

6.7.1 Bone Remodeling and Osteoporosis

Biglycan is predominantly expressed in bone and its genetic ablation results in reduced skeletal growth and bone mass leading to generalized osteopenia (Xu et al. 1998). The mice have less trabecular bone volume and reduced cortical thickness, both important for bone strength and integrity. The important role for biglycan in osteogenesis was confirmed by the fact that its absence following marrow ablation directly impedes bone formation (Chen et al. 2003). Notably, the biglycan gene resides on the X chromosome and patients with Turner syndrome (45,X), a disease characterized by short stature and early-onset osteoporosis, display low levels of biglycan expression. This raises the possibility that bone metabolism in biglycan-deficient mice might be gender dependent. In contrast to male mice, the bone tissue of female mice is less affected, suggesting a gender difference in biglycan skeletal function.

The effects of biglycan deficiency on bone can be linked to collagen fibril abnormalities. These mutant fibrils display an irregular profile, a broader-size range, and reduced packing (Corsi et al. 2002). Interestingly, decorin deficiency also affects the collagen fibril size and shape in bone but in an opposite way: decorin-null mice have smaller average fibril diameter and size range in bone compared to wild-type animals, whereas in the dermis and tendon the fibrils are larger. Thus, it is not surprising that the skeletal phenotypes of the biglycan- and decorin-deficient mice differ from one another and that mice lacking decorin do not feature the marked osteopenia of the biglycan-deficient mice. Double-deficient mice display an almost complete loss of fibril basic geometry, with very few fibrils possessing a predominantly circular cross-sectional profile. The vast majority of the fibrils have a "serrated" fibril morphology observed in many human disorders, including Ehlers–Danlos syndrome. Mice lacking both biglycan and decorin are grossly osteopenic and this abnormality is much more severe (Fig. 6.5) and appears at an earlier age as compared to biglycan-null mice (Bi et al. 2005). Thus, decorin deficiency synergizes with biglycan deficiency in controlling bone mass, although the effects of the individual SLRP deficiencies in bone are quite distinct.

In addition to biglycan and decorin, the SLRP asporin may also play a role in regulating collagen fibril structure in bone (Kalamajski et al. 2009). Asporin competes with decorin, but not biglycan, for binding to collagen where the poly-aspartate in asporin directly regulates collagen mineralization by its collagen- and calcium-binding properties (Heinegård 2009).

While biglycan appears to regulate collagen fibril formation, it is unclear how this could impact the geometric and mechanical properties of mature bone. When bones from biglycan-deficient mice were tested for mechanical strength they had decreased failure load (to bend) and yield energy (to break) at 6 months (Corsi et al. 2002), with biglycan-deficient tibia being the most affected (Wallace et al. 2006).

Wild type   **Biglycan null**   **Biglycan/Decorin null**

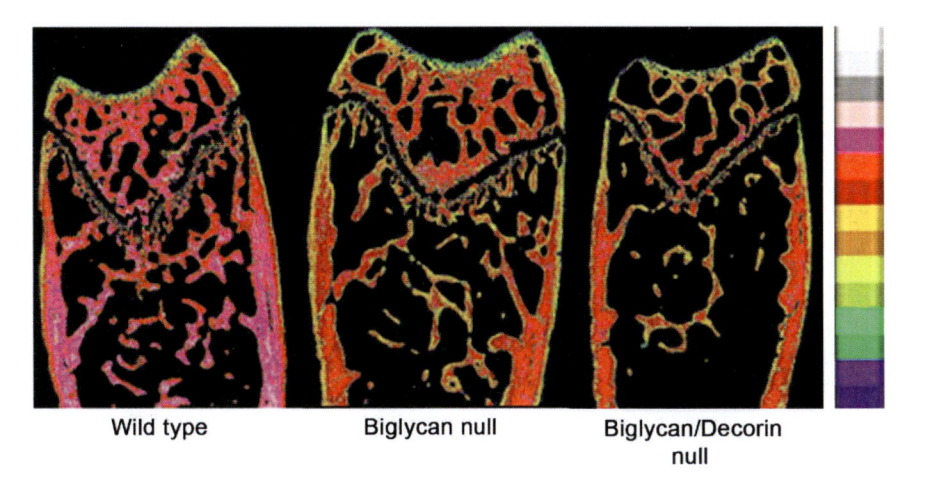

Fig. 6.5 Quantitative backscattered electron imaging of the distal femur from 2-month mice of various genotypes as indicated at the *bottom*. Trabecular bone mass is markedly reduced in double-null animals. A reduced mineral content (as indicated by the *pseudocolor scale*, in which higher values are at the *top*) is observed both in the biglycan-null and in the biglycan/decorin double-deficient animals as compared to the wild-type mice. Reproduced and slightly modified from Corsi et al (2002) with permission of the American Society for Bone and Mineral Research

Biglycan and decorin are highly expressed in the craniofacial bones and cranial sutures of mice and have overlapping yet distinct patterns of expression in the fusing suture and the dura mater (Wadhwa et al. 2007). Mice singly deficient in either biglycan or decorin have no suture formation defects, whereas double-deficient mice have open sutures and severe hypomineralization of both frontal and parietal bones, thus confirming that these two SLRPs have some synergistic effects in regulating craniofacial morphology.

Biglycan-deficient bones exhibit reduced osteoblasts and lower bone formation rates, suggesting that the osteogenic progenitor cells might be affected. Indeed, both the quality and normal activity of bone marrow stromal cells from biglycan-deficient mice are reduced. Specifically, these cells have reduced clonogenic response to TGF-β, produce less type I collagen mRNA and collagen protein, and have enhanced apoptotic rate. Interestingly, when both biglycan and decorin are depleted, the growth factor interplay is also affected but in the opposite way. Specifically, the matrix made by the double-null mice is unable to retain TGF-β and subsequently to maintain proper sequestration (Bi et al. 2005). The excess TGF-β then directly binds to its receptors and over-activates its signaling transduction pathway leading to apoptosis. This results in decreased numbers of osteoprogenitor cells and subsequently reduced bone formation. The class II SLRP PRELP also impairs osteoclastogenesis, thus preventing osteopenia, via its glycosaminoglycan-binding domain which acts as a cell type-specific inhibitor of NF-κB, an established transcriptional inducer of osteoclast-specific gene expression (Rucci et al. 2009). Recently, biglycan and fibromodulin have been found to play key roles in maintaining cartilage integrity and in regulating chondrogenesis and extracellular matrix turnover during the development of temporomandibular osteoarthritis (Embree et al. 2010).

In addition to TGF-β, BMPs also play important roles in regulating osteoblast differentiation and bone formation. The absence of biglycan causes less BMP4 binding to osteoblasts (Chen et al. 2004), resulting in a reduced BMP4-stimulated expression of the osteoblast-specific transcription factor Cbfa1 and ultimately causing a defect in the differentiation process. Interestingly, decorin accumulates to higher levels in biglycan-deficient cell cultures, and thus the distribution ratio of other matrix proteins is changed as a consequence of missing a single protein. This compensation causes additional changes in the extracellular matrix further influencing growth factor activity.

6.7.2 Nonmineralized Musculoskeletal Tissues

Besides controlling bone growth during aging, SLRPs also play important roles in the assembly of normal tendons as well as in the maintenance of articular cartilage. Collagen fibrils in tendons from mice deficient in biglycan and/or fibromodulin are structurally and mechanically altered resulting in unstable joints (Ameye et al. 2002). As a result, these mice develop gait impairment, ectopic ossification in tendon, and severe premature osteoarthritis. At 3 months, both single- and double-deficient mice display torn cruciate ligaments and ectopic ossification in their quadriceps tendon, menisci, and cruciate and patellar ligaments (Kilts et al. 2009). The phenotype is least severe in fibromodulin-null, intermediate in biglycan-null, and the most severe in the double-deficient mice. These problems worsen with age in all three mouse strains and result in the development of large supernumerary sesmoid bones (i.e., bones formed within tendons in regions that wrap around bony prominences). Moderate exercise decreases ectopic ossification in double-deficient mice compared with unchallenged mice, whereas rigorous forced use of the joints further increases ectopic ossification and osteoarthritis. Loss of decorin affects the patellar tendon causing an increase in modulus and stress relaxation, but with little effect on the flexor digitorum longus tendons (Robinson et al. 2005). Conversely, biglycan loss does not significantly affect the patellar tendons, but causes a reduction in both the maximum stress and modulus of the flexor digitorum longus tendon. Thus, biglycan, decorin, and fibromodulin all play critical roles in regulating the structure and function of tendons.

The presence of ectopic ossification in tendons of mice deficient in both biglycan and fibromodulin suggests that tendons could have stem cells that form bone rather than tendon in pathological situations. Human and mouse tendons indeed harbor a unique cell population, termed tendon stem/progenitor cells, which has universal stem cell characteristics such as clonogenicity, multipotency, and self-renewal capacity (Bi et al. 2007). Isolated tendon stem/progenitor cells can regenerate tendon-like tissues after extended expansion in vitro and transplantation in vivo.

As these cells reside within a niche that is surrounded predominantly by extracellular matrix proteins, the latter likely plays a major role in organizing this specialized stem cell niche. Depletion of biglycan and fibromodulin affects the differentiation of tendon stem/progenitor cells by modulating BMP signaling and thereby impairs tendon formation in vivo.

Biglycan- or fibromodulin-null mice exhibit a mild form of knee osteoarthritis, whereas mice doubly deficient in these SLRPs develop severe and premature osteoarthritis in the knee (Ameye et al. 2002). Osteoarthritis appears within the first 3 months of life, and by 6 months the joint is almost completely destroyed. Specifically, the biglycan/fibromodulin-deficient knees display a progressive degeneration of the articular cartilage from early fibrillation to complete erosion, subchondral sclerosis, osteophytes, and bone cysts. Several cellular events are altered during the progression of osteoarthritis, including abnormal expression of aggrecan and type II collagen. Mice deficient in either lumican or lumican and fibromodulin also develop premature knee osteoarthritis, but this occurs more slowly than the mice doubly deficient in biglycan and fibromodulin (Jepsen et al. 2002). Collectively, these data indicate that SLRPs play critical roles in regulating the formation and function of the skeleton, but have unique roles depending on the tissue context.

6.8 Cardiovascular Homeostasis and Diseases

Biglycan is an important regulator of elastogenesis insofar as ectopic expression of a mutant form of biglycan lacking the two glycosaminoglycan-binding sites in vascular smooth muscle cells induces tropoelastin gene expression and increases deposits of cross-linked elastin in vivo (Hwang et al. 2008). The molecular mechanism of this process is not clear. However, it is known that biglycan protein core binds to tropoelastin and elastic fiber microfibrils where it forms a ternary complex with tropoelastin and microfibrillar-associated glyprotein-1 (MAGP-1) (Reinboth et al. 2002), and enhanced biglycan expression coincides with the elastogenic phase of elastic fiber formation during development of nuchal ligament (Reinboth et al. 2000). Thus, it is likely that biglycan could promote physiological interactions among the major components of the elastic fiber assembly (Hwang et al. 2008). This concept is supported by the evidence that biglycan evokes expression of fibrillin-1, a key constituent of microfibrils that form the scaffold on which tropoelastin builds up (Schaefer et al. 2004).

Other studies link biglycan to the development of atherosclerosis by acting as sink for LDL particles. Biglycan confers LDL-binding properties likely mediated by "hyperelongated" glycosaminoglycan side chains (Little et al. 2008), and its accumulation in early human atherosclerosis precedes the inflammatory response (Nakashima et al. 2007), providing strong evidence for a key role for this SLRP in atherogenesis. The inflammatory marker serum amyloid A (SAA) increases TGF-β and, subsequently, biglycan in aorta tissue (Wilson et al. 2008). SAA is elevated in obesity and cardiovascular disease and colocalizes with biglycan and ApoB in the vascular wall (King et al. 2009). Biglycan binds to LDL, accumulates in the subendothelial matrix, and is displaced by endostatin. This proteolytic fragment of collagen XVIII is depleted in vessels with atherosclerosis and thereby unable to inhibit biglycan retention in the disease process (Zeng et al. 2005). Notably, long-term treatment of the pro-atherogenic ApoE-deficient mice with telmisartan, an antagonist of the angiotensin II type 1 receptor, reduces biglycan levels in the atherosclerotic plaques and inhibits atherosclerosis independently of its antihypertensive effects (Nagy et al. 2010). Thus, targeting biglycan could become a therapeutic modality against atherogenesis.

As a regulator of collagen fibrillogenesis, biglycan could play a structural role in maintaining vascular integrity. About half of biglycan-deficient male mice die suddenly within the first 3 months of life due to spontaneous aortic dissection and rupture (Heegaard et al. 2007). This is further supported by the association of low biglycan expression in aneurysms of human abdominal aorta (Theocharis and Karamanos 2002). Thus, biglycan is an essential structural component of the aortic wall and contributes to blood vessel homeostasis. Moreover, these animal studies suggest the possibility of a human disease linked to genetic mutations of biglycan (loss of function) that might lead to aortic dissection in humans.

Decorin has also been involved in regulating elastic fiber formation insofar as it binds to both elastic microfibrils and MAGP-1 (Reinboth et al. 2002; Trask et al. 2000). However, the role for decorin in elastogenesis is less clear. Decorin has been implicated in remodeling of experimentally induced myocardial infarction in mice (Weis et al. 2005). In the decorin-null animals, the infarcted areas were larger than controls as were the right ventricular remote hypertrophy and left ventricular dilatation. Moreover, echocardiography revealed depressed left ventricular systolic function, suggesting that decorin is required for proper fibrotic evolution of myocardial infarction (Weis et al. 2005). In agreement with these studies, adenovirus-mediated gene delivery of decorin in postinfarcted hearts mitigates cardiac remodeling and dysfunction (Li et al. 2009), whereas decorin gene delivery inhibits cardiac fibrosis in spontaneously hypertensive rats (Yan et al. 2009). In a similar vein, biglycan expression increases after myocardial infarction. Mice deficient in biglycan repair poorly after experimentally induced myocardial infarction and have reduced tensile strength, and overall impaired cardiac hemodynamic function (Westermann et al. 2008). A gain-of-function approach showed that the overexpression of biglycan in heart tissue increases numerous genes critical for cardiac remodeling including genes associated with cardiac protection and Ca^{2+} signaling, providing further evidence that it could regulate remodeling after myocardial infarction.

6.9 Cancer and Metastasis

One of the earliest indications that decorin affects cell growth emerged from studies using normal cells: decorin gene expression is greatly enhanced after normal diploid fibroblasts reach confluence and cease to proliferate (Mauviel et al. 1995).

In general, malignant cells do not express decorin and respond to recombinant decorin, with the exception of an osteosarcoma cell line which is resistant to decorin treatment (Zafiropoulos et al. 2008). The first evidence linking decorin to cancer development came from a study utilizing decorin/p53 double knockout mice (Iozzo et al. 1999a). Mutations in the tumor suppressor p53 are found in over half of all human cancers, and mice that lack p53 develop a spectrum of sarcomas, lymphomas, and, less frequently, adenocarcinomas. Remarkably, mice lacking both decorin and p53 genes show a faster rate of tumor development and succumb almost uniformly to a very aggressive form of thymic lymphomas within 6 months. These results indicate that decorin absence is permissive for lymphoma tumorigenesis and suggest that there might be a functional synergism between a secreted extracellular "tumor repressor" (decorin) and an intracellular "tumor suppressor" (p53) (Iozzo et al. 1999a). The second line of evidence arose from an extended analysis of the decorin-null mice. About 30% of these mice develop spontaneous intestinal tumors and a high-risk diet (high fat, low calcium and vitamin D) accelerates this process (Bi et al. 2008). A plausible molecular explanation includes the finding that the intestinal epithelium of the decorin-null mice shows a downregulation of $p21^{WAF1}$ and $p27^{kip1}$, two cyclin-dependent kinase inhibitors, and a concurrent upregulation of β-catenin, a key transcription factor that promotes cell cycle progression. Earlier studies have shown that decorin gene expression is enhanced in the stroma of colon cancer via hypomethylation of its promoter regions (Adany et al. 1990; Adany and Iozzo 1990), suggesting that decorin might be a natural RTK inhibitor. Collectively, this body of literature provides strong direct evidence for a role of decorin as a tumor repressor gene further stressing the role of the tumor microenvironment in cancer progression (Iozzo and Cohen 1993; Iozzo 1995; Friedl 2010).

An anti-oncogenic role for decorin has been documented in various experimental settings including breast cancer cells (Santra et al. 2000), ovarian carcinoma cells (Nash et al. 1999), syngeneic rat gliomas (Biglari et al. 2004), and squamous and colon carcinoma xenografts (Reed et al. 2002; Tralhão et al. 2003; Seidler et al. 2006). Moreover, attenuated decorin expression is associated with poor prognosis in invasive breast cancer (Troup et al. 2003), in aggressive soft tissue tumors (Matsumine et al. 2007), and during mammary gland carcinogenesis in TA2 mice with spontaneous breast cancer (Gu et al. 2010). Collectively, these studies support earlier observations that either ectopic expression of decorin or treatment of various cancer cells of diverse histogenetic backgrounds with exogenous decorin inhibits their growth (Santra et al. 1995). A possible mechanism of action is via a transient activation of the EGFR (Moscatello et al. 1998; Iozzo et al. 1999b), followed by downregulation of the EGFR itself (Csordás et al. 2000; Zhu et al. 2005), and the concomitant induction of $p21^{WAF1}$ (De Luca et al. 1996), which causes the cells to arrest in the G1 phase of the cell cycle (Santra et al. 1997).

Adenovirus-mediated or systemic delivery of decorin prevents metastases in an orthotopic breast carcinoma xenograft model (Reed et al. 2005; Goldoni et al. 2008). Finally, systemic delivery of decorin retards the growth of prostate cancer in a mouse model of prostate carcinogenesis where the tumor suppressor PTEN gene is conditionally deleted in the prostate (Hu et al. 2009). In this study, decorin counteracted not only EGFR but also androgen receptor activity in human prostate cancer cells. An important effect of decorin on cancer cells is the induction of apoptosis via activation of caspase-3 (Seidler et al. 2006; Goldoni et al. 2008). Overexpression of decorin in normal cells also induces apoptosis (Wu et al. 2008). Ultimately, decorin inhibits metastasis formation in breast carcinoma and osteosarcoma tumor models (Goldoni et al. 2008; Araki et al. 2009; Shintani et al. 2008).

Lumican has also been involved in growth control (Nikitovic et al. 2008a, b). It inhibits melanoma progression (Vuillermoz et al. 2004) and anchorage-independent tumor cell growth (Li et al. 2004) and a peptide derived from lumican LRR_9, named "lumcorin," inhibits melanoma cell migration (Zeltz et al. 2009). Low levels of decorin and lumican correlate with a worse prognosis in lymph node-negative invasive breast carcinomas (Troup et al. 2003). Moreover, lumican inhibits melanoma migration by affecting focal adhesion complexes (Brézillon et al. 2009). Thus, targeting constituents of the stable stroma rather than targeting the adaptable cancer cells could be an intelligent therapeutic modality toward solid tumors where tumor stroma is a predominant part of the malignant neoplasm.

6.10 Angiogenesis

The involvement of decorin in angiogenesis and particularly tumor angiogenesis is somewhat controversial. In some experimental settings, decorin seems to be pro-angiogenic (Järveläinen et al. 1992; Nelimarkka et al. 1997, 2001; Schönherr et al. 2001, 2004), whereas in other experimental settings, decorin is anti-angiogenic (de Lange et al. 2001; Grant et al. 2002; Kinsella et al. 2000; Järveläinen et al. 2006). The latter effect occurs via two potential mechanisms: by interfering with thrombospondin-1 (de Lange et al. 2001) and/or by suppressing the endogenous tumor cell production of VEGF (Grant et al. 2002). Another possibility is that decorin inhibition of endothelial cell migration is caused by a decorin-evoked stabilization of pericellular fibrillar matrix, suggesting that balanced decorin expression is important for fine-tuning angiogenesis (Kinsella et al. 2000). In an animal model of angiogenesis, it has been shown that genetic deficiency of decorin markedly increases fibrovascular invasion and enhances the formation of blood vessels in sponge implants (Järveläinen et al. 2006). Recently, peptides derived from the LRR_5 of decorin were shown to be the main mediator of decorin anti-angiogenic activity (Sulochana et al. 2005). Moreover, cleavage of decorin by MT1-MMP favors corneal angiogenesis, further indicating an angiostatic role of decorin (Mimura et al. 2008). All these reports point to a key role for decorin in vasculogenesis and angiogenesis. However, the molecular mechanism of action is unclear. As a general rule, decorin seems to be pro-angiogenic in "normal" endothelial cells (i.e., during development and wound healing), but is anti-angiogenic during pathological angiogenesis such as tumor angiogenesis. This concept has been further supported by a recent study comparing the expression of decorin in malignant and benign vascular tumors (sarcomas versus hemangiomas). Notably, decorin expression was essentially undetectable in the sarcomas whereas there was significant decorin expression in the hemangiomas and surrounding perivascular stroma (Salomäki et al. 2008). Thus, a potential effect of decorin on the tumor microenvironment resides in its ability to modulate angiogenesis.

We hypothesize that all these effects, at time discordant, could be reconciled by a single unifying hypothesis: decorin effect on RTKs such as ErbB and Met receptors is the main mechanism of action. We base this hypothesis on several important pieces of evidence. First, both EGFR and Met are pro-angiogenic and prosurvival receptors, which are enriched in the tumor vasculature (van Cruijsen et al. 2006). Activation of the EGFR pathway increases the production of VEGF, whereas neutralizing antibodies against EGFR and ErbB2 downregulate VEGF production by tumor cells (Petit et al. 1997). Simultaneous blockade of EGFR and VEGFR pathways results in a cooperative antitumor effect (Sini et al. 2005). Activation of the HGF/Met signaling pathway is potently pro-angiogenic by stimulating endothelial cell motility and growth (Bussolino et al. 1992), promoting VEGF secretion (Grant et al. 1993; Zhang et al. 2003; Saucier et al. 2004), or inhibiting the secretion of thrombospondin (Zhang et al. 2003; Saucier et al. 2004), a powerful anti-angiogenic compound. Activation of the HGF/Met axis is also pro-lymphangiogenic (Kajiya et al. 2005; Cao et al. 2006), a quality that could contribute to the metastatic and aggressive behavior of Met-overexpressing tumor cells (Birchmeier et al. 2003). Moreover, tumor angiogenesis is induced by sustained Akt signaling (Phung et al. 2006), a downstream effector of both EGFR and Met signaling pathways.

6.11 Conclusions and Perspectives

When the first two SLRPs were cloned and sequenced nearly 25 years ago (Krusius and Ruoslahti 1986; Day et al. 1987), there was nothing to suggest that this family of proteoglycans and glycoproteins would be implicated in so many biological functions. The SLRPs form an interactive network of extracellular and cell-associated proteins that modulate the activity of key signaling pathways during development and various pathologies. In some cases, SLRPs control bone mass by regulating the number and activity of osteogenic cells and their precursors and affect their ability to utilize growth factors that are critical to skeletal function including TGF-β and BMPs. The SLRP regulation of growth factors is important in many aspects of skeletal cell behavior including proliferation, differentiation, and apoptosis. When these processes are deregulated they cause premature osteopenia leading to an osteoporosis-like phenotype. In other cases, disruption of some SLRP genes causes skin fragility, tendon abnormality, muscular dystrophy, and ocular diseases affecting corneal transparency, visual acuity, or transmission of visual information. The distribution and organization of collagen and other extracellular matrix proteins are compromised in SLRP deficiency, thereby causing abnormal growth factor distribution and function. In cartilage, TGF-β over-activation induces the production of destructive enzymes that cause osteoarthritis. The use of SLRPs as diagnostic biomarkers for these and other diseases will be an exciting and important future development.

SLRPs are critical components of skeletal stem cell niches, and perhaps other niches. Within tendons, SLRP loss leads to over-activation of BMP signaling in stem cells causing them to form bone instead of tendon. The outcome is a "switch in fate" leading to ectopic ossification and tendon malfunction. It is likely that stem cells in other tissues are regulated by SLRPs; however, the nature of the stem cell and exactly which SLRP is involved will need to be determined.

SLRPs are clearly involved in various aspects of cancer development: from the formation of the tumor microenvironment to cancer growth and metastasis. These multifunctional tasks are achieved by their intrinsic (structure-mediated) ability to interact with and downregulate various tyrosine kinase receptors such as the EGFR, Met, and IGF-IR. In most cases, this interaction results in attenuation of prosurvival signaling pathways and the induction of pro-apoptotic pathways. Some of the SLRPs also directly or indirectly affect angiogenesis.

It is upon this wealth of new information that we need to capitalize and focus our next efforts. Much remains to be learned and discovered about the biology of the SLRPs, and hopefully in the next few years, we will be able to see breakthroughs that will put the latest advances onto a firm footing by better defining the SLRP receptor network and their downstream signaling pathways.

Acknowledgments We like to thank all the past and present members of our laboratories for their invaluable contribution to this field of research. We are indebted to Angela McQuillan for her outstanding help with the illustrations. The original work in Iozzo's laboratory was supported in part by NIH grants RO1 CA39481, RO1 CA47282, and RO1 CA120975, and in the Young's laboratory by the Intramural Program of the NIDCR, NIH.

References

Adany R, Iozzo RV (1990) Altered methylation of versican proteoglycan gene in human colon carcinoma. Biochem Biophys Res Commun 171:1402–1413

Adany R, Heimer R, Caterson B, Sorrell JM, Iozzo RV (1990) Altered expression of chondroitin sulfate proteoglycan in the stroma of human colon carcinoma. Hypomethylation of PG-40 gene correlates with increased PG-40 content and mRNA levels. J Biol Chem 265:11389–11396

Ameye L, Young MF (2002) Mice deficient in small leucine-rich proteoglycans: novel in vivo models for osteoporosis, osteoarthritis, Ehlers–Danlos syndrome, muscular dystrophy, and corneal diseases. Glycobiology 12:107R–116R

Ameye L, Aria D, Jepsen K, Oldberg A, Xu T, Young MF (2002) Abnormal collagen fibrils in tendons of biglycan/fibromodulin-deficient mice lead to gait impairment, ectopic ossification, and osteoarthritis. FASEB J 16:673–680

Araki K, Wakabayashi H, Shintani K, Morikawa J, Matsumine A, Kusuzaki K, Sudo A, Uchida A (2009) Decorin suppresses bone metastasis in a breast cancer cell line. Oncology 77:92–99

Babelova A, Moreth K, Tsalastra-Greul W, Zeng-Brouwers J, Eickelberg O, Young MF, Bruckner P, Pfeilschifter J, Schaefer RM, Gröne H-J, Schaefer L (2009) Biglycan, a danger signal that activates the NLRP3 inflammasome via Toll-like and P2X receptors. J Biol Chem 284: 24035–24048

Bech-Hansen NT, Naylor MJ, Maybaum TA, Sparkes RL, Koop B, Birch DG, Bergen AAB, Prinsen CFM, Polomeno RC, Gal A, Drack AV, Musarella MA, Jacobson SG, Young RSL, Weleber RG (2000) Mutations in *NYX*, encoding the leucine-rich proteoglycan nyctalopin, cause X-linked complete congenital stationary night blindness. Nat Genet 26:319–323

Bi Y, Stueltens CH, Kilts T, Wadhwa S, Iozzo RV, Robey PG, Chen X-D, Young MF (2005) Extracellular matrix proteoglycans control the fate of bone marrow stromal cells. J Biol Chem 280:30481–30489

Bi Y, Ehirchiou D, Kilts TM, Inkson CA, Embree MC, Sonoyama W, Li L, Leet AI, Seo B-M, Zhang L, Shi S, Young MF (2007) Identification of tendon stem/progenitor cells and the role of the extracellular matrix in their niche. Nat Med 13:1219–1227

Bi X, Tong C, Dokendorff A, Banroft L, Gallagher L, Guzman-Hartman G, Iozzo RV, Augenlicht LH, Yang W (2008) Genetic deficiency of decorin causes intestinal tumor formation through disruption of intestinal cell maturation. Carcinogenesis 29:1435–1440

Bianco P, Fisher LW, Young MF, Termine JD, Robey PG (1990) Expression and localization of the two small proteoglycans biglycan and decorin in developing human skeletal and non-skeletal tissues. J Histochem Cytochem 38:1549–1563

Biglari A, Bataille D, Naumann U, Weller M, Zirger J, Castro MG, Lowenstein PR (2004) Effects of ectopic decorin in modulating intracranial glioma progression in vivo, in a rat syngeneic model. Cancer Gene Ther 11:721–732

Birchmeier C, Birchmeier W, Gherardi E, Vande Woude GF (2003) MET, metastasis, motility and more. Nat Rev Mol Cell Biol 4:915–925

Brandan E, Retamal C, Cabello-Verrugio C, Marzolo M-P (2006) The low density lipoprotein receptor-related protein functions as an endocytic receptor for decorin. J Biol Chem 281: 31562–31571

Brandan E, Cabello-Verrugio C, Vial C (2008) Novel regulatory mechanisms for the proteoglycans decorin and biglycan during muscle formation and muscular dystrophy. Matrix Biol 27:700–708

Bredrup C, Knappskog PM, Majewski J, Rodahl E, Boman H (2005a) Congenital stromal dystrophy of the cornea caused by a mutation in the decorin gene. Invest Ophthalmol Vis Sci 46:420–426

Bredrup C, Knappskog PM, Majewski J, Rødahl E, Boman H (2005b) Congenital stromal dystrophy of the cornea caused by a mutation in the decorin gene. Invest Ophthalmol Vis Sci 46:420–426

Brézillon S, Radwanska A, Zeltz C, Malkowski A, Ploton D, Bobichon H, Perreau C, Malicka-Blaszkiewicz M, Maquart F-X, Wegrowski Y (2009) Lumican core protein inhibits melanoma cell migration via alterations of focal adhesion complexes. Cancer Lett 283:92–100

Brown EL, Wooten RM, Johnson BJB, Iozzo RV, Smith A, Dolan MC, Guo BP, Weis JJ, Höök M (2001) Resistance to Lyme disease in decorin-deficient mice. J Clin Invest 107:845–852

Bussolino F, Di Renzo MF, Ziche M, Bocchietto E, Olivero M, Naldini L, Gaudino G, Tamagnone L, Coffer A, Comoglio PM (1992) Hepatocyte growth factor is a potent angiogenic factor which stimulates endothelial cell motility and growth. J Cell Biol 119:629–641

Cao R, Björndahl MA, Gallego MI, Chen S, Religa P, Hansen AJ, Cao Y (2006) Hepatocyte growth factor is a lymphangiogenic factor with an indirect mechanism of action. Blood 107:3531–3536

Carlson EC, Lin M, Liu C-Y, Kao WWY, Perez VL, Pearlman E (2007) Keratocan and lumican regulate neutrophil infiltration and corneal clarity in lipopolysaccharide-induced keratitis by direct interaction with CXCL1. J Biol Chem 282:33502–33509

Chakravarti S (2003) Functions of lumican and fibromodulin: lessons from knockout mice. Glycoconj J 19:287–293

Chakravarti S, Magnuson T, Lass JH, Jepsen KJ, LaMantia C, Carroll H (1998) Lumican regulates collagen fibril assembly: skin fragility and corneal opacity in the absence of lumican. J Cell Biol 141:1277–1286

Chakravarti S, Paul J, Roberts L, Chervoneva I, Oldberg A, Birk DE (2003) Ocular and scleral alterations in gene-targeted lumican-fibromodulin double-null mice. Invest Ophthalmol Vis Sci 44:2422–2432

Chen X-D, Allen MR, Bloomfield S, Xu T, Young M (2003) Biglycan-deficient mice have delayed osteogenesis after marrow ablation. Calcif Tissue Int 72:577–582

Chen X-D, Fisher LW, Robey PG, Young MF (2004) The small leucine-rich proteoglycan biglycan modulates BMP-4-induced osteoblast differentiation. FASEB J 18:948–958

Chen ZTY, Wang I-J, Shih Y-F, Lin LLK (2009) The association of haplotype at the lumican gene with high myopia susceptibilty in Taiwanese patients. Ophthalmology 116:1920–1927

Corsi A, Xu T, Chen X-D, Boyde A, Liang J, Mankani M, Sommer B, Iozzo RV, Eichstetter I, Robey PG, Bianco P, Young MF (2002) Phenotypic effects of biglycan deficiency are linked to collagen fibril abnormalities, are synergized by decorin deficiency, and mimic Ehlers–Danlos-like changes in bone and other connective tissues. J Bone Miner Res 17:1180–1189

Csordás G, Santra M, Reed CC, Eichstetter I, McQuillan DJ, Gross D, Nugent MA, Hajnóczky G, Iozzo RV (2000) Sustained down-regulation of the epidermal growth factor receptor by decorin. A mechanism for controlling tumor growth in vivo. J Biol Chem 275:32879–32887

Danielson KG, Fazzio A, Cohen I, Cannizzaro LA, Eichstetter I, Iozzo RV (1993) The human decorin gene: intron–exon organization, discovery of two alternatively spliced exons in the 5′ untranslated region, and mapping of the gene to chromosome 12q23. Genomics 15:146–160

Danielson KG, Baribault H, Holmes DF, Graham H, Kadler KE, Iozzo RV (1997) Targeted disruption of decorin leads to abnormal collagen fibril morphology and skin fragility. J Cell Biol 136:729–743

Day AA, McQuillan CI, Termine JD, Young MR (1987) Molecular cloning and sequence analysis of the cDNA for small proteoglycan II of bovine bone. Biochem J 248:801–805

de Lange DC, Melder RJ, Munn LL, Mouta-Carreira C, Jain RK, Boucher Y (2001) Decorin inhibits endothelial migration and tube-like structure formation: role of thrombospondin-1. Microvasc Res 62:26–42

De Luca A, Santra M, Baldi A, Giordano A, Iozzo RV (1996) Decorin-induced growth suppression is associated with upregulation of p21, an inhibitor of cyclin-dependent kinases. J Biol Chem 271:18961–18965

Du Y, Carlson EC, Funderburgh ML, Birk DE, Pearlman E, Guo N, Kao WW-Y, Funderburgh JL (2009) Stem cell therapy restores transparency to defective murine corneas. Stem Cells 27:1635–1642

Dugan TA, Yang VWC, McQuillan DJ, Höök M (2003) Decorin binds fibrinogen in a Zn^{2+}-dependent interaction. J Biol Chem 278:13655–13662

Elliott DM, Robinson PS, Gimbel JA, Sarver JJ, Abboud JA, Iozzo RV, Soslowsky LJ (2003) Effect of altered matrix proteins on quasilinear viscoelastic properties in transgenic mouse tail tendons. Ann Biomed Eng 31:599–605

Embree MC, Kilts TM, Ono M, Inkson CA, Seyed-Picard F, Karsdal MA, Oldberd Å, Bi Y, Young MF (2010) Biglycan and fibromodulin have essential roles in regulating chondrogenesis and extracellular matrix turnover in temporomandibular joint osteoarthritis. Am J Pathol 176: 812–826

Ezura Y, Chakravarti S, Oldberg Å, Chervoneva I, Birk DE (2000) Differential expression of lumican and fibromodulin regulate collagen fibrillogenesis in developing mouse tendons. J Cell Biol 151:779–787

Fadic R, Mezzano V, Alvarez K, Cabrera D, Holmgren J, Brandan E (2006) Increase in decorin and biglycan in Duchenne muscular dystrophy: role of fibroblasts as cell source of these proteoglycans in the disease. J Cell Mol Med 10:758–769

Ferdous Z, Wei VM, Iozzo RV, Höök M, Grande-Allen KJ (2007) Decorin-transforming growth factor-β interaction regulates matrix organization and mechanical characteristics of three-dimensional collagen matrices. J Biol Chem 282:35887–35898

Ferdous Z, Lazaro LD, Iozzo RV, Höök M, Grande-Allen KJ (2008) Influence of cyclic strain and decorin deficiency on 3D cellularized collagen matrices. Biomaterials 29:2740–2748

Ferdous Z, Peterson SB, Tseng H, Anderson DK, Iozzo RV, Grande-Allen KJ (2010) A role for decorin in controlling proliferation, adhesion, and migration of murine embryonic fibroblasts. J Biomed Mater Res A 93:419–428

Fischer JW, Kinsella MG, Levkau B, Clowes AW, Wight TN (2001) Retroviral overexpression of decorin differentially affects the response of arterial smooth muscle cells to growth factors. Arterioscler Thromb Vasc Biol 21:777–784

Friedl A (2010) Proteoglycans: master modulators of paracrine fibroblast-carcinoma interactions. Semin Cell Dev Biol 21:66–71

Fust A, LeBellego F, Iozzo RV, Roughley PJ, Ludwig MS (2005) Alterations in lung mechanics in decorin deficient mice. Am J Physiol Lung Cell Mol Physiol 288:L159–L166

Geng Y, McQuillan D, Roughley PJ (2006) SLRP interaction can protect collagen fibrils from cleavage by collagenases. Matrix Biol 25:484–491

Gill MR, Oldberd Å, Reinholt FP (2002) Fibromodulin-null murine knee joints display increased incidences of osteoarthritis and alterations in tissue biochemistry. Osteoarthritis Cartilage 10:751–757

Goldberg M, Septier D, Rapoport O, Iozzo RV, Young MF, Ameye LG (2005) Targeted disruption of two small leucine-rich proteoglycans, biglycan and decorin, exerts divergent effects on enamel and dentin formation. Calcif Tissue Int 77:297–310

Goldberg M, Septier D, Oldberd Å, Young MF, Ameye LG (2006) Fibromodulin-deficient mice display impaired collagen fibrillogenesis in predentin as well as altered dentin mineralization and enamel formation. J Histochem Cytochem 54:525–537

Goldoni S, Iozzo RV (2008) Tumor microenvironment: modulation by decorin and related molecules harboring leucine-rich tandem motifs. Int J Cancer 123:2473–2479

Goldoni S, Owens RT, McQuillan DJ, Shriver Z, Sasisekharan R, Birk DE, Campbell S, Iozzo RV (2004) Biologically active decorin is a monomer in solution. J Biol Chem 279:6606–6612

Goldoni S, Seidler DG, Heath J, Fassan M, Baffa R, Thakur ML, Owens RA, McQuillan DJ, Iozzo RV (2008) An anti-metastatic role for decorin in breast cancer. Am J Pathol 173:844–855

Goldoni S, Hunphries A, Nyström A, Sattar S, Owens RT, McQuillan DJ, Ireton K, Iozzo RV (2009) Decorin is a novel antagonistic ligand of the Met receptor. J Cell Biol 185:743–754

Grant DS, Kleinman HK, Goldberg ID, Bhargava MM, Nickoloff BJ, Kinsella JL, Polverini P, Rosen EM (1993) Scatter factor induces blood vessel formation in vivo. Proc Natl Acad Sci USA 90:1937–1941

Grant DS, Yenisey C, Rose RW, Tootell M, Santra M, Iozzo RV (2002) Decorin suppresses tumor cell-mediated angiogenesis. Oncogene 21:4765–4777

Groeneveld TWL, Oroszlán M, Owens RT, Faber-Krol MC, Bakker AC, Arlaud GJ, McQuillan DJ, Kishore U, Daha MR, Roos A (2005) Interactions of the extracellular matrix proteoglycans decorin and biglycan with C1q and collectins. J Immunol 175:4715–4723

Gu Y, Zhang S, Wu Q, Xu S, Cui Y, Yang Z, Zhao X (2010) Differential expression of decorin, EGFR and cyclin D1 during mammary gland carcinogenesis in TA2 mice with spontaneous breast cancers. J Exp Clin Cancer Res 29:6

Gutierrez J, Osses N, Brandan E (2006) Changes in secreted and cell associated proteoglycan synthesis during conversion of myoblasts to osteoblasts in response to bone morphogenetic protein-2: role of decorin in cell response to BMP-2. J Cell Physiol 206:58–67

Häkkinen L, Strassburger S, Kahari VM, Scott PG, Eichstetter I, Iozzo RV, Larjava H (2000) A role for decorin in the structural organization of periodontal ligament. Lab Invest 80: 1869–1880

Haruyama N, Sreenath TL, Suzuki S, Yao X, Wang Z, Wang Y, Honeycutt C, Iozzo RV, Young MF, Kulkarni AB (2009) Genetic evidence for key roles of decorin and biglycan in dentin mineralization. Matrix Biol 28:129–136

Heegaard A-M, Corsi A, Danielsen CC, Nielsen KL, Jorgensen HL, Riminucci M, Young MF, Bianco P (2007) Biglycan deficiency causes spontaneous aortic dissection and rupture in mice. Circulation 115:2731–2738

Heinegård D (2009) Proteoglycans and more – from molecules to biology. Int J Exp Pathol 90:575–586

Henninger HB, Maas SA, Underwood CJ, Whitaker RT, Weiss JA (2006) Spatial distribution and orientation of dermatan sulfate in human medial collateral ligament. J Struct Biol 158:33–45

Hu Y, Sun H, Owens RT, Wu J, Chen YQ, Berquin IM, Perry D, O'Flaherty JT, Edwards IJ (2009) Decorin suppresses prostate tumor growth through inhibition of epidermal growth factor and androgen receptor pathways. Neoplasia 11:1042–1053

Hwang J-Y, Johnson PY, Braun KR, Hinek A, Fischer JW, O'Brien KD, Starcher B, Clowes AW, Merrilees MJ, Wight TN (2008) Retrovirally mediated overexpression of glycosaminoglycan-deficient biglycan in arterial smooth muscle cells induces tropoelastin synthesis and elastic fiber formation in vitro and in neointimae after vascular injury. Am J Pathol 173:1919–1928

Iacob D, Cai J, Tsonis M, Babwah A, Chakraborty RN, Lala PK (2008) Decorin-mediated inhibition of proliferation and migration of the human trophoblast via different tyrosine kinase receptors. Endocrinology 149:6187–6197

Iozzo RV (1995) Tumor stroma as a regulator of neoplastic behavior. Agonistic and antagonistic elements embedded in the same connective tissue. Lab Invest 73:157–160

Iozzo RV (1997) The family of the small leucine-rich proteoglycans: key regulators of matrix assembly and cellular growth. Crit Rev Biochem Mol Biol 32:141–174

Iozzo RV (1998) Matrix proteoglycans: from molecular design to cellular function. Annu Rev Biochem 67:609–652

Iozzo RV (1999) The biology of the small leucine-rich proteoglycans. Functional network of interactive proteins. J Biol Chem 274:18843–18846

Iozzo RV, Cohen I (1993) Altered proteoglycan gene expression and the tumor stroma. Experientia 49:447–455

Iozzo RV, Danielson KG (1999) Transcriptional and post-transcriptional control of proteoglycan gene expression. Prog Nucleic Acid Res Mol Biol 62:19–53

Iozzo RV, Murdoch AD (1996) Proteoglycans of the extracellular environment: clues from the gene and protein side offer novel perspectives in molecular diversity and function. FASEB J 10:598–614

Iozzo RV, Schaefer L (2010) Proteoglycan roles in health and disease: novel regulatory signaling mechanisms evoked by the small leucine-rich proteoglycans. FEBS J 277:3864–3875

Iozzo RV, Chakrani F, Perrotti D, McQuillan DJ, Skorski T, Calabretta B, Eichstetter I (1999a) Cooperative action of germline mutations in decorin and p53 accelerates lymphoma tumorigenesis. Proc Natl Acad Sci USA 96:3092–3097

Iozzo RV, Moscatello D, McQuillan DJ, Eichstetter I (1999b) Decorin is a biological ligand for the epidermal growth factor receptor. J Biol Chem 274:4489–4492

Järveläinen HT, Iruela-Arispe ML, Kinsella MG, Sandell LJ, Sage EH, Wight TN (1992) Expression of decorin by sprouting bovine aortic endothelial cells exhibiting angiogenesis in vitro. Exp Cell Res 203:395–401

Järveläinen H, Puolakkainen P, Pakkanen S, Brown EL, Höök M, Iozzo RV, Sage H, Wight TN (2006) A role for decorin in cutaneous wound healing and angiogenesis. Wound Rep Reg 14:443–452

Jepsen KE, Wu F, Peragallo JH, Paul J, Roberts L, Ezura Y, Oldberg Ä, Birk DE, Chakravarti S (2002) A syndrome of joint laxity and impaired tendon integrity in lumican- and fibromodulin-deficient mice. J Biol Chem 277:35532–35540

Kajiya K, Hirakawa S, Ma B, Drinnenberg I, Detmar M (2005) Hepatocyte growth factor promotes lymphatic vessel formation and function. EMBO J 24:2885–2895

Kalamajski S, Oldberd Å (2010) The role of small leucine-rich proteoglycans in collagen fibrillogenesis. Matrix Biol 29:248–253

Kalamajski S, Oldberg Å (2009) Homologous sequence in lumican and fibromodulin leucine-rich repeat 5–7 competes for collagen binding. J Biol Chem 284:534–539

Kalamajski S, Aspberg A, Lindblom K, Heinegård D, Oldberd Å (2009) Asporin competes with decorin for collagen binding, binds calcium and promotes osteoblast collagen mineralization. Biochem J 423:53–59

Keene DR, San Antonio JD, Mayne R, McQuillan DJ, Sarris G, Santoro SA, Iozzo RV (2000) Decorin binds near the C terminus of type I collagen. J Biol Chem 275:21801–21804

Kilts T, Ameye L, Syed-Picard F, Ono M, Berendsen AD, Oldberg A, Heegaard A-M, Young MF (2009) Potential roles for the small leucine-rich proteoglycans biglycan and fibromodulin in ectopic ossification of tendon induced by exercise and in modulating rotarod performance. Scand J Med Sci Sports 19:536–546

King VL, Hatch NW, Chan H-W, de Beer MC, de Beer FC, Tannock LR (2009) A murine model of obesity with accelerated atherosclerosis. Obesity 18:35–41

Kinsella MG, Fischer JW, Mason DP, Wight TN (2000) Retrovirally mediated expression of decorin by macrovascular endothelial cells. Effects on cellular migration and fibronectin fibrillogenesis in vitro. J Biol Chem 275:13924–13932

Kishioka Y, Thomas M, Wakamatsu J-I, Hattori A, Sharma M, Kambadur R, Nishimura T (2008) Decorin enhances the proliferation and differentiation of myogenic cells through suppressing myostatin activity. J Cell Physiol 215:856–867

Kitaya K, Yasuo T (2009) Dermatan sulfate proteoglycan biglycan as a potential selectin L/CD44 ligand involved in selective recruitment of peripheral blood CD16(−) natural killer cells into human endometrium. J Leukoc Biol 85:391–400

Kizawa H, Kou I, Iida A, Sudo A, Miyamoto Y, Fukuda A, Mabuchi A, Kotani A, Kawakami A, Yamamoto S, Uchida A, Nakamura K, Notoya K, Nakamura Y, Ikegawa S (2005) An aspartic acid repeat polymorphism in asporin inhibits chondrogenesis and increases susceptibility to osteoarthritis. Nat Genet 37:138–144

Kozma EM, Wisowski G, Olczyk K (2009) Platelet derived growth bactor BB is a ligand for dermatan sulfate chain(s) of small matrix proteoglycans from normal and fibrosis affected fascia. Biochimie 91:1394–1404

Krusius T, Ruoslahti E (1986) Primary structure of an extracellular matrix proteoglycan core protein deduced from cloned cDNA. Proc Natl Acad Sci USA 83:7683–7687

Kuriyama S, Lupo G, Ohta K, Ohnuma SI, Harris WA, Tanaka H (2005) *Tsukushi* controls extodermal patterning and neural crest specification in *Xenopus* by direct regulation of BMP4 and X-delta-1 activity. Development 133:75–88

Lee S, Bowrin K, Hamad AR, Chakravarti S (2009) Extracellular matrix lumican deposited on the surface of neutrophils promotes migration by binding to β2 integrin. J Biol Chem 284: 23662–23669

Li Y, Aoki T, Mori Y, Ahmad M, Miyamori H, Takino T, Sato H (2004) Cleavage of lumican by membrane-type matrix metalloprotease-1 abrogates this proteoglycan-mediated suppression of tumor cell colony formation in soft agar. Cancer Res 64:7058–7064

Li L, Okada H, Takemura G, Kosai K-I, Kanamori H, Esaki M, Takahashi T, Goto K, Tsujimoto A, Maruyama R, Kawamura I, Kawaguchi T, Takeyama T, Fujiwara T, Fujiwara H, Minatoguchi S (2009) Postinfarction gene therapy with adenoviral vector expressing decorin mitigates cardiac remodeling and dysfunction. Am J Physiol Heart Circ Physiol 297:H1504–H1513

Liang FT, Wang T, Brown EL, Iozzo RV, Fikrig E (2004) Protective niche for *Borrelia burgdorferi* to evade humoral immunity. Am J Pathol 165:977–985

Little PJ, Osman N, O'Brien KD (2008) Hyperelongated biglycan: the surreptitious initiator of atherosclerosis. Curr Opin Lipidol 19:448–454

Liu C-Y, Birk D, Hassell JR, Kane B, Kao W-Y (2003) Keratocan-deficient mice display alterations in corneal structure. J Biol Chem 278:21672–21677

Maccarana M, Kalamajski S, Kongsgaard M, Magnusson SP, Oldberg Å, Malmström A (2009) Dermatan sulfate epimerase 1-deficient mice have reduced content and changed distribution of iduronic acids in dermatan sulfate and an altered collagen structure in skin. Mol Cell Biol 29:5517–5528

Majava M, Bishop PN, Hägg P, Scott PG, Rice A, Inglehearn C, Hammond CJ, Spector TD, Ala-Kokko L, Männikkö M (2007) Novel mutations in small leucine-rich repeat protein/proteoglycan (SLRP) genes in high myopia. Hum Mutat 28:336–344

Matsumine A, Shintani K, Kusuzaki K, Matsubara T, Satonaka H, Wakabayashi T, Iino T, Uchida A (2007) Expression of decorin, a small leucine-rich proteoglycan, as a prognostic factor in soft tissue tumors. J Surg Oncol 96:411–418

Mauviel A, Santra M, Chen YQ, Uitto J, Iozzo RV (1995) Transcriptional regulation of decorin gene expression. Induction by quiescence and repression by tumor necrosis factor-α. J Biol Chem 270:11692–11700

Mercado ML, Amenta AR, Hagiwara H, Rafii MS, Lechner BE, Owens RT, McQuillan DJ, Froehner SC, Fallon JR (2006) Biglycan regulates the expression and sarcolemmal localization of dystrobrevin, syntrophin and nNOS. FASEB J 20:1724–1726

Merline R, Lazaroski S, Babelova A, Tsalastra-Greul W, Pfeilschifter J, Schluter KD, Gunther A, Iozzo RV, Schaefer RM, Schaefer L (2009) Decorin deficiency in diabetic mice: aggravation of nephropathy due to overexpression of profibrotic factors, enhanced apoptosis and mononuclear cell infiltration. J Physiol Pharmacol 60(suppl 4):5–13

Mimura T, Han KY, Onguchi T, Chang J-H, Kim T-I, Kojima T, Zhou Z, Azar DT (2008) MT1-MMP-mediated cleavage of decorin in corneal angiogenesis. J Vasc Res 46:541–550

Miura T, Kishioka Y, Wakamatsu J, Hattori A, Hennebry A, Berry CJ, Sharma M, Kambadur R, Nishimura T (2006) Decorin binds myostatin and modulates its activity to muscle cells. Biochem Biophys Res Commun 340:675–680

Moreno M, Muñoz R, Aroca F, Labarca M, Brandan E, Larraín J (2005) Biglycan is a new extracellular component of the chordin-BMP4 signaling pathway. EMBO J 24:1397–1405

Moscatello DK, Santra M, Mann DM, McQuillan DJ, Wong AJ, Iozzo RV (1998) Decorin suppresses tumor cell growth by activating the epidermal growth factor receptor. J Clin Invest 101:406–412

Nagy N, Melchior-Becker A, Fischer JW (2010) Long-term treatment with AT1-receptor antagonist telmisartan inhibits biglycan accumulation in murine atherosclerosis. Basic Res Cardiol 105:29–38

Nakashima Y, Fujii H, Sumiyoshi S, Wight TN, Sueishi K (2007) Early human atherosclerosis; accumulation of lipid and proteoglycans in intimal thickenings followed by macrophage infiltration. Arterioscler Thromb Vasc Biol 27:1159–1165

Nash MA, Loercher AE, Freedman RS (1999) In vitro growth inhibition of ovarian cancer cells by decorin: synergism of action between decorin and carboplatin. Cancer Res 59:6192–6196

Nelimarkka L, Kainulainen V, Schönherr E, Moisander S, Jortikka M, Lammi M, Elenius K, Jalkanen M, Järveläinen HT (1997) Expression of small extracellular chondroitin/dermatan sulfate proteoglycans is differentially regulated in human endothelial cells. J Biol Chem 272:12730–12737

Nelimarkka L, Salminen H, Kuopio T, Nikkari S, Ekfors T, Laine J, Pelliniemi L, Järveläinen H (2001) Decorin is produced by capillary endothelial cells in inflammation-associated angiogenesis. Am J Pathol 158:345–353

Nikitovic D, Berdiaki A, Zfiropoulos A, Katonis P, Tsatsakis A, Karamanos N, Tzanakakis GN (2008a) Lumican expression is positively correlated with the differentiation and negatively with the growth of human osteosarcoma cells. FEBS J 275:350–361

Nikitovic D, Berdiaki K, Chalkiadaki G, Karamanos N, Tzanakakis G (2008b) The role of SLRP-proteoglycans in osteosarcoma pathogenesis. Connect Tissue Res 49:235–238

Nili N, Cheema AN, Giordano FJ, Barolet AW, Babaei S, Hickey R, Eskandarian MR, Smeets M, Butany J, Pasterkamp G, Strauss BH (2003) Decorin inhibition of PDGF-stimulated vascular smooth muscle cell function. Potential mechanism for inhibition of intimal hyperplasia after balloon angioplasty. Am J Pathol 163:869–878

Nuka S, Zhou W, Henry SP, Gendron CMSJB, Shinomura T, Johnson J, Wang Y, Keene DR, Ramírez-Solis R, Behringer RR, Young MF, Höök M (2010) Phenotypic characterization of epiphycan-deficient and epiphycan/biglycan double-deficient mice. Osteoarthritis Cartilage 18:88–96

O'Connor E, Eisenhaber B, Dalley J, Wang T, Missen C, Bulleid N, Bishop PN, Trump D (2005) Species specific membrane anchoring of nyctalopin, a small leucine-rich repeat protein. Hum Mol Genet 14:1877–1887

Ohta K, Lupo G, Kuriyama S, Keynes R, Holt CE, Harris WA, Tanaka H, Ohnuma S-I (2004) Tsukushi functions as an organizer inducer by inhibition of BMP activity in cooperation with chordin. Dev Cell 7:347–358

Ohta K, Kuriyama S, Okafuji T, Gejima R, Ohnuma S-I, Tanaka H (2006) Tsukushi cooperates with VG1 to induce primitive streak and Hensen's node formation in the chick embryo. Development 133:3777–3786

Park H, Huxley-Jones J, Boot-Handford RP, Bishop PN, Attwood TK, Bella J (2008) LRRCE: a leucine-rich repeat cysteine capping motif unique to the chordate lineage. BMC Genomics 9:599

Patel S, Santra M, McQuillan DJ, Iozzo RV, Thomas AP (1998) Decorin activates the epidermal growth factor receptor and elevates cytosolic Ca^{2+} in A431 cells. J Biol Chem 273:3121–3124

Pellegata NS, Dieguez-Lucena JL, Joensuu T, Lau S, Montgomery KT, Krahe R, Kivela T, Kucherlapati R, Forsius H, de la Chapelle A (2000) Mutations in KERA, encoding keratocan, cause cornea plana. Nat Genet 25:91–95

Petit AMV, Rak J, Hung M-C, Rockwell P, Goldstein N, Fendly B, Kerbel RS (1997) Neutralizing antibodies against epidermal growth factor and Erb-2/*neu* receptor tyrosine kinases downregulated vascular endothelial growth factor production by tumor cells in vitro and in vivo. Angiogenic implications for signal transduction therapy of solid tumors. Am J Pathol 151:1523–1530

Phung TL, Ziv K, Dabydeen D, Eyah-Mensah G, Riveros M, Peruzzi C, Sun J, Monahan-Early RA, Shiojima I, Nagy JALMI, Walsh K, Dvorak AM, Briscoe DM, Neeman M, Sessa WC, Dvorak HF, Benjamin LE (2006) Pathological angiogenesis is induced by sustained Akt signaling and inhibited by rapamycin. Cancer Cell 10:159–170

Pusch CM, Zeitz C, Brandau O, Pesch K, Achatz H, Feil S, Scharfe C, Maurer J, Jacobi FK, Pinckers A, Andreasson S, Hardcastle A, Wissinger B, Berger W, Meindl A (2000) The complete form of X-linked congenital stationary night blindness is caused by mutations in a gene encoding a leucine-rich repeat protein. Nat Genet 26:324–327

Rafii MS, Hagiwara H, Mercado ML, Seo NS, Xu T, Dugan T, Owens RT, Höök M, McQuillan DJ, Young MF, Fallon JR (2006) Biglycan binds to α- and γ-sarcoglycan and regulates their expression during development. J Cell Physiol 209:439–447

Raspanti M, Vioala M, Forlino A, Tenni R, Gruppi C, Tira ME (2008) Glycosaminoglycans show a specific periodic interaction with type I collagen fibrils. J Struct Biol 164:134–139

Reed CC, Iozzo RV (2002) The role of decorin in collagen fibrillogenesis and skin homeostasis. Glycoconj J 19:249–255

Reed CC, Gauldie J, Iozzo RV (2002) Suppression of tumorigenicity by adenovirus-mediated gene transfer of decorin. Oncogene 21:3688–3695

Reed CC, Waterhouse A, Kirby S, Kay P, Owens RA, McQuillan DJ, Iozzo RV (2005) Decorin prevents metastatic spreading of breast cancer. Oncogene 24:1104–1110

Reinboth BJ, Finnis ML, Gibson MA, Sandberg LB, Cleary EG (2000) Developmental expression of dermatan sulfate proteoglycans in the elastic bovine nuchal ligament. Matrix Biol 19: 149–162

Reinboth B, Hanssen E, Cleary EG, Gibson MA (2002) Molecular interactions of biglycan and decorin with elastic fiber components: biglycan forms a ternary complex with tropoelastin and micrfibril-associated glycoprotein 1. J Biol Chem 277:3950–3957

Robinson PS, Huang TF, Kazam E, Iozzo RV, Birk DE, Soslowsky LJ (2005) Influence of decorin and biglycan on mechanical properties of multiple tendons in knockout mice. J Biomech Eng 127:181–185

Rødahl E, Van Ginderdeuren R, Knappskog PM, Bredrup C, Boman H (2006) A second decorin frame shift mutation in a family with congenital stromal corneal dystrophy. Am J Ophthalmol 142:520–521

Rucci N, Rufo A, Alamanou M, Capulli M, Del Fattore A, Åhrman E, Capece D, Iansante V, Zazzeroni F, Alesse E, Heinegård D, Teti A (2009) The glycosaminoglycan-binding domain of PRELP acts as a cell type-specific NF-κB inhibitor that impairs osteoclastogenesis. J Cell Biol 187:669–683

Rühland C, Schönherr E, Robenek H, Hansen U, Iozzo RV, Bruckner P, Seidler DG (2007) The glycosaminoglycan chain of decorin plays an important role in collagen fibril formation at the early stages of fibrillogenesis. FEBS J 274:4246–4255

Salomäki HH, Sainio AO, Söderström M, Pakkanen S, Laine J, Järveläinen HT (2008) Differential expression of decorin by human malignant and benign vascular tumors. J Histochem Cytochem 56:639–646

Sanches JCT, Jones CJP, Aplin JD, Iozzo RV, Zorn TMT, Oliveira SF (2010) Collagen fibril organization in the pregnant endometrium of decorin-deficient mice. J Anat 216:144–155

Santra M, Danielson KG, Iozzo RV (1994) Structural and functional characterization of the human decorin gene promoter. J Biol Chem 269:579–587

Santra M, Skorski T, Calabretta B, Lattime EC, Iozzo RV (1995) De novo decorin gene expression suppresses the malignant phenotype in human colon cancer cells. Proc Natl Acad Sci USA 92:7016–7020

Santra M, Mann DM, Mercer EW, Skorski T, Calabretta B, Iozzo RV (1997) Ectopic expression of decorin protein core causes a generalized growth suppression in neoplastic cells of various histogenetic origin and requires endogenous p21, an inhibitor of cyclin-dependent kinases. J Clin Invest 100:149–157

Santra M, Eichstetter I, Iozzo RV (2000) An anti-oncogenic role for decorin: downregulation of ErbB2 leads to growth suppression and cytodifferentiation of mammary carcinoma cells. J Biol Chem 275:35153–35161

Santra M, Reed CC, Iozzo RV (2002) Decorin binds to a narrow region of the epidermal growth factor (EGF) receptor, partially overlapping with but distinct from the EGF-binding epitope. J Biol Chem 277:35671–35681

Saucier C, Khoury H, Lai K-M V, Peschard P, Dankort D, Naujokas MA, Holash J, Yancopoulos GD, Muller WJ, Pawson T, Park M (2004) The Shc adaptor protein is critical for VEGF induction by Met/HGF and ErbB2 receptors and for early onset of tumor angiogenesis. Proc Natl Acad Sci USA 101:2345–2350

Schaefer L (2010) Extracellular matrix molecules: endogenous danger signals as new drug targets in kidney diseases. Curr Opin Pharmacol 10:185–190

Schaefer L, Iozzo RV (2008) Biological functions of the small leucine-rich proteoglycans: from genetics to signal transduction. J Biol Chem 283:21305–21309

Schaefer L, Macakova K, Raslik I, Micegova M, Gröne H-J, Schönherr E, Robenek H, Echtermeyer FG, Grässel S, Bruckner P, Schaefer RM, Iozzo RV, Kresse H (2002) Absence of decorin adversely influences tubulointerstitial fibrosis of the obstructed kidney by enhanced apoptosis and increased inflammatory reaction. Am J Pathol 160:1181–1191

Schaefer L, Mihalik D, Babelova A, Krzyzankova M, Grone HJ, Iozzo RV, Young MF, Seidler DG, Lin G, Reinhardt D, Schaefer RM (2004) Regulation of fibrillin-1 by biglycan and decorin is important for tissue preservation in the kidney during pressure-induced injury. Am J Pathol 165:383–396

Schaefer L, Babelova A, Kiss E, Hausser H-J, Baliova M, Krzyzankova M, Marsche G, Young MF, Mihalik D, Götte M, Malle E, Schaefer RM, Gröne H-J (2005) The matrix component biglycan is proinflammatory and signals through Toll-like receptors 4 and 2 in macrophages. J Clin Invest 115:2223–2233

Schaefer L, Tsalastra W, Babelova A, Baliova M, Minnerup J, Sorokin L, Gröne H-J, Reinhardt DP, Pfeilschifter J, Iozzo RV, Schaefer RM (2007) Decorin-mediated regulation of fibrillin-1 in the kidney involves the insulin-like growth factor-1 receptor and mammalian target of rapamycin. Am J Pathol 170:301–315

Scholzen T, Solursh M, Suzuki S, Reiter R, Morgan JL, Buchberg AM, Siracusa LD, Iozzo RV (1994) The murine decorin. Complete cDNA cloning, genomic organization, chromosomal assignment and expression during organogenesis and tissue differentiation. J Biol Chem 269:28270–28281

Schönherr E, Levkau B, Schaefer L, Kresse H, Walsh K (2001) Decorin-mediated signal transduction in endothelial cells. Involvement of Akt/protein kinase B in up-regulation of $p21^{WAF1/CIP1}$ but not $p27^{KIP1}$. J Biol Chem 276:40687–40692

Schönherr E, Sunderkotter C, Schaefer L, Thanos S, Grässel S, Oldberg Å, Iozzo RV, Young MF, Kresse H (2004) Decorin deficiency leads to impaired angiogenesis in injured mouse cornea. J Vasc Res 41:499–508

Schönherr E, Sunderkötter C, Iozzo RV, Schaefer L (2005) Decorin, a novel player in the insulin-like growth factor system. J Biol Chem 280:15767–15772

Scott JE (1988) Proteoglycan-fibrillar collagen interactions. Biochem J 252:313–323

Seidler DG, Schaefer L, Robenek H, Iozzo RV, Kresse H, Schönherr E (2005) A physiologic three-dimensional cell culture system to investigate the role of decorin in matrix organisation and cell survival. Biochem Biophys Res Commun 332:1162–1170

Seidler DG, Goldoni S, Agnew C, Cardi C, Thakur ML, Owens RA, McQuillan DJ, Iozzo RV (2006) Decorin protein core inhibits in vivo cancer growth and metabolism by hindering epidermal growth factor receptor function and triggering apoptosis via caspase-3 activation. J Biol Chem 281:26408–26418

Shintani K, Matsumine A, Kusuzaki K, Morikawa J, Matsubara T, Wakabayashi T, Araki K, Satonaka H, Wakabayashi H, Lino T, Uchida A (2008) Decorin suppresses lung metastases of murine osteosarcoma. Oncol Rep 19:1533–1539

Sini P, Wyder L, Schnell C, O'Reilly T, Littlewood A, Brandt R, Hynes NE, Wood J (2005) The antitumor and antiangiogenic activity of vascular endothelial growth factor receptor inhibition is potentiated by ErbB1 blockade. Clin Cancer Res 11:4521–4532

Ständer M, Naumann U, Dumitrescu L, Heneka M, Löschmann P, Gulbins E, Dichgans J, Weller M (1998) Decorin gene transfer-mediated suppression of TGF-β synthesis abrogates experimental malignant glioma growth in vivo. Gene Ther 5:1187–1194

Ständer M, Naumann U, Wick W, Weller M (1999) Transforming growth factor-β and p-21: multiple molecular targets of decorin-mediated suppression of neoplastic growth. Cell Tissue Res 296:221–227

Sulochana KN, Fan H, Jois S, Subramanian V, Sun F, Kini RM, Ge R (2005) Peptides derived from human decorin leucine-rich repeat 5 inhibit angiogenesis. J Biol Chem 280:27935–27948

Svensson L, Aszódi A, Reinholt FP, Fässler R, Heinegård D, Oldberg Å (1999) Fibromodulin-null mice have abnormal collagen fibrils, tissue organization and altered lumican deposition in tendon. J Biol Chem 274:9636–9647

Theocharis AD, Karamanos NK (2002) Decreased biglycan expression and differential decorin localization in human abdominal aortic aneurysms. Atherosclerosis 165:221–230

Tomoeda M, Yamada S, Shirai H, Ozawa Y, Yanagita M, Murakami S (2008) PLAP-1/asporin inhibits activation of BMP receptor via its leucine-rich repeat motif. Biochem Biophys Res Commun 371:191–196

Tralhão JG, Schaefer L, Micegova M, Evaristo C, Schönherr E, Kayal S, Veiga-Fernandes H, Danel C, Iozzo RV, Kresse H, Lemarchand P (2003) In vivo selective and distant killing of cancer cells using adenovirus-mediated decorin gene transfer. FASEB J 17:464–466

Trask BC, Trask TM, Broekelmann T, Mecham RP (2000) The microfibrillar proteins MAGP-1 and fibrillin-1 form a ternary complex with the chondroitin sulfate proteoglycan decorin. Mol Biol Cell 11:1499–1507

Troup S, Njue C, Kliewer EV, Parisien M, Roskelley C, Chakravarti S, Roughley PJ, Murphy LC, Watson PH (2003) Reduced expression of the small leucine-rich proteoglycans, lumican, and decorin is associated with poor outcome in node-negative invasive breast cancer. Clin Cancer Res 9:207–214

Tufvesson E, Westergren-Thorsson G (2002) Tumor necrosis factor-α interacts with biglycan and decorin. FEBS Lett 530:124–128

van Cruijsen H, Giaccone G, Hoekman K (2006) Epidermal growth factor receptor and angiogenesis:opportunities for combined anticancer strategies. Int J Cancer 118:883–888

Vesentini S, Redaelli A, Montevecchi FM (2005) Estimation of the binding force of the collagen molecule-decorin core protein complex in collagen fibril. J Biomech 38:433–443

Vij N, Roberts L, Joyce S, Chakravarti S (2004) Lumican suppresses cell proliferation and aids Fas–Fas ligand mediated apoptosis: implications in the cornea. Exp Eye Res 78:957–971

Vogel KG, Paulsson M, Heinegård D (1984) Specific inhibition of type I and type II collagen fibrillogenesis by the small proteoglycan of tendon. Biochem J 223:587–597

Vuillermoz B, Khoruzhenko A, D'Onofrio MF, Ramont L, Venteo L, Perreau C, Antonicelli F, Maquart FX, Wegrowski Y (2004) The small leucine-rich proteoglycan lumican inhibits melanoma progression. Exp Cell Res 296:294–306

Wadhwa S, Bi Y, Ortiz AT, Embree MC, Kilts T, Iozzo R, Opperman LA, Young MF (2007) Impaired posterior frontal sutural fusion in the biglycan/decorin double deficient mice. Bone 40:861–866

Wallace JM, Rajachar RM, Chen X-D, Shi S, Allen MR, Bloomfield SA, Les CM, Robey PG, Young MF, Khon DH (2006) The mechanical phenotype of biglycan-deficient mice is bone- and gender-specific. Bone 39:106–116

Wang I-J, Chiang T-H, Shih Y-F, Hsiao CK, Lu S-C, Hou Y-C, Lin LLK (2006) The association of single nucleotide polymorphisms in the 5'-regulatory region of the *lumican* gene with susceptibility to high myopia in Taiwan. Mol Vis 12:852–857

Wegrowski Y, Pillarisetti J, Danielson KG, Suzuki S, Iozzo RV (1995) The murine biglycan: complete cDNA cloning, genomic organization, promoter function and expression. Genomics 30:8–17

Weis SM, Zimmerman SD, Shah M, Covell JW, Omens JH, Ross J Jr, Dalton N, Jones Y, Reed CC, Iozzo RV, McCulloch AD (2005) A role for decorin in the remodeling of myocardial infarction. Matrix Biol 24:313–324

Westermann D, Mersmann J, Melchior A, Freudenberger T, Petrik C, Schaefer L, Lüllmann-Rauch R, Lettau O, Jacoby C, Schrader J, Brand-Herrmann S-M, Young MF, Schultheiss HP, Levkau B, Baba HA, Unger T, Zacharowski K, Tschöpe C, Fischer JW (2008) Biglycan is required for adaptive remodeling after myocardial infarction. Circulation 117:1269–1276

Williams KJ, Qiu G, Usui HK, Dunn SR, McCue P, Bottinger E, Iozzo RV, Sharma K (2007) Decorin deficiency enhances progressive nephropathy in diabetic mice. Am J Pathol 171:1441–1450

Wilson PG, Thompson JC, Webb NR, deBeer FC, King VL, Tannock LR (2008) Serum amyloid A, but not C-reactive protein, stimulates vascular proteoglycan synthesis in a pro-atherogenic manner. Am J Pathol 173:1902–1910

Wu F, Vij N, Roberts L, Lopez-Briones S, Joyce S, Chakravarti S (2007) A novel role of the lumican core protein in bacterial lipopolysaccharide-induced innate immune response. J Biol Chem 282:26409–26417

Wu H, Wang S, Xue A, Liu Y, Liu Y, Liu Y, Wang H, Chen Q, Guo M, Zhang Z (2008) Overexpression of decorin induces apoptosis and cell growth arrest in cultured rat mesangial cells in vitro. Nephrology 13:607–615

Xu T, Bianco P, Fisher LW, Longenecker G, Smith E, Goldstein S, Bonadio J, Boskey A, Heegaard A-M, Sommer B, Satomura K, Dominguez P, Zhao C, Kulkarni AB, Robey PG, Young MF (1998) Targeted disruption of the biglycan gene leads to an osteoporosis-like phenotype in mice. Nat Genet 20:78–82

Yamada S, Tomoeda M, Ozawa Y, Yoneda S, Terashima Y, Ikezawa K, Ikegawa S, Saito M, Toyosawa S, Murakami S (2007) PLAP-1/asporin, a novel negative regulator of periodontal ligament mineralization. J Biol Chem 282:23070–23080

Yamaguchi Y, Mann DM, Ruoslahti E (1990) Negative regulation of transforming growth factor-β by the proteoglycan decorin. Nature 346:281–284

Yan W, Wank P, Zhao CX, Tang J, Xiao X, Wang DW (2009) Decorin gene delivery inhibits cardiac fibrosis in spontaneously hypertensive rats by modulation of transforming growth factor-β/Smad and p38 mitogen-activated protein kinase signaling pathways. Hum Gene Ther 20:1190–1199

Yang VWC, LaBrenz SR, Rosenberg LC, McQuillan D, Höök M (1999) Decorin is a Zn^{2+} metalloprotein. J Biol Chem 274:12454–12460

Zafiropoulos A, Nikitovic D, Katonis P, Tsatsakis A, Karamanos NK, Tzanakakis GN (2008) Decorin-induced growth inhibition is overcome through protracted expression and activation of epidermal growth factor receptors in osteosarcoma cells. Mol Cancer Res 6:785–794

Zeltz C, Brézillon S, Perreau C, Ramont L, Maquart F-X, Wegrowski Y (2009) Lumcorin: a leucine-rich repeat 9-derived peptide from human lumican inhibiting melanoma cell migration. FEBS Lett 583:3027–3032

Zeng X, Chen J, Miller YI, Javaherian K, Moulton KS (2005) Endostatin binds biglycan and LDL and interferes with LDL retention to the subendothelial matrix during atherosclerosis. J Lipid Res 46:1849–1859

Zhang Y-W, Su Y, Volpert OV, Vande Woude GF (2003) Hepatocyte growth factor/scatter factor mediates angiogenesis through positive VEGF and negative thrombospondin 1 regulation. Proc Natl Acad Sci USA 100:12718–12723

Zhang G, Ezura Y, Chervoneva I, Robinson PS, Beason DP, Carine ET, Soslowsky LJ, Iozzo RV, Birk DE (2006) Decorin regulates assembly of collagen fibrils and acquisition of biomechanical properties during tendon development. J Cell Biochem 98:1436–1449

Zhang G, Chen S, Goldoni S, Calder BW, Simpson HC, Owens RT, McQuillan DJ, Young MF, Iozzo RV, Birk DE (2009) Genetic evidence for the coordinated regulation of collagen fibrillogenesis in the cornea by decorin and biglycan. J Biol Chem 284:8888–8897

Zhu J-X, Goldoni S, Bix G, Owens RA, McQuillan D, Reed CC, Iozzo RV (2005) Decorin evokes protracted internalization and degradation of the EGF receptor via caveolar endocytosis. J Biol Chem 280:32468–32479

Zoeller JJ, Pimtong W, Corby H, Goldoni S, Iozzo AE, Owens RT, Ho S-Y, Iozzo RV (2009) A central role for decorin during vertebrate convergent extension. J Biol Chem 284:11728–11737

Chapter 7
Microfibrils and Fibrillin

Dirk Hubmacher and Dieter P. Reinhardt

Abstract Microfibrils are supramolecular structures ubiquitously found in the extracellular matrix of elastic and nonelastic tissues. The three members of the cysteine-rich fibrillin family constitute the core of microfibrils. Mutations in fibrillin-1 and -2 lead to a number of heritable connective tissue disorders termed fibrillinopathies. Clinical symptoms affect blood vessels, bone, the eye, and other organ systems and highlight the importance of fibrillins in development and homeostasis of tissues and organs. Microfibrils have functional significance (1) in conferring mechanical stability and limited elasticity to tissues; (2) in the biogenesis and maintenance of the elastic fiber system; and (3) in the modulation of the activity of growth factors, including transforming growth factor-β and several bone morphogenetic proteins. In this chapter, we provide an overview of the structure, assembly, and functions of fibrillins and microfibrils and also the pathobiology associated with genetic aberrations in the microfibril system. Lessons learned from mouse models will be discussed as well as the emerging role of microfibrils and fibrillins in the regulation of growth factor bioavailability. Due to the large number of articles in the field, we repeatedly cite excellent review articles to which interested readers are referred to for more details.

Abbreviations

BMP   Bone morphogenetic protein
cbEGF Calcium-binding epidermal growth factor like domain

D. Hubmacher and D.P. Reinhardt (✉)
Department of Anatomy and Cell Biology, Faculty of Medicine, McGill University Montreal, Montreal, QC, Canada H3A 2B2
e-mail: dieter.reinhardt@mcgill.ca

CCA   Congenital contractural arachnodactyly
EGF   Epidermal growth factor like domain
FBN1   Human fibrillin-1 gene
Fbn1   Mouse fibrillin-1 gene
FBN2   Human fibrillin-2 gene
Fbn2   Mouse fibrillin-2 gene
LTBP   Latent transforming growth factor-β binding protein
TGF-β   Transforming growth factor-β
Tsk   Tight-skin mutation

7.1 Historical Perspective

Thin extracellular fibers associated with elastic fibers were observed by electron microscopy in various tissues as early as the 1950s. At the beginning of the 1960s, Low coined the term "microfibrils" for fibers 10–12 nm in diameter found primarily in conjunction with elastic fibers or associated with basement membranes (Low 1962). Today, we know that microfibrils are widely distributed in tissues and occur in most multicellular organisms analyzed. Over the past decades, through advances in protein biochemistry, molecular biology, and mouse and human genetics, remarkable discoveries have been made regarding the composition, structure, and function of microfibrils. Seminal work by the Sakai group in 1986 identified the fibrillin protein as a major constituent of microfibrils (Sakai et al. 1986). Cloning and sequencing efforts in the following years revealed that fibrillins constitute a family of large extracellular multidomain proteins. It quickly emerged that microfibrils represent supramolecular protein assemblies not only composed of fibrillins as backbone proteins, but also associated with numerous matrix components. The composition is likely variable depending on the tissue and the developmental stage. An important milestone of microfibril and fibrillin research was reached when their role in human disease was unraveled. In 1990, the genetic defect leading to Marfan syndrome was mapped to chromosome 15 where the fibrillin-1 gene (*FBN1*) is located (Kainulainen et al. 1990). One year later, the first mutation in *FBN1* was identified in patients (Dietz et al. 1991). Today, we know that other genetic disorders are associated with fibrillins. Various mouse models developed in the late 1990s and early 2000s have significantly contributed to our knowledge of the function of microfibrils and fibrillins and to our understanding of pathological processes associated with the microfibril system. These mouse models have been instrumental in discovering a mechanistic link between reduced amounts of functional microfibrils as they occur in affected patients and elevated activity of transforming growth factor-β (TGF-β). Recent work indicates that microfibrils might be involved in the regulation of a larger spectrum of growth factors including some bone morphogenetic proteins (BMPs).

7.2 Structure of Fibrillins

The fibrillins are large (~350 kDa), highly homologous, and extracellular glycoproteins characterized by an unusually high cysteine content of 12–13% (Fig. 7.1). The three fibrillin isoforms, fibrillin-1, -2, and -3, are encoded by different genes on human chromosome 15, 5, and 19, respectively, and are evolutionary conserved (Lee et al. 1991; Maslen et al. 1991; Corson et al. 1993; Zhang et al. 1994; Nagase et al. 2001). Fibrillin genes have been identified in multicellular eukaryotes from primitive coelenterates to the highly developed mammals and, in most cases, the genes are predicted or were shown to be functional. One exception known thus far is the gene for rodent fibrillin-3 which is inactivated due to chromosomal rearrangement (Corson et al. 2004). Fibrillins appear as extended thread-like molecules of about 140–150 nm in length when observed by electron microscopy after rotary shadowing (Sakai et al. 1991; Lin et al. 2002). Like many other extracellular glycoproteins, fibrillins are composed of a combination of individual domains typically containing ~40–80 amino acid residues (see Chap. 1). This domain organization is almost 100% conserved among the three isoforms and between species, while the homology on the amino acid level between isoforms typically ranges from 60 to 70%.

The most frequently occurring domain in fibrillins is the epidermal growth factor-like (EGF) domain, which can also be found in numerous other extracellular proteins and in blood proteins. This domain is present 47 times in fibrillin-1 and -2 and 46 times in fibrillin-3 due to alternative splicing of one N-terminal EGF domain. Most of the EGF domains (42 and 43) contain characteristic amino acid residues that mediate calcium binding (cbEGF domains) (Handford et al. 1991; Downing et al. 1996). While the principal location of the calcium binding site is always situated in the N-terminal pocket of cbEGF domains, the affinities for calcium vary significantly depending on the individual domain context (Handford 2000; Jensen et al. 2005; Whiteman et al. 2007). Calcium binding to fibrillins has been found to be important for protection against proteolysis (Reinhardt et al. 1997a), ligand interaction control (Reinhardt et al. 1996a; Tiedemann et al. 2001; Lin et al. 2002; Rock et al. 2004; Marson et al. 2005), and structural stabilization (Downing et al. 1996; Reinhardt et al. 1997b; Werner et al. 2000). All EGF and cbEGF domains contain six highly conserved cysteine residues that stabilize the structure in the form of three intradomain disulfide bonds arranged in a C1–C3, C2–C4, and C5–C6 pattern. This feature, in combination with interdomain hydrophobic interactions and relatively short linkers between the individual cbEGF domains, renders the tandem arrays of calcium-loaded cbEGF domains relatively stiff and explains the extended rod-like shape of fibrillin and recombinant fibrillin fragments.

236   D. Hubmacher and D.P. Reinhardt

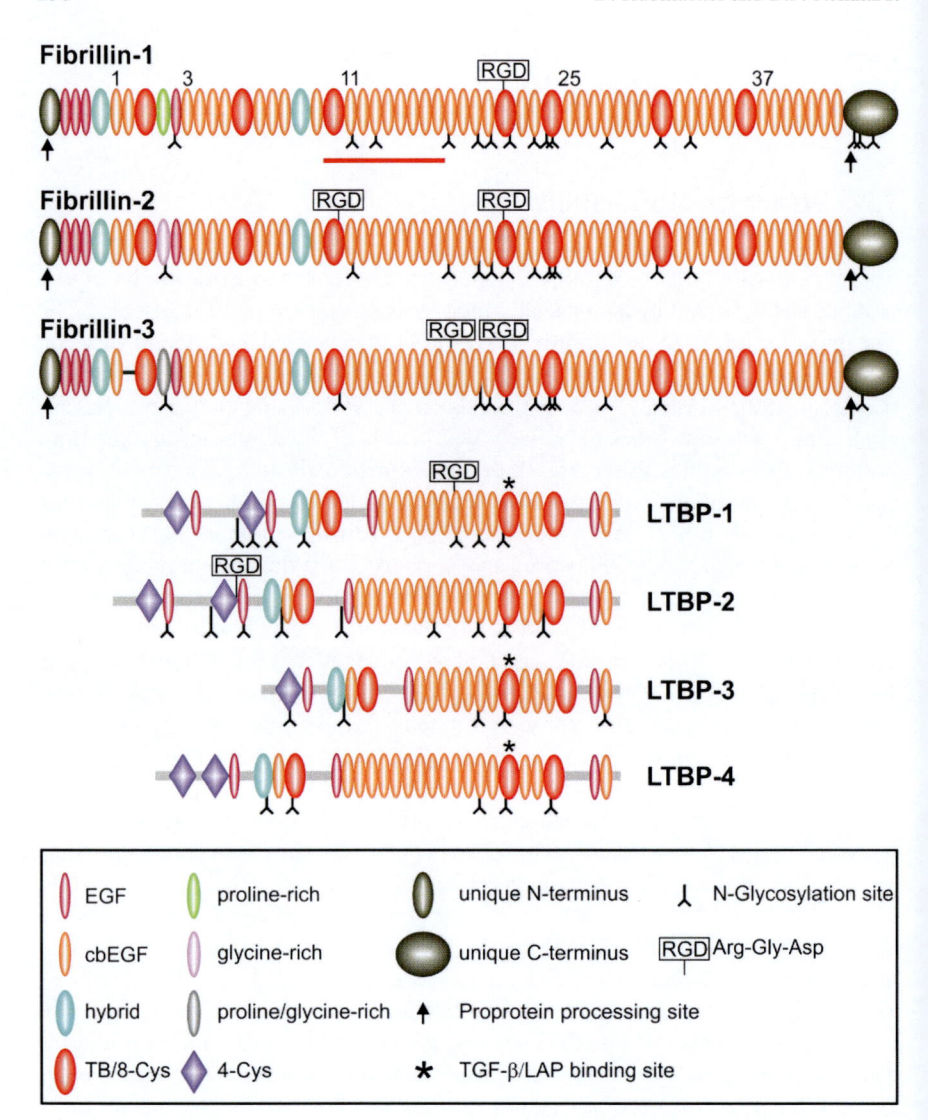

Fig. 7.1 Modular organization of members of the human fibrillin/LTBP superfamily. Schematic representations of the fibrillin and LTBP family are shown with cell binding sites (RGD), predicted N-glycosylation sites, and binding sites for the TGF-β/LAP complex (*asterisk*) as indicated. Numbers above the fibrillin-1 graph indicate the relative numbers of cbEGF domains in the protein and the *red bar* below the graph delineates a region where mutations in fibrillin-1 result in the severe neonatal Marfan syndrome. For simplicity, only the longest splice variant for each LTBP isoform is displayed and suffixes correlated to these variants are omitted in the names. Reprinted from Hubmacher et al. (2006), copyright 2006, with permission from Elsevier

The stretches of EGF and cbEGF domains are interrupted primarily by two other domains unique to fibrillins and to the latent TGF-β binding proteins (LTBPs), providing the rationale to group both protein families into the fibrillin/LTBP superfamily (Fig. 7.1). The eight cysteine-containing TGF-β-binding protein-like (TB) domains occur seven times in fibrillins and three times in LTBPs. They are characterized by four intradomain disulfide bonds in a C1–C3, C2–C6, C4–C7, and C5–C8 pattern. Although the second to last TB domain in LTBPs fulfills a critical function in mediating the covalent interaction with the latent form of TGF-β (see Sect. 7.5), this property is absent in other TB domains in LTBPs and in all fibrillin TB domains. The hybrid (hyb) domain received its name from sequence similarities to both its N-terminus with TB domains and its C-terminus with EGF domains (Corson et al. 1993; Pereira et al. 1993). These similarities in the primary sequence are fully reflected in the three-dimensional structure, emphasizing that the hyb domain is indeed an evolutionary "hybrid" between TB and cbEGF domains stabilized by four disulfide bonds in a C1–C3, C2–C5, C4–C6, and C7–C8 pattern (Jensen et al. 2009). On the basis of the individual domain structures, it can be predicted that almost all cysteine residues in fibrillins are involved in intramolecular disulfide bonds. One exception is the presence of an unpaired ninth cysteine residue in the hyb1 domain, which is conserved between isoforms and between species. This cysteine is surface exposed and has been suggested to participate in the assembly of fibrillins into higher ordered structures (Reinhardt et al. 2000).

The N- and C-terminal fibrillin domains contain four and two conserved cysteine residues respectively, and share minor homology with some of the four cysteine domains in LTBPs and with C-terminal domains of fibulin family members (Giltay et al. 1999). Structurally, the three fibrillin isoforms differ in the following features (Fig. 7.1): a domain with unknown structure close to the N-terminus is rich in proline in fibrillin-1, rich in glycine in fibrillin-2, and rich in both glycine and proline in fibrillin-3. These domains display a relatively low homology between the fibrillin isoforms indicating a diverse functional spectrum. Other structural differences include the position and number of Arg-Gly-Asp (RGD) cell surface integrin binding sites, predicted N- and O-glycosylation sites and predicted tyrosine sulfation sites. Although some of the RGD sites in fibrillins have been shown to interact with various integrins including αvβ3, α5β1, and αvβ6 (Pfaff et al. 1996; Sakamoto et al. 1996; Bax et al. 2003; Jovanovic et al. 2007), the functional contribution of glycosylation remains largely obscure. Sulfation of the predicted sites was ruled out by early metabolic labeling experiments (Sakai et al. 1986).

7.3 Fibrillin-Containing Microfibrils

7.3.1 Properties and Structure of Microfibrils

Fibrillins constitute the core of the multicomponent microfibrils. In association with a number of associated glycoproteins and proteoglycans, microfibrils fulfill multiple tissue-specific physiological roles. Microfibrils are ubiquitously distributed in most tissues where they contribute to structural integrity, provide a scaffold for elastic fiber biogenesis and are involved in the regulation of growth factor bioavailability of the TGF-β/BMP family (Ramirez and Sakai 2010).

In blood vessels, lung, and skin, microfibrils play a crucial role in the formation of elastic fibers, which are essential in conferring resilience to these tissues. Microfibrils provide a scaffold for the deposition of tropoelastin in early phases of elastic fiber formation and primarily, but not exclusively, occupy the surface of mature elastic fibers (Wagenseil and Mecham 2007). Recently, other proteins involved in the assembly of elastic fibers were shown to interact with fibrillin, such as fibulin-4 and -5 and lysyl oxidase, the latter being involved in cross-linking of individual tropoelastin units (El-Hallous et al. 2007; Choudhury et al. 2009). This knowledge manifests a novel concept for the role of microfibrils in elastic fiber biogenesis: in addition to guiding the correct alignment of tropoelastin, microfibrils serve to concentrate accessory proteins, which are involved in the assembly process of elastic fibers including regulation and facilitation of cross-link formation, and possibly other essential molecular events.

In addition to their association with elastic fibers, microfibrils are frequently found in the absence of elastin in organs and tissues including kidney, ciliary zonules of the eye, or superficial regions of the skin (Raviola 1971; Kriz et al. 1990). In these tissues, microfibrils typically intersect with basement membranes and are predicted to act as stress-bearing entities contributing to tissue integrity. Microfibrils may be directly tethered to basement membranes through the proteoglycan perlecan (Tiedemann et al. 2005). Although not entirely clear, basement membranes presumably provide nucleation sites for the formation of microfibrils (Raghunath et al. 1996).

The ultrastructure of microfibrils can be studied by various electron microscopic and X-ray based techniques either directly in tissue or in cell culture. Alternatively, microfibrils can be studied after extraction procedures from tissues including fetal membranes, skin and aorta, or from fibroblast or smooth muscle cell cultures. The appearance of microfibrils in tissues is relatively uniform and they display thread-like structures with 10–12 nm in diameter and frequently arranged in bundles (Low 1962; Fahrenbach et al. 1966; Greenlee et al. 1966). Little is known about the maximum length of microfibrils and whether the ends represent a distinct structure or protein composition. Extraction procedures for microfibrils typically include either collagenase or guanidine–HCl treatments of homogenized tissue followed by gel filtration chromatography (Kielty et al. 1994; Kuo et al. 2007). Extracted microfibrils visualized by electron microscopy after rotary shadowing or negative staining appear significantly different than tissue microfibrils as they display a bead-on-a-string ultrastructure with periodicities of 50–60 nm (Wright and Mayne 1988; Keene et al. 1991; Wallace et al. 1991) (Fig. 7.2). Extraction procedures and mechanical disruption may result in a partial loss of protein components from the interbead regions (Davis et al. 2002). It has been shown that the interbead distance of a bead-on-a-string microfibril can be stretched to more than 100 nm (Keene et al. 1991; Wang et al. 2009). Extensions of up to ~100 nm are fully reversible while irreversible deformation occurs at higher periodicities (Baldock et al. 2001; Eriksen et al. 2001). In addition to this limited elasticity of individual microfibrils, the organization into disulfide-bonded bundles of microfibrils confers

7 Microfibrils and Fibrillin 239

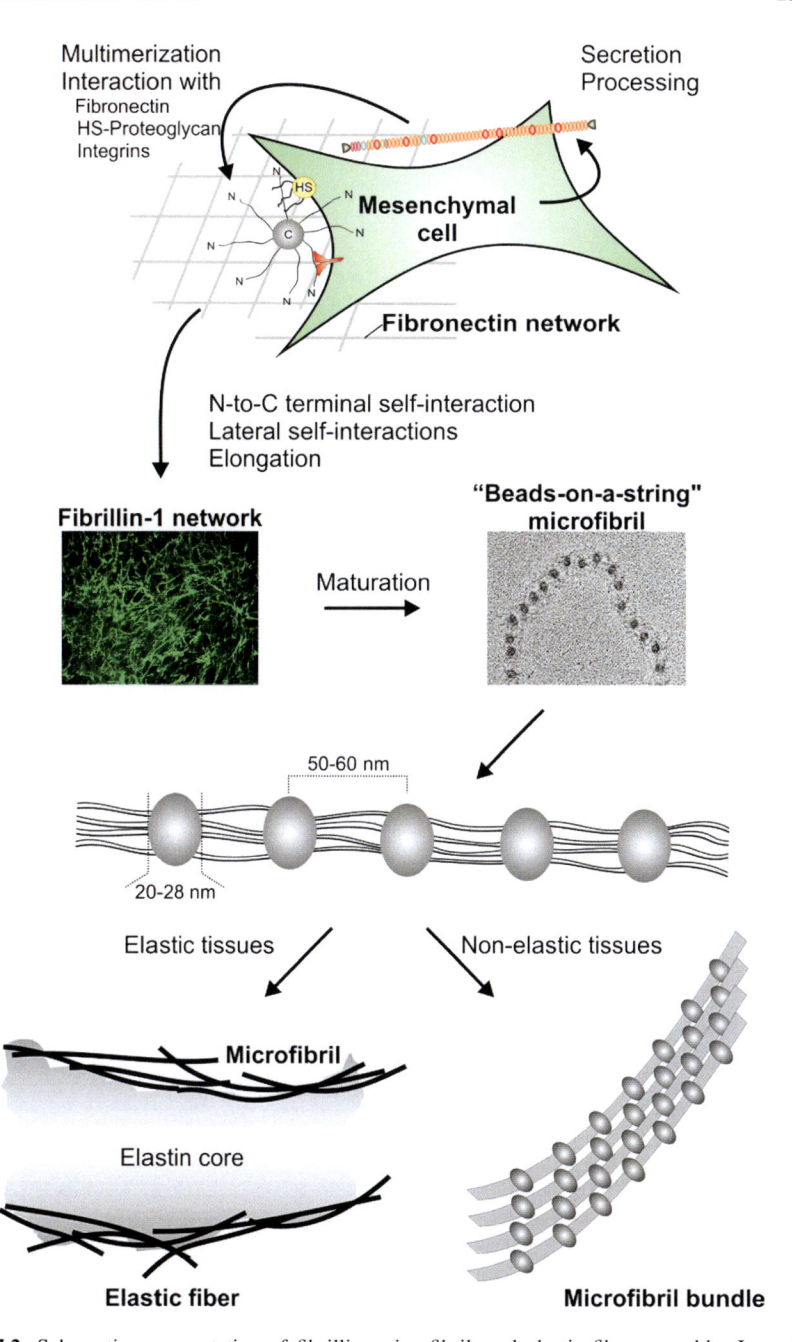

Fig. 7.2 Schematic representation of fibrillin, microfibril, and elastic fiber assembly. In early stages of the assembly process, fibrillins are secreted from mesenchymal cells and the proproteins are cleaved. Likely on the cell surface, fibrillins form disulfide-bonded C-terminal multimers with high affinity for fibronectin and for the fibrillin N-terminus. Interaction with the fibronectin network is essential for fibrillin assembly. Interactions with heparan sulfate proteoglycans

additional elasticity (Thurmond and Trotter 1996; Sherratt et al. 2003). Microfibril elasticity is important in conferring resilience to the low pressure circulatory systems of invertebrates, which do not express elastin (McConnell et al. 1996). In vertebrates, the intrinsic elasticity of microfibrils is likely necessary to prevent microfibril damage as they must stretch together with elastic fibers due to their close association in all elastic tissues.

Despite many efforts and a wide range of techniques used over the past two decades, the molecular organization of individual fibrillin monomers in microfibrils is not completely resolved. Currently, the paradigm exists that only one static alignment of fibrillin in microfibrils is possible. Perhaps the different models put forward in the literature indeed represent different configurations of fibrillins in microfibrils dependent on the demands of specific tissues or the individual composition. Common to all models is a unidirectional head-to-tail orientation of fibrillin molecules in which the ends of the fibrillin molecules reside in or close to the beads (Sakai et al. 1991; Downing et al. 1996; Reinhardt et al. 1996b; Qian and Glanville 1997; Baldock et al. 2001; Lee et al. 2004; Kuo et al. 2007). This unidirectional orientation of the basic microfibril unit (i.e., the fibrillin protein) confers polarity to microfibrils, which may be essential for their molecular function. Differences exist in the alignment models by the degree of overlap (stagger) and by the extent and mechanism of molecular condensation of individual fibrillin molecules (Ramirez and Sakai 2010). Another commonly accepted property of microfibrils is the involvement of 6–8 fibrillin molecules per cross section of the interbead region (Wright and Mayne 1988; Wallace et al. 1991; Baldock et al. 2001; Wang et al. 2009). Currently, it is not known what limits the number of the fibrillin monomers per cross section. Based on multimerization of the recombinant fibrillin C-terminus, which results in a maximum number of 10–12 monomers, it was suggested that steric hindrance may be the limiting factor (Hubmacher et al. 2008).

7.3.2 Biogenesis of Microfibrils

The biogenesis of microfibrils is a multistep process that requires a number of proteins and molecular processes on its way from monomeric fibrillin to mature tissue microfibrils (Hubmacher et al. 2006). Currently, microfibrils cannot be assembled in the test tube because the minimal set of required components is not completely understood and the presence of cells appears to be necessary. In

---

Fig. 7.2 (continued) (HS, *yellow symbol*) and with integrins (*red symbol*) in the pericellular space are likely required, although the mechanistic details and temporal sequences are not known. Elongation and maturation mechanisms including the formation of cross-links follow. A fibrillin network can be detected after a few days in cell culture by immunofluorescence staining. After maturation steps, supramolecular "bead-on-a-string" structures can be extracted from cultured cells or tissues. Microfibrils are involved in the biogenesis of elastic fibers. Alternatively, microfibrils align without elastin to form bundles of microfibrils primarily at basement membranes

addition, full length fibrillin is difficult to purify from cell culture sources or to produce in sufficient quantities with recombinant expression systems. Microfibril assembly assays therefore employ mesenchymal cell-based model systems including fibroblasts and smooth muscle cells. Some of the characterized individual steps of the biogenesis of fibrillin-containing microfibrils described in the following paragraphs are depicted graphically in Fig. 7.2. In most cases, the exact temporal integration of these steps is not known.

During or directly after secretion from cells as proproteins of ~350 kDa, fibrillins are processed to the mature ~320 kDa form by proprotein convertases of the furin/PACE type at the consensus motif (Arg-Xaa-Lys/Arg-Arg) (Milewicz et al. 1992; Milewicz et al. 1995). These recognition sequences are located in both the N- and C-terminal domains and are conserved in all three fibrillin isoforms in all species analyzed to date. Proprotein processing is predicted to result in the release of a small N-terminal (16–48 amino acids) and a larger C-terminal (120–140 amino acids) propeptide depending on the individual fibrillin isoform. Due to its larger size, the processing of the C-terminal propeptide is more amenable to experimental analysis compared to the small N-terminal propeptide. Thus, most of the data available are on C-terminal fibrillin-1 processing. The precise subcellular location for fibrillin processing either within the secretory pathway or directly after secretion from the cell into the extracellular compartment is controversial, although the different cell types used for individual analyses may contribute to the apparent discrepancies (Ritty et al. 1999; Wallis et al. 2003). In fibrillin-secreting fibroblasts, processing of profibrillin-1 is not an intracellular event but occurs during secretion or immediately thereafter (Wallis et al. 2003). The fact that only the C-terminally processed form of fibrillin-1 becomes incorporated into the extracellular matrix suggests that profibrillin-1 conversion to mature fibrillin-1 plays a regulatory role in fibrillin-1 assembly into microfibrils perhaps by preventing premature assembly in the secretory pathway. Presently, the mechanism whereby propeptides would prevent assembly of fibrillin-1 and other fibrillin isoforms is not known. Interestingly, the C-terminal propeptide was detected in proteomic analyses of mature isolated microfibrils (Cain et al. 2006). It was not evident whether the identified peptide represented the cleaved propeptide or originated from the nonprocessed form of fibrillin-1.

After propeptide processing, fibrillin assembly proceeds on or close to the cell surface although the exact sequence of events is unknown. It has been demonstrated that a recombinant C-terminal half of fibrillin-1 forms in close association with cells disulfide-bonded multimers that resemble individual beads of extracted microfibrils (Hubmacher et al. 2008). This multimerization raises the avidity of the relatively low affinity self-interaction sites in the C-terminal monomers to an overall high apparent affinity for the fibrillin-1 N-terminus (Fig. 7.3). These observations suggest that multimerization mediated by the fibrillin C-terminus is a prerequisite for subsequent N-to-C fibrillin-1 self-assembly. Several fibrillin self-interaction mechanisms have been reported on the basis of work with recombinant fibrillin-1 and -2. Full length recombinant fibrillin-1 spontaneously forms multimers in solution and the N- and C-terminal halves of recombinant fibrillin-1 interact with each other with high affinity (Lin et al. 2002; Hubmacher et al. 2008).

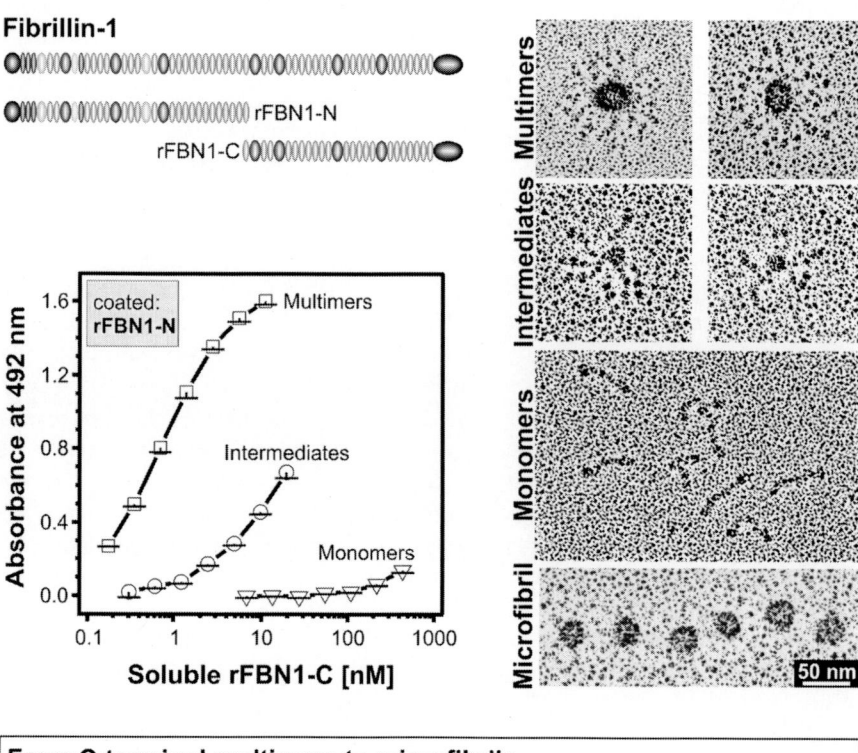

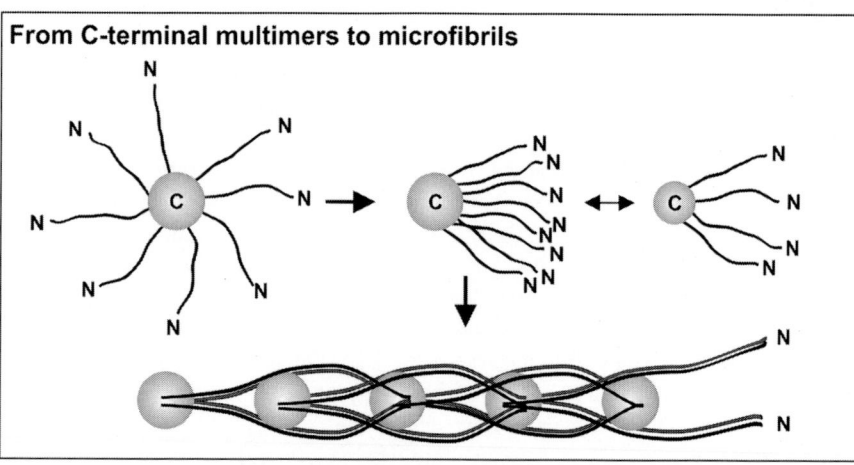

Fig. 7.3 Multimers of the C-terminal half of fibrillin-1 resemble individual beads in microfibrils and interact with the N-terminus. The recombinant fibrillin-1 C-terminal half (rFBN1-C, schematic overview: *top left*) is produced by cells as a mixture of multimers, intermediates, and monomers, which were further separated by size exclusion chromatography. The purified species were visualized by electron microscopy after rotary shadowing (*right panel*). Note the similar appearance of the beaded structures of the multimers (*top*) with the beads of isolated microfibrils (*bottom*). In a solid phase protein–protein interaction assay (*middle left*), the N-terminal half of fibrillin-1 (rFBN1-N) was immobilized and rFBN1-C multimers, intermediates, and monomers

The N-to-C self-interaction epitopes of fibrillin-1 were mapped by the analysis of small overlapping fibrillin-1 fragments in various ligand interaction assays, positioning the interaction sites to the N-terminal region encoded by exons 1–5 (N-terminus to cbEGF2), and the C-terminal region encoded by exons 62–64 (TB7 to processed C-terminus; Fig. 7.4) (Marson et al. 2005; Hubmacher et al. 2008). Collectively, these data explain the exclusive N-to-C arrangement of fibrillin molecules in microfibrils, resulting in polarized arrays of individual fibrillin building blocks. In addition to N-to-C interactions, lateral homotypic interactions in different regions of the fibrillin-1 molecule may play a role in stabilizing initial multimers or lateral associations of individual microfibrils to form microfibrillar bundles (Ashworth et al. 1999; Trask et al. 1999; Marson et al. 2005; Kuo et al. 2007). Mature microfibrils can contain both fibrillin-1 and -2 in the same microfibril and both molecules can heterotypically interact with each other (Lin et al. 2002; Charbonneau et al. 2003). Fibrillin-3 was also located in microfibrils, but it is currently unknown whether it can interact with the other fibrillin isoforms to form heterodimeric or heterotrimeric fibrillin aggregates (Corson et al. 2004).

Three other molecular interactions of fibrillins have been described to play a role in the initial microfibril assembly process. Heparin/heparan sulfate has been demonstrated as a potent inhibitor of microfibril assembly in cell culture and a number of heparin/heparan binding sites have been identified in the center and at the ends of fibrillin-1 (Tiedemann et al. 2001; Ritty et al. 2003; Cain et al. 2005) (Fig. 7.4). From these observations, it was concluded that heparin/heparan sulfate can competitively inhibit the interaction of microfibril proteins with proteoglycans, which may play an active role in microfibril assembly. The heparan sulfate-containing proteoglycan perlecan has been identified as a fibrillin-1 interacting protein with a potential assembly-mediating role for microfibrils in close vicinity of basement membranes (Tiedemann et al. 2005). Recently, it has been shown that the presence of a fibronectin network is absolutely essential for the formation of microfibrils in cell culture (Kinsey et al. 2008; Sabatier et al. 2009). Fibronectin interacts with the same C-terminal fibrillin-1 multimers that initiate the N-to-C self-interaction, suggesting a tightly regulated mechanism. Binding to fibronectin as well as to proteoglycans may support the alignment of fibrillin molecules in the appropriate spatial pattern for subsequent steps in the assembly process or may help to accumulate fibrillin molecules to critical concentrations necessary for self-interaction. A recent study suggested that integrin $\alpha 5\beta 1$, which has been shown to guide fibronectin assembly, is also required for pericellular microfibril assembly (Kinsey et al. 2008). However, it is currently difficult to distinguish whether this

Fig. 7.3 (continued) were tested for interaction. The multimers interacted strongly with the N-terminus, the intermediates showed a reduced binding and the monomeric rFBN1-C interacted very little with rFBN1-N. Multimerization mediated through a C-terminal region raises the avidity of a weak C-terminal self-interaction site of the monomer to enable a strong interaction of the multimer with the N-terminus. On the bottom, a hypothetical model is depicted how the multimerization of the C-terminus of fibrillin-1 could translate into microfibril assembly and polarity. Figure modified after Hubmacher et al. (2008)

Fibrillin-1

Self-assembly and microfibril biogenesis

NC						NC	NC
NN	NN					CC	
He			He		He		He
					Fibronectin		
					Fibrillin-2 (N-terminal half)		
	Fibrillin-2 (C-terminal half)						

Elastic fiber assembly

Fi5	Fi4				
LOX		Te		Te	Te
LTBP2					

Growth factor regulation

LTBP4
LTBP1
BMP7
BMP2,4,10; GDF — BMP10
MAGP1

Proteoglycans

Ve — Decorin — Versican
Perlecan

Integrins

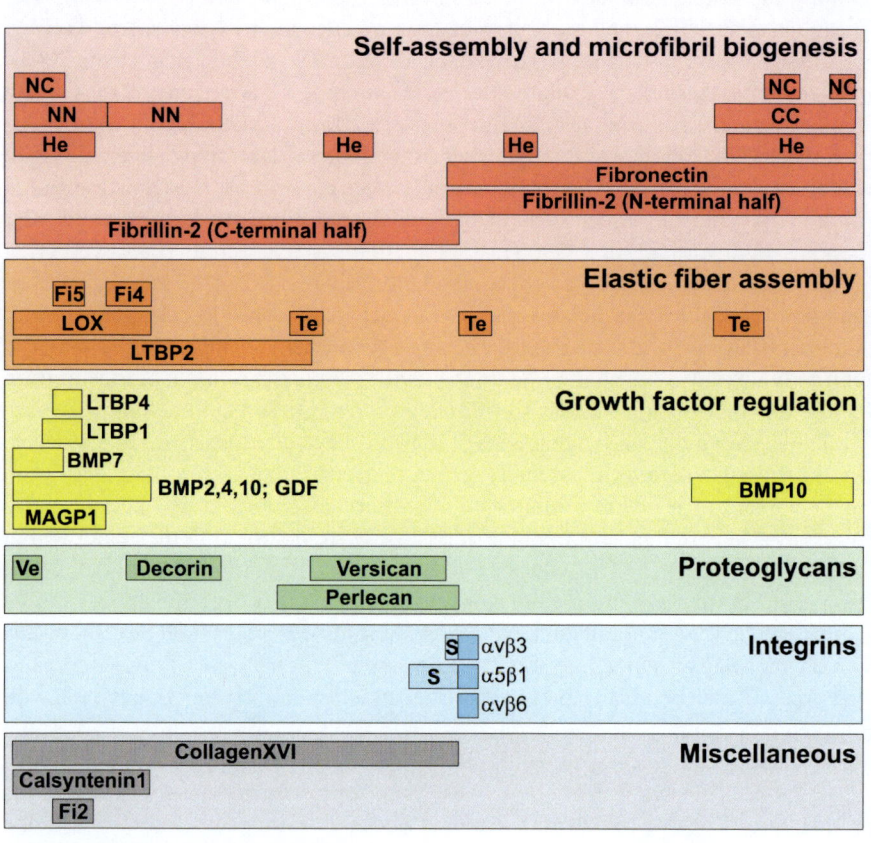

S αvβ3
S α5β1
   αvβ6

Miscellaneous

CollagenXVI
Calsyntenin1
Fi2

Abbreviations:
NC: *N-to-C terminal self-interaction*   Te: *Tropoelastin*
NN: *N-to-N terminal self-interaction*   Fi: *Fibulin*
He: *Heparin/Heparan sulfate*   LOX: *Lysyl oxidase*
GDF: *Growth differentiation factor*   Ve: *Versican*
BMP: *Bone morphogenetic protein*   S: *Synergy site for integrins*
MAGP: *Microfibril associated glycoprotein*
LTBP: *Latent transforming growth factor-β binding protein*

Fig. 7.4 Binding epitopes for fibrillin-1 ligands. More than 20 components have been described to interact with fibrillins or microfibrils. This overview focuses on direct interactions of ligands with fibrillin-1 that have been demonstrated using purified proteins or protein fragments and where mapping data of the binding epitope is available. Other ligands only localized to microfibrils by microscopic techniques were omitted. The proteins are grouped into functional categories and aligned with the fibrillin-1 molecule on top. The width of each colored box, representing the individual ligands, correlates with the smallest interaction region known for the respective ligand. See Fig. 7.1 for the color code of the fibrillin-1 domains and additional structural features

requirement is mediated by a direct fibrillin interaction with integrin α5β1, or indirectly through the dependency of fibronectin network formation on this integrin. The interactions of fibrillins with heparan sulfate, fibronectin, and integrins are likely dynamic in nature and may be designed to transiently "catalyze" and modulate microfibril assembly. Thus, they are important players for microfibril biogenesis and homeostasis in tissues.

It has been demonstrated that mature microfibrils are highly cross-linked. Cross-links between individual fibrillin monomers or between fibrillins and other components likely provide stability required to withstand the mechanical stresses that typically occur in microfibril-containing tissues. Cross-links might further be responsible to maintain a three-dimensional superstructure of microfibrils as bundles of microfibrils display a one-third stagger (Wess et al. 1998). Two types of intermolecular cross-links have been reported for microfibrils, disulfide bonds, and ε(γ-glutamyl)lysine cross-links catalyzed by transglutaminases (Gibson et al. 1989; Qian and Glanville 1997). Intermolecular disulfide bonds between fibrillins or between fibrillin and other proteins form within a few hours in organ cultures of chick aorta (Reinhardt et al. 2000). It is not entirely known which cysteines from the 361 cysteines in fibrillin-1 are involved in intermolecular disulfide bonds. Based on the structures of individual fibrillin domains, almost all cysteines are predicted to be engaged in intradomain disulfide bonds. The first hybrid domain (hyb1) in all fibrillins contains nine conserved cysteine residues and therefore one unpaired cysteine, whose thiol group has been biochemically characterized as solvent accessible (Reinhardt et al. 2000). The data have been further validated by modeling the hyb1 structure based on the high resolution hyb2 structure as template (Jensen et al. 2009). Cys^{204} in fibrillin-1 and Cys^{233} in fibrillin-2 are thus good candidates for intermolecular cross-links in microfibrils. Other candidate cysteines exist in the N- and C-terminal domains. It is presently not known whether intermolecular disulfide bond formation requires the presence of specific enzymes in the extracellular space, or whether they originate from spontaneous oxidation of exposed and properly aligned cysteine residues. It is possible that shuffling mechanisms are in place to generate disulfide bond cross-links. The second type of intermolecular cross-links catalyzed by transglutaminases is nonreducible and is likely formed later during the maturation of microfibrils (Bowness and Tarr 1997; Qian and Glanville 1997; Thurmond et al. 1997). A relatively large portion (10–15%) of lysine residues in microfibrils purified from human amnion was reported to be cross-linked in the interbead region but not in the beads (Qian and Glanville 1997). Transglutaminase as well as disulfide cross-links are potentially critical for correct lateral alignment of fibrillin molecules to facilitate downstream assembly events.

7.3.3 Fibrillins and Microfibrils in Mammalian Tissues

In most tissues of the mouse embryo including blood vessels, lung, bone, and cartilage, the fibrillin-1 (*Fbn1*) and -2 (*Fbn2*) genes are expressed in a diphasic pattern. The onset of *Fbn2* transcription is typically initiated earlier than that of *Fbn1* and is downregulated at later developmental stages. One exception is the cardiovascular system where *Fbn1* expression is detected very early in development at the onset of organ formation and always at higher levels compared to *Fbn2* expression (Yin et al. 1995; Zhang et al. 1995). It was concluded that fibrillin-2 protein expression coincides with early morphogenesis while fibrillin-1 protein expression correlates with late morphogenesis and the development of well-defined organ structures. The original hypothesis that fibrillin-1 microfibrils provide structural support, whereas microfibrils composed of the fibrillin-2 protein regulate early processes of elastic fiber assembly, could not be confirmed by subsequent studies. Elastic fiber formation and blood vessel maturation were apparently normal in fibrillin-2 deficient mice (Arteaga-Solis et al. 2001; Carta et al. 2006). On the other hand, mice completely lacking both *Fbn1* alleles in an *Fbn2*$^{+/+}$ background demonstrated disorganized elastic fibers and impaired matrix maturation in the aortic wall of postnatal animals, whereas complete loss of *Fbn1* in an *Fbn2*$^{-/-}$ background causes embryonic death at mid-gestation with a poorly developed aortic media (Carta et al. 2006). These findings emphasize a critical role for fibrillin-1 in the maturation of the aortic wall and strengthen the idea of a partial functional overlap of both fibrillins. The fibrillin-3 gene is inactivated in the rodent genome likely due to chromosomal rearrangement events during mouse evolution while the gene appears to be functional in humans, cow, chicken, and other organisms (Corson et al. 2004). Fibrillin-1 also exerts a morphogenetic role in lung development. Mice deficient in fibrillin-1 show airspace enlargement of the lung resulting from failure of distal alveolar septation. These perturbations were correlated with overactivation of TGF-β signaling (Neptune et al. 2003).

During early human development, fibrillin-1 and -2 follow a similar temporo-spatial distribution pattern in most developing tissues and organs including aorta, lung, heart, and skin (Zhang et al. 1994; Quondamatteo et al. 2002). Differential expression of both fibrillins was detected in kidney, liver, rib anlagen, and notochord. Similar to the temporal expression pattern of fibrillin-2, fibrillin-3 is found most abundantly in human fetal tissues, suggesting that fibrillin-3 has a role in early development (Corson et al. 2004). The spatial fibrillin-3 expression patterns overlap with those of fibrillin-1 and -2 in some tissues including skeletal elements and skin but differ in other tissues including kidney, lung, blood vessel, and brain. A functional role for fibrillin-3 remains to be established. Other studies in humans focused on skeletal development (Zhang et al. 1994; Keene et al. 1997). Fibrillin-1 is expressed in human fetal limbs in the loose connective tissue around skeletal muscles and tendons and is widely expressed in developing limbs and digits at 16 weeks of gestation, except for the cartilage. At this stage and continuing through adulthood, the perichondrium contains abundant fibrillin-1 microfibrils. In postnatal long bones, fibrillin-1 is colocalized together with LTBP-1 in fibrillar structures in the outer periosteum and in the cartilage fibrillin-1 localizes to the perichondrium (Dallas et al. 2000). From the clinical phenotypes seen in Marfan syndrome (see Sect. 7.4), it is clear that fibrillin-1 plays an important role in the regulation of bone growth, especially in determining the length of the long bones in the extremities.

However, the underlying molecular mechanism is still obscure. Fibrillin containing microfibrils may limit bone growth by exerting tension in the periosteum or perichondrium or may regulate bone deposition in the growth plate.

In summary, fibrillins exhibit broadly overlapping expression patterns during mammalian development with some distinct temporal and tissue-specific differences. Fibrillin-2 and -3 are preferentially expressed during the developmental period whereas fibrillin-1 expression persists throughout life.

7.4 Fibrillinopathies

Mutations in fibrillin-1 cause a number of connective tissue disorders summarized as type 1 fibrillinopathies. Heterozygous mutations in fibrillin-1 have been identified in various forms of Marfan syndrome, isolated ectopia lentis, kyphoscoliosis, familial arachnodactyly, familial thoracic ascending aortic aneurysms and dissections, "MASS" phenotype, Shprintzen–Goldberg syndrome, dominant Weill–Marchesani syndrome, and stiff skin syndrome (Pyeritz 2000; Robinson et al. 2006; Loeys et al. 2010). Fibrillin-1 has been further suggested to be involved in the pathogenesis of systemic sclerosis and homocystinuria albeit further confirmation of such a role is required (Krumdieck and Prince 2000; Lemaire et al. 2006; Glushchenko and Jacobsen 2007). Mutations in fibrillin-2 give rise to congenital contractural arachnodactyly (CCA) or Beals–Hecht syndrome (Putnam et al. 1995; Frederic et al. 2009). It is presently not clear whether fibrillin-3 has a role in human disease.

Autosomal dominant Marfan syndrome occurs with an estimated prevalence of 1 in 5,000–10,000 individuals, while most other fibrillinopathies are relatively rare (Pyeritz 2000). Major clinical symptoms develop in the cardiovascular, skeletal and ocular systems, including mitral valve disease, progressive dilatation of the aortic root, dolichostenomelia, arachnodactyly, scoliosis, and ectopia lentis. Dissection and rupture of the aortic wall represents the major life-threatening clinical complication. More than 1,000 mutations in *FBN1* have been identified in patients with Marfan syndrome (Faivre et al. 2007). The mutations are spread throughout almost the entire gene and thus virtually affect every domain of fibrillin-1. Approximately 12% are recurrent mutations while the majority of the mutations are unique to patients and families. Missense, frameshift, nonsense, and splice site mutations have been observed as well as in-frame deletions and insertions (Collod-Beroud et al. 2003). Mutations in the center of fibrillin-1 encoded by exons 24–32 typically result in a more severe phenotype with a higher probability of ascending aortic dilatations. A significant portion of the mutations in this region (22%) cause a neonatal onset of the disease, which provides the rationale for the terms "neonatal" Marfan syndrome and "neonatal" region of fibrillin-1 (Park et al. 1998; Gupta et al. 2002; Faivre et al. 2007) (Fig. 7.1, red bar). A high degree of inter- and intrafamilial variability is a common feature of Marfan syndrome, suggesting that other genes or environmental factors play a modifying role in the pathogenesis of the disease.

Homocysteine may be one example of such modifiers. Homocystinuria, caused by cystathionine-β-synthase deficiency, and Marfan syndrome overlap in several clinical symptoms of the connective tissue such as ectopia lentis, long bone overgrowth, and scoliosis (Skovby and Kraus 2002). Homocysteine was described as a potential modifier of Marfan syndrome by correlating the severity of aortic aneurysms with elevated homocysteine levels (Giusti et al. 2003). The consequences of elevated homocysteine on the structure and function of fibrillins have been studied with recombinant proteins, in cell culture models, and in a chick model (Hill et al. 2002; Hubmacher et al. 2005; Hutchinson et al. 2005; Hubmacher et al. 2010). In these studies, it was shown that fibrillin-1 is a target for homocysteine, which rendered the protein susceptible to proteolysis and compromised its functional properties. In chick, high homocysteine levels resulted in a reduced amount of fibrillin-2 and microfibrils in the elastic lamina of the aorta.

Other type 1 fibrillinopathies are characterized by various degrees of clinical overlaps with Marfan syndrome. An interesting member of this group of disorders is the autosomal dominant Weill–Marchesani syndrome, which is characterized by short stature, brachydactyly, joint stiffness, and eye abnormalities including ectopia lentis and microspherophakia (Faivre et al. 2003a). Some of these symptoms represent the opposite spectrum of what is observed in patients with Marfan syndrome. An in-frame deletion in exon 41 of the fibrillin-1 gene was identified in a family with autosomal dominant Weill–Marchesani syndrome (Faivre et al. 2003b). It will be of particular interest in the future to unravel the molecular pathogenetic mechanisms that lead to the different clinical symptoms observed in Marfan syndrome, Weill–Marchesani syndrome, and other type 1 fibrillinopathies, and to decipher why certain mutations cause only a subset of symptoms compared to the fully developed Marfan syndrome.

CCA caused by mutations in the fibrillin-2 gene (*FBN2*) is an autosomal dominant connective tissue disorder characterized by overlapping skeletal features with Marfan syndrome such as marfanoid habitus, arachnodactyly, kyphoscoliosis, and camptodactyly (Viljoen 1994). In contrast to Marfan syndrome, individuals with CCA typically present with crumpled appearance of the ear helix and congenital contractures, and do not usually have the ocular and cardiovascular complications seen in Marfan syndrome. Similar to the mutations in *FBN1* causing the severe (neonatal) form of Marfan syndrome, mutations in *FBN2* resulting in CCA are clustered in the central region of fibrillin-2, suggesting that this region has important properties presumably in all fibrillin isoforms (Callewaert et al. 2009).

It is now established that disease progression in Marfan syndrome is the combined consequence of a loss in tissue integrity and perturbed TGF-β signaling (Ramirez et al. 2008). Loss of tissue integrity can result from deficiencies in the biogenesis of microfibrils and associated elastic fibers and/or from excessive degradation of this system (Fig. 7.5). A series of studies using tissues and cells from affected individuals clearly indicate that the biogenesis of microfibrils is often compromised in individuals with Marfan syndrome (Tiedemann et al. 2004). On the other hand, excessive proteolytic fragmentation of the microfibril/elastic

7 Microfibrils and Fibrillin

Unaffected

Genotype: WT / WT

Biogenesis of microfibrils:
A) Normal secretion from cells
B) Normal microfibril formation
C) No degradation of protein
D) Incorporation of normal fibrillin-1 in microfibrils

→ No functional disturbance / No increased degradation of microfibrils

Consequence: NORMAL levels of functional microfibrils → NORMAL levels of active TGF-β

Marfan syndrome

Genotype: WT / MUT *

Biogenesis of microfibrils:
A) Reduced secretion from cells
B) Reduced microfibril formation
C) Enhanced degradation of protein
D) Incorporation of mutant fibrillin-1 in microfibrils

→ Functional disturbance / Increased degradation of microfibrils

Consequence: REDUCED levels of functional microfibrils → ELEVATED levels of active TGF-β

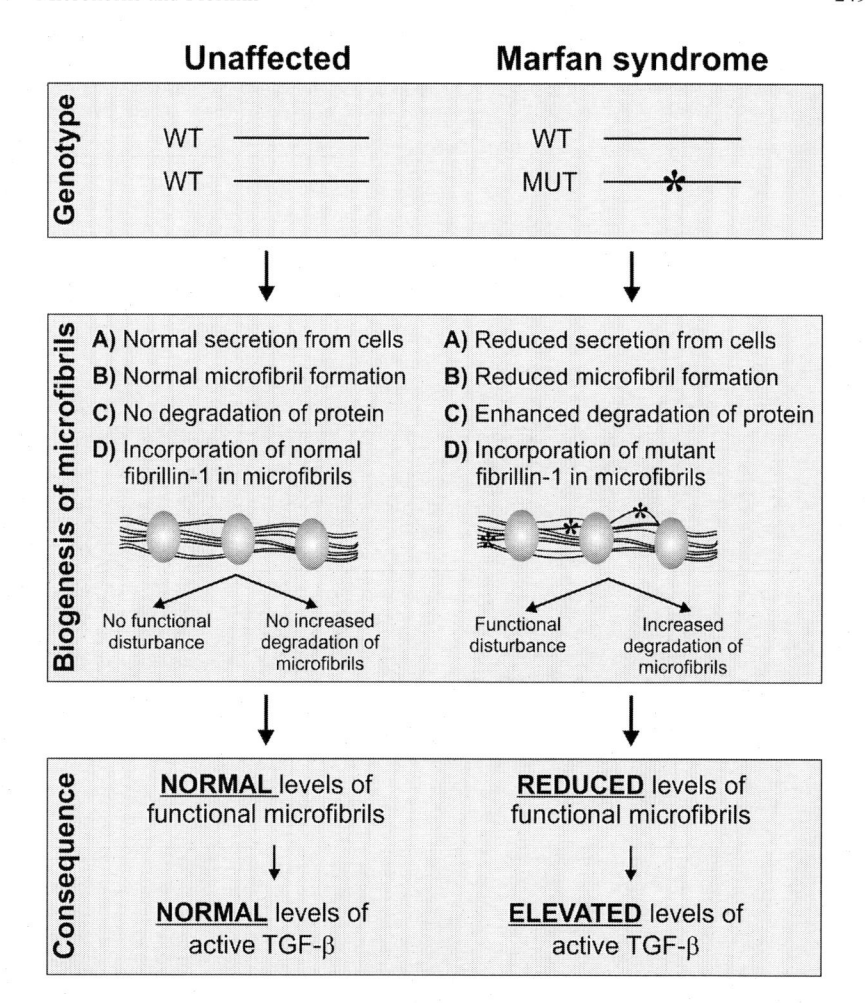

Fig. 7.5 Overview of potential mechanisms involved in the pathogenesis of Marfan syndrome. In unaffected individuals (*left panel*), all aspects of fibrillin-1 and microfibril biology are normal, leading to normal levels of functional microfibrils and normal levels of active TGF-β. In individuals with Marfan syndrome (*right panel*), one fibrillin-1 allele carries a mutation (*asterisk*). This can lead to various molecular malfunctions of fibrillin-1 and microfibrils (A–D). Independent of the underlying mechanism, the common result is a reduced level of functional microfibrils and elevated levels of active TGF-β

fiber system has been observed in the aorta and skin of affected individuals (Halme et al. 1985; Tsuji 1986; Fleischer et al. 1997). Increased matrix metalloproteinase expression was found in aortic tissues from patients with Marfan syndrome (Segura et al. 1998). Independent of the involved mechanism, the common pathway appears to involve a general reduction in the amount of functional microfibrils present in the extracellular matrix (Fig. 7.5). Reduced levels of functional microfibrils cause an elevated level of active TGF-β (see Sect. 7.5). This knowledge has recently led to promising results in a mouse model and in a small cohort of pediatric patients with severe Marfan syndrome using losartan, an angiotensin II type 1 receptor blocker that reduces TGF-β activity by largely unknown mechanisms (Habashi et al. 2006; Brooke et al. 2008). This clinical pilot study requires confirmation and is currently being evaluated in a larger randomized trial.

7.5 Fibrillins and Growth Factors

It was recently demonstrated that fibrillins and microfibrils play a critical role in matrix deposition and potential activation of growth factors of the TGF-β superfamily including TGF-β and BMPs (Ramirez and Rifkin 2009). These growth factors regulate a broad array of developmental and homeostatic processes and are involved in the pathobiology of a variety of tissues including the vascular and pulmonary systems (ten Dijke and Arthur 2007). Furthermore, overactivation of TGF-β has emerged as a central theme in the pathobiochemical mechanism underlying the development of the clinical symptoms observed in Marfan syndrome (see Sect. 7.4).

Mammalian TGF-β1, -2, and -3 are synthesized as proproteins containing mature TGF-β and the latency-associated protein (LAP), which maintains TGF-β in an inactive state (Gentry et al. 1988). In most studied cell lines, the TGF-β/LAP complex is bound to LTBP-1, -3, or -4 during secretion, but not to LTBP-2 (Hyytiäinen et al. 2004; Rifkin 2005). This covalent interaction is mediated by disulfide bonds between the second to last TB domain in these LTBPs and LAP (Fig. 7.1, asterisks). The major fraction of secreted LTBPs does not contain TGF-β, suggesting a dual role for LTBPs as TGF-β targeting molecules and as structural components in the extracellular matrix (Miyazono et al. 1991; Taipale et al. 1994). Since TB domains represent the signature domains of the fibrillin/LTBP family, it was hypothesized that one or more of these domains in fibrillins may also mediate interaction with the TGF-β/LAP complex. However, TB domains in fibrillins and in LTBP-2 are missing the critical residues necessary for the interaction with the LAP protein and are thus unable to directly interact with the TGF-β/LAP complex (Gleizes et al. 1996; Saharinen and Keski-Oja 2000).

It has been demonstrated that nonpolymerized fibrillins and fibrillin-containing microfibrils can interact with LTBPs and therefore indirectly bind and position TGF-β in the extracellular matrix. In cell culture studies, LTBP-1 colocalized with fibrillin-1 and fibronectin (Taipale et al. 1996; Dallas et al. 2000; Dallas et al. 2005). In tissues, LTBP-1 and latent TGF-β1 localization to fibrillin-containing microfibrils was described in skin, the periosteum of the developing long bone, the developing heart, and the cardiovascular system (Nakajima et al. 1997; Raghunath et al. 1998; Dallas et al. 2000; Isogai et al. 2003). LTBP-2, which does not bind TGF-β, has also been localized to microfibrils and it has been speculated that this isoform may target other growth factors to the microfibril system (Gibson et al. 1995; Sinha et al. 2002; Chen et al. 2005). LTBP-1 and -4, but not LTBP-3, interact via their C-terminus with purified fibrillin or with the fibrillin network produced by fibroblasts (Unsöld et al. 2001; Isogai et al. 2003; Koli et al. 2005). While the major interaction sites of LTBPs with extracellular matrix components including fibronectin are located at their N-terminal region and are covalently stabilized by transglutaminase cross-links, the C-terminal interactions with fibrillin are of lower affinity and are mediated by noncovalent interactions (Isogai et al. 2003; Kantola et al. 2008). It is possible that this particular property is important for the physiological role of fibrillin and microfibrils in activation of TGF-β, which depends on various mechanisms including proteolytic cleavage, as well as on interactions with integrins, thrombospondin-1, some fibulins, and fibrillin-1 fragments (Annes et al. 2003; Hyytiäinen et al. 2004; Ge and Greenspan 2006; Chaudhry et al. 2007; Wipff and Hinz 2008; Ono et al. 2009). In fibrillin-1, the interaction site with LTBP-1 has been mapped to a multifunctional N-terminal region spanning EGF2-cbEGF1 using recombinant protein fragments (Isogai et al. 2003; Charbonneau et al. 2004) (Fig. 7.4). Sakai and coworkers suggested a model in which LTBPs are stabilized by their C-terminal interactions with microfibrils in addition to their N-terminal interactions with other matrix components, which maintains the LTBP-bound TGF-β/LAP complex in a quiescent state (Isogai et al. 2003). These investigators speculated that fibrillin-1 deficiency (i.e., decreased amounts of functional microfibrils; Fig. 7.5) may result in TGF-β activation. This original hypothesis is now well supported by several studies with fibrillin-1 deficient mouse models demonstrating overactivation of TGF-β (Neptune et al. 2003; Habashi et al. 2006; Cohn et al. 2007). This overactivation results in typical cardiovascular, pulmonary, and musculoskeletal manifestations resembling those seen in Marfan syndrome. Treatment of these mice with TGF-β neutralizing antibodies or with losartan, an angiotensin II type 1 receptor blocker that interferes with TGF-β signaling, rescued these phenotypes. A recent clinical study using losartan in a small cohort of pediatric patients with severe Marfan syndrome showed the potential of this drug to slow the progression of aortic root dilatation (Brooke et al. 2008). Although it is not clear how deficiencies in fibrillin-1 translate into microfibril malfunction, evidence is accumulating that microfibril formation and incorporation of LTBP-bound TGF-β is a strictly cell coordinated process (Ramirez and Rifkin 2009). It is thus possible that the deficiency of microfibrils to properly regulate the TGF-β bioavailability is determined early in the microfibril assembly process.

Other microfibril-associated proteins may play a modulator role in microfibril-mediated growth factor signaling. MAGP-1 knock-out mice show some phenotypic traits including fat, bone, and muscle phenotypes that are consistent with loss of TGF-β function and are generally opposite those associated with mutations in fibrillin-1 that result in enhanced TGF-β signaling (Weinbaum et al. 2008). The authors demonstrated that MAGP-1, an integral protein of microfibrils, binds the active forms of TGF-β1 and BMP-7. EMILIN-1 (elastin microfibril interface located protein) is localized at the interface between microfibrils and the amorphous core of elastin fibers (Bressan et al. 1993). Mouse EMILIN-1 inhibits TGF-β signaling by binding specifically to the proTGF-β precursor and preventing its maturation by furin convertases in the extracellular space (Zacchigna et al. 2006). Another example for a potential role in fine tuning microfibril-mediated growth factor signaling comes from the observation that fibulin-2, -4, -5 and LTBP-1 and fibulin-4 and LTBP-4 compete for similar binding sites on fibrillin-1, raising the possibility that fibulins may play a role in the sequestration and activation of TGF-β in the extracellular matrix (Ono et al. 2009).

In addition to positioning TGF-β via LTBPs, emerging data point to an even wider role for fibrillins and microfibrils in regulating the bioavailability of other members of the TGF-β superfamily including BMPs. Initially, fibrillin-2 and BMP-7 have been linked to the same genetic pathway by gene targeting experiments in mice (Arteaga-Solis et al. 2001). In this study, homozygous mice deficient for fibrillin-2 ($Fbn2^{-/-}$) are born with temporary joint contractures and a limb patterning defect (syndactyly), whereas the $Fbn2^{+/-}$ animals display no obvious phenotype. Bmp-7 null mice are characterized by several developmental abnormalities including polydactyly, whereas Bmp-$7^{+/-}$ mice are phenotypically silent (Dudley et al. 1995; Luo et al. 1995). Combined heterozygous animals for fibrillin-2 and BMP-7 ($Fbn2^{+/-}$; Bmp-$7^{+/-}$) revealed a limb phenotype that combines the patterning defects (polydactyly and syndactyly) of each homozygous mouse, suggesting that fibrillin-2 and BMP-7 interact with each other in the same signaling pathway during limb development. More recently, BMP-7 was immunolocalized to fibrillin networks in skin and kidney capsules (Gregory et al. 2005). This work demonstrated that the interaction of the BMP-7 prodomain with an N-terminal region of fibrillin-1 polypeptides cannot be attributed to a single TB domain in fibrillin-1, but rather to a larger region of the protein (Fig. 7.4). In addition, direct interactions of an N-terminal region of fibrillin-1 with the prodomain of BMP-2, -4, and -10 and with growth and differentiation factor (GDF)-5 have been established (Sengle et al. 2008a). BMPs are bound to fibrillin by noncovalent forces and can be activated by competitive binding of type II receptors such as BMP receptor II and activin receptors IIA and IIB (Sengle et al. 2008b). These data suggest a molecular replacement mechanism for their activation.

In summary, new concepts have emerged over the past few years pinpointing a role of fibrillin-containing microfibrils in extracellular positioning of growth factors of the TGF-β/BMP superfamily either mediated indirectly through LTBP-1, -3, and -4 or directly through interactions with BMP prodomains. This type of "ready-to-use" storage of microfibril-bound growth factors provides tissues with a tool to rapidly induce growth factor signals in restricted areas. It will be important in the future to identify all growth factors that are localized to microfibrils, to understand the significance of microfibrils in the molecular physiology of these growth factors and to evaluate if and how these growth factor reservoirs can be replenished after activation. We predict that this work will identify a number of new therapeutic targets that can be explored for alternative treatment strategies of fibrillinopathies.

7.6 Mouse Models

The generation of several mouse models over the past years revealed important mechanistic insights into the functional roles of fibrillin-1 and -2 during development and in the homeostasis of microfibril-rich tissues. These animal models significantly contributed to the understanding of pathogenetic mechanisms involved in fibrillinopathies (Table 7.1). In mice, the fibrillin-3 gene is inactive due to chromosomal rearrangements and is thus not accessible to gene targeting experiments (Corson et al. 2004).

The mgΔ (Δ – "deleted") mice express a fibrillin-1 protein at about 10% of the wild-type level with a deletion of cbEGF8-TB3 (exons 19–24) in the center of the molecule (Pereira et al. 1997). The deleted region is positioned upstream of the neonatal region in human fibrillin-1. While heterozygous mgΔ/+ mice show no phenotype, homozygous mgΔ/mgΔ mice die around 3 weeks after birth of cardiovascular complications including dilatation and dissection of the aortic wall. No skeletal abnormalities were observed in homozygous mutant mice. Microfibrils containing the mutant fibrillin-1 still assembled and elastic fibers developed normally, although focal fragmentation of elastic fibers was observed. The mgΔ mice served as a model system for the seminal discovery that defects in the fibrillin-1 gene correlate with enhanced bioavailability of TGF-β (Neptune et al. 2003). In this study, active TGF-β, but not LAP, was greatly enhanced in lung tissue, indicating that TGF-β activation was modulated rather than expression levels or matrix deposition of TGF-β/LAP. Administration of TGF-β-neutralizing antibodies reverted the developmental impairment of distal alveolar septation in these mice.

In the mgR (R – "reduced") model, mice express a normal full length fibrillin-1 protein, but its expression is reduced to about 20–25% compared to the wild-type level (Pereira et al. 1999). Homozygous mgR/mgR mice live significantly longer than the mgΔ/mgΔ mice and die a few months after birth from pulmonary and vascular insufficiency. The mice display phenotypic features in the skeleton including significant kyphosis and overgrowth of the ribs, but not in long bones of the extremities. In the vascular system, medial calcification of elastic lamellae indicates that fibrillin-1 or associated components may be involved in the protection of elastic fibers against calcification. With this model, a threshold theory was suggested for the development of aortic aneurysms and dissection depending on the total amount of functional microfibrils present in a tissue (Pereira et al. 1999).

Complete fibrillin-1 ablation in the mgN/mgN (N – "null") mouse model results in ruptured aortic aneurysms and impaired pulmonary function (Carta et al. 2006). Homozygous animals die within the first 2 weeks of postnatal growth, whereas heterozygous mgN/+ mice are viable and fertile. The elastic lamellar units in the medial layer are disorganized not only in lesions as described for the mgΔ and the mgR mutant mice, but in the entire aorta. These observations suggest a key role for fibrillin-1 in development and maturation of the elastic fibers and elastic

Table 7.1 Overview of fibrillin mouse models

Model	Fibrillin affected	Tissue phenotype	Microfibril/elastic fiber phenotype
mgΔ	Deletion of exons 19–24 (cbEGF8-TB/8-Cys3) in fibrillin-1; mutant fibrillin-1 is expressed at ~10% of wild-type level	Mice die at ~3 weeks after birth due to cardiovascular complications (aortic dilation and dissection); no skeletal phenotype	Fibrillin-1 network from fibroblasts is reduced; mutant fibrillin-1 assembles into microfibrils; focal fragmentation of elastic fibers
mgR	Normal fibrillin-1 is expressed at ~20–25% of the wild-type level	Mice die after 3–4 months of pulmonary and vascular insufficiency; kyphosis and overgrowth of ribs but other long bones not affected	Reduced fibrillin-1 deposition; ~6 weeks after birth onset of focal calcification of aortic elastic lamellae
mgN	Fibrillin-1 null	Mice die within 2 weeks after birth of vascular and pulmonary failure; elongated ribs but no additional bone phenotype; thinner skin; detached endothelial lining	Thin, wavy, and fragmented elastic fibers in whole aorta
C1663R	Transgenic overexpression of human fibrillin-1 containing mutation C1663R in a normal mouse background	No phenotype	No phenotype
C1039G	Mouse fibrillin-1 with missense mutation C1039G	Heterozygous mice live normal life span; aortic wall deterioration (2 months after birth); no death due to aortic dissection. Postnatal development of kyphosis and rib overgrowth Homozygous mice die perinatally due to vascular failure	Reduced microfibril deposition from hetero- and homozygous fibroblasts; late onset of elastic fiber fragmentation
Tsk	In-frame duplication of exons 17–40 in fibrillin-1	Heterozygous mice have thickened skin with decreased elasticity; myocardial fibrosis; emphysema-like condition; increased growth of bone and cartilage; normal life span Homozygous mice die at embryonic days 7–8	Tsk fibrillin-1 incorporates into abnormal microfibrils

(*continued*)

Table 7.1 (continued)

Model	Fibrillin affected	Tissue phenotype	Microfibril/elastic fiber phenotype
Fbn2	Fibrillin-2 null	Mice are viable and fertile; bilateral syndactyly; temporary joint contractures; absence of vascular phenotype	Disorganized microfibrillar patterns in interdigital tissues
sy	Multigene deletion including Fbn2 locus on chromosome 18	Variable fore- and hindlimb syndactyly; deafness; abnormal behavior	Intact microfibrils
sy^{fp}	Frameshift mutation in Fbn2 (5051delA) generates a premature stop codon	Variable fore- and hindlimb syndactyly	Intact microfibrils
sy^{fp-2J}	Exon skipping mutation leading to loss of exon 38 in Fbn2 coding for the second half of the fourth TB/8-Cys domain in fibrillin-2	Variable fore- and hindlimb syndactyly	Intact microfibrils
$Fbn1^{-/-}$ $Fbn2^{-/-}$	Fibrillin-1 and -2 double null	Embryonic lethality around E14.5	Delayed elastic fiber formation in aortic media

Except where specifically mentioned, heterozygous animals or fibroblasts do not show any phenotype and the description is limited to homozygosity
Reprinted from Hubmacher et al. (2006), copyright 2006, with permission from Elsevier. Complete references for all mouse models can be found in that review

lamellae of the aortic wall especially during early postnatal life. Microfibrils isolated from mgN/mgN mice, consisting only of fibrillin-2, showed an asymmetric ultrastructural appearance, which differs from the appearance of other microfibrils containing only fibrillin-1. This suggests differences in the composition and/or assembly process of fibrillin-2 microfibrils compared to fibrillin-1 microfibrils.

Since fibrillin-2 is mainly expressed during development and typically earlier than fibrillin-1, it was surprising that deleting the fibrillin-2 gene generated a relatively mild phenotype (Arteaga-Solis et al. 2001). Homozygous mice ($Fbn2^{-/-}$) are born with temporary joint contractures, mimicking the clinical symptoms observed in patients with CCA. Carta and coworkers proposed that fibrillin-1 may compensate for the loss of fibrillin-2 in $Fbn2^{-/-}$ mice based on the analysis of $Fbn1/Fbn2$ double mutant mice (Carta et al. 2006): complete loss of both fibrillin-1 and -2 are incompatible with embryonic viability since homozygous double mutants ($Fbn1^{-/-}$; $Fbn2^{-/-}$) die around E14.5 due to impaired elastogenesis in the media of the aorta. These results strongly suggested that at least one of the two fibrillin isoforms is necessary for the initial assembly of elastic fibers, although fibrillin-2 is not required for the later steps of elastic fiber development and maintenance. The connective tissue phenotype, including syndactyly of the $Fbn2^{-/-}$ mutant, correlates well with that of the radiation-induced classical mouse mutant shaker-with-syndactylism (sy) caused by a multigene deletion including the locus for *Fbn2* on chromosome 18 (Hertwig 1942; Johnson et al. 1998; Chaudhry et al. 2001). The pathogenesis of Marfan syndrome can be caused by either a dominant negative mechanism, where mutated protein becomes incorporated in microfibrils and compromises their function, or by haploinsufficiency, where mutated fibrillin is expressed at lower levels or is degraded (Fig. 7.5). To address which of the pathogenetic mechanisms may play the predominant role, Judge and coworkers generated transgenic mice models (Judge et al. 2004). In one model, human fibrillin-1 harboring the classical Marfan mutation C1663R in cbEGF24 was expressed in a normal mouse background. Despite expression and integration of the mutated fibrillin1 in microfibrils and deposition in the extracellular matrix, these mice did not show any abnormalities. On the other hand, mice heterozygous for the C1039G mutation in cbEGF11 of mouse fibrillin-1 showed skeletal deformity and progressive deterioration of the aortic wall including elastic fiber fragmentation. In addition, deposition of microfibrils by fibroblasts isolated from these mice showed a diminished fibrillin-1 network. Introduction of a wild-type human *FBN1* transgene in the heterozygous C1039G mouse background rescued the aortic phenotype. These mouse models suggest that haploinsufficiency contributes to the pathogenesis of Marfan syndrome. The C1039G mouse model was used in a seminal study to demonstrate that the angiotensin II type 1 receptor blocker losartan is able to reduce TGF-β signaling and prevent aortic aneurysm formation (Habashi et al. 2006). This work led to the above-mentioned recent clinical study using losartan in a small cohort of pediatric patients with severe Marfan syndrome, where the drug slowed down the rate of progressive aortic root dilatation (Brooke et al. 2008).

Tight-skin (Tsk) is an autosomal dominant mutation that occurred spontaneously (Green et al. 1976). Mice homozygous for the Tsk mutation die in utero at 7–8 days of gestation. Heterozygous Tsk/+ mice are characterized by tight skin, increased growth of cartilage and bone, and large accumulations of microfibrils in the loose connective tissue. The Tsk mutation was identified as a tandem genomic in-frame duplication of the central exons 17–40 of the *Fbn1* gene, resulting in a larger ~420 kDa fibrillin-1 protein compared to the ~350 kDa wild-type protein (Siracusa et al. 1996; Saito et al. 1999). Despite a controversy about the incorporation of both normal and mutated fibrillin-1 in the same or in mutually exclusive microfibrils, it seems clear that the mutant Tsk fibrillin-1 is able to incorporate into microfibrils (Kielty et al. 1998; Gayraud et al. 2000). This is in line with the current view of microfibril assembly, indicating that the critical self-interaction of fibrillin-1 is guided by regions located in the N- and C-terminus and is therefore not disturbed by alterations in the central region (Lin et al. 2002; Marson et al. 2005; Hubmacher et al. 2008). Enhanced proteolytic susceptibility of the Tsk fibrillin-1 may lead to decreased numbers of fully functional microfibrils in tissue, which in turn may destabilize TGF-β/LAP bound to LTBPs leading to activation of TGF-β (see Sect. 7.5) (Gayraud et al. 2000).

7.7 Conclusions

Over the past years, detailed information has emerged about the roles of microfibrils and fibrillins in development and homeostasis of tissues and organs. Important mechanisms have been identified for the biogenesis of microfibrils. It is now evident that microfibrils and fibrillins not only fulfill structural roles, but also function as important extracellular matrix regulators in developmental and signaling processes. Mouse models provided new concepts for pathogenetic mechanisms in fibrillinopathies and offer the possibility to test therapeutic strategies for these disorders. Future research should aim at integrating functions of microfibrils and fibrillins in their specific cellular and organismal context.

Acknowledgments This work was supported by the Canadian Institutes of Health Research (MOP-68836 to DPR) and the German Academic Exchange Service DAAD (postdoctoral fellowship to DH).

References

Annes JP, Munger JS, Rifkin DB (2003) Making sense of latent TGFbeta activation. J Cell Sci 116:217–224
Arteaga-Solis E, Gayraud B, Lee SY, Shum L, Sakai LY, Ramirez F (2001) Regulation of limb patterning by extracellular microfibrils. J Cell Biol 154:275–281
Ashworth JL, Kelly V, Wilson R, Shuttleworth CA, Kielty CM (1999) Fibrillin assembly: dimer formation mediated by amino-terminal sequences. J Cell Sci 112:3549–3558
Baldock C, Koster AJ, Ziese U, Rock MJ, Sherratt MJ, Kadler KE, Shuttleworth CA, Kielty CM (2001) The supramolecular organization of fibrillin-rich microfibrils. J Cell Biol 152:1045–1056
Bax DV, Bernard SE, Lomas A, Morgan A, Humphries J, Shuttleworth A, Humphries MJ, Kielty CM (2003) Cell adhesion to fibrillin-1 molecules and microfibrils is mediated by alpha5 beta1 and alphav beta3 integrins. J Biol Chem 278:34605–34616
Bowness JM, Tarr AH (1997) Epsilon(gamma-Glutamyl)lysine crosslinks are concentrated in a non-collagenous microfibrillar fraction of cartilage. Biochem Cell Biol 75:89–91
Bressan GM, Daga-Gordini D, Colombatti A, Castellani I, Marigo V, Volpin D (1993) Emilin, a component of elastic fibers preferentially located at the elastin-microfibril interface. J Cell Biol 121:201–212
Brooke BS, Habashi JP, Judge DP, Patel N, Loeys B, Dietz HC (2008) Angiotensin II blockade and aortic-root dilation in Marfan's syndrome. N Engl J Med 358:2787–2795
Cain SA, Baldock C, Gallagher J, Morgan A, Bax DV, Weiss AS, Shuttleworth CA, Kielty CM (2005) Fibrillin-1 interactions with heparin: implications for microfibril and elastic fibre assembly. J Biol Chem 280:30526–30537
Cain SA, Morgan A, Sherratt MJ, Ball SG, Shuttleworth CA, Kielty CM (2006) Proteomic analysis of fibrillin-rich microfibrils. Proteomics 6:111–122
Callewaert BL, Loeys BL, Ficcadenti A, Vermeer S, Landgren M, Kroes HY, Yaron Y, Pope M, Foulds N, Boute O, Galan F, Kingston H, Van der Aa N, Salcedo I, Swinkels ME, Wallgren-Pettersson C, Gabrielli O, De BJ, Coucke PJ, De Paepe AM (2009) Comprehensive clinical and molecular assessment of 32 probands with congenital contractural arachnodactyly: report of 14 novel mutations and review of the literature. Hum Mutat 30:334–341

Carta L, Pereira L, Arteaga-Solis E, Lee-Arteaga SY, Lenart B, Starcher B, Merkel CA, Sukoyan M, Kerkis A, Hazeki N, Keene DR, Sakai LY, Ramirez F (2006) Fibrillins 1 and 2 perform partially overlapping functions during aortic development. J Biol Chem 281:8016–8023

Charbonneau NL, Dzamba BJ, Ono RN, Keene DR, Corson GM, Reinhardt DP, Sakai LY (2003) Fibrillins can co-assemble in fibrils, but fibrillin fibril composition displays cell-specific differences. J Biol Chem 278:2740–2749

Charbonneau NL, Ono RN, Corson GM, Keene DR, Sakai LY (2004) Fine tuning of growth factor signals depends on fibrillin microfibril networks. Birth Defects Res C Embryo Today 72:37–50

Chaudhry SS, Gazzard J, Baldock C, Dixon J, Rock MJ, Skinner GC, Steel KP, Kielty CM, Dixon MJ (2001) Mutation of the gene encoding fibrillin-2 results in syndactyly in mice. Hum Mol Genet 10:835–843

Chaudhry SS, Cain SA, Morgan A, Dallas SL, Shuttleworth CA, Kielty CM (2007) Fibrillin-1 regulates the bioavailability of TGF-beta1. J Cell Biol 176:355–367

Chen Y, Ali T, Todorovic V, O'Leary JM, Kristina DA, Rifkin DB (2005) Amino acid requirements for formation of the TGF-beta-latent TGF-beta binding protein complexes. J Mol Biol 345:175–186

Choudhury R, McGovern A, Ridley C, Cain SA, Baldwin A, Wang M-C, Guo C, Mironov AJ, Drymoussi Z, Trump D, Shuttleworth A, Baldock C, Kielty CM (2009) Differential regulation of elastic fiber formation by fibulins-4 and -5. J Biol Chem 284:24553–24567

Cohn RD, van EC H, JP SAA, Klein EC, Lisi MT, Gamradt M, ap Rhys CM, Holm TM, Loeys BL, Ramirez F, Judge DP, Ward CW, Dietz HC (2007) Angiotensin II type 1 receptor blockade attenuates TGF-beta-induced failure of muscle regeneration in multiple myopathic states. Nat Med 13:204–210

Collod-Beroud G, Le Bourdelles S, Ades L, Ala-Kokko L, Booms P, Boxer M, Child A, Comeglio P, De Paepe A, Hyland JC, Holman K, Kaitila I, Loeys B, Matyas G, Nuytinck L, Peltonen L, Rantamaki T, Robinson P, Steinmann B, Junien C, Beroud C, Boileau C (2003) Update of the UMD-FBN1 mutation database and creation of an FBN1 polymorphism database. Hum Mutat 22:199–208

Corson GM, Chalberg SC, Dietz HC, Charbonneau NL, Sakai LY (1993) Fibrillin binds calcium and is coded by cDNAs that reveal a multidomain structure and alternatively spliced exons at the 5' end. Genomics 17:476–484

Corson GM, Charbonneau NL, Keene DR, Sakai LY (2004) Differential expression of fibrillin-3 adds to microfibril variety in human and avian, but not rodent, connective tissues. Genomics 83:461–472

Dallas SL, Keene DR, Bruder SP, Saharinen J, Sakai LY, Mundy GR, Bonewald LF (2000) Role of the latent transforming growth factor beta binding protein 1 in fibrillin-containing microfibrils in bone cells in vitro and in vivo. J Bone Miner Res 15:68–81

Dallas SL, Sivakumar P, Jones CJ, Chen Q, Peters DM, Mosher DF, Humphries MJ, Kielty CM (2005) Fibronectin regulates latent transforming growth factor-beta (TGF beta) by controlling matrix assembly of latent TGF beta-binding protein-1. J Biol Chem 280:18871–18880

Davis EC, Roth RA, Heuser JE, Mecham RP (2002) Ultrastructural properties of ciliary zonule microfibrils. J Struct Biol 139:65–75

Dietz HC, Cutting GR, Pyeritz RE, Maslen CL, Sakai LY, Corson GM, Puffenberger EG, Hamosh A, Nanthakumar EJ, Curristin SM, Stetten G, Meyers DA, Francomano CA (1991) Marfan syndrome caused by a recurrent de novo missense mutation in the fibrillin gene. Nature 352:337–339

Downing AK, Knott V, Werner JM, Cardy CM, Campbell ID, Handford PA (1996) Solution structure of a pair of calcium-binding epidermal growth factor-like domains: implications for the Marfan syndrome and other genetic disorders. Cell 85:597–605

Dudley AT, Lyons KM, Robertson EJ (1995) A requirement for bone morphogenetic protein-7 during development of the mammalian kidney and eye. Genes Dev 9:2795–2807

El-Hallous E, Sasaki T, Hubmacher D, Getie M, Tiedemann K, Brinckmann J, Bätge B, Davis EC, Reinhardt DP (2007) Fibrillin-1 interactions with fibulins depend on the first hybrid domain and provide an adapter function to tropoelastin. J Biol Chem 282:8935–8946

Eriksen TA, Wright DM, Purslow PP, Duance VC (2001) Role of Ca(2+) for the mechanical properties of fibrillin. Proteins 45:90–95

Fahrenbach WH, Sandberg LB, Cleary EG (1966) Ultrastructural studies on early elastogenesis. Anat Rec 155:563–576

Faivre L, Dollfus H, Lyonnet S, Alembik Y, Megarbane A, Samples J, Gorlin RJ, Alswaid A, Feingold J, Le MM, Munnich A, Cormier-Daire V (2003a) Clinical homogeneity and genetic heterogeneity in Weill–Marchesani syndrome. Am J Med Genet A 123:204–207

Faivre L, Gorlin RJ, Wirtz MK, Godfrey M, Dagoneau N, Samples JR, Le MM, Collod-Beroud G, Boileau C, Munnich A, Cormier-Daire V (2003b) In frame fibrillin-1 gene deletion in autosomal dominant Weill–Marchesani syndrome. J Med Genet 40:34–36

Faivre L, Collod-Beroud G, Loeys BL, Child A, Binquet C, Gautier E, Callewaert B, Arbustini E, Mayer K, Arslan-Kirchner M, Kiotsekoglou A, Comeglio P, Marziliano N, Dietz HC, Halliday D, Beroud C, Bonithon-Kopp C, Claustres M, Muti C, Plauchu H, Robinson PN, Ades LC, Biggin A, Benetts B, Brett M, Holman KJ, de Backer J, Coucke P, Francke U, De Paepe A, Jondeau G, Boileau C (2007) Effect of mutation type and location on clinical outcome in 1,013 probands with Marfan syndrome or related phenotypes and FBN1 mutations: an international study. Am J Hum Genet 81:454–466

Fleischer KJ, Nousari HC, Anhalt GJ, Stone CD, Laschinger JC (1997) Immunohistochemical abnormalities of fibrillin in cardiovascular tissues in Marfan's syndrome. Ann Thorac Surg 63:1012–1017

Frederic MY, Monino C, Marschall C, Hamroun D, Faivre L, Jondeau G, Klein HG, Neumann L, Gautier E, Binquet C, Maslen C, Godfrey M, Gupta P, Milewicz D, Boileau C, Claustres M, Beroud C, Collod-Beroud G (2009) The FBN2 gene: new mutations, locus-specific database (Universal Mutation Database FBN2), and genotype-phenotype correlations. Hum Mutat 30:181–190

Gayraud B, Keene DR, Sakai LY, Ramirez F (2000) New insights into the assembly of extracellular microfibrils from the analysis of the fibrillin 1 mutation in the tight skin mouse. J Cell Biol 150:667–680

Ge G, Greenspan DS (2006) BMP1 controls TGFbeta1 activation via cleavage of latent TGFbeta-binding protein 6. J Cell Biol 175:111–120

Gentry LE, Lioubin MN, Purchio AF, Marquardt H (1988) Molecular events in the processing of recombinant type 1 pre-pro-transforming growth factor beta to the mature polypeptide. Mol Cell Biol 8:4162–4168

Gibson MA, Kumaratilake JS, Cleary EG (1989) The protein components of the 12-nanometer microfibrils of elastic and nonelastic tissues. J Biol Chem 264:4590–4598

Gibson MA, Hatzinikolas G, Davis EC, Baker E, Sutherland GR, Mecham RP (1995) Bovine latent transforming growth factor beta 1-binding protein 2: molecular cloning, identification of tissue isoforms, and immunolocalization to elastin-associated microfibrils. Mol Cell Biol 15:6932–6942

Giltay R, Timpl R, Kostka G (1999) Sequence, recombinant expression and tissue localization of two novel extracellular matrix proteins, fibulin-3 and fibulin-4. Matrix Biol 18:469–480

Giusti B, Porciani MC, Brunelli T, Evangelisti L, Fedi S, Gensini GF, ABbate R, Sani G, Yacoub M, Pepe G (2003) Phenotypic variability of cardiovascular manifestations in Marfan Syndrome. Possible role of hyperhomocysteinemia and C677T MTHFR gene polymorphism. Eur Heart J 24:2038–2045

Gleizes PE, Beavis RC, Mazzieri R, Shen B, Rifkin DB (1996) Identification and characterization of an eight-cysteine repeat of the latent transforming growth factor-beta binding protein-1 that mediates bonding to the latent transforming growth factor-beta1. J Biol Chem 271:29891–29896

Glushchenko AV, Jacobsen DW (2007) Molecular targeting of proteins by L-homocysteine: mechanistic implications for vascular disease. Antioxid Redox Signal 9:1883–1898

Green MC, Sweet HO, Bunker LE (1976) Tight-skin, a new mutation of the mouse causing excessive growth of connective tissue and skeleton. Am J Pathol 82:493–512

Greenlee TK, Ross R, Hartman JL (1966) The fine structure of elastic fibers. J Cell Biol 30:59–71

Gregory KE, Ono RN, Charbonneau NL, Kuo CL, Keene DR, Bächinger HP, Sakai LY (2005) The prodomain of BMP-7 targets the BMP-7 complex to the extracellular matrix. J Biol Chem 280:27970–27980

Gupta PA, Putnam EA, Carmical SG, Kaitila I, Steinmann B, Child A, Danesino C, Metcalfe K, Berry SA, Chen E, Delorme CV, Thong MK, Ades LC, Milewicz DM (2002) Ten novel FBN2 mutations in congenital contractural arachnodactyly: delineation of the molecular pathogenesis and clinical phenotype. Hum Mutat 19:39–48

Habashi JP, Judge DP, Holm TM, Cohn RD, Loeys BL, Cooper TK, Myers L, Klein EC, Liu G, Calvi C, Podowski M, Neptune ER, Halushka MK, Bedja D, Gabrielson K, Rifkin DB, Carta L, Ramirez F, Huso DL, Dietz HC (2006) Losartan, an AT1 antagonist, prevents aortic aneurysm in a mouse model of Marfan syndrome. Science 312:117–121

Halme T, Savunen T, Aho H, Vihersaari T, Penttinen R (1985) Elastin and collagen in the aortic wall: changes in the Marfan syndrome and annuloaortic ectasia. Exp Mol Pathol 43:1–12

Handford PA (2000) Fibrillin-1, a calcium binding protein of extracellular matrix. Biochim Biophys Acta 1498:84–90

Handford PA, Mayhew M, Baron M, Winship PR, Campbell ID, Brownlee GG (1991) Key residues involved in calcium-binding motifs in EGF-like domains. Nature 351:164–167

Hertwig P (1942) Neue Mutationen und Koppelungsgruppen bei der Hausmaus. Z Indukt Abstamm Vererbungsl 80:220–246

Hill CH, Mecham R, Starcher B (2002) Fibrillin-2 defects impair elastic fiber assembly in a homocysteinemic chick model. J Nutr 132:2143–2150

Hubmacher D, Tiedemann K, Bartels R, Brinckmann J, Vollbrandt T, Bätge B, Notbohm H, Reinhardt DP (2005) Modification of the structure and function of fibrillin-1 by homocysteine suggests a potential pathogenetic mechanism in homocystinuria. J Biol Chem 280:34946–34955

Hubmacher D, Tiedemann K, Reinhardt DP (2006) Fibrillins: from biogenesis of microfibrils to signaling functions. Curr Top Dev Biol 75:93–123

Hubmacher D, El-Hallous E, Nelea V, Kaartinen MT, Lee ER, Reinhardt DP (2008) Biogenesis of extracellular microfibrils – multimerization of the fibrillin-1 C-terminus into bead-like structures enables self-assembly. Proc Natl Acad Sci USA 105:6548–6553

Hubmacher D, Cirulis JT, Miao M, Keeley FW, Reinhardt DP (2010) Functional consequences of homocysteinylation of the elastic fiber proteins fibrillin-1 and tropoelastin. J Biol Chem 285:1188–1198

Hutchinson S, Aplin RT, Webb H, Kettle S, Timmermans J, Boers GH, Handford PA (2005) Molecular effects of homocysteine on cbEGF domain structure: insights into the pathogenesis of homocystinuria. J Mol Biol 346:833–844

Hyytiäinen M, Penttinen C, Keski-Oja J (2004) Latent TGF-beta binding proteins: extracellular matrix association and roles in TGF-beta activation. Crit Rev Clin Lab Sci 41:233–264

Isogai Z, Ono RN, Ushiro S, Keene DR, Chen Y, Mazzieri R, Charbonneau NL, Reinhardt DP, Rifkin DB, Sakai LY (2003) Latent transforming growth factor beta-binding protein 1 interacts with fibrillin and is a microfibril-associated protein. J Biol Chem 278:2750–2757

Jensen SA, Corbett AR, Knott V, Redfield C, Handford PA (2005) Ca^{2+}-dependent interface formation in fibrillin-1. J Biol Chem 280:14076–14084

Jensen SA, Iqbal S, Lowe ED, Redfield C, Handford PA (2009) Structure and interdomain interactions of a hybrid domain: a disulphide-rich module of the fibrillin/LTBP superfamily of matrix proteins. Structure 17:759–768

Johnson KR, Cook SA, Zheng QY (1998) The original shaker-with-syndactylism mutation (sy) is a contiguous gene deletion syndrome. Mamm Genome 9:889–892

Jovanovic J, Takagi J, Choulier L, Abrescia NG, Stuart DI, van der Merwe PA, Mardon HJ, Handford PA (2007) Alpha v beta 6 is a novel receptor for human fibrillin-1: comparative studies of molecular determinants underlying integrin-RGD affinity and specificity. J Biol Chem 282:6743–6751

Judge DP, Biery NJ, Keene DR, Geubtner J, Myers L, Huso DL, Sakai LY, Dietz HC (2004) Evidence for a critical contribution of haploinsufficiency in the complex pathogenesis of Marfan syndrome. J Clin Invest 114:172–181

Kainulainen K, Pulkkinen L, Savolainen A, Kaitila I, Peltonen L (1990) Location on chromosome 15 of the gene defect causing Marfan syndrome. N Engl J Med 323:935–939

Kantola AK, Keski-Oja J, Koli K (2008) Fibronectin and heparin binding domains of latent TGF-beta binding protein (LTBP)-4 mediate matrix targeting and cell adhesion. Exp Cell Res 314:2488–2500

Keene DR, Maddox BK, Kuo HJ, Sakai LY, Glanville RW (1991) Extraction of extendable beaded structures and their identification as fibrillin-containing extracellular matrix microfibrils. J Histochem Cytochem 39:441–449

Keene DR, Jordan CD, Reinhardt DP, Ridgway CC, Ono RN, Corson GM, Fairhurst M, Sussman MD, Memoli VA, Sakai LY (1997) Fibrillin-1 in human cartilage: developmental expression and formation of special banded fibers. J Histochem Cytochem 45:1069–1082

Kielty CM, Phillips JE, Child AH, Pope FM, Shuttleworth CA (1994) Fibrillin secretion and microfibril assembly by Marfan dermal fibroblasts. Matrix Biol 14:191–199

Kielty CM, Raghunath M, Siracusa LD, Sherratt MJ, Peters R, Shuttleworth CA, Jimenez SA (1998) The tight skin mouse: demonstration of mutant fibrillin-1 production and assembly into abnormal microfibrils. J Cell Biol 140:1159–1166

Kinsey R, Williamson MR, Chaudhry S, Mellody KT, McGovern A, Takahashi S, Shuttleworth CA, Kielty CM (2008) Fibrillin-1 microfibril deposition is dependent on fibronectin assembly. J Cell Sci 121:2696–2704

Koli K, Hyytiäinen M, Ryynanen MJ, Keski-Oja J (2005) Sequential deposition of latent TGF-beta binding proteins (LTBPs) during formation of the extracellular matrix in human lung fibroblasts. Exp Cell Res 310:370–382

Kriz W, Elger M, Lemley K, Sakai T (1990) Structure of the glomerular mesangium: a biomechanical interpretation. Kidney Int Suppl 30:S2–S9

Krumdieck CL, Prince CW (2000) Mechanisms of homocysteine toxicity on connective tissues: implications for the morbidity of aging. J Nutr 130:365S–368S

Kuo CL, Isogai Z, Keene DR, Hazeki N, Ono RN, Sengle G, Bächinger HP, Sakai LY (2007) Effects of fibrillin-1 degradation on microfibril ultrastructure. J Biol Chem 282:4007–4020

Lee B, Godfrey M, Vitale E, Hori H, Mattei MG, Sarfarazi M, Tsipouras P, Ramirez F, Hollister DW (1991) Linkage of Marfan syndrome and a phenotypically related disorder to two different fibrillin genes. Nature 352:330–334

Lee SS, Knott V, Jovanovic J, Harlos K, Grimes JM, Choulier L, Mardon HJ, Stuart DI, Handford PA (2004) Structure of the integrin binding fragment from fibrillin-1 gives new insights into microfibril organization. Structure (Camb) 12:717–729

Lemaire R, Bayle J, Lafyatis R (2006) Fibrillin in Marfan syndrome and tight skin mice provides new insights into transforming growth factor-beta regulation and systemic sclerosis. Curr Opin Rheumatol 18:582–587

Lin G, Tiedemann K, Vollbrandt T, Peters H, Bätge B, Brinckmann J, Reinhardt DP (2002) Homo- and heterotypic fibrillin-1 and -2 interactions constitute the basis for the assembly of microfibrils. J Biol Chem 277:50795–50804

Loeys BL, Gerber EE, Riegert-Johnson D, Iqbal S, Whiteman P, McConnell V, Chillakuri CR, Macaya D, Coucke PJ, De Paepe A, Judge DP, Wigley F, Davis EC, Mardon HJ, Handford P, Keene DR, Sakai LY, Dietz HC (2010) Mutations in fibrillin-1 cause congenital scleroderma: stiff skin syndrome. Sci Transl Med 2:23ra20

Low FN (1962) Microfibrils: fine filamentous components of the tissue space. Anat Rec 142:131–137

Luo G, Hofmann C, Bronckers AL, Sohocki M, Bradley A, Karsenty G (1995) BMP-7 is an inducer of nephrogenesis, and is also required for eye development and skeletal patterning. Genes Dev 9:2808–2820

Marson A, Rock MJ, Cain SA, Freeman LJ, Morgan A, Mellody K, Shuttleworth CA, Baldock C, Kielty CM (2005) Homotypic fibrillin-1 interactions in microfibril assembly. J Biol Chem 280:5013–5021

Maslen CL, Corson GM, Maddox BK, Glanville RW, Sakai LY (1991) Partial sequence of a candidate gene for the Marfan syndrome. Nature 352:334–337

McConnell CJ, Wright GM, DeMont ME (1996) The modulus of elasticity of lobster aorta microfibrils. Experientia 52:918–921

Milewicz D, Pyeritz RE, Crawford ES, Byers PH (1992) Marfan syndrome: defective synthesis, secretion, and extracellular matrix formation of fibrillin by cultured dermal fibroblasts. J Clin Invest 89:79–86

Milewicz DM, Grossfield J, Cao SN, Kielty C, Covitz W, Jewett T (1995) A mutation in FBN1 disrupts profibrillin processing and results in isolated skeletal features of the Marfan syndrome. J Clin Invest 95:2373–2378

Miyazono K, Olofsson A, Colosetti P, Heldin CH (1991) A role of the latent TGF-β1-binding protein in the assembly and secretion of TGF-β1. EMBO J 10:1091–1101

Nagase T, Nakayama M, Nakajima D, Kikuno R, Ohara O (2001) Prediction of the coding sequences of unidentified human genes. XXII. The complete sequences of 100 new cDNA clones from brain which code for large proteins in vitro. DNA Res 8:85–95

Nakajima Y, Miyazono K, Kato M, Takase M, Yamagishi T, Nakamura H (1997) Extracellular fibrillar structure of latent TGF beta binding protein-1: role in TGF beta-dependent endothelial-mesenchymal transformation during endocardial cushion tissue formation in mouse embryonic heart. J Cell Biol 136:193–204

Neptune ER, Frischmeyer PA, Arking DE, Myers L, Bunton TE, Gayraud B, Ramirez F, Sakai LY, Dietz HC (2003) Dysregulation of TGF-beta activation contributes to pathogenesis in Marfan syndrome. Nat Genet 33:407–411

Ono RN, Sengle G, Charbonneau NL, Carlberg V, Bachinger HP, Sasaki T, Lee-Arteaga S, Zilberberg L, Rifkin DB, Ramirez F, Chu ML, Sakai LY (2009) Latent transforming growth factor beta-binding proteins and fibulins compete for fibrillin-1 and exhibit exquisite specificities in binding sites. J Biol Chem 284:16872–16881

Park ES, Putnam EA, Chitayat D, Child A, Milewicz DM (1998) Clustering of FBN2 mutations in patients with congenital contractural arachnodactyly indicates an important role of the domains encoded by exons 24 through 34 during human development. Am J Med Genet 78:350–355

Pereira L, D'Alessio M, Ramirez F, Lynch JR, Sykes B, Pangilinan T, Bonadio J (1993) Genomic organization of the sequence coding for fibrillin, the defective gene product in Marfan syndrome. Hum Mol Genet 2:961–968

Pereira L, Andrikopoulos K, Tian J, Lee SY, Keene DR, Ono RN, Reinhardt DP, Sakai LY, Jensen-Biery N, Bunton T, Dietz HC, Ramirez F (1997) Targeting of fibrillin-1 recapitulates the vascular phenotype of Marfan syndrome in the mouse. Nat Genet 17:218–222

Pereira L, Lee SY, Gayraud B, Andrikopoulos K, Shapiro SD, Bunton T, Biery NJ, Dietz HC, Sakai LY, Ramirez F (1999) Pathogenetic sequence for aneurysm revealed in mice underexpressing fibrillin-1. Proc Natl Acad Sci USA 96:3819–3823

Pfaff M, Reinhardt DP, Sakai LY, Timpl R (1996) Cell adhesion and integrin binding to recombinant human fibrillin-1. FEBS Lett 384:247–250

Putnam EA, Zhang H, Ramirez F, Milewicz DM (1995) Fibrillin-2 (FBN2) mutations result in the Marfan-like disorder, congenital contractural arachnodactyly. Nat Genet 11:456–458

Pyeritz RE (2000) The Marfan syndrome. Annu Rev Med 51:481–510

Qian RQ, Glanville RW (1997) Alignment of fibrillin molecules in elastic microfibrils is defined by transglutaminase-derived cross-links. Biochemistry 36:15841–15847

Quondamatteo F, Reinhardt DP, Charbonneau NL, Pophal G, Sakai LY, Herken R (2002) Fibrillin-1 and fibrillin-2 in human embryonic and early fetal development. Matrix Biol 21:637–646

Raghunath M, Bächi T, Meuli M, Altermatt S, Gobet R, Bruckner-Tuderman L, Steinmann B (1996) Fibrillin and elastin expression in skin regenerating from cultured keratinocyte autografts: morphogenesis of microfibrils begins at the dermo-epidermal junction and precedes elastic fiber formation. J Invest Dermatol 106:1090–1095

Raghunath M, Unsöld C, Kubitscheck U, Bruckner-Tuderman L, Peters R, Meuli M (1998) The cutaneous microfibrillar apparatus contains latent transforming growth factor-beta binding protein-1 (LTBP-1) and is a repository for latent TGF-beta1. J Invest Dermatol 111:559–564

Ramirez F, Rifkin DB (2009) Extracellular microfibrils: contextual platforms for TGF-beta and BMP signaling. Curr Opin Cell Biol 21:616–622

Ramirez F, Sakai LY (2010) Biogenesis and function of fibrillin assemblies. Cell Tissue Res 339:71–82

Ramirez F, Carta L, Lee-Arteaga S, Liu C, Nistala H, Smaldone S (2008) Fibrillin-rich microfibrils – structural and instructive determinants of mammalian development and physiology. Connect Tissue Res 49:1–6

Raviola G (1971) The fine structure of the ciliary zonule and ciliary epithelium. Invest Ophthalmol 10:851–869

Reinhardt DP, Sasaki T, Dzamba BJ, Keene DR, Chu ML, Göhring W, Timpl R, Sakai LY (1996a) Fibrillin-1 and fibulin-2 interact and are colocalized in some tissues. J Biol Chem 271:19489–19496

Reinhardt DP, Keene DR, Corson GM, Pöschl E, Bächinger HP, Gambee JE, Sakai LY (1996b) Fibrillin 1: organization in microfibrils and structural properties. J Mol Biol 258:104–116

Reinhardt DP, Ono RN, Sakai LY (1997a) Calcium stabilizes fibrillin-1 against proteolytic degradation. J Biol Chem 272:1231–1236

Reinhardt DP, Mechling DE, Boswell BA, Keene DR, Sakai LY, Bächinger HP (1997b) Calcium determines the shape of fibrillin. J Biol Chem 272:7368–7373

Reinhardt DP, Gambee JE, Ono RN, Bächinger HP, Sakai LY (2000) Initial steps in assembly of microfibrils. Formation of disulfide-cross-linked multimers containing fibrillin-1. J Biol Chem 275:2205–2210

Rifkin DB (2005) Latent transforming growth factor-beta (TGF-beta) binding proteins: orchestrators of TGF-beta availability. J Biol Chem 280:7409–7412

Ritty TM, Broekelmann T, Tisdale C, Milewicz DM, Mecham RP (1999) Processing of the fibrillin-1 carboxyl-terminal domain. J Biol Chem 274:8933–8940

Ritty TM, Broekelmann TJ, Werneck CC, Mecham RP (2003) Fibrillin-1 and -2 contain heparin-binding sites important for matrix deposition and that support cell attachment. Biochem J 375:425–432

Robinson P, Arteaga-Solis E, Baldock C, Collod-Beroud G, Booms P, De Paepe A, Dietz HC, Guo G, Handford PA, Judge DP, Kielty CM, Loeys B, Milewicz DM, Ney A, Ramirez F, Reinhardt DP, Tiedemann K, Whiteman P, Godfrey M (2006) The molecular genetics of Marfan syndrome and related disorders. J Med Genet 43:769–787

Rock MJ, Cain SA, Freeman LJ, Morgan A, Mellody K, Marson A, Shuttleworth CA, Weiss AS, Kielty CM (2004) Molecular basis of elastic fiber formation. Critical interactions and a tropoelastin-fibrillin-1 cross-link. J Biol Chem 279:23748–23758

Sabatier L, Chen D, Fagotto-Kaufmann C, Hubmacher D, McKee MD, Annis DS, Mosher DF, Reinhardt DP (2009) Fibrillin assembly requires fibronectin. Mol Biol Cell 20:846–858

Saharinen J, Keski-Oja J (2000) Specific sequence motif of 8-Cys repeats of TGF-beta binding proteins, LTBPs, creates a hydrophobic interaction surface for binding of small latent TGF-beta. Mol Biol Cell 11:2691–2704

Saito S, Nishimura H, Brumeanu TD, Casares S, Stan AC, Honjo T, Bona CA (1999) Characterization of mutated protein encoded by partially duplicated fibrillin-1 gene in tight skin (TSK) mice. Mol Immunol 36:169–176

Sakai LY, Keene DR, Engvall E (1986) Fibrillin, a new 350-kD glycoprotein, is a component of extracellular microfibrils. J Cell Biol 103:2499–2509

Sakai LY, Keene DR, Glanville RW, Bächinger HP (1991) Purification and partial characterization of fibrillin, a cysteine-rich structural component of connective tissue microfibrils. J Biol Chem 266:14763–14770

Sakamoto H, Broekelmann T, Cheresh DA, Ramirez F, Rosenbloom J, Mecham RP (1996) Cell-type specific recognition of RGD- and non-RGD-containing cell binding domains in fibrillin-1. J Biol Chem 271:4916–4922

Segura AM, Luna RE, Horiba K, Stetler-Stevenson WG, McAllister HA Jr, Willerson JT, Ferrans VJ (1998) Immunohistochemistry of matrix metalloproteinases and their inhibitors in thoracic aortic aneurysms and aortic valves of patients with Marfan's syndrome. Circulation 98:11331–11337

Sengle G, Charbonneau NL, Ono RN, Sasaki T, Alvarez J, Keene DR, Bachinger HP, Sakai LY (2008a) Targeting of bone morphogenetic protein growth factor complexes to fibrillin. J Biol Chem 283:13874–13888

Sengle G, Ono RN, Lyons KM, Bachinger HP, Sakai LY (2008b) A new model for growth factor activation: type II receptors compete with the prodomain for BMP-7. J Mol Biol 381:1025–1039

Sherratt MJ, Baldock C, Haston JL, Holmes DF, Jones CJ, Shuttleworth CA, Wess TJ, Kielty CM (2003) Fibrillin microfibrils are stiff reinforcing fibres in compliant tissues. J Mol Biol 332:183–193

Sinha S, Heagerty AM, Shuttleworth CA, Kielty CM (2002) Expression of latent TGF-beta binding proteins and association with TGF-beta 1 and fibrillin-1 following arterial injury. Cardiovasc Res 53:971–983

Siracusa LD, McGrath R, Ma Q, Moskow JJ, Manne J, Christner PJ, Buchberg AM, Jimenez SA (1996) A tandem duplication within the fibrillin 1 gene is associated with the mouse tight skin mutation. Genome Res 6:300–313

Skovby F, Kraus JP (2002) The homocystinurias. In: Royce PM, Steinmann B (eds) Connective tissue and its heritable disorders. Wiley-Liss, Inc., New York, pp 627–650

Taipale J, Miyazono K, Heldin CH, Keski-Oja J (1994) Latent transforming growth factor-β1 associates to fibroblast extracellular matrix via latent TGF-β binding protein. J Cell Biol 124:171–181

Taipale J, Saharinen J, Hedman K, Keski-Oja J (1996) Latent transforming growth factor-beta 1 and its binding protein are components of extracellular matrix microfibrils. J Histochem Cytochem 44:875–889

ten Dijke P, Arthur HM (2007) Extracellular control of TGF-beta signalling in vascular development and disease. Nat Rev Mol Cell Biol 8:857–869

Thurmond FA, Trotter JA (1996) Morphology and biomechanics of the microfibrillar network of sea cucumber dermis. J Exp Biol 199:1817–1828

Thurmond FA, Koob TJ, Bowness JM, Trotter JA (1997) Partial biochemical and immunologic characterization of fibrillin microfibrils from sea cucumber dermis. Connect Tissue Res 36:211–222

Tiedemann K, Bätge B, Müller PK, Reinhardt DP (2001) Interactions of fibrillin-1 with heparin/heparan sulfate: implications for microfibrillar assembly. J Biol Chem 276:36035–36042

Tiedemann K, Bätge B, Reinhardt DP (2004) Assembly of microfibrils. In: Robinson PN, Godfrey M (eds) Marfan syndrome: a primer for clinicians and scientists. Landes Bioscience, Georgetown, TX, USA, pp 130–142

Tiedemann K, Sasaki T, Gustafsson E, Göhring W, Bätge B, Notbohm H, Timpl R, Wedel T, Schlötzer-Schrehardt U, Reinhardt DP (2005) Microfibrils at basement membrane zones interact with perlecan via fibrillin-1. J Biol Chem 280:11404–11412

Trask TM, Ritty TM, Broekelmann T, Tisdale C, Mecham RP (1999) N-terminal domains of fibrillin 1 and fibrillin 2 direct the formation of homodimers: a possible first step in microfibril assembly. Biochem J 340:693–701

Tsuji T (1986) Marfan syndrome: demonstration of abnormal elastic fibers in skin. J Cutan Pathol 13:144–153

Unsöld C, Hyytiäinen M, Bruckner-Tuderman L, Keski-Oja J (2001) Latent TGF-beta binding protein LTBP-1 contains three potential extracellular matrix interacting domains. J Cell Sci 114:187–197

Viljoen D (1994) Congenital contractural arachnodactyly. J Med Genet 31:640–643

Wagenseil JE, Mecham RP (2007) New insights into elastic fiber assembly. Birth Defects Res C Embryo Today 81:229–240

Wallace RN, Streeten BW, Hanna RB (1991) Rotary shadowing of elastic system microfibrils in the ocular zonule, vitreous, and ligament nuchae. Curr Eye Res 10:99–109

Wallis DD, Putnam EA, Cretoiu JS, Carmical SG, Cao SN, Thomas G, Milewicz DM (2003) Profibrillin-1 maturation by human dermal fibroblasts: proteolytic processing and molecular chaperones. J Cell Biochem 90:641–652

Wang MC, Lu Y, Baldock C (2009) Fibrillin microfibrils: a key role for the interbead region in elasticity. J Mol Biol 388:168–179

Weinbaum JS, Broekelmann TJ, Pierce RA, Werneck CC, Segade F, Craft CS, Knutsen RH, Mecham RP (2008) Deficiency in microfibril-associated glycoprotein-1 leads to complex phenotypes in multiple organ systems. J Biol Chem 283:25533–25543

Werner JM, Knott V, Handford PA, Campbell ID, Downing AK (2000) Backbone dynamics of a cbEGF domain pair in the presence of calcium. J Mol Biol 296:1065–1078

Wess TJ, Purslow PP, Sherratt MJ, Ashworth J, Shuttleworth CA, Kielty CM (1998) Calcium determines the supramolecular organization of fibrillin-rich microfibrils. J Cell Biol 141:829–837

Whiteman P, Willis AC, Warner A, Brown J, Redfield C, Handford PA (2007) Cellular and molecular studies of Marfan syndrome mutations identify co-operative protein folding in the cbEGF12-13 region of fibrillin-1. Hum Mol Genet 16:907–918

Wipff PJ, Hinz B (2008) Integrins and the activation of latent transforming growth factor beta1 – an intimate relationship. Eur J Cell Biol 87:601–615

Wright DW, Mayne R (1988) Vitreous humor of chicken contains two fibrillar systems: an analysis of their structure. J Ultrastruct Mol Struct Res 100:224–234

Yin W, Smiley E, Germiller J, Sanguineti C, Lawton T, Pereira L, Ramirez F, Bonadio J (1995) Primary structure and developmental expression of Fbn-1, the mouse fibrillin gene. J Biol Chem 270:1798–1806

Zacchigna L, Vecchione C, Notte A, Cordenonsi M, Dupont S, Maretto S, Cifelli G, Ferrari A, Maffei A, Fabbro C, Braghetta P, Marino G, Selvetella G, Aretini A, Colonnese C, Bettarini U, Russo G, Soligo S, Adorno M, Bonaldo P, Volpin D, Piccolo S, Lembo G, Bressan GM (2006) Emilin1 links TGF-beta maturation to blood pressure homeostasis. Cell 124:929–942

Zhang H, Apfelroth SD, Hu W, Davis EC, Sanguineti C, Bonadio J, Mecham RP, Ramirez F (1994) Structure and expression of fibrillin-2, a novel microfibrillar component preferentially located in elastic matrices. J Cell Biol 124:855–863

Zhang H, Hu W, Ramirez F (1995) Developmental expression of fibrillin genes suggests heterogeneity of extracellular microfibrils. J Cell Biol 129:1165–1176

Chapter 8
Elastin

Beth A. Kozel, Robert P. Mecham, and Joel Rosenbloom

Abstract Elastin is the extracellular matrix protein that imparts elasticity to tissue subjected to repeated stretch, such as blood vessels and the lung. It is encoded by a single gene in mammals and is secreted as a 60–70 kDa monomer called tropoelastin that, with the assistance of several fibulins, associates with microfibrils to form the elastic fiber. All tropoelastins share a characteristic domain arrangement of hydrophobic sequences alternating with lysine-containing cross-linking motifs. In the extracellular space, >80% of the lysine residues form covalent cross-links between and within elastin molecules in a process catalyzed by a member of the lysyl oxidase gene family. Mutations in the elastin gene have been linked to supravalvular aortic stenosis and the autosomal dominant form of cutis laxa.

8.1 Introduction

Within the connective tissues of the vertebrate body, rigid materials (bone) and materials of high tensile strength (tendon and ligaments) are prominent components. There is also a need for pliant materials that can stretch, twist, and bend with normal movements, as well as serve certain specialized functions. For example, during systole, the work of the heart is absorbed by expansion of the great vessels, which then recoil elastically during diastole, maintaining the blood pressure and ensuring continuous perfusion of the tissues. Similarly, under normal circumstances

---

B.A. Kozel
Department of Pediatrics, Washington University School of Medicine, 660 South Euclid Avenue, St. Louis, MO 63110, USA

R.P. Mecham (✉)
Department of Cell Biology and Physiology Washington University School of Medicine, 660 South Euclid Avenue, St. Louis, MO 63110, USA
e-mail: bmecham@wustl.edu

J. Rosenbloom
Thomas Jefferson University, Philadelphia, PA, USA

inspiration is an active, energy-requiring process, whereas expiration is a passive one because of the elastic recoil of the respiratory tree. The elastic properties of these tissues are due in large part to the presence of elastic fibers in the extracellular matrix (ECM).

Several unrelated proteins have evolved elasticity, including resilin in arthropods (Anderson 1966), abductin in mollusks (Kelly and Rice 1967), elastomer in octopus (Shadwick and Gosline 1981), and elastin in vertebrates (Gosline 1980). Phylogenetic studies show that elastin appeared after the divergence of the cyclostome and gnathostome lines and is found exclusively in vertebrates, including the cartilaginous fish. Within vertebrate tissues, elastin is found in the ECM as a component of elastic fibers, which may constitute a small (2–4%) but important percentage of the dry weight (as in the skin) or greater than 50% (as in large arteries). Unlike ECM proteins like collagen, which is relatively inelastic (the Young's modulus for collagen is 1×10^6 kPa compared with 300–600 kPa for elastic fibers), mature elastin can stretch to over twice its length and retain full elastic recoil (Gosline 1976).

As visualized in the electron microscope, elastic fibers are composed of two morphologically distinguishable components: (1) an amorphous fraction lacking any apparent regular or repeating structure, which constitutes 90% of the mature fiber and is composed exclusively of elastin; and (2) a microfibrillar component consisting of 10–12 nm diameter fibrils that are located primarily around the periphery of the amorphous component, but also to some extent interspersed within it (Ross 1973; Ross and Bornstein 1969) (Fig. 8.1). The composition and characterization of these microfibrils is discussed in Chap. 7. This chapter will focus on the molecular structure and properties of elastin, regulation of its expression, the assembly of the elastic fiber, and the participation of elastic fibers in heritable and acquired diseases.

8.2 Occurrence and Characterization of Elastin

Consistent with the protein's unique physical properties, the amino acid composition of elastin is peculiar, consisting of approximately 33% glycine, 10–13% proline, more than 40% of other hydrophobic amino acids, but only small amounts of hydrophilic or charged amino acids. The protein also contains hydroxyproline but no hydroxylysine. The functional role of hydroxyproline in elastin is unclear. Unlike the case of collagen, in which hydroxyproline stabilizes the triple helix and inhibition of hydroxylation inhibits secretion (Berg et al. 1973; Rosenbloom et al. 1973), inhibition of hydroxylation has no effect on the rate of elastin secretion and the absence of hydroxyproline has no demonstrated adverse effect on elastin function (Rosenbloom and Cywinski 1976).

An extensive survey of the occurrence and amino acid composition of elastin throughout the animal kingdom was carried out by Sage and Gray (1979). Analyses were performed on samples, mostly from the aorta and related vessels, from

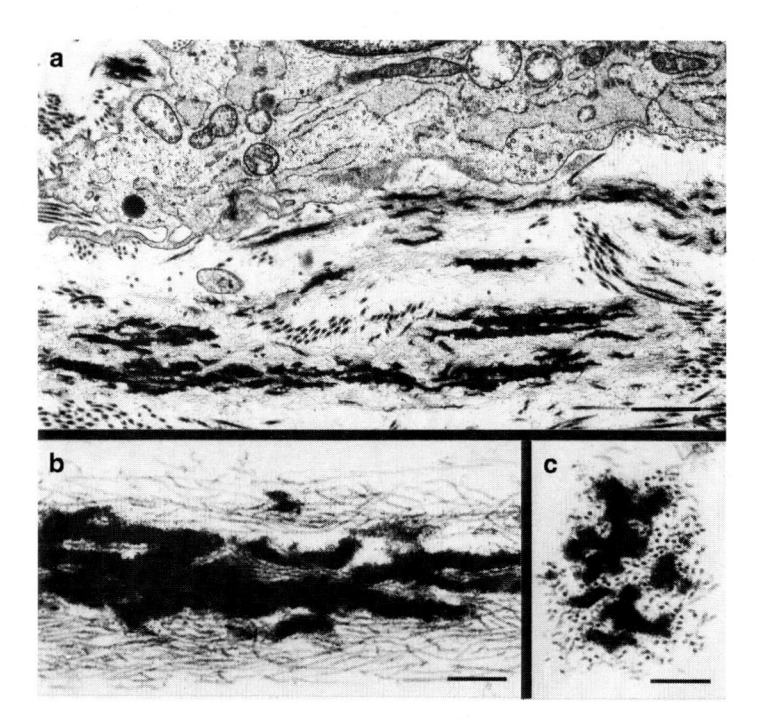

Fig. 8.1 Transmission electron micrograph of developing elastic fiber. (**a**) Elastic fiber forming adjacent to a fibroblast in fetal bovine ligamentum nuchae. *Bar*, 1.0 μm. (**b**) At higher magnification, the elastic fibers are seen to consist of amorphous elastin deposited within a bundle of microfibrils. *Bar*, 0.25 μm. (**c**) Elastic fiber in cross section. *Bar*, 0.25 μm. Taken from Mecham and Davis (1994). Used with permission

representative species of all vertebrates and a number of invertebrate phyla. Results found elastin to be distributed in every vertebrate species examined except in *Agnatha* (jawless fish or cyclostomes) and not in invertebrates. Later studies on the lamprey identified a novel protein, lamprin, which shares some chemical and physiologic features with elastin but is clearly distinct from it (Robson et al. 1993; Bochicchio et al. 2001). Although the amino acid compositions of all elastins have similar general characteristics including the presence of desmosine (a cross-link characteristic of elastin, see below), there is variation in composition among species within a phylum and considerable variation between phyla. In contrast to mammalian and avian elastins that lack histidine, methionine, and cysteine, these amino acids are found in elastin of many reptiles, amphibians, and fish. Consideration of the changes in composition during evolution suggests that the earliest elastin was similar in amino acid composition and cross-linking to that of mammalian elastin, although there has been a progressive increase in hydrophobicity with evolutionary time (see below). These changes in sequence and protein hydrophobicity may be related to a need to accommodate increasing mechanical demands, such as changes in systolic blood pressure, which increases through evolution from a low of 30 mmHg in fish and amphibians to 120–150 mmHg in mammals and birds.

8.3 Coacervation Properties

The hydrophobic nature of elastin imparts interesting physical properties to the molecule, which are important for its elastic recoil properties and for monomer self-association during elastic fiber assembly. Peptides from insoluble elastin obtained from hydrolysis using weak acid (α-elastin) (Partridge et al. 1955) or base–ethanol hydrolysis (κ-elastin) (Moczar et al. 1980) as well as repeat hydrophobic peptides found in tropoelastin and tropoelastin itself manifest the property of coacervation (Vrhovski et al. 1997) in which a solution of the protein or peptide undergoes a phase separation when the temperature is raised. All of these molecules are soluble to a varying extent in aqueous solution at 25°C. On raising the temperature to 37°C the solutions become turbid and, on standing, settle to form two phases: the denser coacervate phase and the equilibrium solution. The protein in the coacervate contains the same percentage of water as fibrous elastin and it is filamentous with periodicities similar to those of native elastin (Cox et al. 1973, 1974; Bressan et al. 1983; Gotte et al. 1974; Partridge 1967).

Detailed studies have suggested that during coacervation, tropoelastin rapidly self-associates to form distinct droplets that possibly facilitate ordered assembly of tropoelastin monomers and specific cross-linking by lysyl oxidase (LOX) (Clarke et al. 2006). The central region including domains encoded in exons 17–27 of human tropoelastin appears to be of particular importance in the coacervation process, determining specific inter-molecular contacts required for proper assembly and cross-linking (Dyksterhuis et al. 2007).

8.4 Elastin Synthesis, Turnover, and Signaling

The extensive cross-linking found in elastin is important for the protein's insolubility and contributes to its longevity. Shapiro et al. (1991) estimated the life span of elastin using aspartic acid racemization and ^{14}C turnover to be ~80 years. Studies using sensitive immunological techniques to measure elastin peptides in the blood or desmosine cross-links excreted in the urine suggest that less than 1% of the total body elastin pool turns over in a year (Starcher 1986). Elastin expression in most tissues occurs over a narrow window of development, beginning in mid-gestation and continuing at high levels through the postnatal period (Fig. 8.2) (Keeley 1979; Berry et al. 1972; Holzenberger et al. 1993). In the aorta, for example, expression decreases rapidly when the physiological rise in blood pressure stabilizes postnatally and there is minimal elastin synthesis in any tissue in the adult animal (Cleary et al. 1967; Bendeck and Langille 1991; Dubick et al. 1981; Wagenseil and Mecham 2009). This explains why damage to elastic fibers during the adult period is so detrimental and why the elastin protein must have a long half-life.

The longevity of mature elastin results from its relative resistance to proteolysis. Because there are few lysine or arginine residues in the cross-linked protein, and few amino acids with large aromatic side chains, elastin is not degraded by trypsin- or

Elastin Expression in Mouse Aorta

Fig. 8.2 Temporal expression profile of elastin in mouse aorta. Expression of elastin was determined by oligonucleotide microarray (median normalized values) in developing mouse aorta. The pattern shows a major increase in expression beginning around embryonic day 14 and continuing through postnatal days 7–10. Thereafter, expression rapidly decreases to low levels that persist into the adult period. Taken from Mecham (2008). Used with permission

chymotrypsin-like proteases. Proteases that do degrade elastin are those that prefer amino acids with small hydrophobic side chains, such as alanine, valine, glycine, and leucine. These proteases, generally termed elastases (Bieth 1986), are predominantly serine proteases, although metalloproteinases and some cathepsins will degrade elastin under appropriate circumstances. Elastases are produced by interstitial and inflammatory cells, and some of the most potent elastases are produced by bacteria (Hajjar et al. 2010; Morihara and Tsuzuki 1967; Hase and Finkelstein 1993).

An interesting property of elastin peptides released from insoluble protein is their ability to act as signaling molecules to a number of cell types, including both interstitial and inflammatory cells (Duca et al. 2005, 2007; Faury et al. 1998; Mochizuki et al. 2002; Senior et al. 1984, 1989; Long et al. 1988). The best-characterized biologically active sequence is VGVAPG, although other sequences conforming to the XGXXPG motif also show activity (Grosso and Scott 1993). These peptides are essentially inactive in the intact cross-linked protein, but have potent signaling activity when the elastin polymer is degraded. Intracellular signaling pathways activated by elastin peptides include protein kinase C (Blood and Zetter 1989), a pertussis toxin-sensitive G-protein pathway that activates RhoA-GTPase (Karnik et al. 2003), a novel Ras-independent ERK1/2 activation system in which p110δ/Raf-1/MEK1/2 and PKA/B-Raf/MEK1/2 cooperate to activate ERK1/2 (Duca et al. 2005, 2007), and a G-protein-associated opening of l-type calcium channels with sequential activation of tyrosine kinases: FAK, c-Src, platelet-derived growth factor-receptor kinase, and then the Ras-Raf-MEK1/2-ERK1/2 phosphorylation cascade (Mochizuki et al. 2002). While the physiological consequences of the elastin-associated "matrikines" (Duca et al. 2004; Antonicelli et al. 2007) are unclear, the generation of such signals associated with tissue damage may provide a means for activating tissue repair or host defense mechanisms.

8.5 Identification of Tropoelastin: The Soluble Elastin Precursor

It has been difficult to characterize mature elastin at the protein level because of its hydrophobicity and insolubility. The major achievements before the late 1960s were the elucidation of the structure of the desmosine cross-links by Partridge and colleagues (Thomas et al. 1963) and the demonstration that these were derived from lysine residues (Partridge et al. 1964, 1966). Nutritional studies involving trace metals showed that animals on a copper-deficient diet suffered aneurysms of the aorta and other defects that could be attributed to a decreased content of the amorphous component in their elastic fibers. This led to the isolation from the aorta of copper-deficient pigs of a soluble protein, tropoelastin, which had an amino acid composition very similar to that of insoluble elastin except for the absence of cross-links and a corresponding increase in lysine residues (Smith et al. 1972; Sandberg et al. 1969). The total lysine content is ~40 residues/mol in tropoelastin compared with 4–6 residues/mol in mature elastin. Tropoelastins isolated from several species share a number of features including similar amino acid composition, a molecular weight of ~65,000–75,000, unusually high solubility in concentrated solutions of short-chain alcohols, and negative temperature coefficient of solubility in salt solutions (Foster et al. 1973; Rucker et al. 1977; Sykes and Partridge 1974; Whiting et al. 1974; Sandberg et al. 1971; Smith et al. 1972).

Initial insight into elastin structure came from sequence analysis of peptides from tryptic digestion of tropoelastin. These peptides are segregated into two classes: (1) small peptides rich in alanine, which are derived from regions destined to form the lysine-derived cross-links; and (2) larger peptides rich in hydrophobic residues, which are derived from the regions responsible for the protein's elastic behavior. As shown later by sequencing of cDNA, the cross-linking peptides are spaced throughout the tropoelastin molecule, being separated by the larger hydrophobic segments. Within some of the larger hydrophobic peptides, smaller limited repeats are discernible, raising the possibility of a secondary structure that contributes to the protein's elastic recoil (Foster et al. 1973; Gray et al. 1973; Urry 1974; Urry et al. 1974) (see Sect. 8.10).

8.6 Insight into Elastin Structure from Elastin cDNA Sequencing

The major advance in our understanding of elastin structure was the characterization of the cDNA for tropoelastin (Bressan et al. 1987; Indik et al. 1987; Pierce et al. 1990; Yeh et al. 1989; Yoon et al. 1984). Analysis of cDNAs from a number of species, including humans, confirmed the basic structural theme suggested from the peptide studies that tropoelastin consists predominantly of alternating hydrophobic and paired lysine-rich domains (Fig. 8.3). In general, there is good agreement at the

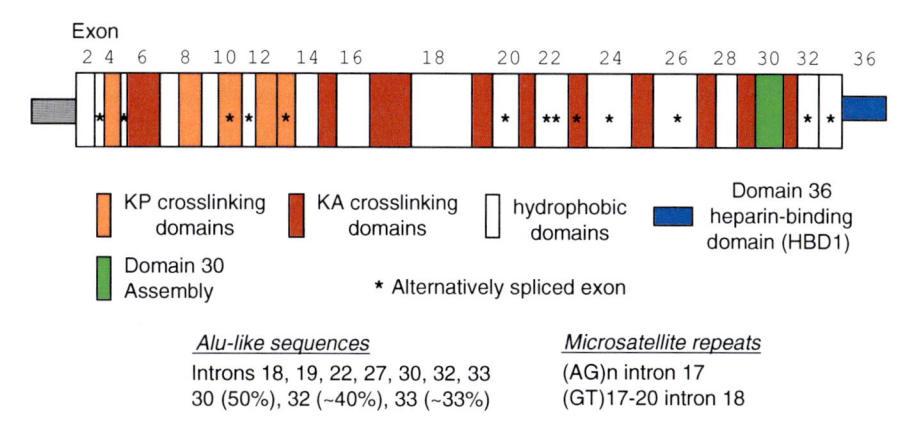

Fig. 8.3 Schematic diagram of the domain structure of human tropoelastin. Human elastin lacks exons 34 and 35 found in other mammalian elastins. Exon number is indicated across the *top*. *Colored squares* represent lysine-cross-linking domains that contain prolines (KP) or are enriched in alanines (KA). *White squares* are hydrophobic sequences. The unique C-terminal sequences important for fiber assembly are indicated in *green* (exon 30) and in *blue* (exon 36). *Stars* indicate exons that are alternatively spliced. Also indicated are introns that contain high numbers of *Alu*-like sequences and microsatellite repeats. Values in parentheses refer to the percentage of the individual intron sequences composed of *Alu* sequences

nucleotide and encoded amino acid sequence levels among the mammalian elastins. These differ, however, in multiple segments in avian, amphibian, and fish. Among mammalian elastins, most amino acid substitutions are of a conservative nature, but some significant differences do exist. For example, near the center of bovine and porcine tropoelastins, a pentapeptide, GVGVP, is repeated 11 times, but this repeat segment is considerably different and more irregular in human tropoelastin and is replaced in rat tropoelastin by GVGIP (Pierce et al. 1990). Similarly, in human tropoelastin, a hexapeptide, GVFVAP, is repeated seven times, but only five times, with conservative substitutions, in bovine tropoelastin, and it is absent altogether from rat and mouse tropoelastin (Rosenbloom et al. 1995). While noticeable variations exist in the length and composition of these hydrophobic sequences, the length of cross-linking segments is highly conserved, indicative of a strong functional requirement.

A comparison of avian, amphibian, and teleost elastins to the mammalian protein shows conservation of the repeating domain structure, but substantial differences occur in the size and number of exons. The chicken sequence is quite homologous to the mammalian sequences for the first 302 and last 57 residues (Bressan et al. 1987). In the central portions, although some segments are homologous, major differences exist, which appear to be due to insertion, duplication, and deletion events, the most striking of which is the occurrence in chicken tropoelastin of the repeating tripeptide (GVP)$_{12}$ not found in mammalian elastins. Zebrafish and *Xenopus* have two elastin genes with gene products larger than avian and mammalian tropoelastins (Chung et al. 2006). These differences are due to both a substantial increase in the size of the hydrophobic domains and an increase in the number of exons in the genes due to extensive replication of a hydrophobic–cross-linking exon pair (Miao et al. 2007; He et al. 2007; Chung et al. 2006).

One region of the protein that shows high homology across all species is the C-terminal region (encoded by the last exon) that contains a tetrabasic sequence motif and a pair of cysteine residues suggested to be important for the structure and function of this region (Brown-Augsburger et al. 1996; Finnis and Gibson 1997; Hsiao et al. 1999; Kozel et al. 2003; Broekelmann et al. 2005; Floquet et al. 2005; Rodgers and Weiss 2004). Interestingly, this region is absent from the second elastin gene in *Xenopus* (Miao et al. 2009). There is also extensive nucleotide homology in the 3′ untranslated region of elastin mRNAs, suggesting that this region may have a function either in stabilizing the mature mRNA or in modulating translation. Within the 3′ untranslated region of several species, a GA-rich segment has been identified that selectively binds proteins and may be involved in regulation of mRNA stability (Hew et al. 1999, 2000).

8.7 Analysis of the Elastin Gene

As mentioned above, the emergence of elastin in evolution is quite recent, appearing coincident with the closed circulatory system and found exclusively in vertebrates (Sage and Gray 1979; Chung et al. 2006). Except in zebrafish and *Xenopus*, tropoelastin is encoded by a single gene [localized to chromosome 7q11.23 in humans (Fazio et al. 1991)]. Analysis of representative fish, frog, avian, and mammalian genes, including human genes, has demonstrated that rather small exons are interspersed between large introns, resulting in a low coding ratio of about 1:20 (Indik et al. 1987). Another important characteristic is that hydrophobic and cross-link domains of the protein are encoded by separate exons, so that the domain structure of the protein is a reflection of the exon organization of the gene. Although all the exons are multiples of three nucleotides and glycine is usually found at the exon/intron junctions, the exons do not exhibit regularity in size as is found in the fibrillar collagen genes. All exon/intron borders have the same phasing. Thus, the second and third nucleotides of a codon are included in the 5′ exon border, while the first nucleotide of a codon is found at the 3′ border of the previous exon. This consistent structure permits extensive alternative splicing of the primary transcript in a cassette-like fashion while maintaining the reading frame. There is also evidence for positive transcriptional regulatory sequences in exon 1 (Pierce et al. 2006) and negative regulatory sequences in intron 1 (Manohar and Anwar 1994).

An extensive comparison of elastin sequences across species identified both highly conserved and taxon and species specific motifs that likely represent important functional and/or structural elements (He et al. 2007). The relative spacing and organization of these elements suggest that exon duplication events have played an important role in the evolution of elastin. Clustering of similarity profiles generated for sets of exons and introns revealed a pattern of putative duplication events involving exons 15–30 in mammalian and chicken elastins, exons 20–31 in both zebrafish elastins, exons 15–20 in fugu elastin, and exons 35–50 in *Xenopus* elastin 1.

The introns of mammalian elastin genes contain an unusual abundance of *Alu* repeat sequences clustered in introns toward the 3' end of the gene (Fig. 8.3). In the human genome, *Alu* repeat sequences occur about once every 4 kb of genomic DNA, but in the elastin gene they occur at about four times that frequency (Indik et al. 1987). In addition to *Alu* repeats, long stretches of alternating purines or pyrimidines occur. The function, if any, of these repetitive elements remains to be determined, but it is clear that the large number of *Alu* elements within primate genomes generates opportunities for unequal homologous recombination events. These events often occur intrachromosomally, resulting in deletion or duplication of exons in a gene, but they also can occur interchromosomally, leading to more complex chromosomal abnormalities and genetic disorders (Batzer and Deininger 2002; Deininger and Batzer 1999). Indeed, the *Alu* elements in the elastin gene have been implicated in a chromosomal recombination that led to gene inactivation, and importantly, facilitated genetic linkage of elastin loss-of-function mutations to the disease supravalvular aortic stenosis (SVAS) (Curran et al. 1993) (see below).

Alu sequences and repetitive elements are also implicated in changes in the structure of the elastin gene in the primate lineage where sequences homologous to exons 34 and 35 in other species are not found in the human gene. An interesting study by Szabo et al. (1999) found that the loss of exon 35 occurred at least 35–45 million years ago, when Catarrhines diverged from Platyrrhines (New World monkeys). The loss of exon 34, in contrast, occurred only about 6–8 million years ago, when Homo separated from the common ancestor shared with chimpanzees and gorillas. The loss of both exons was likely facilitated by *Alu*-mediated recombination events (Szabo et al. 1999). It is unclear what, if any, selective advantage is conferred upon the primate protein by the loss of these two exons and the silencing of a third (exon 22, see below) in primate lineages, but these changes suggest that this relatively new ECM gene is undergoing strong purifying selection (Piontkivska et al. 2004).

8.8 Alternative Splicing of Elastin mRNA

Analysis of elastin cDNAs from all species has demonstrated alternative splicing of the primary transcripts (Indik et al. 1989; Yeh et al. 1987; Pierce et al. 1992; Barrineau et al. 1981). In most cases, splicing occurs in a cassette-like fashion in which an exon is either included or deleted, but occasionally a splicing event may divide an exon. Both hydrophobic and cross-link domains are affected, so that two cross-link domains may be brought into apposition (deletion of exon 22) or the interval between cross-link domains may be increased (deletion of exon 23). The functional significance of these variations is not known, although clearly a tighter or looser fiber network could be produced.

S1 nuclease mapping experiments using elastin mRNA isolated from developing bovine and rat tissues demonstrated that, in the majority of cases, alternative splicing of most exons is infrequent, although in human tissues exon 22 is rarely included in the transcript and exon 32 is spliced out with a frequency of >50%. Numerous experiments indicate that splicing may be developmentally controlled and tissue-specific (Baule and Foster 1988; Heim et al. 1991; Parks and Deak 1990).

8.9 Cross-linking of Elastin and Properties of Elasticity

8.9.1 Elastin Cross-linking

A critical feature of the elastic fiber, crucial to its proper function, is the extensive extracellular cross-linking of tropoelastin mediated by the enzyme protein-lysine-6-oxidase (LOX; EC 1.4.3.13), which oxidizes selective lysine residues in peptide linkage to α-aminoadipic δ-semialdehyde (trivial name allysine). This is the same family of amine oxidases involved in collagen cross-linking (see Chap. 9). There are two major bifunctional cross-links in elastin: dehydrolysinonorleucine, formed through the condensation of one residue of allysine and one of lysine, and allysine aldol formed through the association of two allysine residues (Franzblau et al. 1969; Lent and Franzblau 1967; Lent et al. 1969). These two cross-links can condense with each other, or with other intermediates, to form the tetrafunctional cross-links desmosine or isodesmosine (Partridge et al. 1963; Akagawa and Suyama 2000; Partridge 1963) (Fig. 8.4). Other cross-links that have been identified include a trifunctional cross-link, dehydromerodesmosine (Francis et al. 1973), a cyclopentenosine trifunctional cross-link formed by the condensation of allysine and lysine (Akagawa et al. 1999), and desmopyridine and isodesmopyridine found in trace amounts that form through the condensation of ammonia and allysine (Umeda et al. 2001b). There is also evidence that desmosine/isodesmosine cross-links can be oxidized by reactive oxygen species resulting in dihydrooxopyridine forms (Umeda et al. 2001a).

The lysine residues that serve as cross-link precursors in elastin occur as pairs separated by two or three amino acids, either in alanine sequences (KA domains) or in sequences where the residues are separated by proline residues (KP domains). In contrast to the variability seen in the hydrophobic domains, there is conservation of the cross-linking domains, especially in the number of residues between lysines. The conformation of the alanine-rich cross-linking domains is essentially α-helical (Gray et al. 1973; Wender et al. 1974; Foster et al. 1976), and the lysine residues are always separated by two or three alanines, which results in the lysine side chains being spatially close to one another on the same side of the helix (Fig. 8.4). These positional considerations imply that a critical step in the cross-linking pathway is the formation of a bifunctional "within chain" cross-link intermediate, which then condenses with another bifunctional intermediate on a second chain to form the

8 Elastin

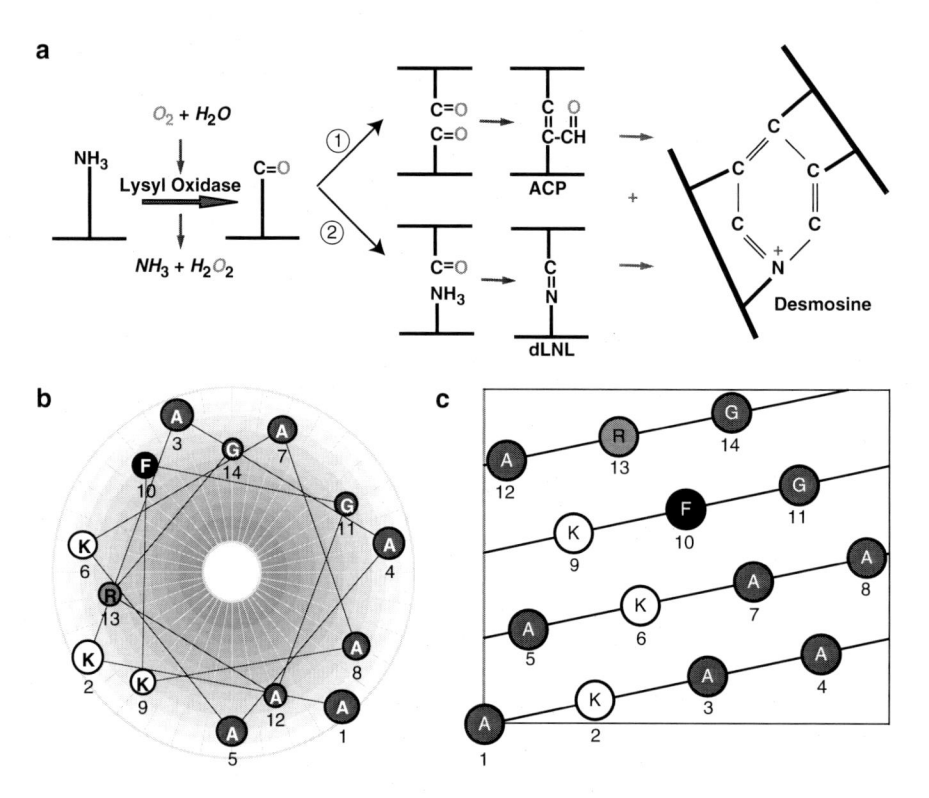

Fig. 8.4 Modification of lysine side chains in cross-linking of elastin. (**a**) Cross-linking of elastin monomers is initiated by the oxidative deamination of lysine side chains by the enzyme lysyl oxidase in a reaction that consumes molecular oxygen and releases ammonia. The aldehyde that is formed can condense with another modified side chain aldehyde (1) to form the bivalent aldol condensation product (ACP) cross-link. Reaction with the amine of an unmodified side chain through a Shiff base reaction (2) produces dehydrolysinonorleucine (dLNL). ACP and dLNL can then condense to form the tetrafunctional cross-link desmosine or isodesmosine. (**b**) *Helical wheel* showing positioning of lysine side chains (K) in sequence encoded by exon 19 around an alpha helix. The view is typical for all lysine residues in alanine-rich sequences (KA cross-linking domains). All three lysine residues are grouped on one side of the helix, as is the arginine residue (R). (**c**) Orientation of lysine side chains on the surface of the alanine-rich helix in KA domains. The lysines are positioned one above the other on the same side of the helix

tetrafunctional desmosine cross-links (Gray 1977; Gray et al. 1973). Insofar as is known, all reactions subsequent to the oxidative deamination of lysine by LOX are spontaneous.

Sequence analysis of cross-link-containing peptides from insoluble elastin found desmosine to be localized exclusively within alanine-rich cross-linking domains (Gerber and Anwar 1974, 1975) and, although desmosine and isodesmosine are tetrafunctional, it is likely that each cross-link normally joins only two chains (Baig et al. 1980; Foster et al. 1974). Brown-Augsburger et al. (1995) went on to show that a major cross-linking site is formed through the association of sequences encoded by exons 10 (KP domain), 19, and 25 (KA domains), with desmosine linking two chains via the alanine-rich cross-linking domains (exons 19 and 25) and a third chain joined via lysinonorleucine cross-links forming from lysines in domain 10 and the third lysine in exons 19 and 25. These findings suggest that the KA domains contribute to tetrafunctional desmosine cross-links, whereas the KP domains contribute to bifunctional cross-links. All but approximately four lysine residues found in tropoelastin can be accounted for in various cross-links and cross-link precursors in mature elastin. The net result is a highly insoluble polymer in which some type of interchain lysine-derived cross-link occurs every 65–70 residues.

8.9.2 Hydrophobic Domains

The fundamental force behind the elastic recoil properties of elastin is entropy, and having a large number of cross-links is important for transferring stress throughout the polymer when the elastic fiber is stretched. Stretching introduces order within the cross-linked protein chains and decreases the entropy of the system. Recoil occurs when the system returns to maximum entropy in the unordered state. Models for elastin structure that account for the entropic behavior vary from a completely isotropic network of random chains (Hoeve 1974; Dyksterhuis et al. 2009) to anisotropic models where regions of order within the hydrophobic domains contribute to the entropic properties of the molecule (Bochicchio and Tamburro 2002; Martino et al. 2000; Urry 1983; Urry and Long 1976). A dynamic blend of the two is probably most realistic.

In the anisotropic models, interactions between water and hydrophobic domains are important to the energetics of elastic function. As stated above, the hydrophobic domains in elastin contain many repeat sequences that are capable of forming β-turns (type I, type II, and type VIII), whose stability depends on both sequence and microenvironment. The first functional model of elastin was the oiled coil proposed by Gray et al. (1973) in which the hydrophobic sequences between the cross-links are rich in β-turns that form a broad coil with buried hydrophobic groups. Upon stretching, the hydrophobic side chains are forced into water and the energy for contraction comes from the return of these groups to a nonpolar environment. While the extensive beta-coil model is not supported by experimental evidence, the concept of secondary structure within these hydrophobic sequences that contributes to the protein's elasticity is generally accepted. Based upon a repeating sequence in one of the hydrophobic domains of elastin, Urry et al. proposed a β-spiral model that consists of a regular array of consecutive type II β-turns subject to "librational" motions responsible for the high entropy of the relaxed state. On applying a stretching force, a damping of librations is produced, and therefore, a decrease of entropy, which will spontaneously increase by removal of the force (Venkatachalam and Urry 1981). Tamburro et al., in turn, proposed a dynamic model in which labile, isolated β-turns form and slide along the chain. The sliding gives rise to an increase in the entropy of the chain, thereby contributing to the elasticity of the protein (Tamburro et al. 2005). This model is supported by experimental findings showing appreciable amounts of poly-L-proline II left-handed helix within the elastin molecule (Martino et al. 2000).

8.10 Tropoelastin Domains Important for Elastic Fiber Assembly

In addition to contributing to the protein's recoil properties, the hydrophobic domains of elastin play an important role in tropoelastin self-interactions during fiber assembly. Important in this regard is the hydrophobic domain encoded by exon 30, which contains repeats of GGLG(V/A) that are similar to sequences found in other proteins such as spider silk and lamprin that aggregate by β-sheet/β-pleat structures (van Beek et al. 2002; Bochicchio et al. 2001; Robson et al. 1993). A peptide with the elastin exon 30 sequence multimerizes and forms amyloid-like super-structures as evidenced by green birefringence when stained with congo red (Kozel et al. 2004). Further work by Tamburro et al. (2005) utilized circular dichroism, nuclear magnetic resonance, and Fourier transform infrared spectroscopy to show that the exon 30 peptide predominately takes on anti-parallel β-sheet structures and aggregates through side-by-side interactions of β-structures.

In vitro studies demonstrated that the aggregation properties of this domain are important for nucleating tropoelastin alignment and subsequent elastic fiber assembly. Expression of tropoelastin cDNAs with deletions of the C-terminal portion of molecule (deletion of exons 16–36 and 29–36) and deletion of exon 30 alone showed that all constructs lacking exon 30 had decreased elastin deposition relative to their wild-type counterparts (Kozel et al. 2003; Sato et al. 2007). Moreover, addback and transfection-based experiments showed that while the C-terminus of tropoelastin containing exon 30 (exons 16–36) is capable of associating with microfibrils in the ECM, constructs containing only the N-terminal portion of the molecule (exons 2–15 or 1–28) cannot (Kozel et al. 2004; Sato et al. 2007). The importance of exon 30 to elastin assembly was confirmed by expressing the cDNA constructs as transgenes in mice. The product of the full-length wild-type transgene (exon 30 included) was able to assemble with pre-existing elastic fibers in the ECM (Kozel et al. 2003). Only minimal incorporation of the protein lacking exon 30 was detected, and there was no incorporation of the protein containing only exons 1–28. Together, these results support a critical role for exon 30 in early stages of elastic fiber assembly.

Like the sequence encoded by exon 30, human exon 16 may also contribute to tropoelastin self-association (Sato et al. 2007; Wachi et al 2007). The sequence of exon 16 is similar to the exon 30 sequence, and constructs of human tropoelastin molecules lacking exons 16/17 show decreased elastin deposition when expressed in vivo. This explains why humans with an acceptor splice site mutation in exon 16, such that exon 16 is spliced out of the mature tropoelastin mRNA/peptide, have haploinsufficiency for elastin and SVAS (Wachi et al. 2007; Urban et al. 1999) even though the mutant transcript is expressed and stable.

Another important sequence for elastic fiber assembly is encoded by exon 36 – the last and most high conserved exon in elastin. Tropoelastin lacking domain 36 forms an elastin polymer that is misassembled with abnormal cross-linking (Hsiao et al. 1999; Kelleher et al. 2005; Kozel et al. 2003). The sequence encoded by exon 36 ends in a cluster of basic amino acids (KxxxRKRK) that defines a heparan binding domain and is important for tropoelastin's binding to cell-surface glycosaminoglycans (Floquet et al. 2005; Broekelmann et al. 2005) and other receptors (Bax et al. 2009). Studies suggest that one or more of the lysine residues in the domain 36 polybasic sequence forms a cross-link when the tropoelastin chain is incorporated into the insoluble polymer (Broekelmann et al. 2008). Given the importance of this domain for proper cross-linking overall, it is likely that the domain 36 sequence functions to facilitate fiber maturation following self-assembly through domain 30 by forming the initial cross-link(s) that serves to help register the multiple tropoelastin cross-linking sites for subsequent oxidation by LOX. As discussed below, mutations that lead to the deletion or alteration of exon 30 and 36 sequences are responsible for several inherited human elastin diseases.

8.11 Elastic Fiber Formation

8.11.1 Microassembly and Macroassembly

The process by which tropoelastin monomers are secreted and deposited into the ECM as the highly cross-linked functional polymer is complicated and the steps are still not completely understood. Numerous studies now support a stepwise model for elastic fiber assembly that involves a number of molecules that assist in the assembly process (Wagenseil and Mecham 2007; Choudhury et al. 2009). Tropoelastin is synthesized on membrane-bound polysomes, transported through the Golgi apparatus and packaged into secretory vesicles (Davis and Mecham 1998; Fahrenbach et al. 1966; Thyberg et al. 1979). At this point, elastin secretion may differ from other ECM proteins by trafficking through an acidic compartment (perhaps a sorting endosome) (Davis and Mecham 1998; Davis and Mecham 1996). There is also evidence that tropoelastin is secreted as a complex with a 67 kDa molecular chaperone that targets the tropoelastin molecule to assembly sites on the cell surface (Privitera et al. 1998).

Electron microscopy and dynamic imaging studies show that tropoelastin is assembled into small globular aggregates on the cell surface that begin the initial stages of cross-linking in a process called microassembly (Fig. 8.5) (Kozel et al. 2004, 2006; Clarke et al. 2006). Accumulating evidence suggests that interactions with proteoglycans on the cell surface facilitate the self-association of tropoelastin monomers. The negatively charged proteoglycans are thought to neutralize the positive charge within the lysine-containing cross-linking domains, thereby reducing

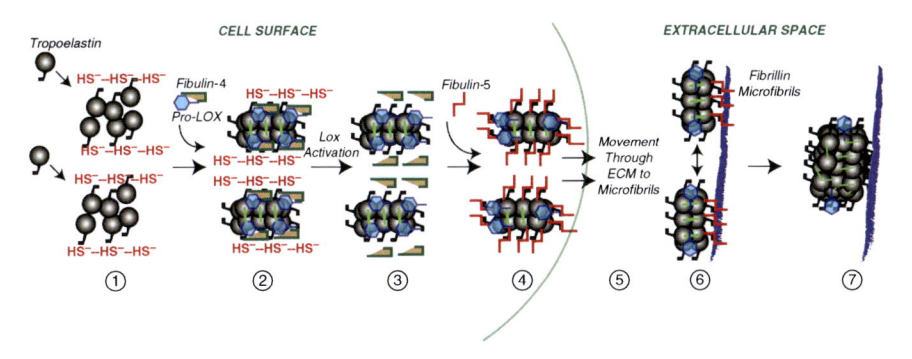

Fig. 8.5 Model of extracellular elastin assembly. ① The first step in elastin assembly (microassembly) involves an interaction between tropoelastin (TE) monomers to form small globules on the cell surface were proteoglycans (heparan sulfate or chondroitin sulfate) may serve to release tropoelastin from its elastin binding protein chaperone and to assist in globule stabilization via an interaction with tropoelastin's C-terminal sequence (domain 36). Cell surface heparan sulfate proteoglycans (HS) could serve two functions: to retain tropoelastin on the cell surface and to facilitate globule formation by lowering the minimum or critical concentration of tropoelastin required for coacervation – a key initiating step of microassembly. ② Fibulin-4 is important in TE chain alignment and in facilitating TE cross-linking by mediating the association of Pro-LOX with the TE coacervate. ③ The oxidation of TE lysine side chains by LOX and subsequent cross-link formation stabilizes the polymer and may serve to catalyze globule release from cell-associated GAGs by reducing the positive charge on the TE chains. ④ Macroassembly involves movement of elastin globules through the ECM where they fuse into larger structures in association with microfibrils. ⑤ Fibulin-5 binds to uncross-linked tropoelastin on the surface of the globules, preventing premature globule aggregation, and neutralizing the positive charge of the uncrosslinked TE so that the globules can move through the negatively charged PG-rich ECM to microfibrils. Through its ability to interact with fibrillin, fibulin-5 could also direct elastin aggregates onto fibrillin-containing microfibrils where globules fuse and are cross-linked ⑥ into larger aggregates ⑦ that eventually constitute the functional fiber

repulsive interactions so that the domains can be brought together for cross-link formation (Wise et al. 2005). Proteoglycans may also participate in tropoelastin alignment prior to cross-linking by interacting with the C-terminus of tropoelastin encoded by exon 36 – a critical assembly domain (Broekelmann et al. 2005). Extracellular glycosaminoglycans are also important for initiating the release of tropoelastin from its chaperone protein (Privitera et al. 1998).

A consequence of tropoelastin cross-linking is a loss of positive charge on the molecule due to oxidative deamination of the lysyl ε-amino group. This is illustrated by the change in the calculated pI from ~11 for tropoelastin to ~6 for crosslinked elastin. Hence, as cross-linking continues, the charge on the tropoelastin aggregate is reduced such that the ionic interactions between it and the negatively charged cell-associated proteoglycans diminish. This reduction of charge could facilitate the release of tropoelastin from the cell and, by neutralizing repulsive charges, assist globular fusion in the presence of microfibrils (macroassembly). Time-lapse video fluorescence imaging shows that once released from the cell surface, globules readily fuse with the developing elastic fiber in the ECM (Kozel et al. 2006). This maturing elastic polymer is simultaneously reorganized by the cells to create fibers with correct orientation that withstand deforming tissue forces (Czirok et al. 2006). All of the molecules that participate in this assembly process are not fully known, but, as discussed below, knocking out candidate genes in mice has provided some important clues.

8.11.2 Fibulin-4 Deficiency Predicts Role of Fibulin-4 in Early Stages of Elastin Assembly

When fibulin-4 is inactivated in mice, elastic fiber assembly largely fails, resulting in dramatically decreased elastin content in elastic tissues. In contrast to the continuous, dense fibers seen in normal animals, electron microscopic analysis of elastic tissues in fibulin-4-deficient mice shows elastic fibers to be unusual aggregates containing rod-like filaments and essentially no desmosine cross-links (McLaughlin et al. 2006). Findings similar to this were seen when inhibitors of LOX were used to decrease elastin cross-linking in the developing chick aorta (Pasquali-Ronchetti et al. 1981; Fornieri et al. 1987). These rod-like structures were shown to be proteoglycans associated with tropoelastin's free ε-amino groups on unoxidized lysines (Baccarani-Contri et al. 1985; Pasquali-Ronchetti et al. 1984).

Studies utilizing a fibulin-4 knockdown mouse where fibulin-4 levels are reduced but not eliminated show progressively less elastin deposited into the ECM with decreasing fibulin-4 content (Horiguchi et al. 2009). Work by Choudhury et al. showed that fibulin-4's binding of LOX, a central cross-linking enzyme of elastin, enhanced fibulin-4's binding to tropoelastin (Choudhury et al. 2009). While LOX is transcribed at normal levels, Lox is also deficient from the elastin bundles in fibulin-4 null animals (Horiguchi et al. 2009; McLaughlin et al. 2006). The absence of Lox suggests that fibulin-4 is responsible for recruiting Lox to the developing bundle so as to promote cross-linking.

8.11.3 Fibulin-5 Acts to Bridge Elastin Between the Matrix and Cells

Animal models of fibulin-5 deficiency reveal large, abnormal elastic fibers consisting of elastin globules that fail to fuse into a uniform structure. Electron microscopy also shows elastin globules trapped at the cell surface, unable to make their way to the microfibril scaffold (Nakamura et al. 2002; Yanagisawa et al. 2002). As a consequence, tissue elastic properties are greatly altered (Nakamura et al. 2002), indicating that while elastin is present in these tissues, it is not organized in a physiologically relevant way. In contrast to the fibulin-4 knockout mouse in which desmosine cross-links fail to develop, fibulin-5 knockout animals show desmosine levels only marginally lower than normal (Choi et al. 2009; Zheng et al. 2007; Yanagisawa et al. 2002). Together, these findings suggest that fibulin-5 acts downstream of the initial elastin globule formation and may participate in their subsequent fusion into larger structures. In support of this idea is evidence from in vitro fibulin-5 knockdown studies suggesting that fibulin-5 controls elastin deposition onto microfibrils (Choudhury et al. 2009; Freeman et al. 2005). In this way, fibulin-5 may serve to link elastin's associations between cell and matrix.

8.11.4 Fibrillin and the Microfibrillar Scaffold

Microfibrils (see Chap. 7) have traditionally been thought to occupy a central role in elastic fiber assembly by serving as inert scaffolds for elastin deposition and cross-linking. Electron microscopy data show that microfibrils are always present at the periphery of the elastic fiber core (Pasquali-Ronchetti and Fornieri 1984; Fahrenbach et al. 1966; Greenlee et al. 1966; Ross and Bornstein 1969; Albert 1972; Ross 1973; Hinek and Thyberg 1977). Moreover, protein studies show direct interactions between elastin and the microfibrillar proteins fibrillin-1 and -2 (Fib1 and Fib2), as well as microfibril-associated glycoprotein-1 (MAGP1) (Brown-Augsburger et al. 1994; Clarke and Weiss 2004; Jensen et al. 2001; Rock et al. 2004; Trask et al. 2000a, b). Interestingly, analyses of mice where individual genes for the fibrillins and MAGP1 have been inactivated, in general, fail to show major abnormalities in elastic fiber development or morphology (Chaudhry et al. 2001; Dietz and Mecham 2000; Pereira et al. 1997, 1999; Weinbaum et al. 2008). MAGP1−/− mice have normal elastic fibers and the functional properties of large elastic vessels and the lung are normal (Weinbaum et al. 2008). Fib1−/− mice die early of ruptured aneurysm but with normal levels of elastin cross-links. These animals display an increase in inflammatory markers, suggesting that elastic fibers in these tissues may have heightened susceptibility to protease activity when stripped of their fibrillin coating. Alternatively, Fib2−/− mice have a relatively preserved vascular integrity. Interestingly, the Fib1−/−; Fib2−/− double knockout mice die in utero with a more severe vascular phenotype than either fibrillin alone. The histology in these animals reveals less inflammation and suggests that, while there may be some built-in redundancy in the fibrillins, some form of fibrillin is required for adequate elastic fiber assembly (Carta et al. 2006).

8.12 Elastin-Associated Diseases

8.12.1 Supravalvular Aortic Stenosis: ELN Loss-of-Function Mutations

The first inherited human disease linked to mutations in the elastin gene was SVAS (MIM #185500). SVAS was first attributed to alterations in the elastin gene by Curran et al. (1993) who identified a translocation in a family with SVAS that disrupted the elastin gene at exon 28 and cosegregated with the disease. In the years that followed, further gene alterations were identified (Ewart et al. 1994; Li et al. 1997; Metcalfe et al. 2000; Pober et al. 2008; Tassabehji et al. 1997; Urban et al. 1999). SVAS is inherited in an autosomal dominant fashion and mutations leading to this condition include small hemizygous deletions removing multiple exons from the elastin gene and nonsense or frameshift mutations. All of these mutations are loss-of-function mutations that lead to haploinsufficiency through nonsense-mediated decay of the mutated transcript (Hinek et al. 1991) or the generation of a protein product that is not capable of assembly (Kozel et al. 2003; Sato et al. 2007; Wachi et al. 2007).

Individuals with SVAS have vascular anomalies including stenosis of the great vessels (most notably the supravalvular aorta) and the large pulmonary vessels. However, the severity and location of stenoses is different in each individual and there is marked intrafamilial variability ranging from asymptomatic carriers to severely affected individuals requiring surgical intervention (Metcalfe et al. 2000; Tassabehji et al. 1997). In familial studies, penetrance for any vascular finding was estimated to be 86% (Chiarella et al. 1989). This variability in penetrance and expressivity predicts the existence of modifiers outside of the elastin gene.

In addition to vascular stenosis, individuals with SVAS have increased risk of inguinal hernia, but do not share the craniofacial features or cognitive phenotype of individuals with Williams–Beuren syndrome (WBS) (see below). Interestingly, there has also been no report of the other connective tissue abnormalities seen in WBS patients in this cohort. Pyloric stenosis, polydactyly, Takyasu's arteritis, cardiac septal defects, and subtle dysmorphic features (Micale et al. 2010; Metcalfe et al. 2000; Park et al. 2006) have been described in individual patients, but there has been no consistent association of these findings in multiple individuals with SVAS.

8.12.2 Williams–Beuren Deletion Syndrome: ELN gene Deletion

The elastin locus on chromosome 7 is in a region flanked by low-copy-repeat blocks known as duplicons. These homologous sequences predispose this region, which includes ~28 genes, to unequal crossing over and deletion, with the majority of the break points within the duplicons (Antonell et al. 2005; Ewart et al. 1993; Pober 2010; Perez Jurado et al. 1996). The result is WBS (MIM # 194050), an autosomal dominant disease characterized by unique facial features, a gregarious personality, and stenosis (SVAS) in major conducting blood vessels. As in individuals with non-syndromic forms of SVAS, haploinsufficiency for the elastin gene is responsible for the vascular abnormalities in these individuals, but not for the other characteristics of the disease. Stenoses are most prevalent in the ascending/supravalvular aorta, but can develop in any vessel, including the pulmonary arteries, cranial vessels causing stroke, and coronary vessels causing myocardial infarction (Eronen et al. 2002; Pober 2010; Pober et al. 2008). Hypertension is a significant concern in this population as 40–55% of WBS patients develop clinically significant hypertension

(Broder et al. 1999; Eronen et al. 2002). Most of these hypertensive individuals have at least one area of detectable stenosis (Rose et al. 2001). Individuals with WBS may also have other connective tissue abnormalities such as hernias, rectal and vaginal prolapse, and joint or skin laxity.

Genotype/phenotype correlations have begun to identify the role of other genes in the WBS deletion region. For example, Lim-kinase1 hemizygositiy has been implicated in the visuospatial abnormalities seen in individuals with this syndrome (Frangiskakis et al. 1996) and deletion of the GTF2-I gene is associated with the Williams cognitive phenotype (Danoff et al. 2004). Individuals whose deletions also affect the NCFI1 gene have decreased risk of hypertension, regardless of their stenosis status (Del Campo et al. 2006).

Histologic analysis of vascular tissue derived from individuals with WBS or SVAS shows increased lamellar number in elastic vessels (Dridi et al. 2005; Li et al. 1998b). In areas of stenosis, subendothelial accumulation of cells occurs, along with hypertrophy of smooth muscle and disruption of the elastic fibers with fibrosis. The skin histopathologic findings in individuals with elastin haploinsufficiency reveal subclinical decreases in the elastin content (Dridi et al. 1999; Urban et al. 2000). Overall, the architecture of vascular elastic fibers in these patients is preserved, but the total elastin content is reduced by approximately 50% (Dridi et al. 2005; Li et al. 1998b), consistent with a haploinsufficiency pathomechanism.

8.12.3 Williams–Beuren Duplication Syndrome: ELN Gene Duplication

Individuals with Williams–Beuren duplication syndrome (MIM# 609757) have three (or more) copies of the elastin gene as well as a duplication of other genes in the Williams region. Cardiopulmonary anomalies other than persistent ductus arteriosus (Van der Aa et al. 2009) have not been reported, but the vascular and pulmonary status of these individuals has not been extensively studied. It is interesting that mice expressing three copies of the elastin gene do not have vascular or pulmonary abnormalities (Hirano et al. 2007), suggesting that too much elastin has few adverse physiological consequences. The nonvascular traits of the duplication syndrome include significant speech delay and autistic features, diaphragmatic hernia, cryptorchidism, and nonspecific brain abnormalities on MRI. A characteristic facies has also been described (Van der Aa et al. 2009). These characteristics undoubtedly result from duplication of other genes in the WBS region.

8.12.4 Autosomal Dominant Cutis Laxa: ELN Dominant-Negative Mutations

Allelic to SVAS in the elastin gene is autosomal dominant cutis laxa (ADCL) (MIM #123700). Mutations associated with ADCL generally arise from single base pair alterations that produce missense sequence, usually near the 3' end of the elastin transcript. Unlike loss-of-function mutations associated with SVAS where the mRNA is frequently degraded through NMD, mutations causing ADCL lead to stable mRNAs and their protein products have been detected in the extracellular space, suggesting a dominant-negative effect for these mutations (Rodriguez-Revenga et al. 2004; Szabo et al. 2006; Tassabehji et al. 1998; Urban et al. 2005; Zhang et al. 1999). Because of the frequency of alternative splicing of exons in the 3' region of the human elastin gene, these mutations often generate multiple mRNA products depending on the splice pattern (Sugitani et al. manuscript submitted). A large number of reported mutations affecting the C-terminus are frame-shift mutations in exon 30, thereby altering the exon 30 self-assembly domain. All of the mutations are expected to modify the sequence of the critical assembly domain in exon 36.

Individuals with ADCL have loose, redundant skin that ages prematurely. Vascular abnormalities that have been reported include mitral and aortic regurgitation, and aortic aneurysms (Szabo et al. 2006). Aortic stenoses have not been noted, except in one case where a child with severe ADCL had a father with the SVAS phenotype (Graul-Neumann et al. 2008). Individuals with ADCL do, however, have increased incidence of inguinal hernia and emphysema (Corbett et al. 1994; Urban et al. 2005). Animal models of ADCL show that the lung can be a major target in this disease and confirm a dominant-negative mechanism associated with these mutations (Hu et al. 2010; Sugitani et al. manuscript submitted).

8.12.5 Autosomal Recessive Cutis Laxa: Mutations in Other Elastic Fiber-Related Genes

Cutis laxa can also arise from mutations in genes other than elastin and can be inherited as X-linked and recessive (ARCL) forms (Milewicz et al. 2000). A common element in these other types of the disease is that causative mutations are in genes for proteins that are involved directly or indirectly in elastic fiber synthesis, secretion, or function. LOX, the enzyme responsible for catalyzing crosslinking of the elastin polymer, requires copper as a cofactor that is transported into the cell by ATP7A, the gene mutated in the X-linked form of the disease. Fibulin-4 and -5 play critical roles in elastic fiber assembly and mutations in the genes for these proteins have been linked to ARCL type I (Claus et al. 2008; Dasouki et al. 2007; Hu et al. 2006; Hucthagowder et al. 2006; Loeys et al. 2002; Lotery et al. 2006). Another interesting set of ARCL mutations point to alterations in the glycosylation and secretion of elastic fiber components. Mutations in the vesicular H^+-ATPase subunit ATP6V0A2 (ARCL type IIA), for example, result in a defect in Golgi and vesicular pH leading to, among other alterations, abnormal glycosylation (Kornak et al. 2008). Mutations in the gene *RIN2*, a ubiquitously expressed protein that interacts with Rab5 and is involved in the regulation of endocytic trafficking, leads to a recessive form of cutis laxa called MACS syndrome (Basel-Vanagaite et al. 2009). Both ATP6V0A2 and *RIN2* mutations target endosomal trafficking, although through different mechanisms. We have previously shown that tropoelastin traffics through an acidic (most likely endosomal) compartment and that modification of intracellular pH has a negative effect on elastin secretion (Davis and Mecham 1998). Interestingly, individuals with ATP6V0A2 mutations accumulate tropoelastin in the Golgi and endoplasmic reticulum and show impaired tropoelastin secretion (Hucthagowder et al. 2009). Cells from individuals with RIN2 mutations also have dilated Golgi and endoplasmic reticulum, and while the secretion of tropoelastin was not specifically studied, RIN2 deficiency was found to be associated with paucity of dermal microfibrils and deficiency of fibulin-5 (Syx et al. 2010). These X-linked and recessive forms of cutis laxa tend to affect a wider spectrum of organ systems with phenotypes that are more severe than the autsomal dominant form of the disease. This difference undoubtedly reflects the wider role the mutated genes play in ECM secretion and cross-linking generally.

8.12.6 Other Genetic Elastin-Associated Diseases

Investigations have been undertaken to look for polymorphisms in the elastin gene that predispose individuals to various conditions such as emphysema or aneurysms. Kelleher et al have reported a mutation in exon 36 that is associated with mild ADCL and early-onset COPD (Kelleher et al. 2005). Other polymorphisms in elastin have been reported to be associated with intracranial aneurysm (Ruigrok et al. 2004) and isolated hypertension (Deng et al. 2009). Further investigation will be required to determine whether these polymorphisms themselves are linked to disease or whether it is the polymorphism-associated haplotype that predisposes to these conditions.

8.13 Animal Models of Elastin Deficiency

8.13.1 Elastin Null Animals

Mouse models of elastin loss-of-function mutations that result in elastin insufficiency are in many ways similar to their human counterparts. Elastin loss-of-function mutations are lethal in the homozygous state ($Eln-/-$), with elastin null mice dying at P0–P4.5 of cardiac failure secondary to aortic stenosis (Li et al. 1998a) and increased left ventricular pressure (Wagenseil et al. 2009). Lung development is also severely affected due to the failure of terminal airway branching (Wendel et al. 2000). The elastin null condition has not been described in humans, suggesting embryonic lethality of the human condition as well. In the mouse, studies of developing $Eln-/-$ embryos reveal aortic development indistinguishable from their wild-type counterparts until approximately E17.5, when elastin deposition increases rapidly under normal conditions (Li et al. 1998a; Wagenseil et al. 2009). After this time point, the $Eln-/-$ conducting vessels have smaller inner and outer diameters and a thicker wall due to smooth muscle over-proliferation (Li et al. 1998a). This sub-endothelial accumulation of cells eventually occludes the vascular lumen, leading to a non-patent vessel. Because of their early demise, pathology in other elastic tissues has not been well described.

8.13.2 Elastin Heterozygous Animals

Elastin heterozygous animals, conversely, are viable. Unlike the null animals, the vessel walls are thinner and, although elastin levels are ~50% of normal levels, the vessel wall has an increased number of elastic layers (i.e., lamellae) (Li et al. 1998b). Elastin heterozygous mice do not exhibit focal stenosis typical of the human disease but do exhibit systemic hypertension that is completely penetrant in the mice but only partially penetrant in humans. The pressure differential in the $Eln+/-$ mice was initially described as a mean arterial pressure 27 mmHg (28%) higher than WT littermate controls (Faury et al. 2003) with the systolic component more affected than the diastolic pressure. This difference leads to a higher pulse pressure in the $Eln+/-$ animals, as would be predicted in a more rigid conduit. The elastin heterozygous mice exhibit reduced vascular compliance (Faury et al. 2003; Li et al. 1998b; Wagenseil et al. 2009). When plotted together, it is apparent that although vessels from elastin heterozygote animals have a smaller inner diameter at any given pressure, at physiological blood pressure the $Eln+/-$ and WT animals have comparable diameters (Faury et al. 2003). Consequently, the increased working blood pressure may be a physiological adaptation required to maintain patency of a stiffer blood vessel so that the system can achieve an appropriate cardiac output for adequate tissue perfusion (Faury et al. 2003). Further analysis of these data suggests that the $Eln+/-$ animal operates closer to the flat upper portion of the compliance curve and may have less overall built-in reserve when stressed. Interestingly, although working at a much higher blood pressure than WT mice, $Eln+/-$ animals have minimal cardiac hypertrophy, heart function is normal, and the animals have a normal life span (Faury et al. 2003).

The lungs of $Eln+/-$ mice contain 45% less elastin than do WT animals. However, lungs from these animals are morphologically similar to WT lungs when evaluated histologically for terminal airspace size and density. Elastin heterozygous animals reveal decreased pulmonary compliance relative to WT animals as evidenced by flattened static pressure–volume curves. And while unchallenged mice do not exhibit emphysema, $Eln+/-$ animals develop increased airspace enlargement when exposed to cigarette smoke, suggesting that the elastin insufficient human population may be at increased risk for toxin-related emphysema (Shifren et al. 2006).

8.13.3 Human Tropoelastin Transgenic Animals

A mouse line has been generated expressing the human elastin gene contained in a bacterial artificial chromosome (BAC). The ELN-BAC expresses as a transgene in mice with a temporal and spatial expression pattern similar to the endogenous mouse elastin gene. The human transgene also retains the human splice pattern and does not alter splicing of the endogenous mouse gene (Hirano et al. 2007). Although expression levels of the human gene are ~70% lower than the mouse gene, sufficient protein is deposited into the ECM to rescue the perinatal lethality of the mouse null ($Eln-/-$) phenotype. With only ~30% of WT elastin levels, this "humanized" elastin mouse (hBAC+; $Eln-/-$) has a more severe phenotype than $Eln+/-$ animals (with 50% elastin). Similar to human patients with SVAS, the ascending aorta in hBAC+; $Eln-/-$ mice shows medial thickening leading to the focal stenosis typical of SVAS. The wall structure of the aorta also reveals discontinuous and fragmented elastic lamellae. hBAC+; $Eln-/-$ animals have a mean arterial pressure of 130 mmHg (50 mmHg higher than their WT littermates). Mechanical studies showed stiff vessels with severely restricted compliance (Hirano et al. 2007). When bred into the $Eln+/-$ background (hBAC+; $Eln+/-$) the hBAC transgene contributes an additional ~20% elastin to the 50% mouse elastin already present in the wall. The consequences are the reversal of the hypertension and cardiovascular changes associated with that $Eln+/-$ phenotype (Fig. 8.6). The results are important in confirming that reestablishing normal elastin levels is a

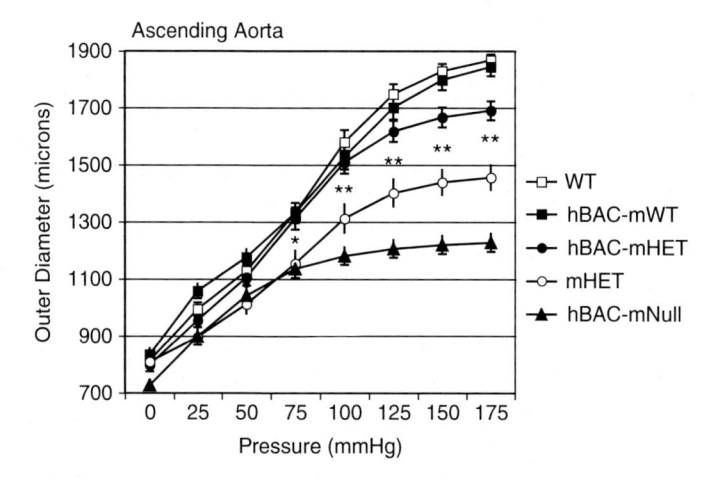

Fig. 8.6 Pressure–diameter analysis of ascending aorta from wild-type and elastin insufficient mice showing a direct relationship between elastin levels and compliance. $Eln+/-$ vessels have decreased compliance compared to WT animals. The aorta in hBAC-mHET mice, where elastin from the hBAC transgene brings elastin levels to ~80% of WT levels, shows mechanical properties intermediate between $Eln+/-$ and WT, suggesting that additional elastin from the human transgene is altering vessel compliance toward normal values. Vessels with only ~30% normal elastin levels (hBac-mNull) are the least compliant. Adapted from Hirano et al. (2007)

logical objective for treating diseases of elastin insufficiency such as SVAS. They also show a strong correlation between disease severity and elastin levels.

8.13.4 Other Animal Models with Elastic Fiber Abnormalities

Two recent mouse models of ADCL have provided insight into the pathogenesis of this disease. Hu et al. (2010) generated a transgenic mouse expressing tropoelastin with a cutis laxa mutation behind a constitutively active promoter. Interestingly, no consistent dermatological or cardiovascular pathologies were observed, but the mice showed increased static lung compliance and decreased stiffness of lung tissue. Markers of transforming growth factor-β signaling and the unfolded protein response were elevated together with increased apoptosis in the lungs of ADCL animals. This evidence suggests that the combined effects of these processes lead to the development of an emphysematous pulmonary phenotype in ADCL.

In the second study, a human ADCL mutation was introduced into the human elastin BAC described in the section above. The mutant BAC was then expressed as a transgene in mice along with the wild-type human gene. RNA stability studies found that alternative exon splicing greatly affects the susceptibility of the mutant transcript to undergo nonsense-mediated decay. Tissue analysis established that the mutant protein is incorporated into elastic fibers in the skin and lung with adverse effects. However, only low levels of mutant protein were found to incorporate into the aorta, which explains why the vasculature is relatively unaffected in individuals with this mutation and in this disease in general. These results also confirm a dominant-negative mechanism for ADCL mutations.

8.14 Conclusions

Elastin is the ECM protein that imparts elasticity to tissues such as the lung, skin, and blood vessels. The emergence of elastin in evolution is quite recent, appearing first in vertebrates. The importance of elastin to the success of vertebrate evolution cannot be understated. Without elastic vessels, it would not be possible to evolve an efficient closed, pulsatile circulatory system that supports efficient distal perfusion and body growth. Similarly, the mechanical function of the vertebrate lung would not be possible without elastin.

Over the past 20 years, we have developed a better understanding of elastin structure, cross-linking, and function, but we still do not know all of the steps involved in elastic fiber assembly. Studies of the phenotypes of mice genetically deficient in elastin and in individual elastic fiber proteins will help us to better understand the assembly process as well as the etiology of human disease involving mutations in elastic fiber genes.

Acknowledgments The original work from the Mecham lab reported in this chapter was supported in part by NIH grants RO1 HL53325, HL74138, and P50-HL84922.

References

Akagawa M, Suyama K (2000) Mechanism of formation of elastin crosslinks. Connect Tissue Res 41:131–141

Akagawa M, Yamazaki K, Suyama K (1999) Cyclopentenosine, major trifunctional crosslinking amino acid isolated from acid hydrolysate of elastin. Arch Biochem Biophys 372:112–120

Albert EN (1972) Developing elastic tissue. An electron microscopic study. Am J Pathol 69:89–102

Anderson SO (1966) Covalent cross-links in a structural protein, resilin. Acta Physiol Scand Suppl 263:1–81

Antonell A, de Luis O, Domingo-Roura X, Perez-Jurado LA (2005) Evolutionary mechanisms shaping the genomic structure of the Williams–Beuren syndrome chromosomal region at human 7q11.23. Genome Res 15:1179–1188

Antonicelli F, Bellon G, Debelle L, Hornebeck W (2007) Elastin-elastases and inflamm-aging. Curr Top Dev Biol 79:99–155

Baccarani-Contri M, Fornieri C, Pasquali Ronchetti I (1985) Elastin-proteoglycans association revealed by cytochemical methods. Connect Tissue Res 13:237–249

Baig KM, Vlaovic M, Anwar RA (1980) Amino acid sequences C-terminal to the cross-links in bovine elastin. Biochem J 185:611–616

Barrineau LL, Rich CB, Foster JA (1981) The biosynthesis of tropoelastin in chick and pig tissues. Connect Tissue Res 8:189–491

Basel-Vanagaite L, Sarig O, Hershkovitz D, Fuchs-Telem D, Rapaport D, Gat A, Isman G, Shirazi I, Shohat M, Enk CD, Birk E, Kohlhase J, Matysiak-Scholze U, Maya I, Knopf C, Peffekoven A, Hennies HC, Bergman R, Horowitz M, Ishida-Yamamoto A, Sprecher E (2009) RIN2 deficiency results in macrocephaly, alopecia, cutis laxa, and scoliosis: MACS syndrome. Am J Hum Genet 85:254–263

Batzer MA, Deininger PL (2002) Alu repeats and human genomic diversity. Nat Rev Genet 3:370–379

Baule VJ, Foster JA (1988) Multiple chick tropoelastin mRNAs. Biochem Biophys Res Commun 154:1054–1060

Bax DV, Rodgers UR, Bilek MM, Weiss AS (2009) Cell adhesion to tropoelastin is mediated via the C-terminal GRKRK motif and integrin $\alpha V \beta 3$. J Biol Chem 284:28616–28623

Bendeck MP, Langille BL (1991) Rapid accumulation of elastin and collagen in the aortas of sheep in the immediate perinatal period. Circ Res 69:1165–1169

Berg RA, Kishida Y, Kobayashi Y, Inouye K, Tonelli AE, Sakakibara S, Prockop DJ (1973) A model for the triple-helical structure of (Pro-Hyp-Gly)10 involving a cis peptide bond and inter-chain hydrogen-bonding to the hydroxyl group of hydroxyproline. Biochim Biophys Acta 328:553–559

Berry CL, Looker T, Germain J (1972) The growth and development of the rat aorta. I. Morphological aspects. J Anat 113:1–16

Bieth JG (1986) Elastases: catalytic and biological properties. In: Mecham RP (ed) Regulation of matrix accumulation. Academic, New York, pp 217–320

Blood CH, Zetter BR (1989) Membrane-bound protein kinase C modulates receptor affinity and chemotactic responsiveness of Lewis lung carcinoma sublines to an elastin-derived peptide. J Biol Chem 264:10614–10620

Bochicchio B, Tamburro AM (2002) Polyproline II structure in proteins: identification by chiroptical spectroscopies, stability, and functions. Chirality 14:782–792

Bochicchio B, Pepe A, Tamburro AM (2001) On (GGLGY) synthetic repeating sequences of lamprin and analogous sequences. Matrix Biol 20:243–250

Bressan GM, Castellani I, Giro MG, Volpin D, Fornieri C, Pasquali-Ronchetti I (1983) Banded fibers in tropoelastin coacervates at physiological temperatures. J Ultrastruct Res 82:335–340

Bressan GM, Argos P, Stanley KK (1987) Repeating structures of chick tropoelastin revealed by complementary DNA cloning. Biochemistry 26:1497–1503

Broder K, Reinhardt E, Ahern J, Lifton R, Tamborlane W, Pober B (1999) Elevated ambulatory blood pressure in 20 subjects with Williams syndrome. Am J Med Genet 83:356–360

Broekelmann TJ, Kozel BA, Ishibashi H, Werneck CC, Keeley FW, Zhang L, Mecham RP (2005) Tropoelastin interacts with cell-surface glycosaminoglycans via its C-terminal domain. J Biol Chem 280:40939–40947

Broekelmann TJ, Ciliberto CH, Shifren A, Mecham RP (2008) Modification and functional inactivation of the tropoelastin carboxy-terminal domain in cross-linked elastin. Matrix Biol 27:631–639

Brown-Augsburger P, Broekelmann T, Mecham L, Mercer R, Gibson MA, Cleary EG, Abrams WR, Rosenbloom J, Mecham RP (1994) Microfibril-associated glycoprotein (MAGP) binds to the carboxy-terminal domain of tropoelastin and is a substrate for transglutaminase. J Biol Chem 269:28443–28449

Brown-Augsburger P, Tisdale C, Broekelmann T, Sloan C, Mecham RP (1995) Identification of an elastin cross-linking domain that joins three peptide chains: possible role in nucleated assembly. J Biol Chem 270:17778–17783

Brown-Augsburger PB, Broekelmann T, Rosenbloom J, Mecham RP (1996) Functional domains on elastin and MAGP involved in elastic fiber assembly. Biochem J 318:149–155

Carta L, Pereira L, Arteaga-Soli E, Lee-Arteaga SY, Lenart B, Starcher B, Merkel CA, Sukoyan M, Kerkis A, Hazeki N, Keene DR, Sakai LY, Ramirez F (2006) Fibrillins 1 and 2 perform partially overlapping functions during aortic development. J Biol Chem 281:8016–8023

Chaudhry SS, Gazzard J, Baldock C, Dixon J, Rock MJ, Skinner GC, Steel KP, Kielty CM, Dixon MJ (2001) Mutation of the gene encoding fibrillin-2 results in syndactyly in mice. Hum Mol Genet 10:835–843

Chiarella F, Bricarelli FD, Lupi G, Bellotti P, Domenicucci S, Vecchio C (1989) Familial supravalvular aortic stenosis: a genetic study. J Med Genet 26:86–92

Choi J, Bergdahl A, Zheng Q, Starcher B, Yanagisawa H, Davis EC (2009) Analysis of dermal elastic fibers in the absence of fibulin-5 reveals potential roles for fibulin-5 in elastic fiber assembly. Matrix Biol 28:211–220

Choudhury R, McGovern A, Ridley C, Cain SA, Baldwin A, Wang MC, Guo C, Mironov Jnr A, Drymoussi Z, Trump D, Shuttleworth A, Baldock C, Kielty CM (2009) Differential regulation of elastic fiber formation by fibulins -4 AND -5. J Biol Chem 284:24553–24567

Chung MI, Ming M, Stahl RJ, Chan E, Parkinson J, Keeley FW (2006) Sequences and domain structures of mammalian, avian, amphibian and teleost tropoelastins: clues to the evolutionary history of elastins. Matrix Biol 25:492–504

Clarke AW, Weiss AS (2004) Microfibril-associated glycoprotein-1 binding to tropoelastin: multiple binding sites and the role of divalent cations. Eur J Biochem 271:3085–3090

Clarke AW, Arnspang EC, Mithieux SM, Korkmaz E, Braet F, Weiss AS (2006) Tropoelastin massively associates during coacervation to form quantized protein spheres. Biochemistry 45:9989–9996

Claus S, Fischer J, Megarbane H, Megarbane A, Jobard F, Debret R, Peyrol S, Saker S, Devillers M, Sommer P, Damour O (2008) A p.C217R mutation in fibulin-5 from cutis laxa patients is associated with incomplete extracellular matrix formation in a skin equivalent model. J Invest Dermatol 128:1442–1450

Cleary EG, Sandberg LB, Jackson DS (1967) The changes in chemical composition during development of the bovine nuchal ligament. J Cell Biol 33:469–479

Corbett E, Glaisyer H, Chan C, Madden B, Khaghani A, Yacoub M (1994) Congenital cutis laxa with a dominant inheritance and early onset emphysema. Thorax 49:836–837

Cox BA, Starcher BC, Urry DW (1973) Coacervation of alpha-elastin results in fiber formation. Biochim Biophys Acta 317:209–213

Cox BA, Starcher BC, Urry DW (1974) Coacervation of tropoelastin results in fiber formation. J Biol Chem 249:997–998

Curran ME, Atkinson DL, Ewart AK, Morris CA, Leppert MF, Keating MT (1993) The elastin gene is disrupted by a translocation associated with supravalvular aortic stenosis. Cell 73: 159–168

Czirok A, Zach J, Kozel BA, Mecham RP, Davis EC, Rongish BJ (2006) Elastic fiber macroassembly is a hierarchical, cell motion-mediated process. J Cell Physiol 207:97–106

Danoff SK, Taylor HE, Blackshaw S, Desiderio S (2004) TFII-I, a candidate gene for Williams syndrome cognitive profile: parallels between regional expression in mouse brain and human phenotype. Neuroscience 123:931–938

Dasouki M, Markova D, Garola R, Sasaki T, Charbonneau NL, Sakai LY, Chu ML (2007) Compound heterozygous mutations in fibulin-4 causing neonatal lethal pulmonary artery occlusion, aortic aneurysm, arachnodactyly, and mild cutis laxa. Am J Med Genet A 143: 2635–2641

Davis EC, Mecham RP (1996) Selective degradation of accumulated secretory proteins in the endoplasmic reticulum. A possible clearance pathway for abnormal tropoelastin. J Biol Chem 271:3787–3794

Davis EC, Mecham RP (1998) Intracellular trafficking of tropoelastin. Matrix Biol 17:245–254

Deininger PL, Batzer MA (1999) Alu repeats and human disease. Mol Genet Metab 67:183–193

Del Campo M, Antonell A, Magano LF, Munoz FJ, Flores R, Bayes M, Perez Jurado LA (2006) Hemizygosity at the NCF1 gene in patients with Williams–Beuren syndrome decreases their risk of hypertension. Am J Hum Genet 78:533–542

Deng L, Huang R, Chen Z, Wu L, Xu DL (2009) A study on polymorphisms of elastin gene in Chinese Han patients with isolated systolic hypertension. Am J Hypertens 22:656–662

Dietz HC, Mecham RP (2000) Mouse models of genetic diseases resulting from mutations in elastic fiber proteins. Matrix Biol 19:481–482

Dridi SM, Ghomrasseni S, Bonnet D, Aggoun Y, Vabres P, Bodemer C, Lyonnet S, de Prost Y, Fraitag S, Pellat B, Sidi D, Godeau G (1999) Skin elastic fibers in Williams syndrome. Am J Med Genet 87:134–138

Dridi SM, Foucault Bertaud A, Igondjo Tchen S, Senni K, Ejeil AL, Pellat B, Lyonnet S, Bonnet D, Charpiot P, Godeau G (2005) Vascular wall remodeling in patients with supravalvular aortic stenosis and Williams Beuren syndrome. J Vasc Res 42:190–201

Dubick MA, Rucker RB, Cross CE, Last JA (1981) Elastin metabolism in rodent lung. Biochim Biophys Acta 672:303–306

Duca L, Floquet N, Alix AJ, Haye B, Debelle L (2004) Elastin as a matrikine. Crit Rev Oncol Hematol 49:235–244

Duca L, Lambert E, Debret R, Rothhut B, Blanchevoye C, Delacoux F, Hornebeck W, Martiny L, Debelle L (2005) Elastin peptides activate extracellular signal-regulated kinase 1/2 via a Ras-independent mechanism requiring both p110gamma/Raf-1 and protein kinase A/B-Raf signaling in human skin fibroblasts. Mol Pharmacol 67:1315–1324

Duca L, Blanchevoye C, Cantarelli B, Ghoneim C, Dedieu S, Delacoux F, Hornebeck W, Hinek A, Martiny L, Debelle L (2007) The elastin receptor complex transduces signals through the catalytic activity of its Neu-1 subunit. J Biol Chem 282:12484–12491

Dyksterhuis LB, Baldock C, Lammie D, Wess TJ, Weiss AS (2007) Domains 17–27 of tropoelastin contain key regions of contact for coacervation and contain an unusual turn-containing crosslinking domain. Matrix Biol 26:125–135

Dyksterhuis LB, Carter EA, Mithieux SM, Weiss AS (2009) Tropoelastin as a thermodynamically unfolded premolten globule protein: the effect of trimethylamine N-oxide on structure and coacervation. Arch Biochem Biophys 487:79–84

Eronen M, Peippo M, Hiippala A, Raatikka M, Arvio M, Johansson R, Kähkönen M (2002) Cardiovascular manifestations in 75 patients with Williams syndrome. J Med Genet 39:554–558

Ewart AK, Morris CA, Atkinson D, Jin W, Sternes K, Spallone P, Stock AD, Leppert M, Keating MT (1993) Hemizygosity at the elastin locus in a developmental disorder, Williams syndrome. Nat Genet 5:11–16

Ewart AK, Jin W, Atkinson D, Morris CA, Keating MT (1994) Supravalvular aortic stenosis associated with a deletion disrupting the elastin gene. J Clin Invest 93:1071–1077

Fahrenbach WH, Sandberg LB, Cleary EG (1966) Ultrastructural studies on early elastogenesis. Anat Rec 155:563–576

Faury G, Usson Y, Robert-Nicoud M, Robert L, Verdetti J (1998) Nuclear and cytoplasmic free calcium level changes induced by elastin peptides in human endothelial cells. Proc Natl Acad Sci USA 95:2967–2972

Faury G, Pezet M, Knutsen RH, Boyle WA, Heximer SP, McLean SE, Minkes RK, Blumer KJ, Kovacs A, Kelly DP, Li DY, Starcher B, Mecham RP (2003) Developmental adaptation of the mouse cardiovascular system to elastin haploinsufficiency. J Clin Invest 112:1419–1428

Fazio MJ, Mattei MG, Passage E, Chu M-L, Black D, Solomon E, Davidson JM, Uitto J (1991) Human elastin gene: new evidence for localization to the long arm of chromosome 7. Am J Hum Genet 48:696–703

Finnis ML, Gibson MA (1997) Microfibril-associated glycoprotein-1 (MAGP-1) binds to the pepsin-resistant domain of the alpha3(VI) chain of type VI collagen. J Biol Chem 272:22817–22823

Floquet N, Pepe A, Dauchez M, Bochicchio B, Tamburro AM, Alix AJ (2005) Structure and modeling studies of the carboxy-terminus region of human tropoelastin. Matrix Biol 24:271–282

Fornieri C, Baccarani-Contri M, Quaglino D Jr, Pasquali-Ronchetti I (1987) Lysyl oxidase activity and elastin/glycosaminoglycan interactions in growing chick and rat aortas. J Cell Biol 105:1463–1469

Foster JA, Burenger E, Gray WR, Sandberg LB (1973) Isolation and amino acid sequence of tropoelastin peptides. J Biol Chem 248:2875–2879

Foster JA, Rubin L, Kagan HM, Franzblau C, Bruenger E, Sandberg LB (1974) Isolation and characterization of cross-linked peptides from elastin. J Biol Chem 249:6191–6196

Foster JA, Bruenger E, Rubin L, Imberman M, Kagan H, Mecham R, Franzblau C (1976) Circular dichroism studies of an elastin cross-linked peptide. Bioplymers 15:833–841

Francis G, John R, Thomas J (1973) Biosynthetic pathway of desmosines in elastin. Biochem J 136:45–55

Frangiskakis JM, Ewart AK, Morris CA, Mervis CB, Bertrand J, Robinson BF, Klein BP, Ensing GJ, Everett LA, Green ED, Proschel C, Gutowski NJ, Noble M, Atkinson DL, Odelberg SJ, Keating MT (1996) LIM-kinase1 hemizygosity implicated in impaired visuospatial constructive cognition. Cell 86:59–69

Franzblau C, Baris B, Lent RW, Salcedo LL, Smith B, Jaffe R, Crombie G (1969) Chemistry and biosynthesis of crosslinks in elastin. In: Balazs EA (ed) Chemistry and molecular biology of the intracellular matrix, vol 1. Academic, New York, pp 617–639

Freeman LJ, Lomas A, Hodson N, Sherratt MJ, Mellody KT, Weiss AS, Shuttleworth A, Kielty CM (2005) Fibulin-5 interacts with fibrillin-1 molecules and microfibrils. Biochem J 388:1–5

Gerber GE, Anwar RA (1974) Structural studies on cross-linked regions of elastin. J Biol Chem 249:5200–5207

Gerber GE, Anwar RA (1975) Comparative studies of the cross-linked regions of elastin from bovine ligamentum nuchae and bovine, porcine, and human aorta. Biochem J 149:685–695

Gosline JM (1976) The physical properties of elastic tissue. Int Rev Connect Tissue Res 7:211–249

Gosline JM (1980) The elastic properties of rubber-like proteins and highly extensible tissues. Symp Soc Exp Biol 34:332–357

Gotte L, Giro MG, Volpin D, Horne RW (1974) The ultrastructural organization of elastin. J Ultrastruct Res 46:23–33

Graul-Neumann LM, Hausser I, Essayie M, Rauch A, Kraus C (2008) Highly variable cutis laxa resulting from a dominant splicing mutation of the elastin gene. Am J Med Genet A 146A:977–983

Gray WR (1977) Some kinetic aspects of crosslink biosynthesis. Adv Exp Med Biol 79:285–290

Gray WR, Sandberg LB, Foster JA (1973) Molecular model for elastin structure and function. Nature 246:461–466

Greenlee TKJ, Ross R, Hartman JL (1966) The fine structure of elastic fibers. J Cell Biol 30:59–71

Grosso LE, Scott M (1993) Peptide sequences selected by BA4, a tropoelastin-specific monoclonal antibody, are ligands for the 67 kDa bovine elastin receptor. Biochemistry 32:13369–13374

Hajjar E, Broemstrup T, Kantari C, Witko-Sarsat V, Reuter N (2010) Structures of human proteinase 3 and neutrophil elastase – so similar yet so different. FEBS J 277:2238–2254

Hase CC, Finkelstein RA (1993) Bacterial extracellular zinc-containing metalloproteases. Microbiol Rev 57:823–837

He D, Chung M, Chan E, Alleyne T, Ha KC, Miao M, Stahl RJ, Keeley FW, Parkinson J (2007) Comparative genomics of elastin: sequence analysis of a highly repetitive protein. Matrix Biol 26:524–540

Heim RA, Pierce RA, Deak SB, Riley DJ, Boyd CD, Stolle CA (1991) Alternative splicing of rat tropoelastin mRNA is tissue-specific and developmentally regulated. Matrix 11:359–366

Hew Y, Grzelczak Z, Lau C, Keeley FW (1999) Identification of a large region of secondary structure in the 3'-untranslated region of chicken elastin mRNA with implications for the regulation of mRNA stability. J Biol Chem 274:14415–14421

Hew Y, Lau C, Grzelczak Z, Keeley FW (2000) Identification of a GA-rich sequence as a protein-binding site in the 3'-untranslated region of chicken elastin mRNA with a potential role in the developmental regulation of elastin mRNA stability. J Biol Chem 275:24857–24864

Hinek A, Thyberg J (1977) Electron microscopic observations on the formation of elastic fibers in primary cultures of aortic smooth muscle cells. J Ultrastruct Res 60:12–20

Hinek A, Botney MD, Mecham RP, Parks WC (1991) Inhibition of tropoelastin expression in fetal bovine auricular chondroblasts by 1,25-dihydroxyvitamin D3. Connect Tissue Res 26:155–166

Hirano E, Knutsen RH, Sugitani H, Ciliberto CH, Mecham RP (2007) Functional Rescue of elastin insufficiency in mice by the human elastin gene: implications for mouse models of human disease. Circ Res 101:523–531

Hoeve CAJ (1974) The elastic properties of elastin. Biopolymers 13:677–686

Holzenberger M, Lievre CA, Robert L (1993) Tropoelastin gene expression in the developing vascular system of the chicken: an in situ hybridization study. Anat Embryol (Berl) 188:481–492

Horiguchi M, Inoue T, Ohbayashi T, Hirai M, Noda K, Marmorstein LY, Yabe D, Takagi K, Akama TO, Kita T, Kimura T, Nakamura T (2009) Fibulin-4 conducts proper elastogenesis via interaction with cross-linking enzyme lysyl oxidase. Proc Natl Acad Sci USA 106: 19029–19034

Hsiao H, Stone PJ, Toselli P, Rosenbloom J, Franzblau C, Schreiber BM (1999) The role of the carboxy terminus of tropoelastin in its assembly into the elastic fiber. Connect Tissue Res 40:83–95

Hu Q, Loeys BL, Coucke PJ, De Paepe A, Mecham RP, Choi J, Davis EC, Urban Z (2006) Fibulin-5 mutations: mechanisms of impaired elastic fiber formation in recessive cutis laxa. Hum Mol Genet 15:3379–3386

Hu Q, Shifren A, Sens C, Choi J, Szabo Z, Starcher BC, Knutsen RH, Shipley JM, Davis EC, Mecham RP, Urban Z (2010) Mechanisms of emphysema in autosomal dominant cutis laxa. Matrix Biol 29:621–628

Hucthagowder V, Sausgruber N, Kim KH, Angle B, Marmorstein LY, Urban Z (2006) Fibulin-4: a novel gene for an autosomal recessive cutis laxa syndrome. Am J Hum Genet 78:1075–1080

Hucthagowder V, Morava E, Kornak U, Lefeber DJ, Fischer B, Dimopoulou A, Aldinger A, Choi J, Davis EC, Abuelo DN, Adamowicz M, Al-Aama J, Basel-Vanagaite L, Fernandez B, Greally MT, Gillessen-Kaesbach G, Kayserili H, Lemyre E, Tekin M, Turkmen S, Tuysuz B, Yuksel-Konuk B, Mundlos S, Van Maldergem L, Wevers RA, Urban Z (2009) Loss-of-function mutations in ATP6V0A2 impair vesicular trafficking, tropoelastin secretion and cell survival. Hum Mol Genet 18:2149–2165

Indik Z, Yoon K, Morrow SD, Cicila G, Rosenbloom J, Rosenbloom J, Ornstein-Goldstein N (1987) Structure of the 3' region of the human elastin gene: great abundance of Alu repetitive sequences and few coding sequences. Connect Tissue Res 16:197–211

Indik Z, Yeh H, Ornstein GN, Kucich U, Abrams W, Rosenbloom JC, Rosenbloom J (1989) Structure of the elastin gene and alternative splicing of elastin mRNA: implications for human disease. Am J Med Genet 34:81–90

Jensen SA, Reinhardt DP, Gibson MA, Weiss AS (2001) Protein interaction studies of MAGP-1 with tropoelastin and fibrillin-1. J Biol Chem 276:39661–39666

Karnik SK, Wythe JD, Sorensen L, Brooke BS, Urness LD, Li DY (2003) Elastin induces myofibrillogenesis via a specific domain, VGVAPG. Matrix Biol 22:409–425

Keeley FW (1979) The synthesis of soluble and insoluble elastin in chicken aorta as a function of development and age. Effect of a high cholesterol diet. Can J Biochem 57:1273–1280

Kelleher CM, Silverman EK, Broekelmann T, Litonjua AA, Hernandez M, Sylvia JS, Stoler J, Reilly JJ, Chapman HA, Speizer FE, Weiss ST, Mecham RP, Raby BA (2005) A functional mutation in the terminal exon of elastin segregates with severe, early onset chronic obstructive pulmonary disease. Am J Respir Cell Mol Biol 33:355–362

Kelly RE, Rice RV (1967) Abductin: a rubber-like protein from the internal triangular hinge ligament of pecten. Science 155:208–210

Kornak U, Reynders E, Dimopoulou A, van Reeuwijk J, Fischer B, Rajab A, Budde B, Nurnberg P, Foulquier F, Lefeber D, Urban Z, Gruenewald S, Annaert W, Brunner HG, van Bokhoven H, Wevers R, Morava E, Matthijs G, Van Maldergem L, Mundlos S (2008) Impaired glycosylation and cutis laxa caused by mutations in the vesicular H+-ATPase subunit ATP6V0A2. Nat Genet 40:32–34

Kozel BA, Wachi H, Davis EC, Mecham RP (2003) Domains in tropoelastin that mediate elastin deposition in vitro and in vivo. J Biol Chem 278:18491–18498

Kozel BA, Ciliberto CH, Mecham RP (2004) Deposition of tropoelastin into the extracellular matrix requires a competent elastic fiber scaffold but not live cells. Matrix Biol 23:23–34

Kozel BA, Rongish BJ, Czirok A, Zach J, Little CD, Davis EC, Knutsen RH, Wagenseil JE, Levy MA, Mecham RP (2006) Elastic fiber formation: a dynamic view of extracellular matrix assembly using timer reporters. J Cell Physiol 207:87–96

Kucik DF, Elson EL, Sheetz MP (1989) Forward transport of glycoproteins on leading lamellipodia in locomoting cells. Nature (Lond) 340:315–317

Lent R, Franzblau C (1967) Studies on the reduction of bovine elastin: evidence for the presence of delta-6,7-dehydrolysinonorleucine. Biochem Biophys Res Commun 26:43–50

Lent RW, Smith B, Salcedo LL, Faris B, Franzblau C (1969) Studies on the reduction of elastin. II. Evidence for the presence of alpha-aminoadipic acid delta-semialdehyde and its aldol condensation product. Biochemistry 8:2837–2845

Li DY, Toland AE, Boak BB, Atkinson DL, Ensing GJ, Morris CA, Keating MR (1997) Elastin point mutations cause an obstructive vascular disease, supravalvular aortic stenosis. Hum Mol Genet 6:1021–1028

Li DY, Brooke B, Davis EC, Mecham RP, Sorensen LK, Boak BB, Eichwald E, Keating MT (1998a) Elastin is an essential determinant of arterial morphogenesis. Nature 393:276–280

Li DY, Faury G, Taylor DG, Davis EC, Boyle WA, Mecham RP, Stenzel P, Boak B, Keating MT (1998b) Novel arterial pathology in mice and humans hemizygous for elastin. J Clin Invest 102:1783–1787

Loeys B, Van Maldergem L, Mortier G, Coucke P, Gerniers S, Naeyaert JM, De Paepe A (2002) Homozygosity for a missense mutation in fibulin-5 (FBLN5) results in a severe form of cutis laxa. Hum Mol Genet 11:2113–2118

Long MM, King VJ, Prasad KU, Urry DW (1988) Chemotaxis of fibroblasts toward the nonapeptide of elastin. Biochim Biophys Acta 968:300–311

Lotery AJ, Baas D, Ridley C, Jones RP, Klaver CC, Stone E, Nakamura T, Luff A, Griffiths H, Wang T, Bergen AA, Trump D (2006) Reduced secretion of fibulin 5 in age-related macular degeneration and cutis laxa. Hum Mutat 27:568–574

Manohar A, Anwar RA (1994) Evidence for a cell-specific negative regulatory element in the first intron of the gene for bovine elastin. Biochem J 300:147–152

Martino M, Bavoso A, Guantieri V, Coviello A, Tamburro AM (2000) On the occurrence of polyproline II structure in elastin. J Mol Struct 519:173–189

McLaughlin PJ, Chen Q, Horiguchi M, Starcher BC, Stanton JB, Broekelmann TJ, Marmorstein AD, McKay B, Mecham R, Nakamura T, Marmorstein LY (2006) Targeted disruption of fibulin-4 abolishes elastogenesis and causes perinatal lethality in mice. Mol Cell Biol 26:1700–1709

Mecham RP (2008) Methods in elastic tissue biology: elastin isolation and purification. Methods 45:32–41

Mecham RP, Davis EC (1994) Elastic fiber structure and assembly. In: Yurchenko PD, Birk DE, Mecham RP (eds) Extracellular matrix assembly and structure. Academic, San Diego, pp 281–314

Metcalfe K, Rucka AK, Smoot L, Hofstadler G, Tuzler G, McKeown P, Siu V, Rauch A, Dean J, Dennis N, Ellis I, Reardon W, Cytrynbaum C, Osborne L, Yates JR, Read AP, Donnai D, Tassabehji M (2000) Elastin: mutational spectrum in supravalvular aortic stenosis. Eur J Hum Genet 8:955–963

Miao M, Bruce AE, Bhanji T, Davis EC, Keeley FW (2007) Differential expression of two tropoelastin genes in zebrafish. Matrix Biol 26:115–124

Miao M, Stahl RJ, Petersen LF, Reintsch WE, Davis EC, Keeley FW (2009) Characterization of an unusual tropoelastin with truncated C-terminus in the frog. Matrix Biol 28:432–441

Micale L, Turturo MG, Fusco C, Augello B, Jurado LA, Izzi C, Digilio MC, Milani D, Lapi E, Zelante L, Merla G (2010) Identification and characterization of seven novel mutations of elastin gene in a cohort of patients affected by supravalvular aortic stenosis. Eur J Hum Genet 18:317–323

Milewicz DM, Urbán Z, Boyd CD (2000) Genetic disorders of the elastic fiber system. Matrix Biol 19:471–480

Mochizuki S, Brassart B, Hinek A (2002) Signaling pathways transduced through the elastin receptor facilitate proliferation of arterial smooth muscle cells. J Biol Chem 277:44854–44863

Moczar M, Moczar E, Robert L (1980) Peptides on partial hydrolysis of elastin with aqueous ethanolic potassium hydroxide. In: Robert AM, Robert L (eds) Biology and pathology of elastic tissues. S. Karger, Basel, pp 174–187

Morihara K, Tsuzuki H (1967) Elastolytic properties of various proteinases from microbial orgin. Arch Biochem Biophys 120:68–78

Nakamura T, Lozano PR, Ikeda Y, Iwanaga Y, Hinek A, Minamisawa S, Cheng CF, Kobuke K, Dalton N, Takada Y, Tashiro K, Ross J Jr, Honjo T, Chien KR (2002) Fibulin-5/DANCE is essential for elastogenesis in vivo. Nature 415:171–175

Park S, Seo EJ, Yoo HW, Kim Y (2006) Novel mutations in the human elastin gene (ELN) causing isolated supravalvular aortic stenosis. Int J Mol Med 18:329–332

Parks WC, Deak SB (1990) Tropoelastin heterogeneity: implications for protein function and disease. Am J Respir Cell Mol Biol 2:399–406

Partridge SM (1963) Elastin. Adv Protein Chem 17:227–302

Partridge SM (1967) Diffusion of solutes in elastin fibres. Biochim Biophys Acta 140:132–141

Partridge SM, Davis HF, Adair GS (1955) The chemistry of connective tissues. 2. Soluble proteins derived from partial hydrolysis of elastin. Biochem J 61:11–21

Partridge SM, Elsden DF, Thomas J (1963) Constitution of the cross-linkages in elastin. Nature (Lond) 197:1297–1298

Partridge SM, Elsden DF, Thomas J, Dorfman A, Telser A, Ho PL (1964) Biosynthesis of the desmosine and isodesmosine cross-bridges in elastin. Biochem J 93:30C–33C

Partridge SM, Elsden DF, Thomas J, Dorfman A, Tesler A, Ho PL (1966) Incorporation of labelled lysine into the desmosine crossbridges in elastin. Nature (Lond) 209:399–400

Pasquali-Ronchetti I, Fornieri C (1984) The ultrastructural organization of the elastin fibre. In: Ruggeri A, Motta PM (eds) Ultrastructure of the connective tissue matrix. Nijhoff, The Hague, pp 126–139

Pasquali-Ronchetti I, Forniere C, Castellani I, Bressan GM, Volpin D (1981) Alterations of the connective tissue components induced by beta-aminopropionitrile. Exp Mol Pathol 35:42–56

Pasquali-Ronchetti I, Bressan GM, Fornieri C, Baccarani-Contri M, Castellani I, Volpin D (1984) Elastin fiber-associated glycosaminoglycans in beta-aminopropionitrile-induced lathyrism. Exp Mol Pathol 40:235–245

Pereira L, Andrikopoulos K, Tian J, Lee SY, Keene DR, Ono R, Reinhardt DP, Sakai LY, Biery NJ, Bunton T, Dietz HC, Ramirez F (1997) Targetting of the gene encoding fibrillin-1 recapitulates the vascular aspect of Marfan syndrome. Nat Genet 17:218–222

Pereira L, Lee SY, Gayraud B, Andrikopoulos K, Shapiro SD, Bunton T, Biery NJ, Dietz HC, Sakai LY, Ramirez F (1999) Pathogenetic sequence for aneurysm revealed in mice underexpressing fibrillin-1. Proc Natl Acad Sci USA 96:3819–3823

Perez Jurado LA, Peoples LA, Kaplan P, Hamel BCJ, Francke U (1996) Molecular definition of the chromosome 7 deletion in Williams syndrome and parent-of-origin effects on growth. Am J Hum Genet 59:781–792

Pierce RA, Deak SB, Stolle CA, Boyd CD (1990) Heterogeneity of rat tropoelastin mRNA revealed by cDNA cloning. Biochemistry 29:9677–9683

Pierce RA, Alatawi A, Deak SB, Boyd CD (1992) Elements of the rat tropoelastin gene associated with alternative splicing. Genomics 12:651–658

Pierce RA, Moore CH, Arikan MC (2006) Positive transcriptional regulatory element located within exon 1 of elastin gene. Am J Physiol Lung Cell Mol Physiol 291:L391–399

Piontkivska H, Zhang Y, Green ED, Elnitski L (2004) Multi-species sequence comparison reveals dynamic evolution of the elastin gene that has involved purifying selection and lineage-specific insertions/deletions. BMC Genomics 5:31

Pober BR (2010) Williams–Beuren syndrome. N Engl J Med 362:239–252

Pober BR, Johnson M, Urban Z (2008) Mechanisms and treatment of cardiovascular disease in Williams–Beuren syndrome. J Clin Invest 118:1606–1615

Privitera S, Prody CA, Callahan JW, Hinek A (1998) The 67-kDa enzymatically inactive alternatively spliced variant of β-galactosidase is identical to the elastin/laminin binding protein. J Biol Chem 273:6319–6326

Robson P, Wright GM, Sitarz E, Maiti A, Rawat M, Youson JH, Keeley FW (1993) Characterization of lamprin, an unusual matrix protein from lamprey cartilage. Implications for evolution, structure, and assembly of elastin and other fibrillar proteins. J Biol Chem 268:1440–1447

Rock MJ, Cain SA, Freeman LJ, Morgan A, Mellody K, Marson A, Shuttleworth CA, Weiss AS, Kielty CM (2004) Molecular basis of elastic fiber formation. Critical interactions and a tropoelastin-fibrillin-1 cross-link. J Biol Chem 279:23748–23758

Rodgers UR, Weiss AS (2004) Integrin alpha v beta 3 binds a unique non-RGD site near the C-terminus of human tropoelastin. Biochimie 86:173–178

Rodriguez-Revenga L, Iranzo P, Badenas C, Puig S, Carrio A, Mila M (2004) A novel elastin gene mutation resulting in an autosomal dominant form of cutis laxa. Arch Dermatol 140: 1135–1139

Rose C, Wessel A, Pankau R, Partsch C-J, Bürsch J (2001) Anomalies of the abdominal aorta in Williams–Beuren syndrome – another cause of arterial hypertension. Eur J Pediatr 160: 655–658

Rosenbloom J, Cywinski A (1976) Inhibition of proline hydroxylation does not inhibit secretion of tropoelastin by chick aorta cells. FEBS Lett 65:246–250

Rosenbloom J, Harsch M, Jimenez S (1973) Hydroxyproline content determines the denaturation temperature of chick tendon collagen. Arch Biochem Biophys 158:478–484

Rosenbloom J, Abrams WR, Indik Z, Yeh H, Ornstein-Goldstein N, Bashir MM (1995) Structure of the elastin gene. Ciba Found Symp 192:59–74
Ross R (1973) The elastic fiber. J Histochem Cytochem 21:199–208
Ross R, Bornstein P (1969) The elastic fiber. I. The separation and partial characterization of its macromolecular components. J Cell Biol 40:366–381
Rucker RB, Murra J, Riemann W, Buckingham K, Tan K, Khoo GS (1977) Putative forms of soluble elastin and their relationship to the synthesis of fibrous elastin. Biochem Biophys Res Commun 75:358–365
Ruigrok YM, Seitz U, Wolterink S, Rinkel GJ, Wijmenga C, Urban Z (2004) Association of polymorphisms and haplotypes in the elastin gene in Dutch patients with sporadic aneurysmal subarachnoid hemorrhage. Stroke 35:2064–2068
Sage H, Gray WR (1979) Studies on the evolution of elastin–I. Phylogenetic distribution. Comp Biochem Physiol B 64:313–327
Sandberg LB, Weissman N, Smith DW (1969) The purification and partial characterization of a soluble elastin-like protein from copper-deficient aorta. Biochemistry 8:2940–2945
Sandberg LB, Weissman N, Gray WR (1971) Structural features of tropoelastin related to the sites of cross-links in aortic elastin. Biochemistry 10:52–56
Sato F, Wachi H, Ishida M, Nonaka R, Onoue S, Urban Z, Starcher BC, Seyama Y (2007) Distinct steps of cross-linking, self-association, and maturation of tropoelastin are necessary for elastic fiber formation. J Mol Biol 369:841–851
Senior RM, Griffin GL, Mecham RP, Wrenn DS, Prasad KU, Urry DW (1984) Val-Gly-Val-Ala-Pro-Gly, a repeating peptide in elastin, is chemotactic for fibroblasts and monocytes. J Cell Biol 99:870–874
Senior RM, Hinek A, Griffin GL, Pipoly DJ, Crouch EC, Mecham RP (1989) Neutrophils show chemotaxis to type IV collagen and its 7 S domain and contain a 67 kD type IV collagen binding protein with lectin properties. Am J Respir Cell Mol Biol 1:479–487
Shadwick RE, Gosline JM (1981) Elastic arteries in invertebrates: mechanics of the octopus aorta. Science 213:759–761
Shapiro SD, Endicott SK, Province MA, Pierce JA, Campbell EJ (1991) Marked longevity of human lung parenchymal elastic fibers deduced from prevalence of D-aspartate and nuclear weapons-related radiocarbon. J Clin Invest 87:1828–1834
Shifren A, Durmowicz AG, Knutsen RH, Hirano E, Mecham RP (2006) Elastin protein levels are a vital modifier affecting normal lung development and susceptibility to emphysema. Am J Physiol Lung Cell Mol Physiol 292:L778–787
Smith DW, Brown DM, Carnes WH (1972) Preparation and properties of salt-soluble elastin. J Biol Chem 247:2427–2432
Starcher BC (1986) Elastin and the lung. Thorax 41:577–585
Sykes BC, Partridge SM (1974) Salt-soluble elastin from lathyritic chicks. Biochem J 141: 657–572
Syx D, Malfait F, Van Laer L, Hellemans J, Hermanns-Le T, Willaert A, Benmansour A, De Paepe A, Verloes A (2010) The RIN2 syndrome: a new autosomal recessive connective tissue disorder caused by deficiency of Ras and Rab interactor 2 (RIN2). Hum Genet 128(1):79–88
Szabo Z, Levi-Minzi SA, Christiano AM, Struminger C, Stoneking M, Batzer MA, Boyd CD (1999) Sequential loss of two neighboring exons of the tropoelastin gene during primate evolution. J Mol Evol 49:664–671
Szabo Z, Crepeau MW, Mitchell AL, Stephan MJ, Puntel RA, Yin Loke K, Kirk RC, Urban Z (2006) Aortic aneurysmal disease and cutis laxa caused by defects in the elastin gene. J Med Genet 43:255–258
Tamburro AM, Bochicchio B, Pepe A (2005) The dissection of human tropoelastin: from the molecular structure to the self-assembly to the elasticity mechanism. Pathol Biol (Paris) 53:383–389

Tassabehji M, Metcalfe K, Donnai D, Hurst J, Reardon W, Burch M, Read AP (1997) Elastin: genomic structure and point mutations in patients with supravalvular aortic stenosis. Hum Mol Genet 6:1029–1036

Tassabehji M, Metcalfe K, Hurst J, Ashcroft GS, Kielty C, Wilmot C, Donnai D, Read AP, Jones CJP (1998) An elastin gene mutation producing abnormal tropoelastin and abnormal elastic fibres in a patient with autosomal dominant cutis laxa. Hum Mol Genet 7:1021–1028

Thomas J, Elsden DF, Partridge SM (1963) Partial structure of two major degradation products from the cross-linkages in elastin. Nature 200:651–652

Thyberg J, Hinek A, Nilsson J, Friberg U (1979) Electron microscopic and cytochemical studies of rat aorta. Intracellular vesicles containing elastin- and collagen-like material. Histochem J 11:1–17

Trask BC, Trask TM, Broekelmann T, Mecham RP (2000a) The microfibrillar proteins MAGP-1 and fibrillin-1 form a ternary complex with the chondroitin sulfate proteoglycan decorin. Mol Biol Cell 11:1499–1507

Trask TM, Crippes Trask B, Ritty TM, Abrams WR, Rosenbloom J, Mecham RP (2000b) Interaction of tropoelastin with the amino-terminal domains of fibrillin-1 and fibrillin-2 suggests a role for the fibrillins in elastic fiber assembly. J Biol Chem 275:24400–24406

Umeda H, Nakamura F, Suyama K (2001a) Oxodesmosine and isooxodesmosine, candidates of oxidative metabolic intermediates of pyridinium cross-links in elastin. Arch Biochem Biophys 385:209–219

Umeda H, Takeuchi M, Suyama K (2001b) Two new elastin cross-links having pyridine skeleton. Implication of ammonia in elastin cross-linking in vivo. J Biol Chem 276:12579–12587

Urban Z, Michels VV, Thibodeau SN, Donis-Keller H, Csiszár K, Boyd CD (1999) Supravalvular aortic stenosis: a splice site mutation within the elastin gene results in reduced expression of two aberrantly spliced transcripts. Hum Genet 104:135–142

Urban Z, Peyrol S, Plauchu H, Zabot MT, Lebwohl M, Schilling K, Green M, Boyd CD, Csiszar K (2000) Elastin gene deletions in Williams syndrome patients result in altered deposition of elastic fibers in skin and a subclinical dermal phenotype. Pediatr Dermatol 17:12–20

Urban Z, Gao J, Pope FM, Davis EC (2005) Autosomal dominant cutis laxa with severe lung disease: synthesis and matrix deposition of mutant tropoelastin. J Invest Dermatol 124: 1193–1199

Urry DW (1974) Studies on the conformation and interactions of elastin. Adv Exp Med Biol 43: 211–243

Urry DW (1983) What is elastin; what is not. Ultrastruct Pathol 4:227–251

Urry DW, Long MM (1976) Conformations of the repeat peptides of elastin in solution. CRC Crit Rev Biochem 4:41–45

Urry DW, Long MM, Cox BA, Ohnishi T, Mitchell LW, Jacobs M (1974) The synthetic polypentapeptide of elastin coacervates and forms filamentous aggregates. Biochim Biophys Acta 371:597–602

van Beek JD, Hess S, Vollrath F, Meier BH (2002) The molecular structure of spider dragline silk: folding and orientation of the protein backbone. Proc Natl Acad Sci USA 99:10266–10271

Van der Aa N, Rooms L, Vandeweyer G, van den Ende J, Reyniers E, Fichera M, Romano C, Delle Chiaie B, Mortier G, Menten B, Destree A, Maystadt I, Mannik K, Kurg A, Reimand T, McMullan D, Oley C, Brueton L, Bongers EM, van Bon BW, Pfund R, Jacquemont S, Ferrarini A, Martinet D, Schrander-Stumpel C, Stegmann AP, Frints SG, de Vries BB, Ceulemans B, Kooy RF (2009) Fourteen new cases contribute to the characterization of the 7q11.23 microduplication syndrome. Eur J Med Genet 52:94–100

Venkatachalam CM, Urry DW (1981) Development of a linear helical conformation from its cyclic correlate. B-spiral model of the elastin poly(pentapeptide) (VPGVG)n. Macromolecules 14:1225–1229

Vrhovski B, Jensen S, Weiss AS (1997) Coacervation characteristics of recombinant human tropoelastin. Eur J Biochem 250:92–98

Wachi H, Sato F, Nakazawa J, Nonaka R, Szabo Z, Urban Z, Yasunaga T, Maeda I, Okamoto K, Starcher BC, Li DY, Mecham RP, Seyama Y (2007) Domains 16 and 17 of tropoelastin in elastic fibre formation. Biochem J 402:63–70

Wagenseil JE, Mecham RP (2007) New insights into elastic fiber assembly. Birth Defects Res C Embryo Today 81:229–240

Wagenseil JE, Mecham RP (2009) Vascular extracellular matrix and arterial mechanics. Physiol Rev 89:957–989

Wagenseil JE, Ciliberto CH, Knutsen RH, Levy MA, Kovacs A, Mecham RP (2009) Reduced vessel elasticity alters cardiovascular structure and function in newborn mice. Circ Res 104:1217–1224

Weinbaum JS, Broekelmann TJ, Pierce RA, Werneck CC, Segade F, Craft CS, Knutsen RH, Mecham RP (2008) Deficiency in microfibril-associated glycoprotein-1 leads to complex phenotypes in multiple organ systems. J Biol Chem 283:25533–25543

Wendel DP, Taylor DG, Albertine KH, Keating MT, Li DY (2000) Impaired distal airway development in mice lacking elastin. Am J Respir Cell Mol Biol 23:320–326

Wender DB, Treiber LR, Bensusan HB, Walton AG (1974) Synthesis and characterization of poly (LysAla3). Biopolymers 13:192919–192941

Whiting AH, Sykes BC, Partridge SM (1974) Isolation of salt-soluble elastin from ligamentum nuchae of copper-deficient calf. Biochem J 179:35–45

Wise SG, Mithieux SM, Raftery MJ, Weiss AS (2005) Specificity in the coacervation of tropoelastin: solvent exposed lysines. J Struct Biol 149:273–281

Yanagisawa H, Davis EC, Starcher BC, Ouchi T, Yanagisawa M, Richardson JA, Olson EN (2002) Fibulin-5 is an elastin-binding protein essential for elastic fibre development in vivo. Nature 415:168–171

Yeh H, Ornstein-Goldstein N, Indik Z, Sheppard P, Anderson N, Rosenbloom JC, Cicila G, Yoon K, Rosenbloom J (1987) Sequence variation of bovine elastin mRNA due to alternative splicing. Coll Relat Res 7:235–247

Yeh H, Anderson N, Ornstein-Goldstein N, Bashir MM, Rosenbloom JC, Abrams W, Indik Z, Yoon K, Parks W, Mecham R, Rosenbloom J (1989) Structure of the bovine elastin gene and S1 nuclease analysis of alternative splicing of elastin mRNA in the bovine nuchal ligament. Biochemistry 28:2365–2370

Yoon K, May M, Goldstein N, Indik Z, Oliver L, Boyd C, Rosenbloom J (1984) Characterization of a sheep elastin cDNA clone contaning translated sequences. Biochem Biophys Res Commun 118:261–269

Zhang M, Pierce RA, Wachi H, Mecham RP, Parks WC (1999) An open reading frame element mediates posttranscriptional regulation of tropoelastin and responsiveness to transforming growth factor β1. Mol Cell Biol 19:7314–7326

Zheng Q, Davis EC, Richardson JA, Starcher BC, Li T, Gerard RD, Yanagisawa H (2007) Molecular analysis of fibulin-5 function during de novo synthesis of elastic fibers. Mol Cell Biol 27:1083–1095

Chapter 9
Lysyl Oxidase and Lysyl Oxidase-Like Enzymes

Herbert M. Kagan and Faina Ryvkin

Abstract Lysyl oxidase (LOX) and its four congeners, lysyl oxidase-like 1 (LOXL1), -2, -3, and -4, have received much investigative attention in recent years. LOX itself, is the prototypic form of these amine oxidase enzymes. LOX has long been considered to function exclusively as the enzyme that oxidizes peptidyl lysine in its collagen and elastin substrates, thereby initiating formation of the covalent cross-linkages that stabilize these fibrous proteins. This view has been greatly expanded in light of the revelations that LOX can function both as an anti-oncogenic agent and as an enhancer of malignancy in selected cancerous conditions. Evidence is also accumulating that points to the roles of specific LOXL members of this family in disease and in biological homeostasis. This chapter reviews structural and catalytic properties as well as the roles in biology of these amine oxidases and presents a computer-generated predicted 3D protein structure of LOX.

9.1 Introduction

Lysyl oxidase (LOX; EC 1. 4. 3.13) has long been recognized as the catalyst that oxidizes peptidyl lysine to peptidyl α-aminoadipic-δ-semialdehyde (AAS) in elastin and collagen, the two major structural proteins of the extracellular matrix (ECM). Once formed, the side chain aldehyde functions can condense with other AAS or intact lysine residues to form a variety of inter- and intrachain cross-linkages that stabilize these connective tissue proteins. The critical role of this enzyme becomes

---

H.M. Kagan (✉)
Department of Biochemistry, Boston University School of Medicine, 715 Albany Street, Boston, MA 02118, USA
e-mail: kagan@biochem.bumc.bu.edu

F. Ryvkin
Emmanuel College, 400 The Fenway, Boston, MA 02115, USA
e-mail: ryvkin@emmanuel.edu

eminently clear if growing animals are treated or fed with β-aminopropionitrile (BAPN), which irreversibly inactivates the catalytic potential of LOX (Barrow et al. 1974; Kagan 1986) and by the perinatal lethality of LOX−/− mice (Mäki et al. 2002; Hornstra et al. 2003). Marked malformation of connective tissues evidenced by fragile skin, aneurysms, and compromised bone formation are consequences of the inhibition, excessive downregulation, or genetic absence of LOX.

This view of the role of LOX has been well supported in recent years. However, this is now recognized to be a limited representation of the multiple and surprising roles in homeostasis and disease that LOX plays in biology beyond its function in stabilizing connective tissue proteins. Moreover, LOX turns out to be just one member of a recently discovered family of amine oxidases whose other members are lysyl oxidase-like (LOXL) enzymes, LOXL1, -2, -3, and -4. Similar to LOX itself, each of these LOXL enzymes has been found to be critical agents involved in normal and diseased biological processes. This chapter summarizes key structural and functional differences as well as similarities that have been documented among these important biological catalysts. Previous reviews on these enzymes have appeared and should be consulted by the interested reader (Mäki 2009; Payne et al. 2007; Lucero and Kagan 2006; Trackman 2005; Molnar et al. 2003; Csiszar 2001).

The chemistry of the LOX-catalyzed reaction is shown in Fig. 9.1. In addition to the production of the peptidyl aldehyde, important features of this reaction include the loss of positive charge at the epsilon carbon of the lysine residue accompanying the oxidative removal of its amino group, the requirement for O_2 as a cosubstrate, and the release of hydrogen peroxide as a product. Specifically, the aldehyde function has the potential to spontaneously form Schiff bases and aldol condensation products, reactions that are required for the formation of the covalent cross-linkages of the structural matrix proteins. While there is no evidence for additional aldehyde reactivity within these proteins, it should be noted that aldehydes can potentially form reversible addition products with oxyanions derived from anionic protein residues, although there are no reports of this effect in proteins. The oxidative deamination of peptidyl lysine, as noted, converts the cationic side chain of lysyl residues to the

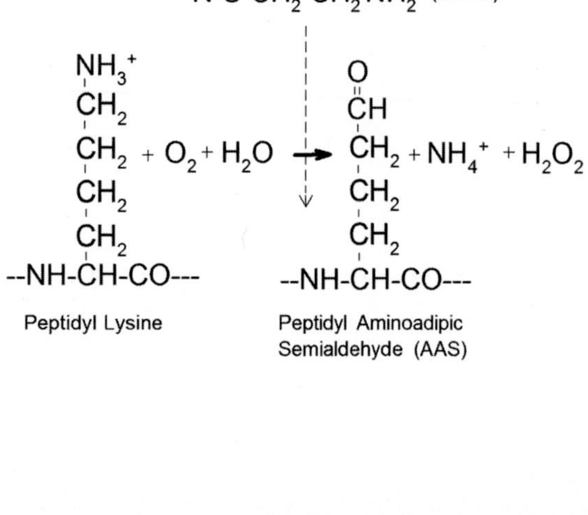

Fig. 9.1 Chemistry of the lysyl oxidase-catalyzed reaction. BAPN irreversibly inhibits this enzymatic reaction

neutral aldehyde function. Since LOX can oxidize lysine in a variety of globular proteins in vitro (Kagan et al. 1984) and in cell membranes in cultured cells (Lucero et al. 2008), the loss of cationic charge within key lysine residues may alter structural and/or functional roles played by the ionized epsilon amino groups of peptidyl lysine. Finally, the hydrogen peroxide product of LOX catalysis cannot be neglected as a possible effector of cell function. Indeed, this peroxide is now known to mediate diverse physiological responses including cell proliferation, differentiation, and migration (Rhee 2006).

9.2 Biosynthesis and Molecular Properties of Lysyl Oxidase

As the properties of LOXL1, -2, -3, and -4 are investigated, the insights gained into these enzymes are conventionally compared with those of LOX, the first and prototypic member of this family to have been explored and which has been subject to extensive analyses of its biosynthesis, catalytic mechanism, specificity, and protein structure (the latter within the limits presented by the lack of success in attempts to crystallize and thereby gain insight into its 3D structure by X-ray crystallography). Sequencing of cloned rat LOX cDNA elucidated the full-length amino acid sequence of the enzyme, which showed that the intracellular protein product of translation was a 46 kDa species (Trackman et al. 1990, 1991). In contrast, SDS-PAGE analysis of purified bovine LOX yielded a single band migrating as a 32 kDa species (Kagan et al. 1979). This anomaly was resolved by biosynthetic studies revealing that LOX first appears intracellularly as a 46 kDa preproenzyme containing an N-terminal signal peptide, a propeptide domain, and a C-terminal catalytic domain arranged sequentially within this preproenzyme. Following signal peptide cleavage and posttranslational N-glycosylation within the propeptide domain and insertion and generation of its copper and quinone cofactors, the resulting 50 kDa proLOX product is secreted and proteolytically cleaved in the extracellular space to release the 32 kDa functional LOX enzyme and the free propeptide (Trackman et al. 1992). Procollagen C-proteinase (BMP-1) has been implicated as the primary catalyst of proLOX cleavage to LOX (Cronshaw et al. 1995; Trackman et al. 1992). Mammalian Tolloid-like (mTLL)-1 and -2, two genetically distinct BMP-1-related proteinases, also have procollagen C-proteinase activity and can process proLOX to LOX in assays in vitro (Uzel et al. 2001).

The proteolytic activation of proLOX has a correlating connection to the processing of procollagen species to the fibrogenic collagen molecule. Thus, LOX does not oxidize type I collagen molecules in vitro until procollagen has been converted to the triple helical collagen molecule, thus permitting spontaneous formation of the quarter-staggered collagen microfibril structure. It is not until the individual collagen molecules aggregate into microfibrils that peptidyl lysine oxidation by LOX can occur, since the microfibrillar structures are the optimal substrate forms of collagen for LOX (Siegel 1974; Nagan and Kagan 1994). Moreover, collagen fibril formation requires the prior proteolytic removal of the C-terminal propeptide segments of procollagen molecules by BMP-1 and of the N-terminal propeptide by procollagen N-proteinase, thus linking activation by BMP-1 of proLOX to the functional LOX catalyst with the BMP-1-catalyzed conversion of procollagen to the collagen molecule, thereby permitting the formation of the microfibrillar substrate.

The characteristics of LOX that serve as benchmarks for comparison with the LOXL enzymes include its (1) biosynthesis as preproLOX; subsequent intracellular posttranslational processing including (2) cleavage of the signal peptide of preproLOX and (3) N-glycosylation; (4) incorporation of its copper cofactor and autocatalytic formation of its lysine tyrosylquinone (LTQ) cofactor; (5) secretion of the posttranslationally processed 50 kDa proLOX species; and (6) extracellular proteolytic processing of proLOX, thereby releasing catalytically functional LOX and the free propeptide.

9.3 Gene and Protein Structures of LOX and LOXL Enzymes

9.3.1 Gene Structure

The chromosomal locations of the five members of the human LOX family have been determined as follows: LOX, 5q23.1; LOXL1, 15q24.1; LOXL2, 8p21.3; LOXL3, 2p13.3; and LOXL4, 10q, 24.2. The LOX gene is distributed into seven exons spanning 15 kb of genomic DNA of which 5.5 kb is the 5' UTR. Approximately half of the coding sequence comprising the signal peptide, propeptide, and 60 residues of the mature enzyme occurs in the first exon along with the last 292 bases of the 5' UTR. The final two amino acids of the coding sequence and a 3.8 kb 3' UTR are encoded in exon 7 (Hämäläinen et al. 1993). LOXL1 protein is the most homologous to that of LOX and, like the LOX gene, the LOXL1 gene also consists of seven exons with coding information distributed among these as in the case of LOX. The LOXL2 gene is composed of 14 exons and 13 introns, distributed through approximately 107 kb of genomic DNA (Fong et al. 2007). Two transcripts of sizes 3.6 and 4.9 kb have been reported, with the smaller transcript much more abundant (Jourdan-Le Saux et al. 1999). At least 17 single-nucleotide polymorphisms (SNPs) occur within the LOXL2 gene (Akagawa et al. 2007). The human LOXL3 gene also has 14 exons spanning more than 21 kb of genomic sequence (Jourdan-Le Saux et al. 2001; Mäki et al. 2001). Similarly, the LOXL4 gene is also composed of 14 exons and intervening introns.

9.3.2 Protein Structure: Comparison of Domains

Figure 9.2 compares the distribution of the domains found in LOX and the four LOXL enzymes. Each of the five members of this family contains conserved components within the C-terminal catalytic domain that are required for the expression of catalytic activity. These include a histidine-rich sequence with similarity to WXWHXCHXHYH, which is known in other proteins to bind Cu(II); conserved lysine and tyrosine residues, which are the two residue components from which the LTQ cofactor is generated; and ten conserved cysteine residues shown in purified bovine aorta LOX to exist as five disulfide cross-linkages within the purified 32 kDa bovine enzyme (Williams and Kagan 1985). In addition, each contains consensus elements characteristic of a cytokine receptor-like domain (CRL), while each of the four LOX-like enzymes also contains a scavenger receptor cysteine-rich (SRCR) domain (Fig. 9.2). The SRCR domain is an ancient and highly conserved domain of ~110 residues and is found in numerous cell surface and secreted proteins implicated in atherosclerosis, adhesion, and host defense (Krieger and Herz 1994; Hohenester et al. 1999).

The complete, predicted amino acid sequences of human preproLOX and those of the human preproLOXL enzymes are shown in Fig. 9.3 in which the four human LOXL sequences have been aligned against that of human preproLOX. It is evident that homology is greatest in the C-terminal catalytic domains of these five species, consistent with the retention of essential structural features required for optimal catalytic expression of these amine oxidases. There is minimal homology between LOX and the four LOXL species in the region between the N-terminal methionine and, approximately, residue 230 of the preproLOX reference sequence. This reflects both significantly different sequences and the greater number of amino acid residues within the putative propeptide regions in the preproLOXL species. The

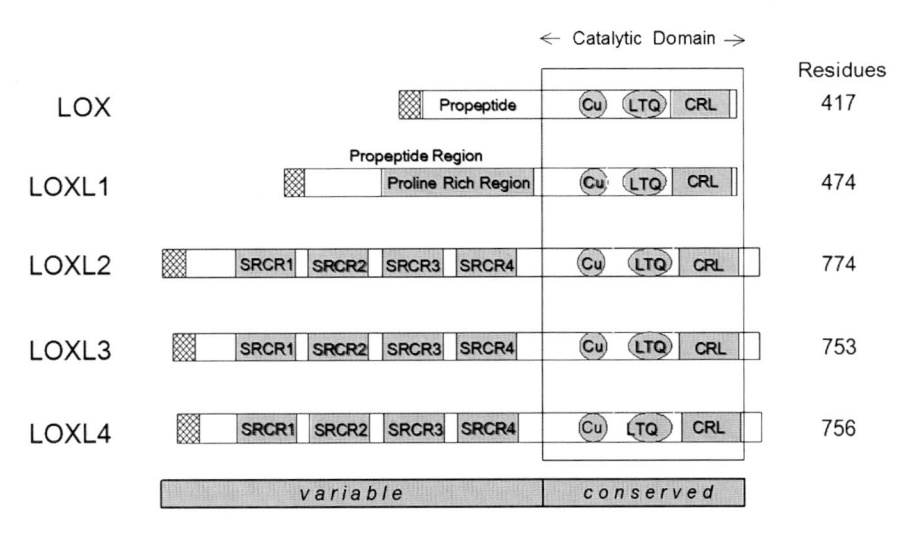

Fig. 9.2 Distribution of domains in LOX and LOXL enzymes. The relative positions of the signal peptides (*crosshatched*), LOX propeptide, LOXL1 propeptide region, copper (Cu) and LTQ sites, and CRL domains are shown. A proline-rich region is unique to LOXL1 within this enzyme family. Putative propeptide regions of LOXL2, -3, and -4 are not designated in view of uncertainties in these cases (see text). This figure is adapted from those of previous reviews (Csiszar 2001; Mäki 2009)

demonstrated or putative sites at which the signal peptides of the preproLOX and preproLOXL enzymes may be cleaved and at which the proenzymes may be cleaved are shown in Fig. 9.3. These sites are based upon specificity determinants shown for mammalian signal peptidase and BMP-1, respectively. In addition, putative BMP-1 sites reflect estimated molecular weights of LOXL proteins secreted by various cell lines after intracellular processing and/or derived from the putative proenzyme forms of these LOXL species by treatment with BMP-1 in vitro (Lemberg and Martoglio 2002). The site of extracellular cleavage of rat proLOX has been shown to be at the Gly162-Asp163 peptide bond (Trackman et al. 1992; Panchenko et al. 1996) as predicted by Cronshaw et al. (1995). This most closely approximates the G168-D169 site for human proLOXL1 as shown in Fig. 9.3. BMP-1 has been shown to cleave procollagen and other protein substrates at the A–D, G–D, N–D, S–D, D–D, Y–D, T–D, and L–D peptide sites with alanine, serine, or glycine the more common P1 residue in that order (Kessler et al. 2001). Where prior reports have indicated uncertainty of BMP-1 sites under different conditions and/or with different cellular sources, the sites of possible BMP-1 cleavage of the proLOXL species have been selected in view of the sequence specificity of BMP-1 and to approximate the molecular weight of the LOXL

a

```
LOX     MRFAWTVL..  .......LLG PLQLC.....  ..........  ..........  ..........  ..........  ..........                 16
LOXL1   *AL*RGSRQL  GALVWGAC..  ....*C....  ..........  ..........  ..........  ..........  ..........                 20
LOXL2   *ERPLCSHLC  SCLAMLA**S  *LS*AQYDSW  PHYPEYFQQP  APEYHQPQAP  ANVAK....I  QLRLAGQKRK  HSEGRVEVYY                 76
LOXL3   ..........  ..........MR  *VSVWQWSPW  GLLLCLLCSS  CLGSPSPSTG  PEKKAGSQGL  RFRLAGFPRK  PYEGRVEIQR                 62
LOXL4   *A........  ..........  ......WSPP  ATLFLFLLL.  .LGQPPPSRP  QSLGT....T  KLRLVGPESK  PEEGRLEVLH                 50

LOX     ..........  ..........  ..........  ..........  ..........  ..........  ..........  ..........                 16
LOXL1   ..........  ..........  ..........  ..........  ..........  ..........  ..........  ..........                 20
LOXL2   DGQWGTVCDD  DFSIHAAHVV  CRELGYVEAK  SWTASSSYGK  GEGPIWLDNL  HCTGNEATLA  ACTSNGWGVT  DCKHTEDVGV                156
LOXL3   AGEWGTICDD  DFTLQAAHIL  CRELGFTEAT  GWTHSAKYGP  GTGRIWLDNL  SCSGTEQSVT  ECASRGWGNS  DCTHDEDAGV                142
LOXL4   QGQWGTVCDD  NFAIQEATVA  CRQLGFEAAL  TWAHSAKYGQ  GEGPIWLDNV  RCVGTESSLD  QCGSNGWGVS  DCSHSEDVGV                130

LOX     ..........  ..........  ..........  ..........  ..........  ..........  ..........  ..........                 16
LOXL1   ..........  ..........  ..........  ..........  ..........  ..........  ..........  ..........                 20
LOXL2   VCSDKRIPGF  KFDNSLINQI  ENLNIQVEDI  RIRAILSTYR  KRTPVMEGYV  EVKEGKTWKQ  ICDKHWTAKN  SRVVCGMFGF                236
LOXL3   ICKDQRLPGF  SDSNVIEVEH  ...HLQVEEV  RIRPAVGWGR  RPLPVTEGLV  EVRLPDGWSQ  VCDKGWSAHN  SHVVCGMLGF                219
LOXL4   ICHPRRHRGY  LSETVSNALG  PQGR.RLEEV  RLKPILASAK  QHSPVTEGAV  EVKYEGHWRQ  VCDQGWTMNN  SRVVCGMLGF                209

LOX     ..........  ..........  ..........  ..........  ..........  ..........  ..........  ...ALVHCAP                 23
LOXL1   ..........  ..........  ..........  ..........  ..........  ..........  ..........  ...V***GQQ                 27
LOXL2   PGERTYNTKV  YKMFASRR..  ..........  .KQRYWPFSM  DCTGTEAHIS  SCKLGPQVSL  DPMKNVTCEN  GLPAV*S*VP                303
LOXL3   PSEKRVNAAF  YRLLAQRQ..  ..........  .QHSFGLHGV  ACVGTEAHLS  LCSLEFYRAN  DTAR...CPG  GGPAV*S*VP                283
LOXL4   PSEVPVDSHY  YRKVWDLKMR  DPKSRLKSLT  NKNSFWIHQV  TCLGTEPHMA  NCQVQVAPAR  GKLRPA.CPG  GMHAV*S*VA                288

LOX     PAAGQQQPPR  EPPAAPGAWR  Q.QIQWENNG  QVFSLLSLGS  YQPQRRRDP   GAAVPGAANA  SAQQPRTPIL  LIRDNRTAAA                102
LOXL1   AQP**GSD..  ......*ARWR  *.*********  **Y***NS**  EYV*AGP...  ..........  ..........  ..........                 66
LOXL2   GQVF..S*DG  PSRFRKAYKP  E.*PL.....  ..........  ......V*LR  *G*YIGEGRV  EVLKNGEWGT  VCD*KWDLVS                359
LOXL3   GPVY..AASS  GQKKQQQSKP  QGEA......  ..........  ....*V*LK   *G*H**EGRV  EVLKAS*WGT  VCD*KWDLH*                340
LOXL4   GPHF..R**K  TK*QRK*S*A  E.EP......  ..........  ....*V*LR   SG*QV*EGRV  EVLMN*QWGT  VCDHRWNLIS                344

LOX     RTRTAGSSGV  TAGRPRPTAR  HWFQAGYSTS  RAREAGASRA  ENQTAPGEV.  ..........  ..........  ..........                151
LOXL1   ..........  ..........  ...*RSE*S*  *VLL***PQ*  QQRRSH*SPR  RRQAPSLPLP  GRVGSDTVRG  QARHPFGFGQ                123
LOXL2   AS........  ..........  ..........  ..........  ..........  ..........  ..........  ..........                361
LOXL3   AS........  ..........  ..........  ..........  ..........  ..........  ..........  ..........                342
LOXL4   AS........  ..........  ..........  ..........  ..........  ..........  ..........  ..........                346

LOX     ..........  ..........  ..........  ..........  ..........  ..........  ..........  ..........                151
LOXL1   VPDNWREVAV  GDSTGMARAR  TSVSQQRHGD  SASSVSASAF  ASTYRQQPSY  PQQFPYPQAP  FVSQYENYDP  ASRTYDQGFV                203
LOXL2   ........V*  CREL*FGS*K  EA*TGS*L**  GIGPIHLNEI  QC*GNEKSI.  ..........  ..........  ..........                402
LOXL3   ........V*  CREL*FGS*R  EALSGA*M*Q  GMGAIHLSEV  RCSGQELSL.  ..........  ..........  ..........                383
LOXL4   ........V*  CRQL*FGS*R  EALFGA*L*Q  GLGPIHLSEV  RCRGYERTL.  ..........  ..........  ..........                387
```

Fig. 9.3 (continued)

b

```
LOX     ..........  ..........  ..........  ..........  ..........  ..........  ..........  .......... 151
LOXL1   YYRPAGGGVG  AGAAAVASAG  VIYPYQPRAR  YEEYGGGEEL  PEYPPQGFYP  APERPYVPPP  PPPPDGLDRR  YSHSLYSEGT 283
LOXL2   ..........  ..........  ..........  ..........  ..........  ..........  ..........  .......... 402
LOXL3   ..........  ..........  ..........  ..........  ..........  ..........  ..........  .......... 383
LOXL4   ..........  ..........  ..........  ..........  ..........  ..........  ..........  .......... 387

LOX     ..........  ..........  .......PAL  SNLRPPSRVD  GMVGDDPYNP  YKYSDDNPYY  NYYDTYERPR  PGGRYRPGY. 203
LOXL1   PGFEQAYPDP  GPEAAQAHGG  DPRLGWY*PY  A*PP*EAYGP  PRALEP**L*  VRS**TP*PG  GERNGAQQG*  LSVGSVYRPN 363
LOXL2   ..........  ..........  ..........  ..........  ..........  ...I*CKFNA  E.SQGCNHEE  DA*VRCNTPA 428
LOXL3   ..........  ..........  ..........  ..........  ..........  ...WKCPHKN  ITAEDCSHSQ  DA*VRCNLPY 410
LOXL4   ..........  ..........  ..........  ..........  ..........  ...**CPALE  GSQNGCQHEN  DAAVRCNVPN 414

LOX     ..........  ..........  ..........  ..........  ..........  ..GT*YFQYG  L.........  .......... 212
LOXL1   QNGR......  ..........  ..........  ..........  ..........  ..........  ........*.  .......... 369
LOXL2   MGLQKKLRLN  GGRNPYEGRV  EVLVERNGSL  VWGMVCGQNW  GIVEAMVVCR  QL*L*FASNA  FQETWYWHGD  VNSNKVVMSG 508
LOXL3   TGAETRIRLS  GGRSQHEGRV  EVQIGGPGPL  RWGLICGDDW  GTLEAMVACR  QL*L**ANH*  *QETWYWDSG  NIT.EVVMSG 489
LOXL4   MGFQNQVRLA  GGRIPEEGLL  EVQVEVNGVP  RWGSVCSENW  GLTEAMVACR  QL*LGFAIHA  YKETWFWSGT  PRAQEVVMGG 494

LOX     ..........  ..........  ..........  ..........  .........P  DLVADPYYIQ  ASTYVQKMSM  YNLRCAAEEN  CLASTAYRAD 253
LOXL1   ..........  ..........  ..........  ..........  ..........* ***P**NYV*  ******RAHL  *S*******K  ******APE 410
LOXL2   VKCSGTELSL  AHCRHDGEDV  ACPQGGVQYG  AGVACSETA*  ***LNAEMV*  QT**LEDRP*  FM*Q**M***  **SAS*AQT* 588
LOXL3   VRCTGTELSL  DQCAHHGTHI  TCKRTGTRFT  AGVICSETAS  **LLHSALV*  ETA*IEDRPL  HM*Y******  ****S*RSAN 569
LOXL4   VRCSGTELAL  QQCQRHGP.V  HCSHGGGRFL  AGVSCMDSA*  ***MNTQLA*  ETA*LEDRPL  SQ*Y**H***  **SKS*DHM* 573

LOX     VRDYDHRVLL  RFPQRVKNQG  TSDFLPSRPR  YSWEWHSCHQ  HYHSMDEFSH  YDLLDANTQR  RVAEGHKASF  CLEDTSCDYG 333
LOXL1   AT***V*VLL  **********  *A****N***  HT*E******  **********  ******A*GK  K****.*.**  ****ST**F* 490
LOXL2   PTT.GY*RLL  **SSQIH*N*  Q***R*KNG*  HA*I**D**R  *****EV*T*  ****NLN.GT  K****.:***  *****E*EGD 666
LOXL3   WP.YG*R**   **SSQIH*L*  RA**R*KAG*  H**V**E**G  ******I*T*  ****TPN.GT  K****.:***  *****E*QED 647
LOXL4   WP.YGY*R**  **STQIY*L*  RT**R*KTG*  D***V**Q**R ****IEV*T*  ****TLN.GS  K****.:***  *****N*PT* 651

LOX     YHR*FACTAH  T.QGLSPGCY  DTYGADIDCQ  WIDITDVKPG  NYILKVSVNP  SYLVPESDYT  NNVVRCDIRY  TGHHAYASGC 412
LOXL1   NLK*Y***S*  *.********  **********  ******Q**   ******H***  K*I*L***F*  *****N*H*   **RYVS*TN* 569
LOXL2   IQKNYE*ANF  GD**ITM**W  *M:RH*****  ******P**   D*LFQ*VI**  NFE*AE***S  **IMK*RS**  D**RIWMYN* 746
LOXL3   VSK*YE*ANF  GE**ITV**W  *L:RH*****  N****Q**    N***Q*VI**  NFE*AE**F*  **AMK*NCK*  D**RIWVHN* 727
LOXL4   LQR*YA*ANF  GE**VTV**W  ***RH*****  *V*****G**  N**FQ*I***  HHE*AE**FS  **MLQ*RCK*  D**RVWLHN* 731

LOX     TISPY.....  ..........  ..........  ..... 417
LOXL1   K*VQ*.....  ..........  ..........  ..... 574
LOXL2   H*GGSFSEET  EKKFEHFSGL  LNNQLSPQ    774
LOXL3   H*GDAFSEEA  NRRFERYPGQ  TSNQII..    753
LOXL4   H*GNSYPANA  ELSLEQEQRL  RNNLI...    756
```

Fig. 9.3 Amino acid sequences of the full lengths of the LOX and LOXL gene products. The sequences of LOXL1, -2, -3, and -4 are aligned against that of human LOX. Areas without residues (*dashed line*); residues identical to corresponding residues in LOX (*asterisk*); predicted signal peptide cleavage sites of preproenzymes (*underlined blue font*); predicted or potential sites of cleavage in proenzymes by BMP-1 (*underlined green font*); histidine-rich Cu binding sequence (*underlined purple font*); highly conserved lysine and tyrosine progenitors of LTQ (*underlined red font*). Sequence positions in human LOX are at lys^{320} and at tyr^{356}

enzyme products with that of LOX, where possible. These selection criteria provide some advantage for comparisons of the chemical properties of regions of the LOXL species, which approximate the corresponding domains of LOX against which the sequences of the LOXL species have been aligned (Table 9.1). Clearly, the putative cleavage sites indicated in Fig. 9.3 wait further experimental testing.

As noted, review of the literature reveals a degree of uncertainty about the precise protease-susceptible sites for the LOXL species. This is particularly seen in the case of LOXL1. Borel et al. (2001) purified bovine LOXL1 by immunoaffinity chromatography as a 56 kDa species. Prediction of the molecular weight of the cloned bovine proenzyme following signal peptide cleavage indicated an expected molecular weight of 61.9 kDa. The possibility was raised that additional cleavage of the released proenzyme might have occurred at one of the two arginine-rich sites (RRR

Table 9.1 Properties of the domains of LOX and LOXL enzymes

Enzyme[a]	Domain	Predicted MW	pI	Lysine content
LOX				
$Q^{28} \to$ C-terminal	Proenzyme	44,113	8.25	6
$Q^{28} \to G^{168}$	Propeptide	15,098	11.81	0
$D^{169} \to$ C-terminal	Enzyme	29,032	5.84	6
LOXL1				
$C^{28} \to$ C-terminal	Proenzyme	60,546	6.8	9
$C^{28} \to G^{303}$	Propeptide	30,054	8.1	0
$D^{304} \to$ C-terminal	Enzyme	30,510	6.42	9
LOXL2				
$Q^{26} \to$ C-terminal	Proenzyme	84,070	5.92	37
$Q^{26} \to G^{497}$	Propeptide	52,670	8.09	28
$D^{498} \to$ C-terminal	Enzyme	69,184	5.06	9
LOXL3				
$S^{26} \to$ C-terminal	Proenzyme	80,349	6.35	23
$S^{26} \to G^{447}$	Propeptide	45,542	8.06	15
$D^{448} \to$ C-terminal	Enzyme	34,825	5.7	8
LOXL4				
$G^{25} \to$ C-terminal	Proenzyme	81,851	7.2	21
$G^{25} \to S^{388}$	Propeptide	40,263	9.05	15
$D^{389} \to$ C-terminal	Enzyme	41,606	5.81	6

The pI and molecular weight values were calculated from the sequences employing the Compute pI/MW program of the Swiss Institute of Bioinformatics

[a]This column indicates the N- and C-terminal residues for the three established or putative domains of these enzymes. Where shown in the column, "C-terminal" refers to the C-terminal residue of the full-length preproenzymes. The sites for signal peptidase cleavage to release the proenzyme and the site for BMP-1 cleavage of the proenzyme to release free enzyme and propeptide domains have been predicted based upon prior demonstrations for the substrate specificities for each of the proteolytic enzymes

at positions 83–85 or RR at positions 89–90) by a furin-like enzyme as also has been suggested for the human enzyme (Kenyon et al. 1993). N-Terminal sequence analysis indicated that the bovine proLOXL1 molecule was cleaved at the Arg91-Gln92 peptide bond yielding a predicted molecular weight of 54.6 kDa (Borel et al. 2002). Incubation of the isolated proenzyme with BMP-1 in vitro yielded two protein bands resolving as ~51 and 28 kDa, respectively, consistent with the occurrence of two potential BMP-1 cleavage sites at G130-D131 and G311-D312 in the bovine enzyme. Similar results have been obtained with the human chondrocyte LOXL1 proenzyme that was cleaved by added BMP-1 at G134-D135 and at S336-D338 releasing ~50 and 30 kDa processed proteins (Jung et al. 2003). In both cases, the mature enzymes resulting from proteolytic processing by BMP-1 of the catalytically quiescent proenzymes exhibited amine oxidase activity as demonstrated with the bovine enzyme against both elastin and collagen substrates (Borel et al. 2001).

Mixed results have also been obtained concerning the issue of extracellular proteolytic processing of the proLOXL2 species (predicted to have a mass of 87 kDa) in that a single 88 kDa band, consistent with the full length of the proenzyme, was expressed by a human gastric cancer cell line (Peng et al. 2009), whereas two bands reactive with anti-LOXL2 resolving as ~100 and 65 kDa, respectively, were secreted by Wilson's disease hepatocytes (Vadasz et al. 2005). Other studies have also observed bands at ~95 and 63 kDa, the latter assumed in this case to be the extracellular proteolytically processed form of LOXL2 (Akiri et al. 2003; Vadasz et al. 2005; Fong et al. 2007; Hollosi et al. 2009). It should be noted that cleavage at the S59-D60 bond in proLOXL2 would be consistent with the specificity of BMP-1 and would yield a 69 kDa protein. Following signal peptide cleavage, the molecular mass of LOXL3 has a calculated value of 81.3 kDa. The LOXL3 protein expressed by and secreted from HT-1080 exhibited a molecular mass of 94.7 kDa (Mäki and Kivirikko 2001). Recombinant expression of LOXL4 in HT-1080 cells also resulted in the secretion of 94.7 kDa, with both the LOXL3 and LOXL4 values corrected for V5 epitope and histidine tags used to identify the expressed proteins. Mäki et al. (2001) suggested that this difference from the calculated value might be due to glycosylation of the expressed protein, but this remains to be established. No evidence of proteolytic processing of the LOXL4 cell product expressed in HT-1080 cells, CHO cells, or in mouse embryonic fibroblasts was seen in this study (Mäki et al. 2001). In contrast, a 67 kDa LOXL3 species was detected in extracts of colon and placental tissues (Lee and Kim 2006). While it is possible that these varied results reflect differences between the specific cell type and/or tissue sources of particular LOXL species, it will be important to reproducibly define the molecular weights of the putative proenzyme forms in specific cases, the question of whether each of the LOXL proenzymes is proteolytically processed, the sequence sites for such processing, and the definition of which of the possible secreted forms exhibit amine oxidase activity.

9.4 Catalytic Properties

9.4.1 Cofactors

Inhibition of LOX activity by chelators with strong affinities for divalent copper ion pointed to the conclusion that LOX activity depends upon the presence of this metal ion. The presence of one tightly bound CuII ion per 32,000 Da of purified bovine aorta LOX has been demonstrated (Gacheru et al. 1990). While there is controversy whether copper plays a role in the expression of catalytic activity (Gacheru et al. 1990) or only in the stabilization of LOX (Tang and Klinman 2001), there is reasonable support at least for the conclusion that the incorporation of copper into the nascent proenzyme within the endoplasmic reticulum is required for the autocatalytic generation of the LTQ cofactor from highly conserved lysine and tyrosine residues within the catalytic domain (Mure 2004; Dubois and Klinman 2005; Bollinger et al. 2005; Kosonen et al. 1997).

Among the variety of chelators and other chemical inhibitors of LOX activity, carbonyl reagents such as dinitrophenylhydrazine and semicarbazide have been

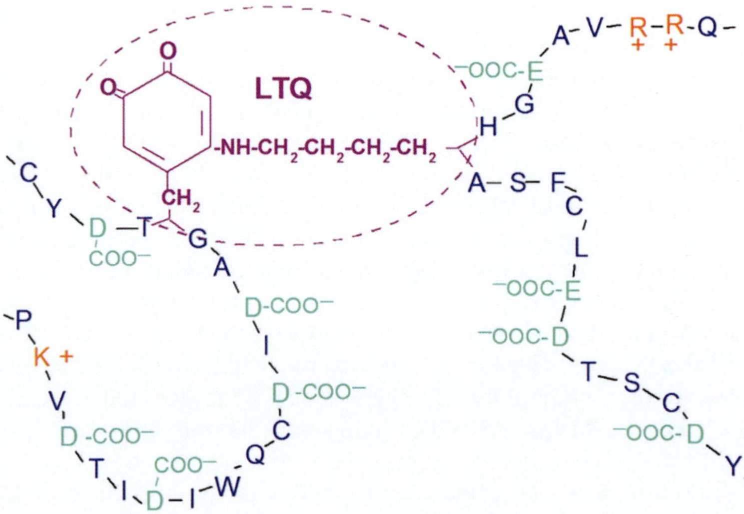

Fig. 9.4 LTQ cofactor (*circled*) and sequences adjacent to its tyrosine and lysine precursors. Anionic residues are in *red* and cationic residues are in *blue*

shown to irreversibly inactivate LOX catalysis, consistent with the presence of a functional carbonyl moiety at the active site. A variety of chemical and spectral studies identified the carbonyl cofactor within LOX as a quinone covalently linked to the enzyme and ultimately identified as LTQ (Wang et al. 1996). LTQ is shown in Fig. 9.4 (top) in peptide-bonded association with the protein sequences vicinal to the lysine and tyrosine progenitors of this cofactor. It appears that LTQ is autocatalytically generated with the assistance of the copper ion, with the synthesis initially requiring hydroxylation of the peptidyl tyrosine ring, followed by Michael addition of the peptidyl lysine component of LTQ to the aromatic nucleus followed by the oxygen-dependent reoxidation of the resulting peptidyl aminoquinol to the oxidized LTQ cofactor, the latter required for catalytic oxidation of amine substrates by the functional enzyme (Mure 2004). Quinones are chemically susceptible to reduction to quinols. In turn, quinols can be reoxidized to the quinone, thus possessing the properties of a redox-active cofactor mediating the oxidation of amine substrates, followed by its oxygen-dependent reoxidation to the quinone, thus regenerating the functional state for further catalytic events. The proposed participation of LTQ in the mechanism of action of LOX has been previously reviewed (Kagan and Li 2003).

9.4.2 Assays and Substrate Specificity

Initial studies of LOX relied heavily upon assay of its enzymatic activity against its elastin and collagen substrates that had been pulse labeled in organ culture with L-4,5-[3H]lysine (Kagan et al. 1974). Oxidation of tritiated lysine to the peptidyl aldehyde product releases free tritium ions into solution that then can be isolated as tritiated water and quantified. It was subsequently observed that LOX will also oxidize various cationic globular proteins in addition to its elastin and collagen substrates including histone H1 and can also oxidize a variety of monoamines and diamines of various carbon chain lengths (Trackman and Kagan 1979; Kagan et al. 1984). In view of the apparent flexibility of substrate requirements, a convenient continuous, enzyme-coupled assay was developed (Trackman et al. 1981) [and subsequently modified (Palamakumbura and Trackman 2002)] that monitored activity by fluorescence spectroscopy to quantify the accumulation of hydrogen peroxide as the enzyme oxidizes 1,5-diaminopentane or other alkylamines or protein substrates.

Many enzymes catalyzing proteolysis or posttranslational modifications of target residues of their protein substrates commonly display specificity for the target residue and also toward side chains of residues vicinal to the target residue. It is surprising, therefore, that purified LOX readily oxidizes lysine found in largely nonpolar residue regions, such as ...AAAKAAKAA... as in elastin substrates but also oxidizes specific lysine residues in collagen where surrounding sequences are highly polar as seen in the N-telopeptide of the α1 chain of type I collagen (...SYGYDEKSTG...). It is possible, of course, that the substrate specificity in vivo may be more restricted than is apparent in vitro, while the individual LOX and LOXL enzymes may have distinctly different specificity requirements yet to be established.

9.4.3 Electrostatic Factors Affecting LOX Substrate Specificity

Comparison of the chemistry of LOX substrates revealed that this enzyme is distinctly responsive to the net charge of various globular protein substrates, with proteins whose isoelectric points are less than 8 being resistant to oxidation by LOX, whereas a variety of basic proteins with isoelectric points greater than 8 are readily oxidized by this enzyme. As noted, the susceptible lysine in the N-telopeptide domain of the collagen α1(I) chain is immediately preceded by two dicarboxylic amino acid residues (...DEKSTG...), whereas this sequence is resistant to oxidation when presented to the enzyme within the context of a synthetic peptide. In contrast, a synthetic peptide in which glycine was substituted for the aspartic acid residue in this sequence was readily oxidized by LOX (Nagan and Kagan 1994). Since individual collagen molecules are not oxidized by LOX until quarter-staggered, microfibrillar aggregates have been formed, a possible basis to these effects was observed in structural models of interacting collagen molecules within microfibrillar, quarter-staggered arrays. These models indicate that the aspartic acid moiety of this N-telopeptide sequence occurs in close proximity to an arginine residue projecting from the triple helical segment of the neighboring collagen molecule within the quarter-staggered array, thereby potentially neutralizing the anionic character of the unfavorable aspartate residue. The quarter-staggered, overlapping neighboring collagen molecule also contains a lysine residue in its triple helical domain that occurs in the immediate vicinity of the LOX-susceptible lysine in the adjacent collagen molecule, thus introducing additional cationic charge within this microenvironment (Nagan and Kagan 1994). Such interchain interactions between ionic residues would be consistent with the requirement by LOX for microfibril formation of its type I collagen substrate.

Clearly, detailed information about the orientation of residues of the active site of LOX would be essential to more completely understand the factors that influence the catalytic mechanism of action as well as the substrate specificity of this enzyme. Thus far, however, attempts to gain insight into the 3D structure of the enzyme have been unsuccessful, since the purified enzyme has a strong tendency to form amorphous aggregates, thus making the determination of its secondary and tertiary structure by X-ray crystallography presently untenable. Nevertheless, close inspection of the primary sequence of LOX suggests the nature of potential interactions of the propeptide with the catalytic domain and points to plausible bases of the electrostatic component of its substrate specificity.

It has been previously noted that the sequence in the immediate vicinity of the tyrosine residue precursor (Y345 in human LOX) of the LTQ cofactor contains an abundance of aspartic acid residues as shown in Fig. 9.4, suggesting that this anionic microenvironment could account for the preference of LOX for proteins possessing a net cationic charge by providing an attractive charge field complementary to those in favorable substrates (Kagan and Li 2003). The negative charge density of the active site of LOX is also likely to be strongly influenced by the additional anionic residues within the sequence microenvironment of the conserved lysine residue (K320 in human LOX) that becomes a covalent component of the LTQ cofactor (Fig. 9.4).

In view of the participation of electrostatic factors in catalysis by LOX, it was deemed of interest to compare the pI values of the proLOX and proLOXL species as well as those of their putative propeptides and processed enzymes. These data along with the corresponding molecular weights computed from the predicted sequences as well as the lysine content of each of these protein forms are shown in Table 9.1. It was assumed that each of the preproLOX and preproLOXL species is cleaved by signal peptidase and that each of the proLOX and proLOXL species is cleaved by BMP-1 at sites which are consistent with possible or previously demonstrated specificity of these two proteases. Predicted molecular weights derived exclusively from amino acid sequence data can differ from those seen in western blots of proLOX species by virtue of the fact that at least two N-glycosylation sites exist within the various proenzyme forms. The degree of N-glycosylation has not been established in all of the proLOX or proLOXL species. Moreover, the LTQ cofactor is, in reality, a cross-link between two different sequence regions and, therefore, is likely to prevent complete unfolding of LOX or LOXL enzymes. In turn, this factor as well as the degree of glycosylation would likely affect the observed protein molecular weight as determined from SDS-PAGE electrophoretograms, potentially affecting estimates made here of the BMP-1 cleavage sites based upon available molecular weights of cleaved species.

Considering the predicted pI values of the preproproteins, only preproLOX among the five enzyme species is a distinctly basic protein with a pI of 8.43. Interestingly, the basicity of this protein primarily derives from the strong basic character (pI 11.86) of the propeptide region of the full sequence. In contrast, the proteolytically derived, free enzyme domain is acidic with a pI of 5.84. As noted above, the sequence regions surrounding the lysine and tyrosine components giving rise to the LTQ cofactor are rich in dicarboxylic amino acids (Fig. 9.4). Given these complementary properties of the propeptide and catalytic domains, it seems possible that the cationic propeptide domain of proLOX may be attracted to the grouped anionic sites in the proLOX catalytic domain, thus blocking access of protein substrates to the active site in the intact proenzyme and prevention of LOX catalysis intracellularly and during secretion.

It has been reported that the propeptide domains of proLOX and proLOXL1 are required for deposition of these enzymes onto elastic fibers (Thomassin et al. 2005), based partly upon the findings that the products of full-length proLOX and pro-LOXL1 constructs localized to elastic fibers in cultured cells while the processed enzymes were secreted, but did not associate with the matrix. Ligand blot and mammalian two-hybrid assays also confirmed an interaction between tropoelastin and the pro-regions of both LOX and LOXL (Thomassin et al. 2005; Liu et al. 2004). Thus, the propeptide domains appear to be important elastin substrate recognition sites for LOX and LOXL1. It is of interest that rat and bovine tropoelastins, the soluble precursors of elastin fiber formation, are themselves highly cationic proteins with calculated pI values of 10.43 and 10.64, respectively. It is notable that rat tropoelastin contains no ionizable dicarboxylic amino acids (asp and glu), while the bovine protein contains only three aspartyl and two glutamyl residues within its 741 amino acid polypeptide chains. Moreover, grouping of proximal lysine residues in similar steric orientation relative to the tropoelastin backbone results in localized fields of positive charge density at sites of LOX oxidation (Brown-Augsburger et al. 1995), consistent with the demonstrated preference of LOX for global cationic features in its protein substrates. At least in the case of the highly cationic proLOX propeptide (pI 11.86), one might assume that charge–charge repulsion between tropoelastin and the propeptide would hinder binding of the propeptide to this substrate. A possible clue to this seemingly discrepant result is suggested by the studies demonstrating that proLOXL1 avidly binds to fibulin-5 in vitro and that each of these proteins colocalizes with the elastic matrix produced by vascular smooth muscle cells (VSMCs) (Liu et al. 2004; Choi et al. 2009). Fibulin-5 appears to play an important role as a component of the protein scaffold upon and in which elastogenesis occurs. Interestingly, fibulin-5 is a strong acidic protein (pI 4.58) and, therefore, might mediate the proximal relationship of cationic tropoelastin units and the cationic propeptides of proLOXL1 and proLOX within the elastogenic scaffold complexes (see Chaps. 8 and 10). The LOXL1 propeptide moiety (pI 8.1), which is considerably less basic than the LOX propeptide (pI 11.86), was abundantly present in fibulin-5−/− dermis, but was not immunologically detectable in wild-type dermis, suggesting that another role of fibulin-5 in elastogenesis may be to facilitate the proteolytic activation of proLOXL1 (Choi et al. 2009). It has also been suggested that LOX may be more involved in the oxidation and cross-linking of elastin and its precursor, while LOXL1 may have a more structural role in the development of previously oxidized and partially cross-linked insoluble elastin (Liu et al. 2004). Certainly, as oxidation and cross-linking of elastin proceeds, possibly initially by LOX, the basicity of the individual tropoelastin chains in the elastic fiber would decrease, potentially favoring interaction with LOXL1. It may also be relevant to note that fibulin-5 contains only 3 lysines per 448 total residues, which, coupled with its basicity, would make fibulin-5 an unlikely substrate of LOX.

Table 9.1 also notes the total lysine content of the propeptide and of the catalytic domains of the various LOX and LOXL species. Note that the fewest total lysines are found in the intact proproteins of LOX and LOXL1 and that the propeptide domains of only these two species are unique among the five family members in that they contain no lysine residues. In contrast, proLOXL2, -3, and -4 contain significantly greater quantities of lysine residues, with the bulk of these residues occurring within the propeptide domains of these three species. These properties agree with the assignment of the former pair as a genetic subfamily and of the other three LOXL enzymes as a second genetic subfamily (Csiszar 2001). Considering possible consequences of the presence or absence of lysine within the propeptide domains raises the speculation that nature may be attempting to avoid intramolecular, autocatalytic oxidation of lysine(s) within the propeptides of LOX and LOXL1 that could lead to the cross-linking of the propeptides to the catalytic domain, thus blocking access of substrates to the active site. The contrast between the lysine contents of LOX and LOXL1 versus those of LOXL2, -3, and -4 might also affect the substrate specificities of these two groups of amine oxidases. A similar possibility is raised by the report that the proteolytic activation of proLOX requires the interaction of the proenzyme with fibronectin, apparently mediated by the propeptide domain of proLOX (Fogelgren et al. 2005). Interestingly, fibronectin, like fibulin, is a strong basic protein (pI 5.29), and might then favorably accommodate the binding of the cationic propeptide domain of LOX.

The substrate specificities of the four LOXL enzymes have not been explored in much detail, although all are catalytically active. Recombinant forms of LOXL1 express BAPN-inhibitable amine oxidizing enzyme activity against benzylamine and an elastin substrate (Borel et al. 2001; Jung et al. 2003). Recombinant LOXL2 has been shown to oxidize a type I collagen substrate, although this activity was not inhibited by concentrations of BAPN that readily inhibit LOX, but was inhibited by a copper chelator (Vadasz et al. 2005). BAPN-inhibitable catalysis by recombinant LOXL3 has been noted against collagen types I, IV, VIII, X, and VI substrates (Lee and Kim 2006). LOXL4 displayed BAPN-inhibitable enzymatic activity against benzylamine (Kim et al. 2003).

9.5 Computer-Assisted Modeling of the Catalytic Domain of Human LOX

As noted, there is no crystallographic data presently available concerning the three-dimensional structure of LOX. In addition, there is no structural information as might be obtained from NMR analyses of this enzyme, while only limited EPR and

CD characterization has been possible. The absence of this information reflects both the difficulty in obtaining sufficient quantities of the purified enzyme and the tendency of concentrated preparations of LOX to form amorphous aggregates in urea-free solution. Hence, little is known about the location and role of copper in the catalytic mechanism, nor is information available concerning specific amino acid residues that might function in LOX catalysis and inhibition. In view of the critical importance of gaining some insight into the detailed structure of this enzyme, we have undertaken computer-based molecular modeling in an effort to progress toward these goals (Ryvkin et al. submitted). This method offers a powerful alternative to traditional methods for the elucidation of the three-dimensional structure of this enzyme and can contribute to our understanding of structural bases of those physical and chemical properties of LOX that have been reported, such as its stability at high temperature, retention of catalytic function in high concentrations of urea, and possibly elucidate aspects of its mechanism of its action.

Given the tendency of LOX to aggregate in aqueous solution, it is surprising that its hydrophilic side chain content is more than 60% greater than its hydrophobic content. We hypothesize that the limited solubility of LOX may reflect secondary and tertiary levels of its protein structure, resulting in its hydrophobic regions being exposed.

9.5.1 Model Construction and Refinement

The strategy for structure prediction took into consideration the following procedures: (1) sequence alignment and secondary structure prediction, (2) backbone construction, (3) loop structural determination and refinement, (4) LTQ and copper placement, and (5) solvation and energy minimization. All computer simulations were performed using Schrödinger, Inc. Molecular Modeling Software specifically employing the use of the Prime, Macromodel, and Impact packages. The refined structure of the model was subjected to a series of tests of its internal consistency and reliability. The proposed model of LOX, shown in Fig. 9.5a, had 88% of its amino acids with Ψ and Φ angles in allowed and generously allowed regions in the Ramachandran plot (data not shown), indicating reliable molecular geometry.

Referring to Fig. 9.5a, the overall structure of human LOX approximates that of a globular protein and contains six regions of beta chains (about 25%), 20% α-helix, and 65% random coil and is in good agreement with existing CD (Ryvkin and Greenaway 2004) and fluorescence data (Ryvkin et al. submitted), indicating that there is a high content of random coil and that its three tryptophan residues are at the surface and exposed to solvent. The model also predicts the presence of a "freely moving" random coil fragment at the N-terminal end of the structure projecting from the bottom as shown in Fig. 9.5a. The sequence of this N-terminal fragment consists of 35 residues enriched with tyrosine residues. In view of the abundance of hydrophobic tyrosine residues, it is conceivable that this tail might act as an initiator of LOX polymerization and consequent self-aggregation of this enzyme.

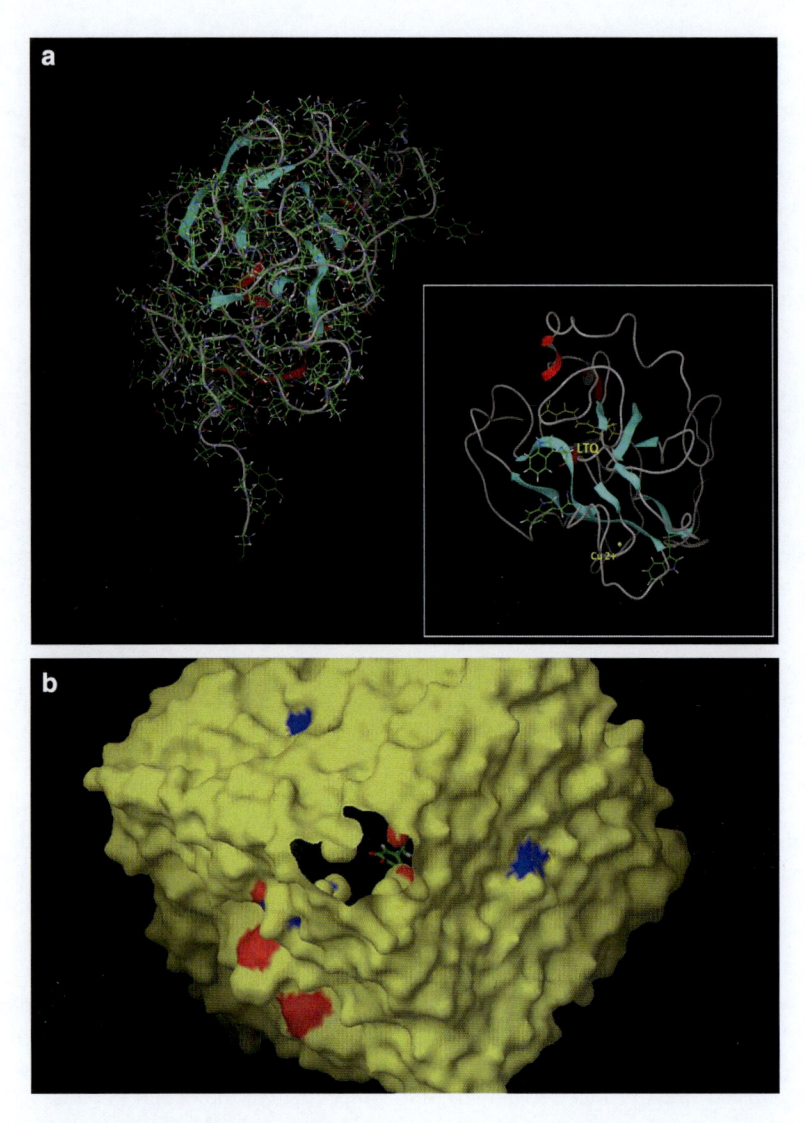

Fig. 9.5 (a) Computer generated model of human lysyl oxidase. *Inset*: model rotated to reveal three surface tryptophan residues and the bound copper ion at 20.019 Å distance from LTQ. (b) View of predicted structure of LOX showing LTQ within a solvent accessible pore of the enzyme and sites of anionic (*red*) and cationic (*blue*) charge density

Although these results must be considered with caution, they predict the copper ion cofactor to be at a distance of about 20 Å from the LTQ cofactor at the active site (Fig. 9.5a inset). This distance would be consistent with the report that the copper ion has a structural but not a catalytic role in LOX (Tang and Klinman 2001). However, the LTQ-copper distance in the predicted LOX structure contrasts with the expectation of proximity between these two cofactors as observed with other copper amine oxidases that contain tyrosine-derived trihydroxyphenylalanine quinone cofactors but which are not covalently linked to a lysine residue (Dubois and Klinman 2005; Cai and Klinman 1994; Ruggiero and Dooley 1999; Kim et al. 2002). The copper atom in those instances has been proposed to participate in the autocatalytic hydroxylation and oxidation of the tyrosine precursor of the TPQ cofactor as the proenzyme is being constructed in the intracellular compartment. This putative discrepancy would be resolved by the possibility that the copper in nascent proLOX is proximal to the developing LTQ during its synthesis and then reorients to the site seen in Fig. 9.5a (inset) due to a significant conformational change as the lysine residue is added to the tyrosyl quinone component of LTQ.

Consistent with evidence for the role of charge complementarity in the substrate specificity of this enzyme, negative charge density is also seen in the view looking down into the passageway between the surface of the enzyme and the LTQ (Fig. 9.5b). This view suggests that the peptidyl lysine side chain of a protein substrate projects into the active site pore, while its adjacent sequence interacts with the surface of the enzyme. The distribution of the prominent sites of anionic charge and the lesser sites of cationic charge density seen on the enzyme surface (Fig. 9.5b) could underlie the observation that the -Gly_n-Glu-Lys-Gly_n- sequence is the optimal substrate, while the -Gly_n-Lys-Glu-Gly_n- is the poorest substrate sequence among several peptides in which anionic residues are placed at varying orientations and at varying residue distances from the susceptible lysine residue (Nagan and Kagan 2000). The suggestion proposed in that study that such peptides bind to LOX in a preferred N- to C-terminal directional sense would be consistent with the influence of the orientation of anionic and cationic sites at the surface of the LOX molecule seen in Fig. 9.5b.

9.6 LOX Effects on Cell Phenotype

9.6.1 LOX and Oncogenesis

The early view that the biological function of LOX was restricted to its oxidation of lysine within its collagen and elastin substrates was dramatically altered with the publication in 1991 documenting that the cDNA sequence of rat LOX is 95+% homologous with that of a mouse ras-rescission gene previously shown to be expressed in NIH 3T3 cells, markedly reduced in carcinogenic RS485 cells derived from the 3T3 fibroblasts, and fully reexpressed upon reversion of the RS485 cells to the noncarcinogenic PR4 cells (Contente et al. 1990; Kenyon et al. 1991). The issue was raised, therefore, that LOX may act as an anti-oncogenic agent. Oncogenic ras mediates transformation in part through the activation of the transcription factor NF-kappa B. Indeed, the expression of LOX in ras-transformed NIH 3T3 cells led to decreased NF-kappa B binding and activity, decreased expression of the NF-kappa B target gene *c-myc*, and a dramatic decrease in colony formation. These phenomena apparently resulted from LOX-induced downregulation of the PI3K and Akt kinase components of signal transduction pathways through which ras mediates the induction of NF-kappa B (Jeay et al. 2003). Consistent with the oncogenic potential of LOX, LOX expression is decreased in ductal breast carcinoma cells (Peyrol et al. 1997), in prostate tumors (Ren et al. 1998), and in bronchogenic carcinoma (Woznick et al. 2005). LOX promoter methylation and loss of heterozygosity have been found in human gastric cancers (Kaneda et al. 2004).

Further insight into putative mechanisms whereby LOX suppresses tumor formation was made by the finding that BAPN, the irreversible inhibitor of LOX activity, was unable to block suramin-induced reversion of RS485 cells to the nontransformed phenotype, indicating that the catalytic activity of LOX may not underlie its antioncogenic potential. Surprisingly, these results were made explicable by the finding that the 18 kDa LOX propeptide and not the LOX enzyme domain functions to inhibit ras-dependent cell transformation (Palamakumbura et al. 2004). A further series of studies emphasized the significant effect of the LOX propeptide in cell biology. Thus, the LOX propeptide has been reported to inhibit smooth muscle cell signaling and proliferation (Hurtado et al. 2008), reverse the invasive phenotype of Her-2/neu-driven breast cancer (Min et al. 2007), attenuate fibronectin-mediated activation of focal adhesion kinase (FAK) and p130Cas in breast cancer cells (Zhao et al. 2009), inhibit prostate cancer cell growth by mechanisms that target FGF-2 cell binding and signaling (Palamakumbura et al. 2009), and repress expression of BCL2, thus inhibiting the transformed phenotype of lung and pancreatic cancer cells (Wu et al. 2007a; Min et al. 2009). It is of further interest that an SNP resulting in an Arg158Gln substitution in a highly conserved region within the proLOX propeptide occurs in several breast cancer cell lines examined. Min et al. (2009) have now found that the Arg-to-Gln substitution profoundly impairs the ability of the LOX propeptide to inhibit the invasive phenotype and tumor formation of NF639 cells in a xenograft model. Moreover, a potential association of the Gln-encoding A allele was seen with increased risk of estrogen receptor (ER)-alpha-negative invasive breast cancer in African American women (Min et al. 2009). The possibility that the LOX propeptide might exert its influence on cells at the intracellular level was raised by the observation that it was principally associated with the Golgi and endoplasmic reticulum, while mature LOX epitopes were found principally in the nucleus and perinuclear region. In differentiating cells, LOX propeptide and the mature LOX enzyme colocalized with the microtubule network (Guo et al. 2007). This follows upon evidence that the functional 32 kDa rhodamine-labeled catalyst is readily taken up by fibroblasts than to be concentrated within the nucleus (Nellaiappan et al. 2000), while the immunoreactive 32 kDa processed catalyst has been observed within the nuclei of VSMC and fibroblasts (Li et al. 1997). The reports that the purified enzyme readily oxidizes histone H1 (Kagan et al. 1983) and can bind to isolated forms of histones H1 and H2 (Giampuzzi et al. 2003) have led to the speculation that the mature enzyme might have an intranuclear function. Di Donato et al. (1997) have reported evidence that microinjected, intracellular LOX resulted in the blocking of oncogenic p21-Ha-Ras and of progesterone effects on *Xenopus laevis* oocyte maturation, although these effects were independent of enzyme activity (Di Donato et al. 1997) and seem consistent with the role of the propeptide as the regulating agent. It should be pointed out that propeptide-specific antibody developed against specific sequences within proLOX detected the 50 kDa proenzyme in both cytoplasmic and nuclear extracts of nontransformed mouse fibroblasts, although the free 18 kDa propeptide was not detectable under these conditions. A 30 kDa protein, assumed to be the processed LOX enzyme, was immunologically detected in both transformed and nontransformed mouse fibroblasts, although the cellular content of this protein was not affected by RNA interference directed against mature LOX suggesting that the 30 kDa species may be an unrelated protein that cross-reacts with the anti-LOX used in this instance (Contente et al. 2009).

The possibility that LOX may exert anti-oncogenic effects in specific cases by virtue of its enzymatic activity has been supported by the finding that bFGF is a productive substrate for this enzyme and, once oxidized by LOX, no longer induces a proliferative response of cells to this growth factor. IgBNM 6–1 cells overexpress bFGF which participates in an autocrine mechanism accounting for the transformation of these cells into a tumorigenic state. Exposure of these cells to nanomolar concentrations of LOX in culture oxidized lysine and generated cross-linkages in bFGF within the cell and markedly reduced proliferative rates (Li et al. 2003). It has also been reported that an FGF-2 autocrine pathway inhibits LOX transcription in the tumorigenic, transformed RS485 cell line (Palamakumbura et al. 2003).

9.6.2 LOX as a Chemokine

Among the experimental results that have revealed unexpected roles of LOX was the observation that nanomolar concentrations of purified bovine LOX induced significant directional cell migration of human peripheral blood monocytes, a response that was fully prevented by prior heat denaturation of the enzyme or by the presence of BAPN, thus implicating LOX catalysis and its native structure in the chemotactic response (Lazarus et al. 1995). Subsequent studies established that VSMCs are also chemotactically attracted to LOX and that VSMC migration was fully inhibited not only by BAPN but also by the presence of catalase, thereby pointing to the H_2O_2 product of LOX catalysis as a mediator of LOX-dependent chemotaxis. The chemotactic response appeared to require direct access between LOX and a substrate molecule (or molecules) tightly associated with the VSMC and was accompanied by LOX-dependent elevation of intracellular levels of H_2O_2, enhanced stress fiber, and focal adhesion assembly (Li et al. 2000). Lucero et al. (2008) have shown that, in addition to the intrinsic chemotactic property of exogenous LOX, endogenous LOX appears to "prime" VSMC for chemotactic attraction not only to LOX but also to PDGF. The affinity of VSMC for the binding of PDGF was significantly reduced by their prior incubation with BAPN, while the chemotactic response to exogenous LOX was markedly decreased in LOX−/− cells. The inhibition of LOX-induced chemotaxis by BAPN pointed to the possibility that cellular proteins involved in the chemotactic response might have been oxidized by LOX. Since the aldehyde function of the peptidyl AAS product of LOX catalysis covalently reacts with dinitrophenylhydrazine (DNPH), it was possible to identify such oxidized proteins by reaction with DNPH followed by probing of western blots with anti-DNP for DNP-proteins which had been resolved by SDS-PAGE. Using this technique, it was found that endogenous as well as exogenous LOX oxidized lysine residues within the PDGF receptor as well as other unspecified plasma membrane proteins of LOX competent cells, raising the possibility that such LOX-dependent oxidation of membrane proteins may contribute to enhanced receptor–chemokine interactions (Lucero et al. 2008). Levels of active LOX correlate with FAK/paxillin activation and migration of invasive astrocytes pointing toward a spectrum of cell types that may chemotactically respond to LOX (Laczko et al. 2007).

LOX-dependent chemotaxis has been seen to be involved in certain disease processes (Payne et al. 2007). The expression of LOX in breast cancer has been associated with tumor cell migration, estrogen receptor negative status, and reduced patient survival. Notably, LOX regulates breast cancer cell migration and adhesion through a hydrogen peroxide-mediated mechanism (Payne et al. 2005) consistent with the previous findings with VSMC of Li et al. (2000), and thus can play an active role in malignant spread of these transformed cells. Induction of migration by LOX involved the activation of the FAK/Src signaling complex leading to changes in actin filament polymerization via activation of the p130^{Cas}/Crk/DOCK180 signaling pathway (Payne et al. 2006). Similarly, inhibition of LOX expression or activity prevents in vitro migration of melanoma and pancreatic cancer cells (Kirschmann et al. 2002). It has also been established that tumor hypoxia is associated with increased invasion and metastasis (Cairns et al. 2003; Erler et al. 2006a; Sion and Figg 2006). A correlation between hypoxia and increased levels of LOX in lung tissue has been previously demonstrated (Brody et al. 1976). Consistent with this early finding, hypoxia markedly increased LOX protein expression leading to the transformation of poorly invasive breast cancer cells toward a more aggressive phenotype. This study further indicated that both hypoxia and reoxygenation are necessary for full expression of LOX catalytic activity, thus facilitating LOX-dependent, hydrogen peroxide-mediated breast cancer cell migration (Postovit et al. 2008). Sahlgren et al. (2008) have shown that Notch signaling is required to convert the hypoxic stimulus into epithelial–mesenchymal transition (EMT), increased motility, and invasiveness. Notch potentiated hypoxia-inducible factor 1alpha (HIF-1alpha) recruitment to the LOX promoter, leading to the hypoxia-induced upregulation of LOX. LOX catalytic activity and therefore BAPN-inhibitable hydrogen peroxide production was significantly reduced under hypoxic conditions (Postovit et al. 2008), consistent with its requirement for its oxygen substrate (Fig. 9.1). However, the increased levels of LOX protein generated under hypoxia might be expected to regain full catalytical competency to produce H_2O_2 upon tissue reoxygenation, consistent with the findings noted above. The interaction of LOX with cell membranes to initiate cell migration activates FAK/Src signal transduction components (Postovit et al. 2008; Li et al. 2000; Laczko et al. 2007), and, as has been shown, results in the oxidation by cell-bound LOX of peptidyl lysine residues of specific cell membrane proteins (Lucero et al. 2008). In view of the apparent requirement for the H_2O_2 product of LOX catalysis to mediate cell migration, it is possible that the production of this peroxide directly at the cell surface facilitates the uptake of H_2O_2 into the cells to activate specific signals regulating the chemotactic response to LOX. It is also possible that the LOX-catalyzed oxidation of cell membrane proteins, including, as has been shown, the PDGF receptor (Lucero et al. 2008), directly results in activation of those receptors that can regulate chemotaxis. LOX-induced cell migration presents a rational basis for chemotherapeutic control of malignancy as exemplified by the prevention or inhibition of LOX-induced cell migration by BAPN, LOX antisense oligonucleotides, and by LOX shRNA (Erler and Giaccia 2006b).

The LOX 3′ UTR contains a binding site for a single miRNA, mir-145, which is downregulated in many cancers and has been reported to be deleted in prostate cancers (Iorio et al. 2005). Low mir-145 is also part of a poor prognosis signature in lung cancer (Yanaihara et al. 2006). The LOX gene is also a target for the anti-oncogenic transcription factor IRF-1, which contributes to the process of malignant transformation (Tan et al. 1996). Examination of the role of LOX in human basal and squamous cell carcinomas revealed the lack of LOX expression in epidermal tumor cells, consistent with other reports documenting the lack of LOX expression in transformed cells (Kuivaniemi et al. 1986; Csiszar et al. 2002; Hämäläinen et al. 1995). Interestingly, LOX was upregulated in association with the stromal reaction surrounding invading tumor cells (Bouez et al. 2006). The presence of LOX in normal skin cells surrounding the tumor tissue raises the possibility that this exogenous LOX source could have been inducing the migratory response of the tumor cells.

9.6.3 LOX and Fibrosis

There is abundant evidence for the upregulation of LOX protein and activity in a variety of fibrotic diseases (Smith-Mungo and Kagan 1998; Kagan 1986). In view of the critical role of LOX in stabilizing collagen fibers in such disease states, there has been interest in the use of LOX-specific inhibitors as potential anti-fibrotic agents. Varied pathologies of connective tissues result from the inhibition of LOX activity, notably by BAPN administered to animal models especially during developmental stages of growth, as reviewed (Kagan 1986; Smith-Mungo and Kagan 1998; Rodríguez et al. 2008), and by the deletion of the LOX gene (Mäki et al. 2002). Characteristically, skeletal, cardiovascular, pulmonary, and dermal deformities result from the lack of LOX oxidation of peptidyl lysine to the peptidyl aldehyde precursor of the stabilizing cross-linkages in collagen and elastin. Noting that increased LOX expression has been correlated with increased metastasis and decreased survival in breast cancer patients, Bondareva et al. (2009) found that administration of BAPN significantly reduced the metastatic colonization potential of the human breast cancer cell line, MDA-MB-231 mice, consistent with the established role of LOX activity in migration of breast cancer cells (Payne et al. 2005, 2006).

While increased LOX activity can have adverse effects as seen in fibrosis and breast cancer, LOX may also exert a potentially beneficial role in specific disease states (Ovchinnikova et al. 2009; Fernández-Hernando et al. 2009). For example, the early stages of atherosclerotic plaque formation include the deposition of new collagen and elastin fibers in the developing arterial lesions, stabilized by the increased levels of LOX as seen in arterial plaques of rabbits rendered atherosclerotic by high lipid diet (Kagan et al. 1981). However, LOX-dependent cross-linking and stabilizing of collagen in fibrous caps overlying atherosclerotic lesions would be expected to limit aneurysmatic disruption of the affected artery at lesion sites. Similarly, inhibition of LOX activity may be beneficial by, for example, limiting the development of left ventricular stiffness in failing hearts due to excess collagen deposition at myocardial lesions (López et al. 2009). In view of these contrasting possibilities, the use of chemical inhibitors of LOX enzyme activity is not a simple matter with potentially untoward or beneficial effects on health. Inhibition of LOX activity should not affect those biological responses dependent exclusively upon the noncatalytic LOX propeptide. There is at least one example of the modulation of LOX activity by a chemotherapeutic agent, which is not directed at LOX itself. Thus, pharmacological concentrations of statins (atorvastatin and simvastatin) modulated LOX transcriptional activity, counteracting the downregulation of LOX at the mRNA, protein, and activity levels caused by tumor necrosis factor-alpha in porcine, bovine, and human aortic endothelial cells. Statins also counteracted the decrease in LOX expression produced by atherogenic concentrations of LDL to partially prevent the increase in endothelial permeability elicited by this lipoprotein (Rodríguez et al. 2009). An additional relationship between LOX and vascular endothelium was supported by the finding that pathophysiological concentrations (35 μM) of homocysteine inhibited LOX activity in porcine aortic endothelial cells, while higher concentrations (250 μM) inhibited the expression of LOX protein. Raposo et al. (2004) attributed the inhibition of LOX activity to a free radical mechanism and suggested that LOX inhibition contributes to endothelial dysfunction which is associated with hyperhomocysteinemia. Other studies have shown that homocysteine thiolactone, derived spontaneously from homocysteine, is a potent irreversible inhibitor of LOX activity (Liu et al. 1997).

9.7 LOXL Enzymes in Disease and Homeostasis

9.7.1 *LOXL1*

As noted, the sequence of LOXL1 is the most similar to that of LOX among the four LOXL species especially in their C-terminal catalytic domains, although significant sequence differences exist between these two enzymes within their propeptide domains. There appears to be specific relationships of individual members of these five LOX and LOXL amine oxidases with specific disease states. For example, selected mutants of LOXL1 appear to predispose affected individuals to exfoliation glaucoma in which abnormal deposits are found on the surface of the lens and other structures within the eye. This results in insufficient recirculation of the fluid of the eye and increased intraocular pressure (Jonasson 2009). Decreased levels of LOXL1 have been implicated in the development of abnormal elastic fibers appearing in venous insufficiency (Pascual et al. 2008). Lower urogenital tract anatomical and functional phenotype in LOXL1 knockout mice resembles female pelvic floor dysfunction in humans (Lee et al. 2008). LOXL1 and LOXL4 are epigenetically silenced and can inhibit ras/extracellular signal-regulated kinase signaling pathway in human bladder cancer (Wu et al. 2007b).

9.7.2 LOXL2

As previously noted, hypoxia enhances LOX expression, thus enhancing the opportunity for LOX-dependent migration of malignant cells from localized tumors. A variety of reports have shed light on possible molecular bases of this relationship of LOX to hypoxia and to the EMT seen in carcinogenesis. The invasive and metastatic phenotype is associated with downregulation of E-cadherin, a cell adhesion molecule with anti-invasive properties in numerous epithelial-derived cancers, while both LOXL2 and LOXL3 have been shown to physically interact with Snail, an important repressor of E-cadherin. Peinado et al. (2005) have demonstrated that the interaction of LOXL2 with Snail stabilizes this transcription factor by counteracting the protein phosphorylating action of glycogen synthase kinase-3, thus leading to E-cadherin repression and EMT.

There is evidence that HIFα, a growth promoter for cancer cells and which is stimulated in hypoxia, and/or von Hippel Landau factor, a tumor suppressor that targets HIFα for degradation, can regulate E-cadherin expression, the former negatively, thus favoring EMT, and the latter indirectly positively. As previously noted, HIFα has been shown to upregulate expression of the LOX gene (Erler and Giaccia 2006) and, as recently reported, the LOXL2 gene is also upregulated as a direct target of HIF-1 (Schietke et al 2010). Deficient levels of VHL are associated with multiple tumor type diseases including retinal and central nervous system hemangioblastomas, pheochromocytoma, and clear-cell renal cell carcinoma. Recent studies have shown that E-cadherin expression can depend on the VHL status of cells, where hypoxic incubation was able to suppress E-cadherin protein expression (Esteban et al. 2006; Evans et al. 2007; Russell and Ohh 2007). These data indicate that VHL and/or HIFα are capable of regulating E-cadherin and, thereby, potentially influencing the process of EMT. It is of further interest that VHL exerts inhibitory effects on the invasive and migratory capacity of patient-derived human breast cancer cells in vitro, with the lowest levels of VHL expression occurring in the most aggressive breast tumors (Zia et al. 2007). Thus, malignant breast cancer is accompanied by decreased expression of VHL, increased expression of HIF-1 (van der Groep et al. 2008), and HIF-1-assisted induction of LOXL2 (Higgins et al. 2007) leading to LOXL2-dependent migration of breast cancer cells (Hollosi et al. 2009).

LOXL2 has been implicated in the initiation and/or progression of other pathologic conditions in addition to breast cancer, as recently summarized (Fong et al. 2009). LOXL2 has been observed in hepatocytes from patients with Wilson's disease or primary biliary cirrhosis (Vadasz et al. 2005), renal tubulointerstitial fibrosis associated with diabetic and IgA nephropathies, and hypertensive nephrosclerosis (Higgins et al. 2007). The expression of LOXL2 mRNA was reduced in human pelvic organ prolapse (Klutke et al. 2008) and is increased in intracranial aneurysms (Akagawa et al. 2007).

9.7.3 LOXL3 and LOXL4

The properties and functions of LOXL3 (Szauter and Csiszar 2008) and LOXL4 (Szauter et al. 2007) have been recently reviewed. LOXL3 was expressed in highly invasive but not in poorly invasive and nonmetastatic breast cancer cell lines (Hollosi et al. 2009) and, as noted, participates as does LOXL2 in the downregulation of cadherin with potentially consequent effects on metastasis (Peinado 2005). Sebban et al. (2009) have reported the finding of two new alternative splice variants of LOXL4, one of which was associated with ovarian carcinoma while the other was elevated in breast carcinoma. The specific roles of these enzymes remain to be determined. LOXL4 has also been found to be the only member of the LOX family whose expression is induced by TGF-beta1 in PLC/PRF/5 hepatoma cells. Moreover, expression of LOXL4 in these cells resulted in the inhibition of cell motility in the presence of TGF-beta1 and suppressed the expression of laminins and alpha3 integrin as well as the activity of MMP2. The authors suggest that LOXL4 may function as a negative feedback regulator of TGF-beta1 in cell invasion by inhibiting the metabolism of ECM components (Kim et al. 2008). LOXL4 is overexpressed in head and neck squamous cell carcinoma (HNSCC) compared with normal squamous epithelium (Holtmeier et al. 2003). The degree of expression was significantly correlated with local lymph node metastases. These findings and related data point to LOXL4 expression as a distinctive trait of HNSCC and suggest that it plays a functional role in the pathogenesis of this disease (Görögh et al. 2007).

9.8 Summary and Prospects

It is clear that investigative interest in LOX and its amine oxidase isomers has increased considerably in recent times, stimulated by the new evidence for the surprising and multiple roles of these enzymes in health and disease. The participation of LOX in various types and stages of carcinogenesis seems well supported, especially in the case of breast cancer, while changes in levels of expression of individual LOXL species in cancerous tissue and cells indicate that these enzymes as well as LOX may prove to be desirable therapeutic targets in efforts to stem tumor growth and malignant spread of transformed cells. At least in the case of LOX, it is evident that its effects on cellular homeostasis may derive from the catalytic function of the processed enzyme and/or, surprisingly, from the propeptide released from proLOX during maturation of the proenzyme. In turn, the catalytic function of LOX may contribute to the observed changes in cell phenotypes by oxidation of cell membrane proteins including specific receptors and/or intracellular protein targets and/or inactivation by oxidation of exocrine- or autocrine-derived growth factors and/or by stimulating cell migration. These possibilities require further investigation of function, and especially in those instances in which specific members of this enzyme family are associated with alterations in cell phenotype.

The highly conserved sequences of the active site regions of the LOXL species and the available demonstrations in vitro of their catalytic potential make a strong argument for the need for detailed studies of their substrate specificities and for possible cellular protein substrates. The use of the dinitrophenylhydrazine probe (Lucero et al. 2008) should prove to be useful in this regard when coupled with immunodetection of tissue, cellular, and subcellular sites at which the activity of these enzymes might be expressed as well as for the identification of specific protein substrates of these enzymes. Such efforts must include controls preventing the expression of the enzyme protein and/or enzyme activity. Efforts to identify which of the presently known five members of the LOX family are relevant in each case would also be critical in such studies. Several reports noting the association of LOX or specific LOXL enzymes with specific disease states focused on the control of expression of these enzymes quantified by measurements of mRNA species but not of the LOX protein levels. It is now evident that more complete understanding of molecular bases of these effects must differentiate between the contribution of the free propeptide moieties and that of the catalytic domain in the effects seen. Since the relative positions of the lysine and tyrosine precursors of LTQ in LOX are highly homologous within the LOXL species, the reasonable assumption has been made, although not chemically proven, that this unusual cofactor is present and underlies the catalytic ability of the LOXL enzymes as well as LOX. As noted, the issues of whether proteolytic processing of the proenzyme forms of selected LOXL species occurs and, if so, the identification of cleavage sites and molecular weights of the processed enzymes are yet to be definitively resolved in specific cases.

The predicted structure of the catalytic domain of LOX presented here will obviously have to await analyses obtained by direct physical–chemical approaches for full verification. Nevertheless, the present results point to the possible role of charged residues in the expression of LOX catalysis. Moreover, the prominently hydrophobic surface of this enzyme appearing in the predicted model of its structure offers insight into the possible basis of the tendency of LOX to undergo intermolecular amorphous aggregation that strongly hinders efforts at crystallization of this enzyme. Clearly, future studies may lead to the solution of this problem as well as to the revelation of additional roles that these unusual enzymes may play in health and disease.

References

Akagawa H, Narita A, Yamada H, Tajima A, Krischek B, Kasuya H, Hori T, Kubota M, Saeki N, Hata A, Mizutani T, Inoue I (2007) Systematic screening of lysyl oxidase-like (LOXL) family genes demonstrates that LOXL2 is a susceptibility gene to intracranial aneurysms. Hum Genet 121:377–387

Akiri G, Sabo E, Dafni H, Vadasz Z, Kartvelishvily Y, Gan N, Kessler O, Cohen T, Resnick M, Neeman M, Neufeld G (2003) Lysyl oxidase-related protein-1 promotes tumor fibrosis and tumor progression in vivo. Cancer Res 63:1657–1666

Barrow MV, Steffek AJ (1974) Teratologic and other embryotoxic effects of beta-aminopropionitrile in rats. Teratology 10:165–172

Bollinger JA, Brown DE, Dooley DM (2005) The formation of lysine tyrosylquinone (LTQ) is a self-processing reaction. Expression and characterization of a *Drosophila* lysyl oxidase. Biochemistry 44:11708–11714

Bondareva A, Downey CM, Ayres F, Liu W, Boyd SK, Hallgrimsson B, Jirik FR (2009) The lysyl oxidase inhibitor, beta-aminopropionitrile, diminishes the metastatic colonization potential of circulating breast cancer cells. PLoS ONE 4:e5620

Borel A, Eichenberger D, Farjanel J, Kessler E, Gleyzal C, Hulmes DJ, Sommer P, Font B (2001) Lysyl oxidase-like protein from bovine aorta. Isolation and maturation to an active form by bone morphogenetic protein-1. J Biol Chem 276:48944–48949

Bouez C, Reynaud C, Noblesse E, Thépot A, Gleyzal C, Kanitakis J, Perrier E, Damour O, Sommer P (2006) The lysyl oxidase LOX is absent in basal and squamous cell carcinomas and its knockdown induces an invading phenotype in a skin equivalent model. Clin Cancer Res 12:1463–1469

Brody JS, Kagan HM, Manalo AD, Hu CA, Franzblau C (1976) Lung lysyl oxidase and elastin synthesis during compensatory lung growth. Chest 69:271–272

Brown-Augsburger P, Tisdale C, Broekelmann T, Sloan C, Mecham RP (1995) Identification of an elastin cross-linking domain that joins three peptide chains. Possible role in nucleated assembly. J Biol Chem 270:17778–177783

Cai D, Klinman JP (1994) Evidence of a self-catalytic mechanism of 2,4,5-trihydroxy-phenylalanine quinone biogenesis in yeast copper amine oxidase. J Biol Chem 269:32039–32042

Cairns RA, Khokha R, Hill RP (2003) Molecular mechanisms of tumor invasion and metastasis: an integrated view. Curr Mol Med 3:659–671

Choi J, Bergdahl A, Zheng Q, Starcher B, Yanagisawa H, Davis EC (2009) Analysis of dermal elastic fibers in the absence of fibulin-5 reveals potential roles for fibulin-5 in elastic fiber assembly. Matrix Biol 28:211–220

Contente S, Kenyon K, Rimoldi D, Friedman RM (1990) Expression of gene rrg is associated with reversion of NIH 3T3 transformed by LTR-c-H-ras. Science 249:796–798

Contente S, Yeh TJ, Friedman RM (2009) Tumor suppressive effect of lysyl oxidase proenzyme. Biochim Biophys Acta 1793:1272–1278

Cronshaw AD, Fothergill-Gilmore LA, Hulmes DJ (1995) The proteolytic processing site of the precursor of lysyl oxidase. Biochem J 306:279–284

Csiszar K (2001) Lysyl oxidases: a novel multifunctional amine oxidase family. Prog Nucleic Acid Res Mol Biol 70:1–32

Csiszar K, Fong SF, Ujfalusi A, Krawetz SA, Salvati EP, Mackenzie JW, Boyd CD (2002) Somatic mutations of the lysyl oxidase gene on chromosome 5q23.1 in colorectal tumors. Int J Cancer 97:636–642

Di Donato A, Lacal JC, Di Duca M, Giampuzzi M, Ghiggeri G, Gusmana R (1997) Microinjection of recombinant lysyl oxidase blocks oncogenic p21-Ha-Ras and progesterone effects on *Xenopus laevis* oocyte maturation. FEBS Lett 419:63–68

Dubois JL, Klinman JP (2005) Mechanism of post-translational quinone formation in copper amine oxidases and its relationship to the catalytic turnover. Arch Biochem Biophys 433:255–265

Erler JT, Bennewith KL, Nicolau M, Dornhöfer N, Kong C, Le QT, Chi JT, Jeffrey SS, Giaccia AJ (2006a) Lysyl oxidase is essential for hypoxia-induced metastasis. Nature 440:1222–1226

Erler JT, Giaccia AJ (2006b) Lysyl oxidase mediates hypoxic control of metastasis. Cancer Res 66:10238–10241

Esteban MA, Tran MG, Harten SK, Hill P, Castellanos MC, Chandra A, Raval R, O'Brien TS, Maxwell PH (2006) Regulation of E-cadherin expression by VHL and hypoxia-inducible factor. Cancer Res 66:3567–3575

Evans AJ, Russell RC, Roche O, Burry TN, Fish JE, Chow VW, Kim WY, Saravanan A, Maynard MA, Gervais ML, Sufan RI, Roberts AM, Wilson LA, Betten M, Vandewalle C, Berx G, Marsden PA, Irwin MS, Teh BT, Jewett MA, Ohh M (2007) VHL promotes E2 box-dependent E-cadherin transcription by HIF-mediated regulation of SIP1 and snail. Mol Cell Biol 27:157–169

Fernández-Hernando C, József L, Jenkins D, Di Lorenzo A, Sessa WC (2009) Absence of Akt1 reduces vascular smooth muscle cell migration and survival and induces features of plaque vulnerability and cardiac dysfunction during atherosclerosis. Arterioscler Thromb Vasc Biol 29:2033–2040

Fogelgren B, Polgár N, Szauter KM, Ujfaludi Z, Laczkó R, Fong KS, Csiszar K (2005) Cellular fibronectin binds to lysyl oxidase with high affinity and is critical for its proteolytic activation. J Biol Chem 280:24690–24697

Fong SF, Dietzsch E, Fong KS, Hollosi P, Asuncion L, He Q, Parker MI, Csiszar K (2007) Lysyl oxidase-like 2 expression is increased in colon and esophageal tumors and associated with less differentiated colon tumors. Genes Chromosom Cancer 46:644–655

Fong SFT, Fong KSK, Csiszar K (2009) LOXL2 (lysyl oxidase-like 2). Atlas Genet Cytogenet Oncol Haematol. http://AtlasGeneticsOncology.org/Genes/LOXL2ID41192ch8p21.html

Gacheru SN, Trackman PC, Shah MA, O'Gara CY, Spacciapoli P, Greenaway FT, Kagan HM (1990) Structural and catalytic properties of copper in lysyl oxidase. J Biol Chem 265:19022–19027

Giampuzzi M, Oleggini R, Di Donato A (2003) Demonstration of in vitro interaction between tumor suppressor lysyl oxidase and histones H1 and H2: definition of the regions involved. Biochim Biophys Acta 1647:245–251

Görögh T, Weise JB, Holtmeier C, Rudolph P, Hedderich J, Gottschlich S, Hoffmann M, Ambrosch P, Csiszar K (2007) Selective upregulation and amplification of the lysyl oxidase like-4 (LOXL4) gene in head and neck squamous cell carcinoma. J Pathol 212:74–82

Guo Y, Pischon N, Palamakumbura AH, Trackman PC (2007) Intracellular distribution of the lysyl oxidase propeptide in osteoblastic cells. Am J Physiol Cell Physiol 292:C2095–C2102

Hämäläinen ER, Kemppainen R, Pihlajaniemi T, Kivirikko KI (1993) Structure of the human lysyl oxidase gene. Genomics 17:544–548

Hämäläinen ER, Kemppainen R, Kuivaniemi H, Tromp G, Vaheri A, Pihlajaniemi T, Kivirikko KI (1995) Quantitative polymerase chain reaction of lysyl oxidase mRNA in malignantly transformed human cell lines demonstrates that their low lysyl oxidase activity is due to low quantities of its mRNA and low levels of transcription of the respective gene. J Biol Chem 270:21590–21593

Higgins DF, Kimura K, Bernhardt WM, Shrimanker N, Akai Y, Hohenstein B, Johnson SY, RS KM, Cohen CD, Eckardt KU, Iwano M, Haase VH (2007) Hypoxia promotes fibrogenesis in vivo via HIF-1 stimulation of epithelial-to-mesenchymal transition. J Clin Invest 117:3810–3820

Hohenester E, Sasaki T, Timpl R (1999) Crystal structure of a scavenger receptor cysteine-rich domain sheds light on an ancient superfamily. Nat Struct Biol 6:228–232

Hollosi P, Yakushiji JK, Fong KS, Csiszar K, Fong SF (2009) Lysyl oxidase-like 2 promotes migration in noninvasive breast cancer cells but not in normal breast epithelial cells. Int J Cancer 15:318–327

Holtmeier C, Görögh T, Beier U, Meyer J, Hoffmann M, Gottschlich S, Heidorn K, Ambrosch P, Maune S (2003) Overexpression of a novel lysyl oxidase-like gene in human head and neck squamous cell carcinomas. Anticancer Res 23:2585–2591

Hornstra IK, Birge S, Starcher B, Bailey AJ, Mecham RP, Shapiro SD (2003) Lysyl oxidase is required for vascular and diaphragmatic development in mice. J Biol Chem 278: 14387–14393

Hurtado PA, Vora S, Sume SS, Yang D, St. Hilaire C, Guo Y, Palamakumbura AH, Schreiber BM, Ravid K, Trackman PC (2008) Lysyl oxidase propeptide inhibits smooth muscle cell signaling and proliferation. Biochem Biophys Res Commun 366:156–161

Iorio MV, Ferracin M, Liu CG, Veronese A, Spizzo R, Sabbioni S, Magri E, Pedriali M, Fabbri M, Campiglio M, Ménard S, Palazzo JP, Rosenberg A, Musiani P, Volinia S, Nenci I, Calin GA, Querzoli P, Negrini M, Croce CM (2005) MicroRNA gene expression deregulation in human breast cancer. Cancer Res 65:7065–7070

Jeay S, Pianetti S, Kagan HM, Sonenshein GE (2003) Lysyl oxidase inhibits ras-mediated transformation by preventing activation of NF-kappa B. Mol Cell Biol 23:2251–2263

Jonasson F (2009) From epidemiology to lysyl oxidase like one (LOXL1) polymorphisms discovery: phenotyping and genotyping exfoliation syndrome and exfoliation glaucoma in Iceland. Acta Ophthalmol 87:478–487

Jourdan-Le Saux C, Tronecker H, Bogic L, Bryant-Greenwood GD, Boyd CD, Csiszar K (1999) The LOXL2 gene encodes a new lysyl oxidase-like protein and is expressed at high levels in reproductive tissues. J Biol Chem 274:12939–12944

Jourdan-Le Saux C, Tomsche A, Ujfalusi A, Jia L, Csiszar K (2001) Central nervous system, uterus, heart, and leukocyte expression of the LOXL3 gene, encoding a novel lysyl oxidase-like protein. Genomics 74:211–218

Kagan HM (1986) Characterization and regulation of lysyl oxidase. In: Mecham RP (ed) Biology of the extracellular matrix, vol 1, Regulation of matrix accumulation. Academic, Orlando, FL, pp 321–398

Kagan HM, Li W (2003) Lysyl oxidase: properties, specificity, and biological roles inside and outside of the cell. J Cell Biochem 88:660–672

Kagan HM, Hewitt NA, Salcedo LL, Franzblau C (1974) Catalytic activity of aortic lysyl oxidase in an insoluble enzyme–substrate complex. Biochim Biophys Acta 365:223–234

Kagan HM, Sullivan KA, Olsson TA 3rd, Cronlund AL (1979) Purification and properties of four species of lysyl oxidase from bovine aorta. Biochem J 177:203–214

Kagan HM, Raghavan J, Hollander W (1981) Changes in aortic lysyl oxidase activity in diet-induced atherosclerosis in the rabbit. Arteriosclerosis 1:287–291

Kagan HM, Williams MA, Calaman SD, Berkowitz EM (1983) Histone H1 is a substrate for lysyl oxidase and contains endogenous sodium borotritide-reducible residues. Biochem Biophys Res Commun 115:186–192

Kagan HM, Williams MA, Williamson PR, Anderson JM (1984) Influence of sequence and charge on the specificity of lysyl oxidase toward protein and synthetic peptide substrates. J Biol Chem 259:11203–11207

Kaneda A, Wakazono K, Tsukamoto T, Watanabe N, Yagi Y, Tatematsu M, Kaminishi M, Sugimura T, Ushijima T (2004) Lysyl oxidase is a tumor suppressor gene inactivated by methylation and loss of heterozygosity in human gastric cancers. Cancer Res 64:6410–6415

Kenyon K, Contente S, Trackman PC, Tang J, Kagan HM, Friedman RM (1991) Lysyl oxidase and rrg messenger RNA. Science 253:802

Kenyon K, Modi WS, Contente S, Friedman RM (1993) A novel human cDNA with a predicted protein similar to lysyl oxidase maps to chromosome 15q24–q25. J Biol Chem 268: 18435–18437

Kessler E, Fichard A, Chanut-Delalande H, Brusel M, Ruggiero F (2001) Bone morphogenetic protein-1 (BMP-1) mediates C-terminal processing of procollagen V homotrimer. J Biol Chem 276:27051–27057

Kim M, Okajima T, Kishishita S, Yoshimura M, Kawamori A, Tanizawa K, Yamaguchi H (2002) X-ray snapshots of quinone cofactor biogenesis in bacterial copper amine oxidase. Nat Struct Biol 9:591–596

Kim MS, Kim SS, Jung ST, Park JY, Yoo HW, Ko J, Csiszar K, Choi SY, Kim Y (2003) Expression and purification of enzymatically active forms of the human lysyl oxidase-like protein 4. J Biol Chem 278:52071–52074

Kim DJ, Lee DC, Yang SJ, Lee JJ, Bae EM, Kim DM, Min SH, Kim SJ, Kang DC, Sang BC, Myung PK, Park KC, Yeom YI (2008) Lysyl oxidase like 4, a novel target gene of TGF-beta1 signaling, can negatively regulate TGF-beta1-induced cell motility in PLC/PRF/5 hepatoma cells. Biochem Biophys Res Commun 373:521–527

Kirschmann DA, Seftor EA, Fong SF, Nieva DR, Sullivan CM, Edwards EM, Sommer P, Csiszar K, Hendrix MJ (2002) A molecular role for lysyl oxidase in breast cancer invasion. Cancer Res 62:4478–4483

Klutke J, Ji Q, Campeau J, Starcher B, Felix JC, Stanczyk FZ, Klutke C (2008) Decreased endopelvic fascia elastin content in uterine prolapse. Acta Obstet Gynecol Scand 87:111–115

Kosonen T, Uriu-Hare JY, Clegg MS, Keen CL, Rucker RB (1997) Incorporation of copper into lysyl oxidase. Biochem J 327:283–289

Krieger M, Herz J (1994) Structures and functions of multiligand lipoprotein receptors: macrophage scavenger receptors and LDL receptor-related protein (LRP). Annu Rev Biochem 63:601–637

Kuivaniemi H, Korhonen RM, Vaheri A, Kivirikko KI (1986) Deficient production of lysyl oxidase in cultures of malignantly transformed human cells. FEBS Lett 195:261–264

Laczko R, Szauter KM, Jansen MK, Hollosi P, Muranyi M, Molnar J, Fong KS, Hinek A, Csiszar K (2007) Active lysyl oxidase (LOX) correlates with focal adhesion kinase (FAK)/paxillin activation and migration in invasive astrocytes. Neuropathol Appl Neurobiol 33:631–643

Lazarus HM, Cruikshank WW, Narasimhan N, Kagan HM, Center DM (1995) Induction of human monocyte motility by lysyl oxidase. Matrix Biol 14:727–731

Lee J-E, Kim Y (2006) A tissue specific variant of the human lysyl oxidase-like protein 3 (LOXL3) functions as an amine oxidase with substrate specificity. J Biol Chem 281:37282–37290

Lee UJ, Gustilo-Ashby AM, Daneshgari F, Kuang M, Vurbic D, Lin DL, Flask CA, Li T, Damaser MS (2008) Lower urogenital tract anatomical and functional phenotype in lysyl oxidase like-1 knockout mice resembles female pelvic floor dysfunction in humans. Am J Physiol Renal Physiol 295:F545–555

Lemberg MK, Martoglio B (2002) Requirements for signal peptide peptidase-catalyzed intramembrane proteolysis. Mol Cell 10:735–744

Li W, Nellaiappan K, Strassmaier T, Graham L, Thomas KM, Kagan HM (1997) Localization and activity of lysyl oxidase within nuclei of fibrogenic cells. Proc Natl Acad Sci USA 94:12817–12822

Li W, Liu G, Chou IN, Kagan HM (2000) Hydrogen peroxide-mediated, lysyl oxidase-dependent chemotaxis of vascular smooth muscle cells. J Cell Biochem 78:550–557

Li W, Nugent MA, Zhao Y, Chau AN, Li SJ, Chou IN, Liu G, Kagan HM (2003) Lysyl oxidase oxidizes basic fibroblast growth factor and inactivates its mitogenic potential. J Cell Biochem 88:152–164

Liu G, Neillaiappan K, Kagan HM (1997) Irreversible inhibition of lysyl oxidase by homocysteine thiolactone and its selenium and oxygen analogues: implications for homocystinuria. J Biol Chem 272:32370–32377

Liu X, Zhao Y, Gao J, Pawlyk B, Starcher B, Spencer JA, Yanagisawa H, Zuo J, Li T (2004) Elastic fiber homeostasis requires lysyl oxidase-like 1 protein. Nat Genet 36:178–182

López B, Querejeta R, González A, Beaumont J, Larman M, Díez J (2009) Impact of treatment on myocardial lysyl oxidase expression and collagen cross-linking in patients with heart failure. Hypertension 53:236–242

Lucero HA, Kagan HM (2006) Lysyl oxidase: an oxidative enzyme and effector of cell function. Cell Mol Life Sci 63:2304–2316

Lucero HA, Ravid K, Grimsby JL, Rich CB, DiCamillo SJ, Mäki JM, Myllyharju J, Kagan HM (2008) Lysyl oxidase oxidizes cell membrane proteins and enhances the chemotactic response of vascular smooth muscle cells. J Biol Chem 283:24103–24117

Mäki JM (2009) Lysyl oxidases in mammalian development and certain pathological conditions. Histol Histopathol 24:651–660

Mäki J, Kivirikko KI (2001) Cloning and characterization of a fourth human lysyl oxidase isoenzyme. Biochem J 355:381–387

Mäki JM, Tikkanen H, Kivirikko KI (2001) Cloning and characterization of a fifth human lysyl oxidase isoenzyme: the third member of the lysyl oxidase-related subfamily with four scavenger receptor cysteine-rich domains. Matrix Biol 20:493–496

Mäki JM, Räsänen J, Tikkanen H, Sormunen R, Mäkikallio K, Kivirikko KI, Soininen R (2002) Inactivation of the lysyl oxidase gene Lox leads to aortic aneurysms, cardiovascular dysfunction, and perinatal death in mice. Circulation 106:2503–2509

Min C, Kirsch KH, Zhao Y, Jeay S, Palamakumbura AH, Trackman PC, Sonenshein GE (2007) The tumor suppressor activity of the lysyl oxidase propeptide reverses the invasive phenotype of Her-2/neu-driven breast cancer. Cancer Res 67:1105–1112

Min C, Yu Z, Kirsch KH, Zhao Y, Vora SR, Trackman PC, Spicer DB, Rosenberg L, Palmer JR, Sonenshein GE (2009) A loss-of-function polymorphism in the propeptide domain of the LOX gene and breast cancer. Cancer Res 69:6685–6693

Molnar J, Fong KS, He QP, Hayashi K, Kim Y, Fong SF, Fogelgren B, Szauter KM, Mink M, Csiszar K (2003) Structural and functional diversity of lysyl oxidase and the LOX-like proteins. Biochim Biophys Acta 1647:220–224

Mure M (2004) Tyrosine-derived quinone cofactors. Acc Chem Res 37:131–139

Nagan N, Kagan HM (1994) Modulation of lysyl oxidase activity toward peptidyl lysine by vicinal dicarboxylic amino acid residues. Implications for collagen cross-linking. J Biol Chem 269:22366–22371

Nagan N, Kagan HM (2000) Modulation of lysyl oxidase activity toward peptidyl lysine by vicinal dicarboxylic amino acid residues. Implications for collagen cross-linking. J Biol Chem 269:22366–22371

Nellaiappan K, Risitano A, Liu G, Nicklas G, Kagan HM (2000) Fully processed lysyl oxidase catalyst translocates from the extracellular space into nuclei of aortic smooth-muscle cells. J Cell Biochem 79:576–582

Ovchinnikova O, Gylfe A, Bailey L, Nordström A, Rudling M, Jung C, Bergström S, Waldenström A, Hansson GK, Nordström P (2009) Osteoprotegerin promotes fibrous cap formation in atherosclerotic lesions of ApoE-deficient mice – brief report. Arterioscler Thromb Vasc Biol 29:1478–1480

Palamakumbura AH, Trackman PC (2002) A fluorometric assay for detection of lysyl oxidase enzyme activity in biological samples. Anal Biochem 300:245–251

Palamakumbura AH, Sommer P, Trackman PC (2003) Autocrine growth factor regulation of lysyl oxidase expression in transformed fibroblasts. J Biol Chem 278:30781–30787

Palamakumbura AH, Jeay S, Guo Y, Pischon N, Sommer P, Sonenshein GE, Trackman PC (2004) The propeptide domain of lysyl oxidase induces phenotypic reversion of ras-transformed cells. J Biol Chem 279:40593–40600

Palamakumbura AH, Vora SR, Nugent MA, Kirsch KH, Sonenshein GE, Trackman PC (2009) Lysyl oxidase propeptide inhibits prostate cancer cell growth by mechanisms that target FGF2 cell binding and signaling. Oncogene 28:3390–3400

Panchenko MV, Stetler-Stevenson WG, Trubetskoy OV, Gacheru SN, Kagan HM (1996) Metalloproteinase activity secreted by fibrogenic cells in the processing of prolysyl oxidase. Potential role of procollagen C-proteinase. J Biol Chem 271:7113–7119

Pascual G, Mendieta C, Mecham RP, Sommer P, Bellón JM, Buján J (2008) Down-regulation of lysyl oxydase-like in aging and venous insufficiency. Histol Histopathol 23:179–186

Payne SL, Fogelgren B, Hess AR, Seftor EA, Wiley EL, Fong SF, Csiszar K, Hendrix MJ, Kirschmann DA (2005) Lysyl oxidase regulates breast cancer cell migration and adhesion through a hydrogen peroxide-mediated mechanism. Cancer Res 65:11429–11436

Payne SL, Hendrix MJ, Kirschmann DA (2006) Lysyl oxidase regulates actin filament formation through the p130(Cas)/Crk/DOCK180 signaling complex. J Cell Biochem 98:827–837

Payne SL, Hendrix MJ, Kirschmann DA (2007) Paradoxical roles for lysyl oxidases in cancer – a prospect. J Cell Biochem 101:1338–1354

Peinado H, Iglesias-de DC, la Cruz M, Olmeda D, Csiszar K, Fong KS, Vega S, Nieto MA, Cano A, Portillo F (2005) A molecular role for lysyl oxidase-like 2 enzyme in snail regulation and tumor progression. EMBO J 24:3446–3458

Peng L, Ran YL, Hu H, Yu L, Liu Q, Zhou J, Sun YM, Sun LC, Pan J, Sun LX, Zhao P, Yang ZH (2009) Secreted LOXL2 is a novel therapeutic target that promotes gastric cancer metastasis via the Src/FAK pathway. Carcinogenesis 30:1660–1669

Peyrol S, Raccurt M, Gerard F, Gleyzal C, Grimaud JA, Sommer P (1997) Lysyl oxidase gene expression in the stromal reaction to in situ and invasive ductal breast carcinoma. Am J Pathol 150:497–507

Postovit LM, Abbott DE, Payne SL, Wheaton WW, Margaryan NV, Sullivan R, Jansen MK, Csiszar K, Hendrix MJ, Kirschmann DA (2008) Hypoxia/reoxygenation: a dynamic regulator of lysyl oxidase-facilitated breast cancer migration. J Cell Biochem 103:1369–1378

Raposo B, Rodríguez C, Martínez-González J, Badimon L (2004) High levels of homocysteine inhibit lysyl oxidase (LOX) and downregulate LOX expression in vascular endothelial cells. Atherosclerosis 177:1–8

Ren C, Yang G, Timme TL, Wheeler TM, Thompson TC (1998) Reduced lysyl oxidase messenger RNA levels in experimental and human prostate cancer. Cancer Res 58:1285–1290

Rhee SG (2006) Cell signaling. H_2O_2, a necessary evil for cell signaling. Science 312:1882–1883

Rodríguez C, Martínez-González J, Raposo B, Alcudia JF, Guadall A, Badimon L (2008) Regulation of lysyl oxidase in vascular cells: lysyl oxidase as a new player in cardiovascular diseases. Cardiovasc Res 79:7–13

Rodríguez C, Alcudia JF, Martínez-González J, Guadall A, Raposo B, Sánchez-Gómez S, Badimon L (2009) Statins normalize vascular lysyl oxidase down-regulation induced by proatherogenic risk factors. Cardiovasc Res 83:595–603

Rückert F, Joensson P, Saeger HD, Grützmann R, Pilarsky C (2010) Functional analysis of LOXL2 in pancreatic carcinoma. Int J Colorectal Dis 25:303–311

Ruggiero CE, Dooley DM (1999) Stoichiometry of the topa quinone biogenesis reaction in copper amine oxidases. Biochemistry 38:2892–2898

Russell RC, Ohh M (2007) The role of VHL in the regulation of E-cadherin: a new connection in an old pathway. Cell Cycle 6:56–59

Ryvkin F, Greenaway FT (2004) A peptide model of the copper-binding region of lysyl oxidase. J Inorg Biochem 98:1427–1435

Sahlgren C, Gustafsson MV, Jin S, Poellinger L, Lendahl U (2008) Notch signaling mediates hypoxia-induced tumor cell migration and invasion. Proc Natl Acad Sci USA 105:6392–6397

Saito H, Papaconstantinou J, Sato H, Goldstein S (1997) Regulation of a novel gene encoding a lysyl oxidase-related protein in cellular adhesion and senescence. J Biol Chem 272:8157–8160

Schietke RE, Warnecke C, Wacker I, Schodel J, Mole DR, Campean V, Amann K, Goppelt-Struebe M, Behrens J, Eckardt KU, Wiesener MS (2010) The lysyl oxidases LOX and LOXL2 are necessary and sufficient to repress E-cadherin in hypoxia – insights into cellular transformation processes mediated by HIF-1. J Biol Chem 285(9):6658–6669

Sebban S, Davidson B, Reich R (2009) Lysyl oxidase-like 4 is alternatively spliced in an anatomic site-specific manner in tumors involving the serosal cavities. Virchows Arch 454:71–79

Siegel RC (1974) Biosynthesis of collagen crosslinks: increased activity of purified lysyl oxidase with reconstituted collagen fibrils. Proc Natl Acad Sci USA 71:4826–4830

Sion AM, Figg WD (2006) Lysyl oxidase (LOX) and hypoxia-induced metastases. Cancer Biol Ther 5:909–911

Smith-Mungo LI, Kagan HM (1998) Lysyl oxidase: properties, regulation and multiple functions in biology. Matrix Biol 16:387–398

Szauter KM, Csiszar K (2008) LOXL3 (lysyl oxidase-like 3). Atlas Genet Cytogenet Oncol Haematol. http://AtlasGeneticsOncology.org/Genes/LOXL3ID44000ch2p13.html

Szauter KM, Gorogh T, Csiszar K (2007) LOXL4 (lysyl oxidase-like 4). Atlas Genet Cytogenet Oncol Haematol. http://AtlasGeneticsOncology.org/Genes/LOXL4ID41193ch10q24.html

Tan RS, Taniguchi T, Harada H (1996) Identification of the lysyl oxidase gene as target of the antioncogenic transcription factor, IRF-1, andits possible role in tumor suppression. Cancer Res 56:2417–2421

Tang C, Klinman JP (2001) The catalytic function of bovine lysyl oxidase in the absence of copper. J Biol Chem 276:30575–30578

Thomassin L, Werneck CC, Broekelmann TJ, Gleyzal C, Hornstra IK, Mecham RP, Sommer P (2005) The pro-regions of lysyl oxidase and lysyl oxidase-like-1 are required for deposition onto elastic fibers. J Biol Chem 280:42848–42855

Trackman PC (2005) Diverse biological functions of extracellular collagen processing enzymes. J Cell Biochem 96:927–937

Trackman PC, Kagan HM (1979) Nonpeptidyl amine inhibitors are substrates of lysyl oxidase. J Biol Chem 254:7831–7836

Trackman PC, Zoski CG, Kagan HM (1981) Development of a peroxidase-coupled fluorometric assay for lysyl oxidase. Anal Biochem 113:336–342

Trackman PC, Pratt AM, Wolanski A, Tang SS, Offner GD, Troxler RF, Kagan HM (1990) Cloning of rat aorta lysyl oxidase cDNA: complete codons and predicted amino acid sequence. Biochemistry 29:4863–4870

Trackman PC, Pratt AM, Wolanski A, Tang SS, Offner GD, Troxler RF, Kagan HM (1991) Cloning of rat aorta lysyl oxidase cDNA: complete codons and predicted amino acid sequence. Biochemistry 30:8282 (erratum)

Trackman PC, Bedell-Hogan D, Tang J, Kagan HM (1992) Post-translational glycosylation and proteolytic processing of a lysyl oxidase precursor. J Biol Chem 267:8666–8671

Uzel MI, Scott IC, Babakhanlou-Chase H, Palamakumbura AH, Pappano WN, Hong HH, Greenspan DS, Trackman PC (2001) Multiple bone morphogenetic protein 1-related mammalian met alloproteinases process pro-lysyl oxidase at the correct physiological site and control lysyl oxidase activation in mouse embryo fibroblast cultures. J Biol Chem 276:22537–22543

Vadasz Z, Kessler O, Akiri G, Gengrinovitch S, Kagan HM, Baruch Y, Izhak OB, Neufeld G (2005) Abnormal deposition of collagen around hepatocytes in Wilson's disease is associated with hepatocyte specific expression of lysyl oxidase and lysyl oxidase like protein-2. J Hepatol 43:499–507

Van der Groep P, Bouter A, Menko FH, van der Wall E, van Diest PJ (2008) High frequency of HIF-1alpha overexpression in BRCA1 related breast cancer. Breast Cancer Res Treat 111:475–480

Wang SX, Mure M, Medzihradszky KF, Burlingame AL, Brown DE, Dooley DM, Smith AJ, Kagan HM, Klinman JP (1996) A crosslinked cofactor in lysyl oxidase: redox function for amino acid side chains. Science 273:1078–1084

Williams MA, Kagan HM (1985) Assessment of lysyl oxidase variants by urea gel electrophoresis: evidence against disulfide isomers as bases of the enzyme heterogeneity. Anal Biochem 149:430–437

Woznick AR, Braddock AL, Dulai M, Seymour ML, Callahan RE, Welsh RJ, Chmielewski GW, Zelenock GB, Shanley CJ (2005) Lysyl oxidase expression in bronchogenic carcinoma. Am J Surg 189:297–301

Wu G, Guo Z, Chang X, Kim MS, Nagpal JK, Liu J, Maki JM, Kivirikko KI, Ethier SP, Trink B, Sidransky D (2007a) LOXL1 and LOXL4 are epigenetically silenced and can inhibit ras/

extracellular signal-regulated kinase signaling pathway in human bladder cancer. Cancer Res 67:4123–4129

Wu M, Min C, Wang X, Yu Z, Kirsch KH, Trackman PC, Sonenshein GE (2007b) Repression of BCL2 by the tumor suppressor activity of the lysyl oxidase propeptide inhibits transformed phenotype of lung and pancreatic cancer cells. Cancer Res 67:6278–6285

Yanaihara N, Caplen N, Bowman E, Seike M, Kumamoto K, Yi M, Stephens RM, Okamoto A, Yokota J, Tanaka T, Calin GA, Liu CG, Croce CM, Harris CC (2006) Unique microRNA molecular profiles in lung cancer diagnosis and prognosis. Cancer Cell 9:189–198

Zhao Y, Min C, Vora SR, Trackman PC, Sonenshein GE, Kirsch KH (2009) The Lysyl oxidase propeptide attenuates fibronectin-mediated activation of focal adhesion kinase and p130Cas in breast cancer cells. J Biol Chem 284:1385–1393

Zia MK, Rmali KA, Watkins G, Mansel RE, Jiang WG (2007) The expression of the von Hippel-Lindau gene product and its impact on invasiveness of human breast cancer cells. Int J Mol Med 20:605–611

Chapter 10
The Fibulins

Marion A. Cooley and W. Scott Argraves

Abstract This year, 2010, marked 21 years of fibulin research. Over these two decades, findings reported in nearly 400 manuscripts have shown a family of eight fibulin genes that serve a variety of critical extracellular matrix (ECM)-related functions. In particular, phenotypic analysis of humans, mice, and worms carrying mutations in genes of the fibulin family has led to great advances in our understanding of the roles that these proteins play in physiological and pathological processes. One of the most significant roles to emerge for the fibulins as a group is their ability to coordinate the assembly of elastic fibers. This chapter will convey our understanding of the key roles that the fibulins play in the processes of elastogenesis, cell adhesion, and motility, as well as pathological processes including eye and cardiovascular disease and cancer.

10.1 Introduction

Fibulins are a family of glycoproteins that in mammals are encoded by eight genes (Table 10.1). The prototypic member of the family, fibulin-1, is an ancient gene, found in organisms as evolutionarily primitive as nematodes (Barth et al. 1998). Homologs of fibulin-1 have been identified in zebrafish (Zhang et al. 1997), chicken (Barth et al. 1998), mouse (Pan et al. 1993), and man (Argraves et al. 1990). The conservation of fibulin-1 throughout evolution suggests a conservation of functional features. Indeed, fibulin-1 has retained several common activities between worms and humans including its ability to interact with nidogen (Kubota et al. 2008; Sasaki et al. 1995b) and members of the ADAMTS (a disintegrin and MMP with thrombospondin motifs) family of matrix metalloproteinases (MMP) [i.e., GON-1 (Hesselson et al. 2004) and its human homolog ADAMTS9 (McCulloch

---

M.A. Cooley and W.S. Argraves (✉)
Department of Regenerative Medicine and Cell Biology, Medical University of South Carolina, 173 Ashley Avenue, Charleston, SC 29425, USA
e-mail: argraves@musc.edu

Table 10.1 Fibulin family nomenclature

Name	Synonymous names	Gene symbol	Human chromosome location	References
Fibulin-1	BM-90	FBLN1	22q13.31	Argraves et al. (1990), Kluge et al. (1990)
Fibulin-2		FBLN2	3p24–p25	Pan et al. (1993)
Fibulin-3	S1-5, T16, EFEMP1	FBNL3	2p16	Tran et al. (1997)
Fibulin-4	MBP1, EFEMP2, UPH1, H411	EFEMP2	11q13	Gallagher et al. (1999), Giltay et al. (1999)
Fibulin-5	DANCE, EVEC, UP50	FBLN5	14q32.1	Kowal et al. (1999), Nakamura et al. (1999)
Fibulin-6	Hemicentin-1	HMCN1	1q25.3	Vogel and Hedgecock (2001)
Fibulin-7	TM-14	FBLN7	2q13	de Vega et al. (2007)
Fibulin-8	Hemicentin-2	HMCN2	9q34.11	Xu et al. (2007)

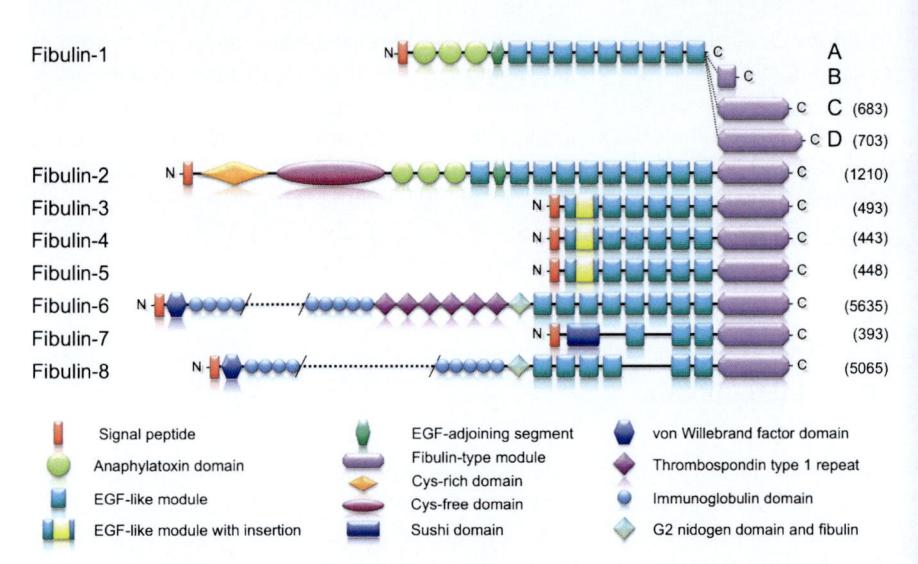

Fig. 10.1 Polypeptide structure of members of the fibulin family. Shown are the four variants of human fibulin-1 produced as a result of alternative splicing. The worm, chicken, and mouse fibulin-1 genes produce only the C and D variants. Three of the smaller fibulins, 3, 4, and 5 contain an unusual interrupted EGF-like module. The number of amino acids in the human forms of each family member are indicated at the right in *parentheses*. The string of immunoglobulin domains in fibulin-6 and fibulin-8 is interrupted in the diagram. Fibulin-6 contains 43 immunoglobulin domains (42 IG-C type and 1 IG type). Fibulin-8 contains 42 immunoglobulin domains (40 IG-C type and 2 IG type)

et al. 2009)]. Of the remaining members of the fibulin family, only fibulin-6 (also known as hemicentin-1) is found in the nematode (Vogel and Hedgecock 2001), suggesting that the other fibulins have emerged later in evolution (see Chap. 1).

Similar to many other extracellular matrix (ECM) proteins, the fibulins have repeated domain structures (Fig. 10.1), which likely form the basis for the array of

Table 10.2 Ligands of fibulin family members

Category	Ligand name	Fibulin-1	Fibulin-2	Fibulin-3	Fibulin-4	Fibulin-5	Fibulin-6	Fibulin-7
Basement membrane proteins	Laminin-1 ($\alpha1\beta1\gamma1$)	± (Brown et al. 1994)						
	Laminin-4 ($\alpha2\beta2\gamma1$)	+ (Brown et al. 1994)						
	Laminin-5 ($\alpha3\beta3\gamma2$)	+ (Sasaki et al. 2001)	+ (Sasaki et al. 2001)					
	Laminin $\alpha1$ chain		+ (Utani et al. 1997)					
	Laminin $\alpha2$ chain	+ (Talts et al. 1999)	+ (Talts et al. 1999)					
	Laminin $\alpha4$ chain	+ (Talts et al. 2000)	+ (Talts et al. 2000)					
	Laminin $\gamma2$ chain	+ (Sasaki et al. 2001)	+ (Sasaki et al. 2001; Utani et al. 1997)					
	Nidogen-1	+ (Adam et al. 1997)	+ (Salmivirta et al. 2002; Sasaki et al. 1995b)					
	Nidogen-2		+ (Kobayashi et al. 2007; Salmivirta et al. 2002)		+ (Kobayashi et al. 2007)			
	Perlecan		+ (Hopf et al. 1999; Sasaki et al. 1995a)					
	Type IV collagen		+ (Kobayashi et al. 2007; Sasaki et al. 1995a)		+ (Kobayashi et al. 2007)			
Loose connective tissue matrix proteins	Fibronectin	+ (Balbona et al. 1992)	+ (Sasaki et al. 1995a)	− (Kobayashi et al. 2007)	− (Kobayashi et al. 2007)	− (Kobayashi et al. 2007)		
	Tropoelastin	+ (Sasaki et al. 1999)	+ (Sasaki et al. 1999)	+ (Kobayashi et al. 2007)	+ (Kobayashi et al. 2007)	+ (Choudhury et al. 2009; Nakamura et al. 2002; Yanagisawa et al. 2002)		+ (de Vega et al. 2007)
	Lysyl oxidase (Lox)				+ (Choudhury et al. 2009)	+ (Choudhury et al. 2009)		
	Lysyl oxidase-like protein-1,2 and 4					+ (Hirai et al. 2007b)		

(*continued*)

Table 10.2 (continued)

Category	Ligand name	Fibulin-1	Fibulin-2	Fibulin-3	Fibulin-4	Fibulin-5	Fibulin-6	Fibulin-7
	Fibrillin-1		+ (Reinhardt et al. 1996)		+ (Choudhury et al. 2009)	+ (Choudhury et al. 2009) + (Hirai et al. 2007a) + (Zanetti et al. 2004)		
	Latent TGF-β-binding protein 2 (LTBP-2)							
	Emilin-1							+ (de Vega et al. 2007)
	Aggrecan	+ (Aspberg et al. 1999)	+ (Olin et al. 2001)					
	Versican	+ (Aspberg et al. 1999)	+ (Olin et al. 2001)					
	Brevican		+ (Olin et al. 2001)					
	Dentin sialoprotein (Dsp)			+ (Klenotic et al. 2004)				
	ADAMTS1	+ (Lee et al. 2005)						
	Tissue inhibitor of metallo-proteinases-3 (TIMP-3)							
	Other							
	Extracellular matrix protein 1 (ECM1)	+ (Kortvely et al. 2010)	+ (Fujimoto et al. 2005)		+ (Sercu et al. 2009)			
ARMS2							+ (Kortvely et al. 2010)	
Sex hormone binding globulin (SHBG)		+ (Ng et al. 2006)	+ (Ng et al. 2006)					
NOVH (CCN3)		+ (Perbal et al. 1999)						
Connective tissue growth factor (CCN2/CTGF)		+ (Perbal et al. 1999)						
HB-EGF		+ (Brooke et al. 2002)						

10 The Fibulins

Papillomavirus E6 protein	+ (Du et al. 2002)		
β-Amyloid precursor protein	+ (Ohsawa et al. 2001)		
DA41		+ (Ozaki et al. 1997)	
Mutant p53			+ (Gallagher et al. 1999)
Angiogenin	+ (Zhang et al. 2008)		
Collagen XV-derived endostatin	+ (Kobayashi et al. 2007)	+ (Kobayashi et al. 2007)	
Collagen XVIII endostatin	+ (Kobayashi et al. 2007)	+ (Kobayashi et al. 2007)	
Lipoprotein(a) [Lp(a)]			+ (Kapetanopoulos et al. 2002)
Extracellular superoxide dismutase (ecSOD)			+ (Nguyen et al. 2004)
Receptors			
α4β1			+ (Lomas et al. 2007)
α5β1		−(Pfaff et al. 1995)	+ (Lomas et al. 2007)
α9β1			(Nakamura et al. 2002)
αvβ3	+ (Kobayashi et al. 2007; Pfaff et al. 1995)		+ (Nakamura et al. 2002)
αvβ5			
αIIbβ3	+ (Pfaff et al. 1995)		
Serum opacity factor (SOF)	+ (Courtney et al. 2009)		+ (Nakamura et al. 2002)

At this time, no proteins have been found to bind fibulin-8

ligands that individual fibulins have been found to bind (Table 10.2). Members of the fibulin family all share a common architectural signature, namely a series of epidermal growth factor (EGF)-like modules followed by a carboxy terminal fibulin-type module (Fig. 10.1). A subset of the fibulins, fibulin-3, 4, and 5, are referred to as short fibulins and have structures entirely comprising tandemly repeated EGF modules followed by a fibulin-type module. Fibulin-7 is structurally similar to these other short fibulins but additionally possesses an amino terminal sushi domain (de Vega et al. 2007). The remaining members of the fibulin family possess longer amino terminal segments comprising various types of repeated structural modules.

10.2 The Fibulin-Type Module

The fibulin-type module is a unique structural domain found at the carboxy terminus of all fibulin proteins (Fig. 10.2). The module averages 118 amino acids in length and is preceded by a variable number of repeated EGF-like modules. Key to the architecture of the fibulin-type module is the inclusion of a cysteine moiety that acts to satisfy

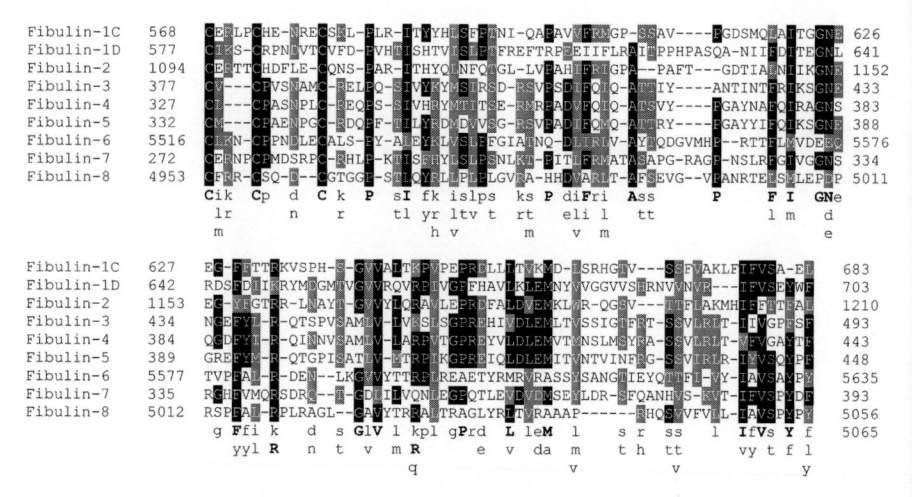

Fig. 10.2 Alignment of fibulin-type module sequences from the human fibulins. The carboxy terminal regions of the fibulins were aligned using Multialign (Corpet 1988) and the alignments refined manually. Identical residues are highlighted with a *black background* and chemically similar residues with a *gray background*. Under the aligned sequences is the consensus. A *bolded capital letter* in the consensus sequence indicates that the residue was the same in at least six of the eight fibulins. The GenBank accession numbers used for the sequences depicted are as follows: fibulin-1C, BC022497; fibulin-1D, AF126110; fibulin-2, NM_001998; fibulin-3, NM_001039349; fibulin-4, NM_016938; fibulin-5, NM_006329; fibulin-6, NM_031935; fibulin-7, NM_001128165; and fibulin-8, XP_002347119

disulfide bonding requirements of the preceding EGF-like module, whose exon encodes an EGF-like module with an odd number of cysteines. Thus far, a fibulin-type module has not been reported to be present in proteins other than the fibulins. The three-dimensional structure of a fibulin-type module has not yet been determined.

Functionally, fibulin-type modules have been implicated in mediating a number of protein–protein interactions. For example, the fibulin-type modules of fibulin-1D and fibulin-2 bind to sex hormone-binding globulin (SHBG) (Ng et al. 2006), thus providing a mechanism by which the steroid-binding protein is sequestered in the ECM. The fibulin-type modules of the fibulin-1C and fibulin-1D variants bind to the protein ECM1 (Fujimoto et al. 2005), and the fibulin-type module of fibulin-5 has been shown to bind to extracellular superoxide dismutase (ecSOD), lysyl oxidase-like-1 (Loxl1), and Loxl2 (Hirai et al. 2007b; Nguyen et al. 2004).

Several studies have mapped protein binding sites within fibulin to regions that include the fibulin-type modules, but left open the possibility for involvement of the adjacent EGF domains. For example, the TIMP-3 binding site in fibulin-3 has been mapped to amino residues 256–493, which includes the fibulin-type module plus an upstream EGF-like domain (Klenotic et al. 2004). Similarly, the NOVH (CCN3) and CTGF (CCN2) binding sites within fibulin-1 have been localized to the region containing the fibulin-type module and two upstream EGF-like domains (Perbal et al. 1999). Recently, the binding site for ARMS2 (ARMD8) has been mapped to the fibulin-type module and upstream EGF-like domain of both fibulin-1D and fibulin-6 (Kortvely et al. 2010).

Findings from several studies indirectly implicate fibulin-type modules as mediating specific ECM binding interactions. For example, the fibulin-1C variant has been shown to bind to nidogen with 30-fold higher affinity than fibulin-1D (Sasaki et al. 1995b), and in *C. elegans*, the assembly of fibulin-1D into an ECM is dependent on perlecan-M, whereas the fibulin-1C variant is not (Muriel et al. 2006).

10.3 Receptors for the Fibulins

Most of the fibulins, including fibulin-1, 2, 3, 5, and 7, have been shown to interact with cell surface receptors. Fibulin-1 was originally identified by virtue of its ability to interact with the integrin $\beta 1$ cytoplasmic domain, although the significance of this interaction remains to be established (Argraves et al. 1989). Fibulin-1 has recently been found to bind to serum opacity factor (SOF), a surface protein of group A streptococci (Courtney et al. 2009). This interaction is believed to be involved in the adhesion of bacteria to host ECMs, the first stage in establishing bacterial infections.

Murine fibulin-2, which contains an integrin-binding, arginine-glycine-aspartic acid (RGD) sequence, mediates cell adhesion via binding to the integrins $\alpha IIb\beta 3$ and $\alpha v\beta 3$ (Pfaff et al. 1995). The RGD sequence is not conserved between mouse and human fibulin-2, and human fibulin-2 only weakly interacts with $\alpha IIb\beta 3$ and does not mediate cell adhesion (Pfaff et al. 1995).

Fibulin-3 binds and activates the EGF receptor (EGFR) leading to downstream activation of intermediates of the mitogen-activated protein kinase and Akt pathways (Camaj et al. 2009). In fact, the fibulin-3 binding site on EGFR appears to be overlapping with that of EGF since fibulin-3 can compete for EGF binding to EGFR. These findings highlight the potential for fibulin-3 to serve directly as a growth factor.

Fibulin-5 binds to a subset of integrins, including $\alpha 5\beta 1$, $\alpha 4\beta 1$, $\alpha v\beta 3$, and $\alpha v\beta 5$ (Kobayashi et al. 2007; Lomas et al. 2007; Nakamura et al. 2002), through an evolutionarily conserved RGD sequence located in the N-terminal region of fibulin-5 (Nakamura et al. 1999). Interestingly, there is evidence that the RGD binding site within fibulin-5 requires unmasking, and even after doing so, the affinity of chemically unmasked fibulin-5 is tenfold lower than integrin binding to vitronectin (Kobayashi et al. 2007). Although fibulin-5 binds to the fibronectin-binding integrins, $\alpha 5\beta 1$ and $\alpha 4\beta 1$, and facilitates cell adhesion [i.e., human aortic smooth muscle cells (SMC)], it fails to activate downstream signaling (Lomas et al. 2007). This, together with evidence that it can modulate fibronectin-mediated cell spreading and morphology, leads to speculation that fibulin-5 may act to antagonize fibronectin–integrin adhesive signaling (Lomas et al. 2007). Recently, fibulin-5 has been shown to suppress fibronectin–integrin signaling that promotes production of reactive oxygen species (ROS) (Schluterman et al. 2010).

Fibulin-7 also mediates cell adhesive interactions, although it remains to be established whether these interactions involve direct or indirect binding to cell surface receptors. Fibulin-7 has been shown to mediate mesenchymal cell binding in a manner that can be blocked by antibody to integrin $\beta 1$ (de Vega et al. 2007). Furthermore, the fact that cell adhesion to fibulin-7 can be inhibited by heparin suggests that heparan sulfate-containing receptors may also bind to fibulin-7 (de Vega et al. 2007).

10.4 Regulation of the Expression of Fibulins

There is rather little known about the physiological and pathological regulation of the fibulins. The steroid hormones, estrogen and progesterone, have been shown to regulate expression of several of the fibulins. The expression of fibulin-1 is greatly increased by estradiol in estrogen-responsive ovarian cancer cell lines [i.e., estrogen receptor (ER)-positive cells] (Clinton et al. 1996). The fibulin-1 promoter region contains a binding site for the transcription factor, Sp1 (Bardin et al. 2004), which is involved in mediating transcriptional responses to estrogens through recruitment of ER-α. Estrogen has been found not only to upregulate the transcription of the fibulin-1 gene but also to decrease the stability of the fibulin-1D transcript and thereby influence the ratio of fibulin-1C and fibulin-1D mRNAs (Bardin et al. 2004). This may underlie the findings that the fibulin-1C to fibulin-1D ratio is greater in estrogen-responsive ovarian cancers than in normal tissues (Moll et al. 2002). In endometrial stromal cells, progesterone has also been shown to upregulate the expression of fibulin-1 (Nakamoto et al. 2005); however, the mechanism has not been defined. The fibulin-2 promoter contains ER half sites and Sp1 binding sites (Grassel et al. 1999), although it remains uncertain whether this gene is regulated by estrogen. Similarly, the fibulin-3 promoter contains Sp1 binding sites, as well as an estrogen response element. However, in contrast to the positive effects of estrogen on fibulin-1 expression, estrogen suppresses the transcription of the fibulin-3 gene (Blackburn et al. 2003). Fibulin-1 and several other fibulins (i.e., fibulins 4 and 5) are apparently not regulated by androgens, at least in prostate cancer cells (Wlazlinski et al. 2007).

To date only a few growth factors have been shown to regulate the expression of the fibulins. Fibroblast growth factor 18 (Fgf-18), a critical growth factor in lung morphogenesis and in postnatal lung elastogenesis, increases the expression of fibulin-1 and fibulin-5 in fibroblasts isolated from fetal and postnatal lungs (Chailley-Heu et al. 2005). The Fgf-18-mediated increase in fibulin-1 and fibulin-5 in lung fibroblasts is accompanied by increases in the expression of tropoelastin and lysyl oxidase (Chailley-Heu et al. 2005). The expression of fibulin-5 by fibroblasts and endothelial cells has been shown to be stimulated by transforming growth factor-β (TGF-β) (Kuang et al. 2006; Schiemann et al. 2002). In endothelial cells, the stimulatory effects of TGF-β on fibulin-5 expression can be blocked by vascular endothelial growth factor (VEGF), which itself is a negative regulator of fibulin-5 expression (Albig and Schiemann 2004). In lung fibroblasts, TGF-β-mediated upregulation of fibulin-5 occurs via Smad2/3-dependent binding to two Smad binding sites located in the proximal region of the fibulin-5 promoter (Kuang et al. 2006). Consistent with these findings is evidence that TGF-β-mediated induction of fibulin-5 in lung fibroblasts involves the PI3-kinase/Akt pathway (Kuang et al. 2006). By contrast, TGF-β-stimulated expression of fibulin-5 in 3T3-L1 fibroblasts acts through a Smad2/3-independent pathway (Schiemann et al. 2002).

10.5 Fibulins as Regulators of Cell Adhesion and Motility

A major function ascribed to fibulin-1 is regulation of cell motility and guidance. Evidence for this role initially came from cell culture studies that showed that fibulin-1 could suppress the fibronectin-stimulated motility (i.e., migration velocity and persistence time) of many types of cancer cells (Hayashido et al. 1998; Lee et al. 2005; Qing et al. 1997; Twal et al. 2001). From studies of *C. elegans* mutants, fibulin-1 was found to be required for proper guidance of migrating distal tip cells (DTCs) engaged in gonad morphogenesis (Hesselson et al. 2004; Kubota et al. 2004; Kubota and Nishiwaki 2003). In fibulin-1-deficient nematodes, there is a distention/widening of sheets of gonadal cells during development and failure of DTCs to complete migration to the midline (Hesselson et al. 2004). Suppressor mutation studies in *C. elegans* showed that fibulin-1 point mutations suppress DTC migration defects that occur in worms with mutations in two secreted MMPs belonging to the ADAMTS family, Mig-17 and GON-1 (Hesselson et al. 2004;

Hesselson et al. 2003; Kubota et al. 2004; Kubota and Nishiwaki 2003). ADAMTS family members are a group of zinc-dependent MMPs that mainly degrade ECM components such as proteoglycans and collagens. GON-1 function is essential for the motility of distal tip cells, whereas Mig-17 is required for directed migration of distal tip cells (Blelloch and Kimble 1999; Nishiwaki et al. 2000).

A relationship between fibulin-1 and ADAMTS MMPs also exists in vertebrates where fibulin-1 has been shown to bind to ADAMTS-1 and act as a positive cofactor for ADAMTS-1-mediated cleavage of the fibulin-1-binding proteoglycan, aggrecan, to liberate its amino terminal G1 domain (Lee et al. 2005). The significance of this activity remains to be established, but it may relate to regulation of the inhibitory effects of aggrecan on neural crest cell (NCC) migration (Perissinotto et al. 2000; Perris et al. 1996). Another fibulin-1 binding proteoglycan, versican, is also a substrate for ADAMTS-1 (Sandy et al. 2001). Similar to the consequence of ADAMTS-1 cleavage of aggrecan, ADAMTS-1 cleavage of versican results in production of an amino terminal globular domain-containing fragment, G1. During vertebrate development, not only are the expression patterns of versican, fibulin-1, and ADAMTS-1 closely similar in tissues such as endocardial cushions, but G1 domain-containing cleavage products of versican are also detectable at these sites as well (Henderson and Copp 1998; Kern et al. 2006). Together, the findings support the possibility that in vertebrates, fibulin-1 may promote ADAMTS-1-mediated proteolysis of versican and aggrecan and thereby regulate the motility and guidance activities of these proteoglycans. In support of this is the finding that fibulin-1 can also act as a cofactor for ADAMTS-5-mediated cleavage of versican (McCulloch et al. 2009).

This hypothesis is not applicable to invertebrates such as *C. elegans* and *Drosophila* whose genomes do not encode the proteoglycans aggrecan and versican (Hesselson et al. 2004). However, these organisms express other ECM constituents that may serve similar motility regulating functions. For example, the ECM within which nematode DTCs migrate is basement membrane-like, composed of laminins, type IV collagen, nidogen, and the proteoglycan perlecan (Kramer 1997). Both laminin and perlecan are required for DTC migration (Merz et al. 2003) and interact with fibulin-1 (Brown et al. 1994; Muriel et al. 2006). While laminin and perlecan are cleaved by MMPs in vertebrate systems, there are no reports that they are substrates for ADAMTS family members. It remains to be determined whether they are the substrates for GON-1 or Mig-17 involved in DTC migration or whether fibulin-1 influences their proteolysis. Collectively, fibulin-1C, nidogen, and Mig-17 have also been found to be required both for nidogen incorporation into the nematode basement membrane and for directional migration of DTCs (Kubota et al. 2008). Given that in vertebrates, ADAMTS1 cleaves nidogen (Canals et al. 2006), it is possible that nidogen in worms may be a substrate for the ADAMTS relatives, Mig-17 or GON-1.

Fibulin-1 is also implicated as a regulator of cell motility during vertebrate development based on the fact that it is expressed in association with migrating mesenchymal cells, including endocardial cushion mesenchymal cells and NCCs (Bouchey et al. 1996; Kern et al. 2006; Spence et al. 1992; Zhang et al. 1995; Zhang et al. 1993). Findings from characterization of fibulin-1-deficient mouse embryos demonstrate that fibulin-1 is required for guidance of cranial NCCs (Cooley et al. 2008). This is evident by abnormalities in the patterns of streams of NCCs emanating from rhombomeres 6 and 7 as well as by structural defects in the cranial nerves derived from these NCC streams, cranial nerves IX and X (Cooley et al. 2008) (Fig. 10.3). Consistent with these findings is the fact that fibulin-1-binding ECM components including fibronectin, versican, and laminin α5 have also been implicated in regulating the migration and guidance of cranial NCCs derived from rhombomeres 6 and 7 (Bronner-Fraser 1993; Costell et al. 2002; Henderson et al. 1997; Perris and Perissinotto 2000). For example, in mice deficient in laminin α5,

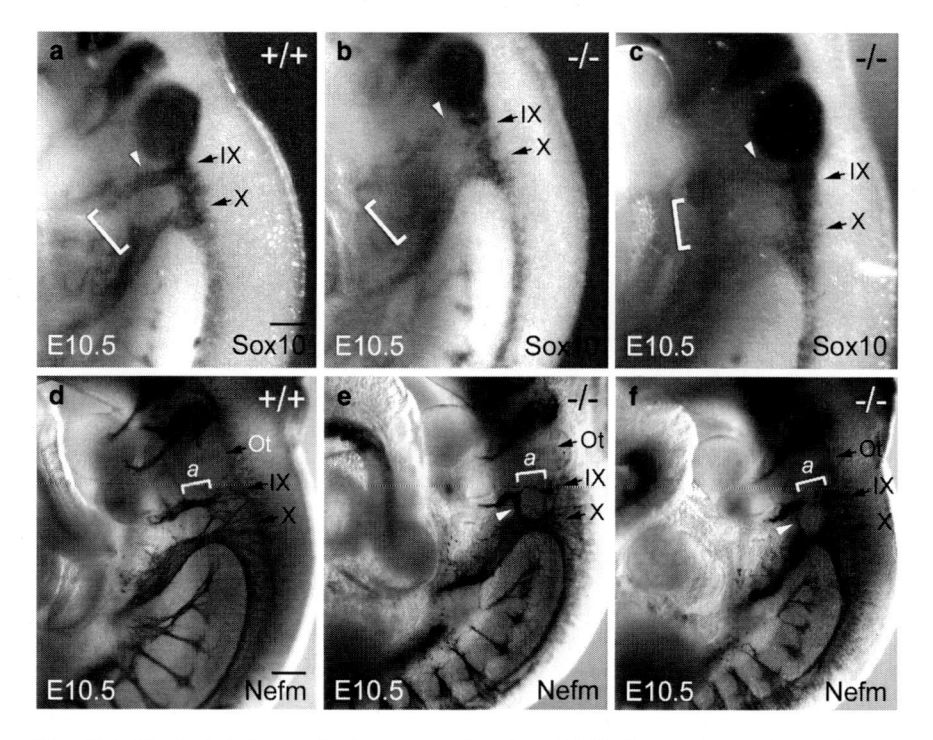

Fig. 10.3 Fibulin-1 deficiency leads to anomalies of cranial NCC patterning and cranial nerve morphogenesis. (a–c) Lateral views of E10.5 *Fbln1*[+/+] (A) and *Fbln1*[−/−] embryos (b and c) after whole mount in situ hybridization with antisense riboprobe for Sox-10, a transcription factor expressed by migratory NCCs that form the cranial nerves. *Arrows* in panels a–c indicate the forming glossopharyngeal and vagus nerves (cranial nerves IX and X). Note the decreased level of Sox-10 expression in the forming glossopharyngeal nerves (*white arrowheads* in panels b and c). Abnormal fusion of glossopharyngeal ganglia with the distal ganglia of cranial nerve X is apparent in the nulls (*brackets in* panels b and c). In (c), cranial nerve X has a decreased level of Sox-10 expression relative to the control. (d–f) Lateral views of wild-type (d) and *Fbln1*[−/−] (e and f) E10.5 embryos with cranial nerves immunolabeled with monoclonal neurofilament-M antibody. *Brackets* labeled *a* in panels d–f indicate the proximal portion of the cranial nerve IX, which is hypoplastic and absent in the Fbln1 null embryos shown in panels e and f, respectively. *White arrowheads* in panels e and f indicate bridging in the epibranchial placode-derived regions of cranial nerves IX and X. *Ot* otic vesicle. *Bar* in panel **a** = 300 μm and applies to panels **a–c**. *Bar* in **d** = 300 μm and applies to panels **d–f**

there are abnormalities in the migration of NCCs from rhombomeres 6 and 7. This results in improper condensation of the NCC-derived primordia of cranial nerves IX and X (Coles et al. 2006). Similarly, deficiency of integrin β1, a subunit of integrins that bind both laminin α5 and fibronectin (Kikkawa et al. 2000), leads to defective NCC migration from rhombomeres 6 and 7 resulting in abnormal formation of cranial nerves IX and X (Pietri et al. 2004). The role of fibulin-1 as a regulator of cranial NCC motility is also important to the morphogenesis of the heart, pharyngeal glands, and other NCC-derived structures of the head. Indeed, mice deficient in fibulin-1 display malformations of the aortic arch, heart, thymus, thyroid, and bones of the head, all similar to congenital defects known as neurocristopathies, which result from defective NCCs (Cooley et al. 2008).

Evidence for other members of the fibulin family playing roles in cell adhesion and motility is rather limited. Similar to fibulin-1, fibulin-5 inhibits fibronectin-mediated spreading and motility (Lomas et al. 2007). Consistent with fibulin-5 having an antimigratory activity are findings showing that vascular SMCs from fibulin-5-deficient mice have enhanced migratory responses to stimulation with PDGF (Spencer et al. 2005). Unlike fibulin-1 and fibulin-5, several of the other fibulin family members promote cell adhesion and motility. For example, fibulin-2 promotes vascular SMC migration and its interaction with versican is potentially important to the process (Strom et al. 2006). Similarly, fibulin-3 has been shown to enhance adhesion and promote the motility of glioma cells (Hu et al. 2009a) and olfactory ensheathing cells (Vukovic et al. 2009).

10.6 Fibulins in Elastogenesis

Most of the fibulins, including fibulin-2–5, have been shown to be critical for elastogenesis in various tissue settings. In this section, the order in which fibulins are discussed is related to the relative severity of elastogenic anomalies observed as a result of loss-of-function mutations in the fibulin genes.

Fibulin-4: Mice deficient in fibulin-4 exhibit the most severe defects in elastogenesis and widest range of elastic tissue abnormalities among the fibulin gene mutants (see Chap. 7). Fibulin-4 null mice are perinatal lethal and display severe lung and vascular defects (McLaughlin et al. 2006). These mice lack elastic fibers in elastin-containing tissues including the skin, lungs, and aorta. For example, while amorphous elastin is deposited in the extracellular spaces of the aorta of fibulin-4 nulls, nothing structurally resembling normal elastic lamina are formed (McLaughlin et al. 2006). Instead, small aggregates of elastin are present that contain rod-like filaments not seen within normal elastic laminae.

Biochemical analysis of elastin in fibulin-4-deficient mice showed a 94% and 88% decrease in desmosine levels in the aorta and lung, respectively, as compared with wild-type mice (McLaughlin et al. 2006). These findings, taken together with evidence that levels of tropoelastin mRNA were not reduced in the nulls, point to there being a defect in the process of elastin cross-linking in fibulin-4 nulls

(McLaughlin et al. 2006). This conclusion is supported by evidence that fibulin-4 binds to lysyl oxdase (LOX) (Horiguchi et al. 2009), the prototypic member of a family of lysyl oxidases that mediate the polymerization of tropoelastin to insoluble elastic fibers. Mice deficient in LOX share features with fibulin-4-deficient mice in that both display perinatal lethality and tortuous aorta and both develop aortic aneurysms (Hornstra et al. 2003). In both LOX and fibulin-4 null models, elastic fiber formation is disrupted resulting in decreased levels of the cross-linked amino acid desmosine, although fibulin-4-deficient mice appear to have more severe defects (Hornstra et al. 2003; McLaughlin et al. 2006).

Horiguchi et al. (2009) report that fibulin-4 mediates the formation of a complex between proLOX and tropoelastin. This group concluded that proLOX is unable to bind tropoelastin except in the presence of fibulin-4 (Horiguchi et al. 2009). However, Choudhury et al. (2009) suggest that proLOX can bind tropoelastin in the absence of fibulin-4. The fact that fibulin-4 nulls have decreased desmosine levels (McLaughlin et al. 2006) suggests that fibulin-4 may indeed act to mediate cross-linking of tropoelastin. It is possible that fibulin-4 may play a role in the activation of proLOX. In this regard, it is interesting to note that following activation of LOX by cleavage of the propeptide, fibulin-4 binding to LOX cannot occur (Horiguchi et al. 2009). At present, it is unknown whether fibulin-4 binds other members of the LOX family.

Studies of fibulin-4 hypomorphs have led to different perspectives on the role of fibulin-4 in elastogenesis. Reduction of the level of fibulin-4, achieved through transcriptional interference (Hanada et al. 2007), results not only in tortuosity and stiffening of the mouse aorta but also dilatation of the ascending aorta resembling that which is seen in human aortic aneurisms. Quantitative PCR analysis of elastin mRNA in the aortas of fibulin-4 hypomorphs shows a decrease in elastin transcripts suggesting a regulatory role for fibulin-4. The ability of fibulin-4 to promote elastin transcription has also been reported in studies in which skin fibroblasts displaying haploinsufficiency for elastin were treated with conditioned culture medium containing fibulin-4 (Chen et al. 2009). The mechanism by which fibulin-4 may positively regulate elastin gene expression is not known. However, in fibulin-4 hypomorphs, the levels of several intermediates in the TGF-β signaling pathway have been shown to be upregulated (Hanada et al. 2007).

Several mutations in the human fibulin-4 gene have been reported in association with cutis laxa, as well as an array of elastic tissue abnormalities. For example, a recessive homozygous missense mutation (169G>A, Glu57Lys) has been defined in a subject having congenital cutis laxa, vascular tortuosity, ascending aortic aneurysm, developmental emphysema, inguinal and diaphragmatic hernias, joint laxity, pectus excavatum, and multiple bone fractures (Hucthagowder et al. 2006). In another subject, compound heterozygous mutations in the fibulin-4 gene [c.835C>T (p.R279C)/c.1070_1073dupCCGC] were associated with aortic aneurysm, tortuous pulmonary arteries, and mild generalized lax skin (Dasouki et al. 2007).

Fibulin-5: Mice deficient in fibulin-5 show defects in elastogenesis in the skin, lungs, aorta, and pelvic organs (Choi et al. 2009; Nakamura et al. 2002; Yanagisawa et al. 2002). For example, fibulin-5-deficient mice display a tortuous aorta, severe emphysema, and loose skin (cutis laxa) (Nakamura et al. 2002). The relative severity of the elastic fiber abnormalities in these mutants appears less than that seen in fibulin-4 nulls. For example, unlike fibulin-4 nulls, the aortas of fibulin-5 nulls have elastic laminae albeit they display abnormalities that include interruptions in elastic laminae continuity. These abnormalities include the presence of microfibrils devoid of elastin and large aggregates of elastin occurring outside of microfibril bundles (Choi et al. 2009; Nakamura et al. 2002; Yanagisawa et al. 2002). These in vivo findings as well as findings from in vitro studies point to a role for fibulin-5 in controlling elastin aggregation by slowing the kinetics of coalescence of cross-linked elastin onto microfibrils (i.e., elastin fiber maturation) (Choi et al. 2009; Choudhury et al. 2009; Cirulis et al. 2008).

In contrast to fibulin-4 nulls that have ~90% decrease in desmosine levels (McLaughlin et al. 2006), the fibulin-5 nulls have only a 16% decrease (Yanagisawa et al. 2002). Fibulin-5 binds to LOX with low affinity (K_D = 304 nM) as compared with fibulin-4, which binds with relatively high affinity (K_D = 33 nM) (Choudhury et al. 2009). Fibulin-5 also binds to several of the LOX-like proteins including lysyl oxidase-like-1 (Loxl1), -2 (Loxl2), and -4 (Loxl4) (Hirai et al. 2007b). While little is known as to the roles of Loxl2 and Loxl4 in elastogenesis, Loxl1 appears to have a prominent role in the process of elastogenesis in tissues other than the aorta including the skin and pelvic fascia (Choi et al. 2009; Liu et al. 2004). Indeed, mice deficient in Loxl1 display connective tissue abnormalities similar to those observed in fibulin-5 nulls including loose skin and pelvic organ prolapse and enlarged airspaces in the lungs, each associated with elastic fiber defects (Drewes et al. 2007; Nakamura et al. 2002; Yanagisawa et al. 2002).

Binding site mapping studies have shown that fibulin-5 binds via its carboxy terminal fibulin-type module to Loxl1, Loxl2, and Loxl4 (Hirai et al. 2007b). This is in contrast to fibulin-4, which binds via its amino terminal region to LOX (Horiguchi et al. 2009). The significance of the fibulin-5 interaction with lysyl oxidase-like proteins is not certain; however, the finding that skin samples from fibulin-5-deficient mice display elevated levels of inactive Loxl1 suggests that fibulin-5 may promote the activation of Loxl1 (Choi et al. 2009).

Fibulin-3: Mice deficient in fibulin-3 show defects in elastin fibers of the pelvic fascia that include disruptions in elastic fibers. Fibulin-3-deficient mice are prone to pelvic organ prolapse similar to fibulin-5 null and Loxl1 null mice (Liu et al. 2004), but display a higher incidence of rectal prolaspe and inguinal hernias (McLaughlin et al. 2007; Rahn et al. 2009). Indeed, there is an absence of elastic fibers in the facia covering the myopectinal orifice in fibulin-3 nulls allowing the content of the abdominal cavity to herniate at the myopectineal orifice. Interestingly, genetic background impacts the degree to which fibulin-3 deficiency effects elastic fiber formation in myopectinal orifice facia such that fibulin-3$^{-/-}$ mice on a C57BL6 background are 100% penetrant for hernia formation whereas on a Balb/c background, fibulin-3 null mice do not develop inguinal hernias.

Similar to the phenotype of fibulin-4 and fibulin-5 nulls, fibulin-3 nulls exhibit abnormalities of elastic fibers in the skin, although these abnormalities are rather subtle. Nonetheless, the nonuniformities of elastin in the dermis likely underlie the relatively small degree of skin laxity seen in fibulin-3 mutants (McLaughlin et al. 2007; Rahn et al. 2009). Mice deficient in fibulin-3 do not display overt abnormalities of the elastic fibers in the aorta or lungs as seen in fibulin-4 and fibulin-5 nulls.

Fibulin-2: Fibulin-2 binds tropoelastin with relatively high affinity ($K_D = 0.6$ nM) (Sasaki et al. 1999); however, its expression in elastin-containing tissues is more restricted than that of other fibulin family members. For example, fibulin-2 is not found in elastic fiber-rich lung parenchyma as are fibulins 1 and 4, but it is found in elastin-containing blood vessels of the lung and aorta in association with the internal elastic lamina (IEL) (Chapman et al. 2009; Sicot et al. 2008). During postnatal periods of vascular elastogenesis, fibulin-2 is prominently expressed in the subendothelial region in proximity to where the IEL is forming (Tsuda et al. 2001). However, mice deficient in fibulin-2 do not display apparent morphological or biochemical abnormalities in elastic fiber formation in the aorta or other elastin-containing tissues (Sicot et al. 2008). Functional redundancy with fibulin-1 has been suggested to account for the lack of elastin defects in fibulin-2 nulls given that fibulin-1 expression is elevated in the IEL of aortas of fibulin-2 nulls (Sicot et al. 2008). Evidence that fibulin-2 indeed plays a role in IEL formation comes from studies of mice with compound loss-of-function mutations in fibulin-2 and fibulin-5 genes (Chapman et al. 2009). Although the IEL is overtly normal in fibulin-5 nulls, the IEL is thinner in mice doubly deficient for both fibulin-5 and fibulin-2 (Chapman et al. 2009). Thus far, mutations in the human fibulin-2 gene have not yet been directly associated with elastin abnormalities. Reduced expression of fibulin-2 has been observed in a patient with cutis laxa although associated mutations in the fibulin-2 gene were not found (Markova et al. 2003).

Fibulin-1: The role of fibulin-1 with respect to elastogenesis is still uncertain. Clearly, fibulin-1 is a component of elastin fibers in tissues including skin (Fig. 10.4), lung, and muscular arteries (Roark et al. 1995). Ultrastructural analysis shows that fibulin-1 is localized within the elastic-containing cores of elastic fibers (Roark et al. 1995). This is in contrast to other fibulins such as fibulin-2, fibulin-4, and fibulin-5 that are found along the surface of elastic lamina (Nakamura et al. 2002). The localization of these fibulins on the surface of elastic fibers is consistent with the fact that they bind to fibrillin-1 (Choudhury et al. 2009; Reinhardt et al.

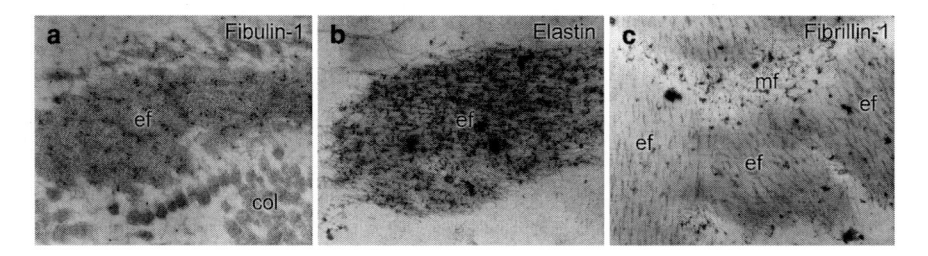

Fig. 10.4 Electron microscopic immunolocalization of fibulin-1 to the core of elastic fibers in skin. Immunogold staining of human skin was performed using (**a**) monoclonal fibulin-1 antibody, (**b**) monoclonal elastin antibody, and (**c**) polyclonal fibrillin-1 antibody. *ef* elastic fiber, *col* collagen bundles, *mf* microfibrils. Image adapted from Roark et al. (1995)

1996) which is a component of the microfibrils that surround elastic fibers. By contrast, fibulin-1 apparently does not bind to the fibrillins (El-Hallous et al. 2007) but binds to tropoelastin (Sasaki et al. 1999).

The perinatal lethality of the fibulin-1 nulls (Cooley et al. 2008; Kostka et al. 2001) has limited the scope of studies related to determining its role in elastogenesis. In fibulin-1-deficient embryos having a C57BL6 background, moderate to severe morphological abnormalities in the aorta have been reported (Kostka et al. 2001), but the relationship of these abnormalities to elastin/elastogenesis was not evaluated. In fibulin-1-deficient mouse embryos having a mixed 129/C57BL6 background, aortic narrowing was apparent in some mutants (Cooley et al. 2008), but no biochemical or histological analysis of the aorta was reported. In aortic tissue from patients with acute aortic dissection (Mohamed et al. 2009), the level of fibulin-1 mRNA is downregulated.

In *C. elegans*, a species that lacks a gene orthologous to elastin, fibulin-6 has been implicated in the formation of flexible linear matrix structures that allow the animal to flex during feeding and which are also involved in uterine attachment (Muriel et al. 2005). These structures may be evolutionarily primitive precursors to elastic fibers of vertebrates. Deficiency of fibulin-6 in *C. elegans* results in loss of these structures resulting in defective feeding movement and prolapse of the uterus. Fibulin-1 is also critically important to the assembly of these fibulin-6-containing structures, with the two splice variants of fibulin-1 each playing distinct roles in the assembly of these structures in different regions of the animal (Muriel et al. 2005; Muriel et al. 2006). For example, fibulin-1D, but not C, is required for assembly of the flexible elastic-like structures that surround the pharynx (Muriel et al. 2005). By contrast, fibulin-1C, but not D, is required for normal formation of the structures that specifically mediate uterine attachment to the body wall (Muriel et al. 2005).

10.7 Fibulins in Disease

10.7.1 Fibulins in Cancer

Evidence that the expression of several members of the fibulin family is downregulated in a variety of cancers has lead to speculation that these glycoproteins might be tumor suppressors. For example, fibulin-1 expression is low or undetectable in many malignant cell lines (Qing et al. 1997) and is downregulated in primary gastric carcinoma tissues (Cheng et al. 2008), prostate tumors (Wlazlinski et al. 2007), and ovarian tumors relative to normal surrounding tissues (Clinton et al. 1996). Other experimental evidence also supports fibulin-1 as being a tumor suppressor and includes findings that fibulin-1 inhibits the adhesion and motility of various carcinoma cell lines (Twal et al. 2001), the growth of human fibrosarcomas in nude mice (Qing et al. 1997), and papillomavirus E6-mediated transformation (Du et al. 2002). Fibulin-2 is also downregulated in a number of breast cancer cell lines and in breast cancer tissue samples (Yi et al. 2007). The fibulin-2 gene shows frequent methylation in childhood acute lymphoblastic leukemia (B-ALL), and its expression can be restored in methylated leukemia cell lines treated with 5-aza-2′-deoxycytidine (Dunwell et al. 2009). Similarly, the downregulation of fibulin-3 expression in non-small cell lung tumors as compared with normal lung parenchyma located both adjacent and distal to the tumors (Yue et al. 2007) has been attributed to promoter hypermethylation (Yue et al. 2007). Finally, the expression of both fibulin-4 and fibulin-5 is downregulated in prostate cancer (Wlazlinski et al. 2007), and fibulin-5 is downregulated in urothelial carcinoma (Hu et al. 2009a, b) and lung cancers (Yue et al. 2009). Promoter hypermethylation is responsible for suppression of fibulin-5 expression in lung cancer cell lines and primary lung tumors (Yue et al. 2009). Lung cancer cells in which fibulin-5 was overexpressed displayed reduced metastasis (Yue et al. 2009).

There are findings that appear to be discordant with the view that fibulin-1 and fibulin-3 are tumor suppressors. For example, fibulin-3 is upregulated in malignant gliomas (Hu et al. 2009a) and pancreatic adenocarcinomas (Seeliger et al. 2009). Furthermore, overexpression of fibulin-3 in glioblastoma cells augments the invasiveness of tumors in vivo, an outcome that is consistent with evidence that fibulin-3 promotes adhesion and migration of cells in culture (Hu et al. 2009a). Findings from several studies also show elevated expression of fibulin-1 in association with certain cancers. Elevated fibulin-1 expression has been reported in human breast cancers (Greene et al. 2003) and in the stroma of human ovarian epithelial tumors (Roger et al. 1998). Such findings do not necessarily undermine the notion that fibulin-1 is a tumor suppressor. Upregulation of fibulin-1 expression in stroma surrounding tumors may be part of a reactive stromal response (Ronnov-Jessen et al. 1996) which is typified in many cancers by alterations in stromal ECM (Rowley 1998) and may negatively regulate tumor growth and invasion. Indeed, low stromal expression of fibulin-1 is correlated with a higher proliferation of breast cancer epithelial cells (Sadlonova et al. 2009). There is also speculation that the two major spice variants of fibulin-1 may have opposing effects on tumorigenesis (Moll et al. 2002). Evidence for this comes from studies showing that the ratio of fibulin-1C to fibulin-1D mRNA expression is higher in ovarian cancer cells than normal ovarian cells (Moll et al. 2002). Human fibrosarcoma tumor cell lines also show a reduction or absence of fibulin-1D expression (Qing et al. 1997). Furthermore, fibrosarcoma cells engineered to express fibulin-1D display reduced growth in vivo, as well as a lowered growth capacity in soft agar and a reduced ability to invade reconstituted basement membranes (Qing et al. 1997). Therefore, augmented expression of fibulin-1D in certain cancers may be part of an antitumor response.

Steroid hormone regulation appears to underlie the observed differential expression of fibulin-1 variants in ovarian cancers. In estrogen-responsive ovarian cancer cell lines [i.e., estrogen receptor (ER) positive], the expression of fibulin-1 is greatly increased by estradiol (Clinton et al. 1996). In particular, fibulin-1C mRNA is induced by estradiol in ER-α, but not ER-β expressing breast cancer cells, indicating that fibulin-1C induction is mediated through ER-α (Moll et al. 2002). This response may be significant to ovarian carcinogenesis given that ER-α levels are greater than ER-β in ovarian cancer cells as compared with normal ovaries (Pujol et al. 1998).

In one breast cancer study (Pupa et al. 2004), fibulin-1 expression has been associated with improved survival in patients with lymphoid infiltrate at tumor sites. On the basis of these findings, as well as evidence of fibulin-1-specific humoral and cellular immune responses in a few patients with breast cancer (Pupa et al. 2004), fibulin-1 has been proposed to play a role in triggering protective antitumor immune responses.

Of particular therapeutic importance are recent findings showing that fibulin-1 acts to promote resistance of breast cancer cells to the antitumor drug doxorubicin (Pupa et al. 2007). Indeed, in doxorubicin-treated breast cancer cell lines, there is a marked increase in fibulin-1 mRNA and protein levels. Furthermore, suppression of fibulin-1 expression in breast cancer cells has been shown to result in a tenfold increase in doxorubicin-induced apoptosis as compared with control cells (Pupa et al. 2007). These findings are consistent with those from other studies showing that a number of ECM components also provide prosurvival signals to neoplastic cells.

Fibulins 1 and 5 have been implicated as inhibitors of tumor blood vessel formation. Fibrosarcoma tumor cells (HT1080 cells) engineered to express either fibulin-1 or fibulin-5 produce tumors having greatly reduced vascularization as compared with tumors arising from HT1080 cells not carrying these transgenes (Xie et al. 2008). Mice deficient in fibulin-5 display an increase in branching of blood vessels such as the long thoracic artery (Sullivan et al. 2007). Furthermore, vascular invasion is increased within polyvinyl sponges implanted into fibulin-5 null mice (Sullivan et al. 2007), and pancreatic tumor growth is attenuated in fibulin-5 null mice as compared with tumor growth in wild-type animals (Schluterman et al. 2010).

The mechanisms by which fibulin-1 and fibulin-5 exert inhibitory effects on angiogenesis are not yet clear. The antiangiogenic activity of fibulin-1 has been attributed to a proteolytically derived ~35-kDa fragment of fibulin-1 that bears similarity in size and sequence to full-length fibulin-5 and that inhibits endothelial cell proliferation (Xie et al. 2008). The antiangiogenic effects of fibulin-5 are consistent with evidence that it too inhibits endothelial cell proliferation (Xie et al. 2008). Mechanistically, this appears to relate to its ability to suppress expression of the angiopoietins and VEGF (Sullivan et al. 2007) and inhibit VEGF-stimulated activation of p38 MAPK and ERK1/2 in endothelial cells (Albig and Schiemann 2004).

In contrast with evidence that fibulin-5 is antiangiogenic are recent findings showing decreased angiogenesis in tumors grown in fibulin-5 null mice (Schluterman et al. 2010). The apparent proangiogenic activity of fibulin-5 on tumor angiogenesis is attributed to its ability to regulate the oxidative environment within the tumor (Schluterman et al. 2010). Specifically, it was found that there was an increase in the level of ROS production in tumors grown in fibulin-5 nulls (Schluterman et al. 2010). Furthermore, fibulin-5 was shown to mediate its negative effects on ROS production by inhibiting fibronectin–integrin-induced ROS generation (Schluterman et al. 2010). One consequence of this activity would be to protect endothelial cells from death due to oxidative stress.

Fibulin-3 also appears to be proangiogenic. Tumors in mice arising from pancreatic cancer cells engineered to express fibulin-3 display increased growth and a greater number of blood vessels as compared with tumors from control cancer cells (Seeliger et al. 2009). Fibulin-3 expressing tumor cells also showed a marked increase in VEGF secretion as compared with controls (Seeliger et al. 2009). There is also a reduction of intratumoral necroses in the xenografts formed by fibulin-3-transfected cells (Seeliger et al. 2009). Diminished intratumoral necrosis may relate to findings showing that tumor cells transfected to express fibulin-3 display inactivation of apoptotic signals and a shift from G_0–G_1 phase toward S phase and mitosis (Seeliger et al. 2009). These findings, together with evidence that the expression of fibulin-3 is significantly upregulated in human pancreatic adenocarcinomas (Seeliger et al. 2009), make fibulin-3 a potentially important target for antipancreatic tumor drug design.

10.7.2 Fibulins in Cardiovascular Disease

Fibulin-1 is deposited within human coronary artery atherosclerotic lesions in association with fibrinogen (Argraves et al. 2009). Whether it is playing a role in the etiology of atherosclerosis has not yet been established. Fibulin-1 binds to fibrinogen (Tran et al. 1995) that binds to lipoprotein(a) [Lp(a)] and mediates the accumulation of the atherogenic lipoprotein in blood vessel walls (Lou et al. 1998). The association of fibulin-1 and fibrinogen in atherosclerotic lesions may also regulate thrombus formation at these sites because fibulin-1 can influence fibrin formation (Tran et al. 1995). Finally, the ability of fibulin-1 to inhibit motility of cells may influence vascular SMC migration during lesion remodeling.

Fibulin-2 is expressed at relatively low levels in medial layers of blood vessels such as the aorta (Strom et al. 2006). By contrast, in SMC-rich regions of atherosclerotic aortic lesions, its expression is high. Fibulin-2 has also been detected in mechanically injured mouse carotid arteries, colocalizing with versican and hyaluronan (Strom et al. 2006). In response to ligation-induced injury, blood vessels in mice that are deficient in fibulin-2 do not display abnormalities (Chapman et al. 2009). However, mice carrying compound deletion of fibulin-2 and fibulin-5 genes show a reduced level of neointima formation as compared with fibulin-5 null mice, suggesting that fibulin-2 does play a role in injury-induced vascular SMC proliferation (Chapman et al. 2009).

Vascular tortuosity and ascending aortic aneurysm have been observed in a patient with a missense mutation in the human fibulin-4 gene coupled with reduced expression of fibulin-4 (Hucthagowder et al. 2006). These findings are consistent with those showing that fibulin-4 deficiency in mice results in enlarged and tortuous aortas with intramural bleeding, aneurysm formation, aortic stiffening, and aortic dissection (Hanada et al. 2007). In addition to disruption of the elastic laminae in regions of the aortas of mice with fibulin-4 deficiency, there is also increased vascular SMC proliferation in the tunica adventitia and increased collagen deposition in the media (Hanada et al. 2007). This is consistent with findings showing that fibulin-4 deficiency initiates excessive TGF-β signaling in aorta tissue (Hanada et al. 2007). The finding that pulse pressure in fibulin-4-deficient mice was twofold to threefold higher than in wild-type mice is an indication that the changes in the arterial wall lead to increased aortic stiffness, which in humans is important in predicting cardiovascular disease risk (Mackenzie et al. 2002).

Since deficiency of elastin does not result in aortic aneurysms (Li et al. 1998), defective elastogenesis may not account for aneurysm formation in fibulin-4-deficient mice. Conditional ablation of fibulin-4 expression in vascular SMCs has shown that fibulin-4 is critical for SMC differentiation as evidenced by reduction in expression of SM myosin heavy chain and α-SM-actin (Huang et al. 2009). The importance of proper expression and function of SMC contractile proteins to the etiology of aneurysms is underscored by the fact that thoracic aortic aneurysms in humans have been linked to mutations in genes encoding SMC contractile proteins (i.e., *MYH11* and *ACTA2*) (Guo et al. 2007; Pannu et al. 2007; Zhu et al. 2007). Alteration in the expression of SMC contractile proteins in SMCs from fibulin-4 null mice is also accompanied by increased expression of the ERK1/2 signaling pathway (Huang et al. 2009). While it remains to be established how such changes in SMC contractility and signaling contribute to aneurysm development, compromised SMC contractility has been proposed to cause familial thoracic aortic aneurysms (Milewicz et al. 2008).

Unlike fibulin-4-deficient mice, fibulin-5-deficient mice as well as humans with fibulin-5 deficiency do not develop aneurysms. Despite this, there does appear to be a role for fibulin-5 in vascular injury response. For example, fibulin-5 is normally expressed at very low levels in normal carotid arteries of mice; however, in injured carotid arteries, its expression is significantly upregulated in the neointima and to a lesser extent the adventitia (Spencer et al. 2005). Its upregulation occurs after a relatively long time period (~28 days) following the vascular injury (Spencer et al. 2005). This may indicate that fibulin-5 plays a role in ameliorating the injury response. Indeed, genetic deficiency of fibulin-5 in mice affects the vascular response to injury in that blood vessel ligation leads to development of a complex lesion having an organized thrombus and SMC proliferation adjacent to the ligature. Analysis of vascular SMCs from fibulin-5-deficient mice has revealed that fibulin-5 normally acts as a negative regulator of proliferation and migration. The observation that ligated blood vessels in fibulin-5 null mice frequently develop thrombi (Spencer et al. 2005) suggests that fibulin-5 may normally also be a positive regulator of the fibrinolysis pathway.

10.7.3 Fibulins in Eye Disease

Several of the fibulins have been implicated in the pathobiology of age-related macular degeneration (AMD). An A16,263G transition in exon 104 of the fibulin-6 gene (hemicentin-1, *HMCN1*) producing a Gln5346Arg mutation has been found to segregate exclusively with AMD in members of a large family (Schultz et al. 2003). However, a subsequent study determined that the Gln5346Arg variant is not likely to be causally related to AMD and may simply be a rare polymorphism (Fisher et al. 2007). Furthermore, this study concluded that common variants of the fibulin-6 gene do not account for a substantial proportion of AMD cases (Fisher et al. 2007). Despite these findings, a recent study has identified fibulin-6 as being able to bind ARMS2, a product of a gene identified through linkage analysis to be associated with AMD (Fritsche et al. 2008; Kortvely et al. 2010).

In earlier-onset macular dystrophies including Malattia Leventinese (ML) and Doyne honeycomb retinal dystrophy (DHRD; OMIM #126600), a single mutation (a C→T transition, Arg345Trp) has been discovered in the fibulin-3 gene (*EFEMP1*) (Stone et al. 1999). The mutation occurs within the last EGF domain of the fibulin-3 polypeptide. Further evidence that this mutation is pathogenic came from studies of mice engineered to carry the Arg345Trp mutation. Mice carrying this mutation developed deposits of material located between Bruch's membrane and the retinal pigment epithelium (RPE), which resemble basal drusen deposits in humans with inherited macular degeneration disease (Fu et al. 2007; Marmorstein et al. 2007). Fibulin-3 was found to accumulate in these basal RPE deposits (Marmorstein et al. 2007).

The fact that fibulin-3 binds to TIMP3 (Klenotic et al. 2004) and that TIMP3 mutations are associated with AMD (Weber et al. 1994) prompted evaluation of the impact of fibulin-3 deficiency on TIMP3 deposition in basal deposits (Fu et al. 2007). As a result, TIMP3 was found to be deposited basal to the RPE of fibulin-3 Arg345Trp (Fu et al. 2007). Mice homozygous for the Arg345Trp fibulin-3 mutation also were found to display elevated levels of activated complement C3 in Bruch's membrane and the RPE (Fu et al. 2007). These findings suggest that fibulin-3 may normally have a suppressive role in complement activation. Impairment of this activity may underlie the pathogenesis of AMD.

Mutations in the human fibulin-5 gene have also been linked to AMD. In one study of 402 patients with AMD, seven were found to have *FBLN5* missense mutations (Stone et al. 2004). All seven displayed drusen formation and most had detachment of the RPE, findings consistent with the fact that fibulin-5 is normally expressed in the retina and the RPE (Stone et al. 2004). Given that fibulin-5 is an adhesion protein, the observed retinal detachment in the seven patients with ADM may relate to defective cell adhesion. Since elastin is a component of Bruch's membrane in which drusen is formed, it is also possible that the missense mutations in the fibulin-5 gene could also lead to alterations in elastin assembly in Bruch's membrane.

10.8 Perspectives

Studies of the phenotypes of mice genetically deficient in individual members of the fibulin gene family have led to great advances in our understanding of the roles that these proteins play in physiological and pathological processes. Knockout mouse models for several of the fibulin genes including fibulin-6 and fibulin-8 have yet to be created. Moreover, only one of the fibulin genes, fibulin-4 (Horiguchi et al. 2009), has thus far been floxed so as to permit the generation of conditional knockout mice. Floxing of all of the genes in the fibulin family should be a major goal for the field. This will be particularly important for understanding the role of fibulin-1 in the adult given that mice that are nonconditionally deficient in fibulin-1 die perinatally. Since there appears to be a measure of redundancy among several members of the fibulin gene family, development of compound deletion mutants is also an important direction for future studies. The value of this approach was demonstrated by study of mice having compound deletion of fibulin-2 and fibulin-5 genes (Chapman et al. 2009), which showed the role of fibulin-2 in elastogenesis of the IEL.

Defining the underlying mechanisms of the various fibulins in elastogenesis will be a primary avenue for continued investigation in the field of fibulin research. Particularly interesting in this regard will be further elucidation of the nonoverlapping, but essential roles of fibulin-4 and fibulin-5 in elastic fiber formation. We herein present a refinement (Fig. 10.5) of elastogenesis models by Choudhury et al. (2009) and Horiguchi et al. (Horiguchi et al. 2009) based on the biochemical evidence that (1) fibulin-4 binds the N-terminal propeptide region of proLOX (Horiguchi et al. 2009), (2) the fibulin-4-proLOX complex interacts with tropoelastin via binding sites on proLOX and on fibulin-4 (Horiguchi et al. 2009), (3) activation of proLOX by the removal of the propeptide region leads to cross-linking of the coacervated tropoelastin monomers to form aggregates (Horiguchi et al. 2009), (4) fibulin-5, in concert with fibulin-4, regulates the process of elastin aggregration (Choudhury et al. 2009), (5) fibulin-5 chaperones cross-linked elastin aggregates onto microfibrils (Choi et al. 2009; Choudhury et al. 2009; Nakamura et al. 2002; Yanagisawa et al. 2002), and (6) elastin aggregates interact with fibrillin-1, a major microfibril component, thereby weakening the interaction of the aggregates with fibulin-4 and fibulin-5 (Choudhury et al. 2009). A major tenet of the model is that

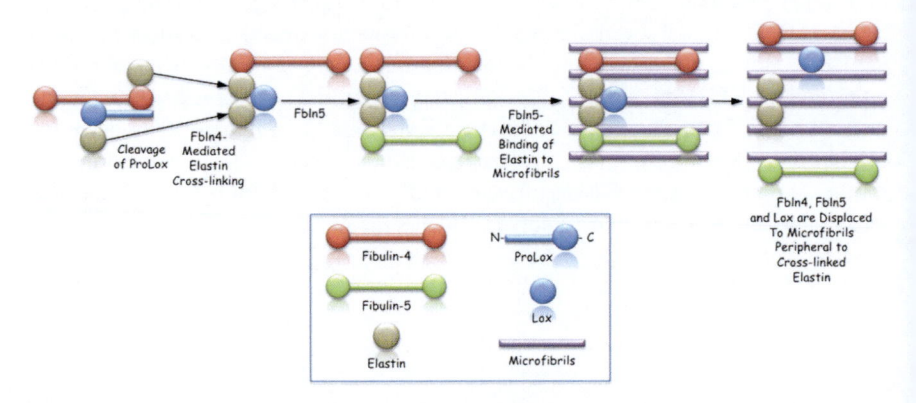

Fig. 10.5 Model of the roles of fibulin-4 and fibulin-5 in elastogenesis. According to the model, fibulin-4 regulates early events of tropoelastin cross-linking and fibulin-5 chaperones cross-linked elastin onto microfibrils

fibulin-4 regulates early events of elastin cross-linking and that fibulin-5 chaperones cross-linked elastin onto microfibrils. In light of the similarities in the elastin abnormalities of LOX-like protein-1 nulls and mice deficient in fibulin-3, future studies will need to define the interrelationship between LOX-like protein-1 and fibulin-3. Given the total lack of elastic fiber formation in the pelvic fascia of fibulin-3 nulls (McLaughlin et al. 2007), it will also be of interest to determine whether fibulin-3 functions in a manner similar to fibulin-4, but in a connective tissue-restricted manner (i.e., facilitating tropoelastin cross-linking in the pelvic fascia).

As is the case with other ECM proteins, the fibulins do not serve purely structural functions. For example, fibulin-1 is critical in regulating motility, guidance, and survival required for tissue morphogenesis, which has remained conserved between worms and man (Hesselson et al. 2004; McCulloch et al. 2009). A key aspect of this regulatory role is the involvement of MMPs of the ADAMTS family. It remains to be established whether other members of the fibulin family besides fibulin-1 also act in concert with MMPs to mediate any of their biological activities.

References

Adam S, Gohring W, Wiedemann H, Chu ML, Timpl R, Kostka G (1997) Binding of fibulin-1 to nidogen depends on its C-terminal globular domain and a specific array of calcium-binding epidermal growth factor-like (EG) modules. J Mol Biol 272:226–236

Albig AR, Schiemann WP (2004) Fibulin-5 antagonizes vascular endothelial growth factor (VEGF) signaling and angiogenic sprouting by endothelial cells. DNA Cell Biol 23:367–379

Argraves WS, Dickerson K, Burgess WH, Ruoslahti E (1989) Fibulin, a novel protein that interacts with the fibronectin receptor beta subunit cytoplasmic domain. Cell 58:623–629

Argraves WS, Tanaka A, Smith EP, Twal WO, Argraves KM, Fan D, Haudenschild CC (2009) Fibulin-1 and fibrinogen in human atherosclerotic lesions. Histochem Cell Biol 132:559–565

Argraves WS, Tran H, Burgess WH, Dickerson K (1990) Fibulin is an extracellular matrix and plasma glycoprotein with repeated domain structure. J Cell Biol 111:3155–3164

Aspberg A, Adam S, Kostka G, Timpl R, Heinegard D (1999) Fibulin-1 is a ligand for the C-type lectin domains of aggrecan and versican. J Biol Chem 274:20444–20449

Balbona K, Tran H, Godyna S, Ingham KC, Strickland DK, Argraves WS (1992) Fibulin binds to itself and to the carboxyl-terminal heparin-binding region of fibronectin. J Biol Chem 267:20120–20125

Bardin A, Moll F, Margueron R, Delfour C, Chu M, Maudelonde T, Cavailles V, Pujol P (2004) Transcriptional and post-transcriptional regulation of fibulin-1 by estrogens leads to differential induction of mRNA variants in ovarian and breast cancer cells. Endocrinology 146: 760–768

Barth JL, Argraves KM, Roark EF, Little CD, Argraves WS (1998) Identification of chicken and C. elegans fibulin-1 homologs and characterization of the C. elegans fibulin-1 gene. Matrix Biol 17:635–646

Blackburn J, Tarttelin EE, Gregory-Evans CY, Moosajee M, Gregory-Evans K (2003) Transcriptional regulation and expression of the dominant drusen gene FBLN3 (EFEMP1) in mammalian retina. Invest Ophthalmol Vis Sci 44:4613–4621

Blelloch R, Kimble J (1999) Control of organ shape by a secreted metalloprotease in the nematode *Caenorhabditis elegans*. Nature 399:586–590

Bouchey D, Argraves WS, Little CD (1996) Fibulin-1, vitronectin, and fibronectin expression during avian cardiac valve and septa development. Anat Rec 244:540–551

Bronner-Fraser M (1993) Mechanisms of neural crest cell migration. Bioessays 15:221–230
Brooke JS, Cha JH, Eidels L (2002) Latent transforming growth factor beta-binding protein-3 and fibulin-1C interact with the extracellular domain of the heparin-binding EGF-like growth factor precursor. BMC Cell Biol 3:2
Brown JC, Wiedemann H, Timpl R (1994) Protein binding and cell adhesion properties of two laminin isoforms (AmB1eB2e, AmB1sB2e) from human placenta. J Cell Sci 107:329–338
Camaj P, Seeliger H, Ischenko I, Krebs S, Blum H, De Toni EN, Faktorova D, Jauch KW, Bruns CJ (2009) EFEMP1 binds the EGF receptor and activates MAPK and Akt pathways in pancreatic carcinoma cells. Biol Chem 390:1293–1302
Canals F, Colome N, Ferrer C, Plaza-Calonge Mdel C, Rodriguez-Manzaneque JC (2006) Identification of substrates of the extracellular protease ADAMTS1 by DIGE proteomic analysis. Proteomics 6(Suppl 1):S28–S35
Chailley-Heu B, Boucherat O, Barlier-Mur AM, Bourbon JR (2005) FGF-18 is upregulated in the postnatal rat lung and enhances elastogenesis in myofibroblasts. Am J Physiol Lung Cell Mol Physiol 288:L43–L51
Chapman SL, Sicot FX, Davis EC, Huang J, Sasaki T, Chu ML, Yanagisawa H (2009) Fibulin-2 and Fibulin-5 cooperatively function to form the internal elastic lamina and protect from vascular injury. Arterioscler Thromb Vasc Biol 30:68–74
Chen Q, Zhang T, Roshetsky JF, Ouyang Z, Essers J, Fan C, Wang Q, Hinek A, Plow EF, Dicorleto PE (2009) Fibulin-4 regulates expression of the tropoelastin gene and consequent elastic-fibre formation by human fibroblasts. Biochem J 423:79–89
Cheng YY, Jin H, Liu X, Siu JM, Wong YP, Ng EK, Yu J, Leung WK, Sung JJ, Chan FK (2008) Fibulin 1 is downregulated through promoter hypermethylation in gastric cancer. Br J Cancer 99:2083–2087
Choi J, Bergdahl A, Zheng Q, Starcher B, Yanagisawa H, Davis EC (2009) Analysis of dermal elastic fibers in the absence of fibulin-5 reveals potential roles for fibulin-5 in elastic fiber assembly. Matrix Biol 28:211–220
Choudhury R, McGovern A, Ridley C, Cain SA, Baldwin A, Wang MC, Guo C, Mironov A Jr, Drymoussi Z, Trump D, Shuttleworth A, Baldock C, Kielty CM (2009) Differential regulation of elastic fiber formation by fibulin-4 and -5. J Biol Chem 284:24553–24567
Cirulis JT, Bellingham CM, Davis EC, Hubmacher D, Reinhardt DP, Mecham RP, Keeley FW (2008) Fibrillins, fibulins, and matrix-associated glycoprotein modulate the kinetics and morphology of in vitro self-assembly of a recombinant elastin-like polypeptide. Biochemistry 47:12601–12613
Clinton GM, Rougeot C, Derancourt J, Roger P, Defrenne A, Godyna S, Argraves WS, Rochefort H (1996) Estrogens increase the expression of fibulin-1, an extracellular matrix protein secreted by human ovarian cancer cells. Proc Natl Acad Sci USA 93:316–320
Coles EG, Gammill LS, Miner JH, Bronner-Fraser M (2006) Abnormalities in neural crest cell migration in laminin alpha5 mutant mice. Dev Biol 289:218–228
Cooley MA, Kern CB, Fresco VM, Wessels A, Thompson RP, McQuinn TC, Twal WO, Mjaatvedt CH, Drake CJ, Argraves WS (2008) Fibulin-1 is required for morphogenesis of neural crest-derived structures. Dev Biol 319:336–345
Corpet F (1988) Multiple sequence alignment with hierarchical clustering. Nucleic Acids Res 16:10881–10890
Costell M, Carmona R, Gustafsson E, Gonzalez-Iriarte M, Fassler R, Munoz-Chapuli R (2002) Hyperplastic conotruncal endocardial cushions and transposition of great arteries in perlecan-null mice. Circ Res 91:158–164
Courtney HS, Li Y, Twal WO, Argraves WS (2009) Serum opacity factor is a streptococcal receptor for the extracellular matrix protein fibulin-1. J Biol Chem 284:12966–12971
Dasouki M, Markova D, Garola R, Sasaki T, Charbonneau NL, Sakai LY, Chu ML (2007) Compound heterozygous mutations in fibulin-4 causing neonatal lethal pulmonary artery occlusion, aortic aneurysm, arachnodactyly, and mild cutis laxa. Am J Med Genet A 143A:2635–2641

de Vega S, Iwamoto T, Nakamura T, Hozumi K, McKnight DA, Fisher LW, Fukumoto S, Yamada Y (2007) TM14 is a new member of the fibulin family (fibulin-7) that interacts with extracellular matrix molecules and is active for cell binding. J Biol Chem 282:30878–30888

Drewes PG, Yanagisawa H, Starcher B, Hornstra I, Csiszar K, Marinis SI, Keller P, Word RA (2007) Pelvic organ prolapse in fibulin-5 knockout mice: pregnancy-induced changes in elastic fiber homeostasis in mouse vagina. Am J Pathol 170:578–589

Du M, Fan X, Hong E, Chen JJ (2002) Interaction of oncogenic papillomavirus E6 proteins with fibulin-1. Biochem Biophys Res Commun 296:962–969

Dunwell TL, Hesson LB, Pavlova T, Zabarovska V, Kashuba V, Catchpoole D, Chiaramonte R, Brini AT, Griffiths M, Maher ER, Zabarovsky E, Latif F (2009) Epigenetic analysis of childhood acute lymphoblastic leukemia. Epigenetics 4:185–193

El-Hallous E, Sasaki T, Hubmacher D, Getie M, Tiedemann K, Brinckmann J, Batge B, Davis EC, Reinhardt DP (2007) Fibrillin-1 interactions with fibulins depend on the first hybrid domain and provide an adaptor function to tropoelastin. J Biol Chem 282:8935–8946

Fisher SA, Rivera A, Fritsche LG, Keilhauer CN, Lichtner P, Meitinger T, Rudolph G, Weber BH (2007) Case-control genetic association study of fibulin-6 (FBLN6 or HMCN1) variants in age-related macular degeneration (AMD). Hum Mutat 28:406–413

Fritsche LG, Loenhardt T, Janssen A, Fisher SA, Rivera A, Keilhauer CN, Weber BH (2008) Age-related macular degeneration is associated with an unstable ARMS2 (LOC387715) mRNA. Nat Genet 40:892–896

Fu L, Garland D, Yang Z, Shukla D, Rajendran A, Pearson E, Stone EM, Zhang K, Pierce EA (2007) The R345W mutation in EFEMP1 is pathogenic and causes AMD-like deposits in mice. Hum Mol Genet 16:2411–2422

Fujimoto N, Terlizzi J, Brittingham R, Fertala A, McGrath JA, Uitto J (2005) Extracellular matrix protein 1 interacts with the domain III of fibulin-1C and 1D variants through its central tandem repeat 2. Biochem Biophys Res Commun 333:1327–1333

Gallagher WM, Argentini M, Sierra V, Bracco L, Debussche L, Conseiller E (1999) MBP1: a novel mutant p53-specific protein partner with oncogenic properties. Oncogene 18:3608–3616

Giltay R, Timpl R, Kostka G (1999) Sequence, recombinant expression and tissue localization of two novel extracellular matrix proteins, fibulin-3 and fibulin-4. Matrix Biol 18:469–480

Grassel S, Sicot FX, Gotta S, Chu ML (1999) Mouse fibulin-2 gene. Complete exon-intron organization and promoter characterization. Eur J Biochem 263:471–477

Greene LM, Twal WO, Duffy MJ, McDermott EW, Hill AD, O'Higgins NJ, McCann AH, Dervan PA, Argraves WS, Gallagher WM (2003) Elevated expression and altered processing of fibulin-1 protein in human breast cancer. Br J Cancer 88:871–878

Guo DC, Pannu H, Tran-Fadulu V, Papke CL, Yu RK, Avidan N, Bourgeois S, Estrera AL, Safi HJ, Sparks E, Amor D, Ades L, McConnell V, Willoughby CE, Abuelo D, Willing M, Lewis RA, Kim DH, Scherer S, Tung PP, Ahn C, Buja LM, Raman CS, Shete SS, Milewicz DM (2007) Mutations in smooth muscle alpha-actin (ACTA2) lead to thoracic aortic aneurysms and dissections. Nat Genet 39:1488–1493

Hanada K, Vermeij M, Garinis GA, de Waard MC, Kunen MG, Myers L, Maas A, Duncker DJ, Meijers C, Dietz HC, Kanaar R, Essers J (2007) Perturbations of vascular homeostasis and aortic valve abnormalities in fibulin-4 deficient mice. Circ Res 100:738–746

Hayashido Y, Lucas A, Rougeot C, Godyna S, Argraves WS, Rochefort H (1998) Estradiol and fibulin-1 inhibit motility of human ovarian- and breast-cancer cells induced by fibronectin. Int J Cancer 75:654–658

Henderson DJ, Copp AJ (1998) Versican expression is associated with chamber specification, septation, and valvulogenesis in the developing mouse heart. Circ Res 83:523–532

Henderson DJ, Ybot-Gonzalez P, Copp AJ (1997) Over-expression of the chondroitin sulphate proteoglycan versican is associated with defective neural crest migration in the Pax3 mutant mouse (splotch). Mech Dev 69:39–51

Hesselson D, Newman C, Kim KW, Kimble J (2004) GON-1 and fibulin have antagonistic roles in control of organ shape. Curr Biol 14:2005–2010

Hesselson D, Newman C, Kimble J (2003) Control of gonad size and shape by distal tip cell migration, the GON-1 metalloproteinase and fibulin in *C. elegans*. Mol Biol Cell 14:140a

Hirai M, Horiguchi M, Ohbayashi T, Kita T, Chien KR, Nakamura T (2007a) Latent TGF-beta-binding protein 2 binds to DANCE/fibulin-5 and regulates elastic fiber assembly. EMBO J 26:3283–3295

Hirai M, Ohbayashi T, Horiguchi M, Okawa K, Hagiwara A, Chien KR, Kita T, Nakamura T (2007b) Fibulin-5/DANCE has an elastogenic organizer activity that is abrogated by proteolytic cleavage in vivo. J Cell Biol 176:1061–1071

Hopf M, Gohring W, Kohfeldt E, Yamada Y, Timpl R (1999) Recombinant domain IV of perlecan binds to nidogens, laminin-nidogen complex, fibronectin, fibulin-2 and heparin. Eur J Biochem 259:917–925

Horiguchi M, Inoue T, Ohbayashi T, Hirai M, Noda K, Marmorstein LY, Yabe D, Takagi K, Akama TO, Kita T, Kimura T, Nakamura T (2009) Fibulin-4 conducts proper elastogenesis via interaction with cross-linking enzyme lysyl oxidase. Proc Natl Acad Sci USA 106:19029–19034

Hornstra IK, Birge S, Starcher B, Bailey AJ, Mecham RP, Shapiro SD (2003) Lysyl oxidase is required for vascular and diaphragmatic development in mice. J Biol Chem 278:14387–14393

Hu B, Thirtamara-Rajamani KK, Sim H, Viapiano MS (2009a) Fibulin-3 is uniquely upregulated in malignant gliomas and promotes tumor cell motility and invasion. Mol Cancer Res 7:1756–1770

Hu Z, Ai Q, Xu H, Ma X, Li HZ, Shi TP, Wang C, Gong DJ, Zhang X (2009) Fibulin-5 is down-regulated in urothelial carcinoma of bladder and inhibits growth and invasion of human bladder cancer cell line 5637. Urol Oncol.

Huang J, Davis EC, Chapman SL, Budatha M, Marmorstein LY, Word RA, Yanagisawa H (2009) Fibulin-4 deficiency results in ascending aortic aneurysms. A potential link between abnormal smooth muscle cell phenotype and aneurysm progression. Circ Res 106:583–592

Hucthagowder V, Sausgruber N, Kim KH, Angle B, Marmorstein LY, Urban Z (2006) Fibulin-4: a novel gene for an autosomal recessive cutis laxa syndrome. Am J Hum Genet 78:1075–1080

Kapetanopoulos A, Fresser F, Millonig G, Shaul Y, Baier G, Utermann G (2002) Direct interaction of the extracellular matrix protein DANCE with apolipoprotein(a) mediated by the kringle IV-type 2 domain. Mol Genet Genomics 267:440–446

Kern CB, Twal WO, Mjaatvedt CH, Fairey SE, Toole BP, Iruela-Arispe ML, Argraves WS (2006) Proteolytic cleavage of versican during cardiac cushion morphogenesis. Dev Dyn 235:2238–2247

Kikkawa Y, Sanzen N, Fujiwara H, Sonnenberg A, Sekiguchi K (2000) Integrin binding specificity of laminin-10/11: laminin-10/11 are recognized by alpha 3 beta 1, alpha 6 beta 1 and alpha 6 beta 4 integrins. J Cell Sci 113(Pt 5):869–876

Klenotic PA, Munier FL, Marmorstein LY, Anand-Apte B (2004) Tissue inhibitor of metalloproteinases-3 (TIMP-3) is a binding partner of epithelial growth factor-containing fibulin-like extracellular matrix protein 1 (EFEMP1). Implications for macular degenerations. J Biol Chem 279:30469–30473

Kluge M, Mann K, Dziadek M, Timpl R (1990) Characterization of a novel calcium-binding 90-kDa glycoprotein (BM-90) shared by basement membranes and serum. Eur J Biochem 193:651–659

Kobayashi N, Kostka G, Garbe JH, Keene DR, Bachinger HP, Hanisch FG, Markova D, Tsuda T, Timpl R, Chu ML, Sasaki T (2007) A comparative analysis of the fibulin protein family. Biochemical characterization, binding interactions, and tissue localization. J Biol Chem 282:11805–11816

Kortvely E, Hauck SM, Duetsch G, Gloeckner CJ, Kremmer E, Alge-Priglinger CS, Deeg C, Ueffing M (2010) ARMS2 is a constituent of the extracellular matrix providing a link between familial and sporadic age-related macular degenerations. Invest Ophthalmol Vis Sci 51:79–88

Kostka G, Giltay R, Bloch W, Addicks K, Timpl R, Fassler R, Chu ML (2001) Perinatal lethality and endothelial cell abnormalities in several vessel compartments of fibulin-1-deficient mice. Mol Cell Biol 21:7025–7034

Kowal RC, Richardson JA, Miano JM, Olson EN (1999) EVEC, a novel epidermal growth factor-like repeat-containing protein upregulated in embryonic and diseased adult vasculature [see comments]. Circ Res 84:1166–1176

Kramer JM (1997) Extracellular matrix. In: Riddle DL, Blumenthal T, Meyer BJ, Priess JR (eds) *C. elegans* II. Cold Spring Harbor Laboratory Press, Cold Spring Harbor, NY, pp 471–500

Kuang PP, Joyce-Brady M, Zhang XH, Jean JC, Goldstein RH (2006) Fibulin-5 gene expression in human lung fibroblasts is regulated by TGF-beta and phosphatidylinositol 3-kinase activity. Am J Physiol Cell Physiol 291:C1412–C1421

Kubota Y, Kuroki R, Nishiwaki K (2004) A Fibulin-1 homolog interacts with an ADAM protease that controls cell migration in *C. elegans*. Curr Biol 14:2011–2018

Kubota Y, Nishiwaki K (2003) Mutations in a fibulin-1 homolog bypass the requirement of the MIG-17 ADAM protease for cell migration in *C. elegans*. Mol Biol Cell 14:261a

Kubota Y, Ohkura K, Tamai KK, Nagata K, Nishiwaki K (2008) MIG-17/ADAMTS controls cell migration by recruiting nidogen to the basement membrane in *C. elegans*. Proc Natl Acad Sci USA 105:20804–20809

Lee NV, Rodriguez-Manzaneque JC, Thai SN, Twal WO, Luque A, Lyons KM, Argraves WS, Iruela-Arispe ML (2005) Fibulin-1 acts as a cofactor for the matrix metalloprotease ADAMTS-1. J Biol Chem 280:34796–34804

Li DY, Brooke B, Davis EC, Mecham RP, Sorensen LK, Boak BB, Eichwald E, Keating MT (1998) Elastin is an essential determinant of arterial morphogenesis. Nature 393:276–280

Liu X, Zhao Y, Gao J, Pawlyk B, Starcher B, Spencer JA, Yanagisawa H, Zuo J, Li T (2004) Elastic fiber homeostasis requires lysyl oxidase-like 1 protein. Nat Genet 36:178–182

Lomas AC, Mellody KT, Freeman LJ, Bax DV, Shuttleworth CA, Kielty CM (2007) Fibulin-5 binds human smooth-muscle cells through alpha5beta1 and alpha4beta1 integrins, but does not support receptor activation. Biochem J 405:417–428

Lou XJ, Boonmark NW, Horrigan FT, Degen JL, Lawn RM (1998) Fibrinogen deficiency reduces vascular accumulation of apolipoprotein(a) and development of atherosclerosis in apolipoprotein(a) transgenic mice. Proc Natl Acad Sci USA 95:12591–12595

Mackenzie IS, Wilkinson IB, Cockcroft JR (2002) Assessment of arterial stiffness in clinical practice. QJM 95:67–74

Markova D, Zou Y, Ringpfeil F, Sasaki T, Kostka G, Timpl R, Uitto J, Chu ML (2003) Genetic heterogeneity of cutis laxa: a heterozygous tandem duplication within the fibulin-5 (FBLN5) gene. Am J Hum Genet 72:998–1004

Marmorstein LY, McLaughlin PJ, Peachey NS, Sasaki T, Marmorstein AD (2007) Formation and progression of sub-retinal pigment epithelium deposits in Efemp1 mutation knock-in mice: a model for the early pathogenic course of macular degeneration. Hum Mol Genet 16: 2423–2432

McCulloch DR, Nelson CM, Dixon LJ, Silver DL, Wylie JD, Lindner V, Sasaki T, Cooley MA, Argraves WS, Apte SS (2009) ADAMTS metalloproteases generate active versican fragments that regulate interdigital web regression. Dev Cell 17:687–698

McLaughlin PJ, Bakall B, Choi J, Liu Z, Sasaki T, Davis EC, Marmorstein AD, Marmorstein LY (2007) Lack of fibulin-3 causes early aging and herniation, but not macular degeneration in mice. Hum Mol Genet 16:3059–3070

McLaughlin PJ, Chen Q, Horiguchi M, Starcher BC, Stanton JB, Broekelmann TJ, Marmorstein AD, McKay B, Mecham R, Nakamura T, Marmorstein LY (2006) Targeted disruption of fibulin-4 abolishes elastogenesis and causes perinatal lethality in mice. Mol Cell Biol 26:1700–1709

Merz DC, Alves G, Kawano T, Zheng H, Culotti JG (2003) UNC-52/perlecan affects gonadal leader cell migrations in C. elegans hermaphrodites through alterations in growth factor signaling. Dev Biol 256:173–186

Milewicz DM, Guo DC, Tran-Fadulu V, Lafont AL, Papke CL, Inamoto S, Kwartler CS, Pannu H (2008) Genetic basis of thoracic aortic aneurysms and dissections: focus on smooth muscle cell contractile dysfunction. Annu Rev Genomics Hum Genet 9:283–302

Mohamed SA, Sievers HH, Hanke T, Richardt D, Schmidtke C, Charitos EI, Belge G, Bullerdiek J (2009) Pathway analysis of differentially expressed genes in patients with acute aortic dissection. Biomark Insights 4:81–90

Moll F, Katsaros D, Lazennec G, Hellio N, Roger P, Giacalone PL, Chalbos D, Maudelonde T, Rochefort H, Pujol P (2002) Estrogen induction and overexpression of fibulin-1C mRNA in ovarian cancer cells. Oncogene 21:1097–1107

Muriel JM, Dong C, Hutter H, Vogel BE (2005) Fibulin-1C and Fibulin-1D splice variants have distinct functions and assemble in a hemicentin-dependent manner. Development 132: 4223–4234

Muriel JM, Xu X, Kramer JM, Vogel BE (2006) Selective assembly of fibulin-1 splice variants reveals distinct extracellular matrix networks and novel functions for perlecan/UNC-52 splice variants. Dev Dyn 235:2632–2640

Nakamoto T, Okada H, Nakajima T, Ikuta A, Yasuda K, Kanzaki H (2005) Progesterone induces the fibulin-1 expression in human endometrial stromal cells. Hum Reprod 20:1447–1455

Nakamura T, Lozano PR, Ikeda Y, Iwanaga Y, Hinek A, Minamisawa S, Cheng CF, Kobuke K, Dalton N, Takada Y, Tashiro K, Ross J Jr, Honjo T, Chien KR (2002) Fibulin-5/DANCE is essential for elastogenesis in vivo. Nature 415:171–175

Nakamura T, Ruiz-Lozano P, Lindner V, Yabe D, Taniwaki M, Furukawa Y, Kobuke K, Tashiro K, Lu Z, Andon NL, Schaub R, Matsumori A, Sasayama S, Chien KR, Honjo T (1999) DANCE, a novel secreted RGD protein expressed in developing, atherosclerotic, and balloon-injured arteries. J Biol Chem 274:22476–22483

Ng KM, Catalano MG, Pinos T, Selva DM, Avvakumov GV, Munell F, Hammond GL (2006) Evidence that fibulin family members contribute to the steroid-dependent extravascular sequestration of sex hormone-binding globulin. J Biol Chem 281:15853–15861

Nguyen AD, Itoh S, Jeney V, Yanagisawa H, Fujimoto M, Ushio-Fukai M, Fukai T (2004) Fibulin-5 is a novel binding protein for extracellular superoxide dismutase. Circ Res 95:1067–1074

Nishiwaki K, Hisamoto N, Matsumoto K (2000) A metalloprotease disintegrin that controls cell migration in *Caenorhabditis elegans*. Science 288:2205–2208

Ohsawa I, Takamura C, Kohsaka S (2001) Fibulin-1 binds the amino-terminal head of beta-amyloid precursor protein and modulates its physiological function. J Neurochem 76: 1411–1420

Olin AI, Morgelin M, Sasaki T, Timpl R, Heinegard D, Aspberg A (2001) The proteoglycans aggrecan and Versican form networks with fibulin-2 through their lectin domain binding. J Biol Chem 276:1253–1261

Ozaki T, Kondo K, Nakamura Y, Ichimiya S, Nakagawara A, Sakiyama S (1997) Interaction of DA41, a DAN-binding protein, with the epidermal growth factor-like protein, S(1–5). Biochem Biophys Res Commun 237:245–250

Pan TC, Sasaki T, Zhang RZ, Fassler R, Timpl R, Chu ML (1993) Structure and expression of fibulin-2, a novel extracellular matrix protein with multiple EGF-like repeats and consensus motifs for calcium binding. J Cell Biol 123:1269–1277

Pannu H, Tran-Fadulu V, Papke CL, Scherer S, Liu Y, Presley C, Guo D, Estrera AL, Safi HJ, Brasier AR, Vick GW, Marian AJ, Raman CS, Buja LM, Milewicz DM (2007) MYH11 mutations result in a distinct vascular pathology driven by insulin-like growth factor 1 and angiotensin II. Hum Mol Genet 16:2453–2462

Perbal B, Martinerie C, Sainson R, Werner M, He B, Roizman B (1999) The C-terminal domain of the regulatory protein NOVH is sufficient to promote interaction with fibulin 1C: a clue for a role of NOVH in cell-adhesion signaling. Proc Natl Acad Sci USA 96:869–874

Perissinotto D, Iacopetti P, Bellina I, Doliana R, Colombatti A, Pettway Z, Bronner-Fraser M, Shinomura T, Kimata K, Morgelin M, Lofberg J, Perris R (2000) Avian neural crest cell migration is diversely regulated by the two major hyaluronan-binding proteoglycans PG-M/versican and aggrecan. Development 127:2823–2842

Perris R, Perissinotto D (2000) Role of the extracellular matrix during neural crest cell migration. Mech Dev 95:3–21

Perris R, Perissinotto D, Pettway Z, Bronner-Fraser M, Morgelin M, Kimata K (1996) Inhibitory effects of PG-H/aggrecan and PG-M/versican on avian neural crest cell migration. FASEB J 10:293–301

Pfaff M, Sasaki T, Tangemann K, Chu ML, Timpl R (1995) Integrin-binding and cell-adhesion studies of fibulins reveal a particular affinity for alpha IIb beta 3. Exp Cell Res 219:87–92

Pietri T, Eder O, Breau MA, Topilko P, Blanche M, Brakebusch C, Fassler R, Thiery JP, Dufour S (2004) Conditional beta1-integrin gene deletion in neural crest cells causes severe developmental alterations of the peripheral nervous system. Development 131:3871–3883

Pujol P, Rey JM, Nirde P, Roger P, Gastaldi M, Laffargue F, Rochefort H, Maudelonde T (1998) Differential expression of estrogen receptor-alpha and -beta messenger RNAs as a potential marker of ovarian carcinogenesis. Cancer Res 58:5367–5373

Pupa SM, Argraves WS, Forti S, Casalini P, Berno V, Agresti R, Aiello P, Invernizzi A, Baldassari P, Twal WO, Mortarini R, Anichini A, Menard S (2004) Immunological and pathobiological roles of fibulin-1 in breast cancer. Oncogene 23:2153–2160

Pupa SM, Giuffre S, Castiglioni F, Bertola L, Cantu M, Bongarzone I, Baldassari P, Mortarini R, Argraves WS, Anichini A, Menard S, Tagliabue E (2007) Regulation of breast cancer response to chemotherapy by fibulin-1. Cancer Res 67:4271–4277

Qing J, Maher VM, Tran H, Argraves WS, Dunstan RW, McCormick JJ (1997) Suppression of anchorage-independent growth and matrigel invasion and delayed tumor formation by elevated expression of fibulin-1D in human fibrosarcoma-derived cell lines. Oncogene 15:2159–2168

Rahn DD, Acevedo JF, Roshanravan S, Keller PW, Davis EC, Marmorstein LY, Word RA (2009) Failure of pelvic organ support in mice deficient in fibulin-3. Am J Pathol 174:206–215

Reinhardt DP, Sasaki T, Dzamba BJ, Keene DR, Chu ML, Gohring W, Timpl R, Sakai LY (1996) Fibrillin-1 and fibulin-2 interact and are colocalized in some tissues. J Biol Chem 271: 19489–19496

Roark EF, Keene DR, Haudenschild CC, Godyna S, Little CD, Argraves WS (1995) The association of human fibulin-1 with elastic fibers: an immunohistological, ultrastructural, and RNA study. J Histochem Cytochem 43:401–411

Roger P, Pujol P, Lucas A, Baldet P, Rochefort H (1998) Increased immunostaining of fibulin-1, an estrogen-regulated protein in the stroma of human ovarian epithelial tumors. Am J Pathol 153:1579–1588

Ronnov-Jessen L, Petersen OW, Bissell MJ (1996) Cellular changes involved in conversion of normal to malignant breast: importance of the stromal reaction. Physiol Rev 76:69–125

Rowley DR (1998) What might a stromal response mean to prostate cancer progression? Cancer Metastasis Rev 17:411–419

Sadlonova A, Bowe DB, Novak Z, Mukherjee S, Duncan VE, Page GP, Frost AR (2009) Identification of molecular distinctions between normal breast-associated fibroblasts and breast cancer-associated fibroblasts. Cancer Microenviron 2:9–21

Salmivirta K, Talts JF, Olsson M, Sasaki T, Timpl R, Ekblom P (2002) Binding of mouse nidogen-2 to basement membrane components and cells and its expression in embryonic and adult tissues suggest complementary functions of the two nidogens. Exp Cell Res 279:188–201

Sandy JD, Westling J, Kenagy RD, Iruela-Arispe ML, Verscharen C, Rodriguez-Mazaneque JC, Zimmermann DR, Lemire JM, Fischer JW, Wight TN, Clowes AW (2001) Versican V1 proteolysis in human aorta in vivo occurs at the Glu441-Ala442 bond, a site that is cleaved by recombinant ADAMTS-1 and ADAMTS-4. J Biol Chem 276:13372–13378

Sasaki T, Gohring W, Mann K, Brakebusch C, Yamada Y, Fassler R, Timpl R (2001) Short arm region of laminin-5 gamma2 chain: structure, mechanism of processing and binding to heparin and proteins. J Mol Biol 314:751–763

Sasaki T, Gohring W, Miosge N, Abrams WR, Rosenbloom J, Timpl R (1999) Tropoelastin binding to fibulins, nidogen-2 and other extracellular matrix proteins. FEBS Lett 460:280–284

Sasaki T, Gohring W, Pan TC, Chu ML, Timpl R (1995a) Binding of mouse and human fibulin-2 to extracellular matrix ligands. J Mol Biol 254:892–899

Sasaki T, Kostka G, Gohring W, Wiedemann H, Mann K, Chu ML, Timpl R (1995b) Structural characterization of two variants of fibulin-1 that differ in nidogen affinity. J Mol Biol 245:241–250

Schiemann WP, Blobe GC, Kalume DE, Pandey A, Lodish HF (2002) Context-specific effects of fibulin-5 (DANCE/EVEC) on cell proliferation, motility, and invasion. Fibulin-5 is induced by transforming growth factor-beta and affects protein kinase cascades. J Biol Chem 277:27367–27377

Schluterman MK, Chapman SL, Korpanty G, Ozumi K, Fukai T, Yanagisawa H, Brekken RA (2010) Loss of fibulin-5 binding to beta1 integrins inhibits tumor growth by increasing the level of ROS. Dis Model Mech 3:333–342

Schultz DW, Klein ML, Humpert AJ, Luzier CW, Persun V, Schain M, Mahan A, Runckel C, Cassera M, Vittal V, Doyle TM, Martin TM, Weleber RG, Francis PJ, Acott TS (2003) Analysis of the ARMD1 locus: evidence that a mutation in HEMICENTIN-1 is associated with age-related macular degeneration in a large family. Hum Mol Genet 12:3315–3323

Seeliger H, Camaj P, Ischenko I, Kleespies A, De Toni EN, Thieme SE, Blum H, Assmann G, Jauch KW, Bruns CJ (2009) EFEMP1 expression promotes in vivo tumor growth in human pancreatic adenocarcinoma. Mol Cancer Res 7:189–198

Sercu S, Lambeir AM, Steenackers E, El Ghalbzouri A, Geentjens K, Sasaki T, Oyama N, Merregaert J (2009) ECM1 interacts with fibulin-3 and the beta 3 chain of laminin 332 through its serum albumin subdomain-like 2 domain. Matrix Biol 28:160–169

Sicot FX, Tsuda T, Markova D, Klement JF, Arita M, Zhang RZ, Pan TC, Mecham RP, Birk DE, Chu ML (2008) Fibulin-2 is dispensable for mouse development and elastic fiber formation. Mol Cell Biol 28:1061–1067

Spence SG, Argraves WS, Walters L, Hungerford JE, Little CD (1992) Fibulin is localized at sites of epithelial-mesenchymal transitions in the early avian embryo. Dev Biol 151:473–484

Spencer JA, Hacker SL, Davis EC, Mecham RP, Knutsen RH, Li DY, Gerard RD, Richardson JA, Olson EN, Yanagisawa H (2005) Altered vascular remodeling in fibulin-5-deficient mice reveals a role of fibulin-5 in smooth muscle cell proliferation and migration. Proc Natl Acad Sci USA 102:2946–2951

Stone EM, Braun TA, Russell SR, Kuehn MH, Lotery AJ, Moore PA, Eastman CG, Casavant TL, Sheffield VC (2004) Missense variations in the fibulin 5 gene and age-related macular degeneration. N Engl J Med 351:346–353

Stone EM, Lotery AJ, Munier FL, Heon E, Piguet B, Guymer RH, Vandenburgh K, Cousin P, Nishimura D, Swiderski RE, Silvestri G, Mackey DA, Hageman GS, Bird AC, Sheffield VC, Schorderet DF (1999) A single EFEMP1 mutation associated with both Malattia Leventinese and Doyne honeycomb retinal dystrophy. Nat Genet 22:199–202

Strom A, Olin AI, Aspberg A, Hultgardh-Nilsson A (2006) Fibulin-2 is present in murine vascular lesions and is important for smooth muscle cell migration. Cardiovasc Res 69:755–763

Sullivan KM, Bissonnette R, Yanagisawa H, Hussain SN, Davis EC (2007) Fibulin-5 functions as an endogenous angiogenesis inhibitor. Lab Invest 87:818–827

Talts JF, Andac Z, Gohring W, Brancaccio A, Timpl R (1999) Binding of the G domains of laminin alpha1 and alpha2 chains and perlecan to heparin, sulfatides, alpha-dystroglycan and several extracellular matrix proteins. EMBO J 18:863–870

Talts JF, Sasaki T, Miosge N, Gohring W, Mann K, Mayne R, Timpl R (2000) Structural and functional analysis of the recombinant G domain of the laminin alpha4 chain and its proteolytic processing in tissues. J Biol Chem 275:35192–35199

Tran H, Mattei M, Godyna S, Argraves WS (1997) Human fibulin-1D: Molecular cloning, expression and similarity with S1-5 protein, a new member of the fibulin gene family. Matrix Biol 15:479–493

Tran H, Tanaka A, Litvinovich SV, Medved LV, Haudenschild CC, Argraves WS (1995) The interaction of fibulin-1 with fibrinogen. A potential role in hemostasis and thrombosis. J Biol Chem 270:19458–19464

Tsuda T, Wang H, Timpl R, Chu ML (2001) Fibulin-2 expression marks transformed mesenchymal cells in developing cardiac valves, aortic arch vessels, and coronary vessels. Dev Dyn 222:89–100

Twal WO, Czirok A, Hegedus B, Knaak C, Chintalapudi MR, Okagawa H, Sugi Y, Argraves WS (2001) Fibulin-1 suppression of fibronectin-regulated cell adhesion and motility. J Cell Sci 114:4587–4598

Utani A, Nomizu M, Yamada Y (1997) Fibulin-2 binds to the short arms of laminin-5 and laminin-1 via conserved amino acid sequences. J Biol Chem 272(5):2814–2820

Vogel BE, Hedgecock EM (2001) Hemicentin, a conserved extracellular member of the immunoglobulin superfamily, organizes epithelial and other cell attachments into oriented line-shaped junctions. Development 128:883–894

Vukovic J, Ruitenberg MJ, Roet K, Franssen E, Arulpragasam A, Sasaki T, Verhaagen J, Harvey AR, Busfield SJ, Plant GW (2009) The glycoprotein fibulin-3 regulates morphology and motility of olfactory ensheathing cells in vitro. Glia 57:424–443

Weber BH, Vogt G, Pruett RC, Stohr H, Felbor U (1994) Mutations in the tissue inhibitor of metalloproteinases-3 (TIMP3) in patients with Sorsby's fundus dystrophy. Nat Genet 8:352–356

Wlazlinski A, Engers R, Hoffmann MJ, Hader C, Jung V, Muller M, Schulz WA (2007) Downregulation of several fibulin genes in prostate cancer. Prostate 67:1770–1780

Xie L, Palmsten K, MacDonald B, Kieran MW, Potenta S, Vong S, Kalluri R (2008) Basement membrane derived fibulin-1 and fibulin-5 function as angiogenesis inhibitors and suppress tumor growth. Exp Biol Med (Maywood) 233:155–162

Xu X, Dong C, Vogel BE (2007) Hemicentins assemble on diverse epithelia in the mouse. J Histochem Cytochem 55:119–126

Yanagisawa H, Davis EC, Starcher BC, Ouchi T, Yanagisawa M, Richardson JA, Olson EN (2002) Fibulin-5 is an elastin-binding protein essential for elastic fibre development in vivo. Nature 415:168–171

Yi CH, Smith DJ, West WW, Hollingsworth MA (2007) Loss of fibulin-2 expression is associated with breast cancer progression. Am J Pathol 170:1535–1545

Yue W, Dacic S, Sun Q, Landreneau R, Guo M, Zhou W, Siegfried JM, Yu J, Zhang L (2007) Frequent inactivation of RAMP2, EFEMP1 and Dutt1 in lung cancer by promoter hypermethylation. Clin Cancer Res 13:4336–4344

Yue W, Sun Q, Landreneau R, Wu C, Siegfried JM, Yu J, Zhang L (2009) Fibulin-5 suppresses lung cancer invasion by inhibiting matrix metalloproteinase-7 expression. Cancer Res 69:6339–6346

Zanetti M, Braghetta P, Sabatelli P, Mura I, Doliana R, Colombatti A, Volpin D, Bonaldo P, Bressan GM (2004) EMILIN-1 deficiency induces elastogenesis and vascular cell defects. Mol Cell Biol 24:638–650

Zhang H, Gao X, Weng C, Xu Z (2008) Interaction between angiogenin and fibulin 1: evidence and implication. Acta Biochim Biophys Sin (Shanghai) 40:375–380

Zhang HY, Chu ML, Pan TC, Sasaki T, Timpl R, Ekblom P (1995) Extracellular matrix protein fibulin-2 is expressed in the embryonic endocardial cushion tissue and is a prominent component of valves in adult heart. Dev Biol 167:18–26

Zhang HY, Kluge M, Timpl R, Chu ML, Ekblom P (1993) The extracellular matrix glycoproteins BM-90 and tenascin are expressed in the mesenchyme at sites of endothelial-mesenchymal conversion in the embryonic mouse heart. Differentiation 52:211–220

Zhang HY, Lardelli M, Ekblom P (1997) Sequence of zebrafish fibulin-1 and its expression in developing heart and other embryonic organs. Dev Genes Evol 207:340–351

Zhu L, Bonnet D, Boussion M, Vedie B, Sidi D, Jeunemaitre X (2007) Investigation of the MYH11 gene in sporadic patients with an isolated persistently patent arterial duct. Cardiol Young 17:666–672

Chapter 11
Matricellular Proteins

David D. Roberts and Lester F. Lau

Abstract In addition to its major structural elements, extracellular matrix contains a number of factors that are important for orchestrating developmental morphogenesis, maintaining tissue homeostasis in adults, and regenerating tissue following injury. Several proteins that serve these functions share a complex modular structure that enables them to interact with specific components of the matrix while engaging specific cell surface receptors through which they control cell behavior. These have been named matricellular proteins. Matricellular proteins, including the thrombospondins, some thrombospondin-repeat superfamily members, tenascins, SPARC, CCN proteins, and SIBLING proteins, are increasingly recognized to play important roles in inherited disorders, responses to injury and stress, and the pathogenesis of several chronic diseases of aging. Improved understanding of the functions and mechanisms of action of matricellular proteins is beginning to yield novel therapeutic strategies for prevention or treatment of these diseases.

11.1 Introduction

In 1995, Paul Bornstein coined the term *matricellular proteins* and defined them as "a group of modular, extracellular proteins whose functions are achieved by binding to matrix proteins as well as to cell surface receptors, or to other molecules such as cytokines and proteases that interact, in turn, with the cell surface" and thereby modulate cellular function (Bornstein 1995). Because this definition is functional rather than structural, the boundaries of this family are subject to individual interpretation. ECM proteins that serve structural roles were excluded in his original definition, but Bornstein acknowledged that some classic structural ECM proteins share the modulatory properties of matricellular proteins to regulate cell function. Conversely, matricellular proteins can have some properties of conventional ECM proteins such as serving, at least transiently, as structural components of ECM and exhibiting pro-adhesive activities for some cell types. Thus, matricellular proteins can be seen as falling in the middle of a spectrum of secreted proteins bounded at one end by purely structural ECM components (some collagens and proteoglycans) and at the other end by secreted proteins such as proteases, TGF-β, and fibroblast growth factors, which also modulate cell function and bind to other ECM components but generally lack the modular complexity of matricellular proteins and function mainly via a defined enzymatic activity or interaction with a specific cell surface signaling receptor.

Despite the above ambiguity, the term matricellular has proven popular and has been used to date in over 400 publications. The original members of the matricellular protein family were thrombospondin-1, tenascin C, SPARC, and osteopontin (Bornstein 1995). In addition to some close relatives of the above proteins, the CCN family and several members of the thrombospondin repeat (TSR) superfamily are now generally regarded as also belonging to this family.

Matricellular proteins share several common features (Fig. 11.1). Constitutive expression in adult tissues tends to be limited, contrasting with their high

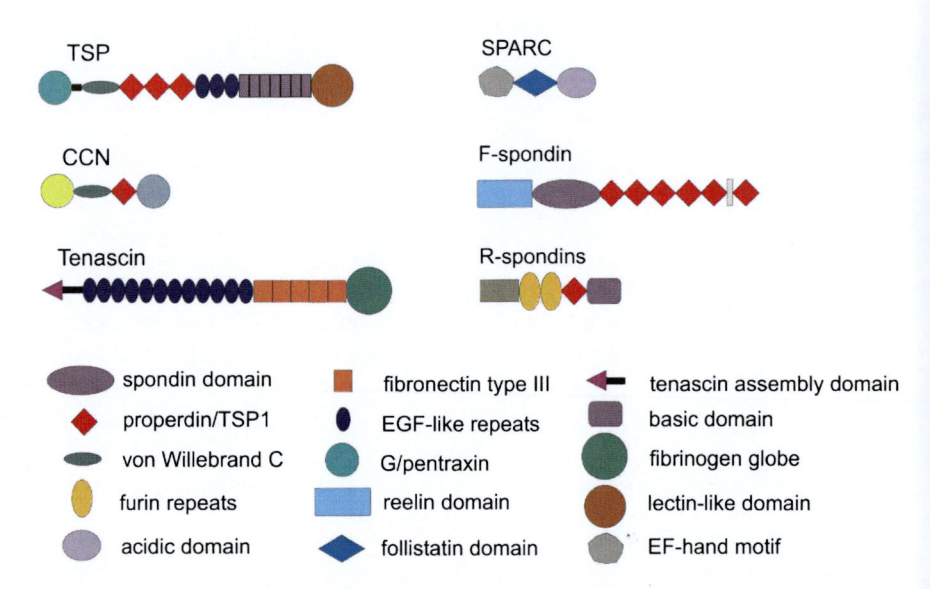

Fig. 11.1 Schematics depicting the structural organization of the matricellular proteins discussed in this chapter. Several domains are present in multiple matricellular proteins including the thrombospondin repeats, EGF-like repeats, and von Willebrand C motifs

expression during development and in response to acute injuries or in chronic disease states. Although they are expressed with tissue and temporal specificity during development, some mice with targeted disruption of individual matricellular genes are viable and appear grossly normal. Interesting phenotypes have been more frequently found when adult transgenic mice with disrupted matricellular protein genes are subjected to stresses that replicate aspects of disease states where their expression is normally induced. A second general feature is that each interacts specifically with several cell surface receptors, components of the extracellular matrix, growth factors, cytokines, or proteases. Historically, antiadhesive activities were considered characteristic of matricellular proteins (Sage and Bornstein 1991; Bornstein 1995). While some have this activity and also inhibit cell motility and proliferation in certain contexts, another emerging shared characteristic of matricellular proteins is the complexity of their functions. The integration of signals from their different receptors and their regulation of the activities of other cytokines and proteases result in responses that tend to be cell type specific, and even within a given cell type the functions of a given matricellular protein tend to be context dependent. Specific examples of this context dependence will be elaborated for several of the proteins discussed in this chapter.

11.2 Thrombospondins

The thrombospondin family contains five members in vertebrates (Fig. 11.2). Current evidence indicates that this family of matricellular proteins originated as a single gene in invertebrates. *Drosophila* thrombospondin (Tsp) is expressed in developing tendon cells and is required for their attachment of muscle cells to form a functional contractile unit (Chanana et al. 2007). Genetic evidence in this study suggested that Tsp serves a structural role in *Drosophila* as a ligand for the αPS2 integrin (Chanana et al. 2007), and this activity was demonstrated using purified Tsp C-terminal domain polypeptide, which mediated adhesion of *Drosophila* S2 cells in an integrin-dependent manner (Subramanian et al. 2007). Gene duplication of this ancestral gene gave rise to two subfamilies of thrombospondins in modern vertebrates. TSP3–5 are pentameric proteins that are prominently expressed in cartilage, muscle, and bone reminiscent of *Drosophila* Tsp. TSP1 and TSP2 are trimeric proteins that share the insertion of a repeated module that occurs in complement components (Fig. 11.2). These are known as properdin or TSRs. Studies of mice with disruption of multiple *Thbs* genes have shown only limited evidence for redundant function within or between either subfamily (Agah et al. 2002; Christopherson et al. 2005; Posey et al. 2008a), implying that the five TSPs generally have distinct functions in vertebrates.

11.2.1 Thrombospondin-1

11.2.1.1 Structure and Interactions

TSP1 was the first member of this family to be studied based on its availability as a soluble protein that is released when mammalian platelets are activated. Digestion with various proteases yielded stable fragments that first revealed the modular structure of TSP1, and the different ligand binding and functional activities of these proteolytic fragments were instrumental to map specific functions to each domain (Fig. 11.3). Sequencing of partial and full-length cDNAs showed that the modular structure of TSP1 is based on several repeated protein modules (Fig. 11.2). The N-terminal module forms a globular domain that is derived from the laminin-G/pentraxin superfamily but structurally more resembles the concanavalin-A family and contains binding sites for heparin, sulfated glycolipids, several β1 integrins, and calreticulin/LRP1 (Elzie and Murphy-Ullrich 2004; Carlson et al. 2008). This is followed by a coiled-coil motif containing a pair of cysteines that mediate trimer formation and a von Willebrand factor C (vWC) module. Three TSRs mediate

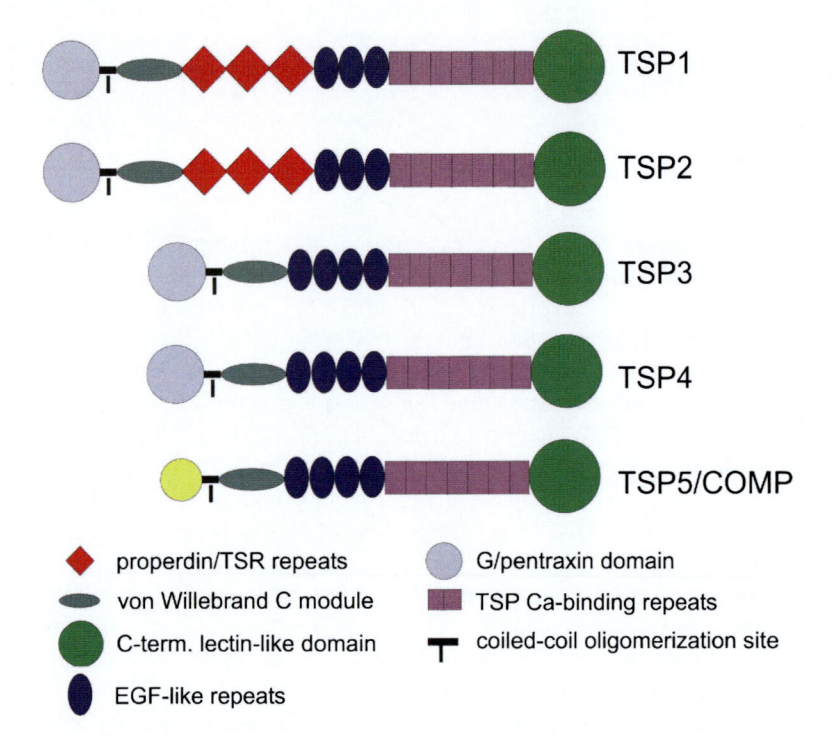

Fig. 11.2 Organization of the five vertebrate thrombospondins. Structures for individual subunits are depicted. TSP1 and TSP2 form trimers mediated by pairs of conserved disulfides in the coiled-coil motif. TSP3, TSP4, and TSP5/COMP form disulfide-bonded trimers via the same motif. TSRs are found only in TSP1 and TSP2

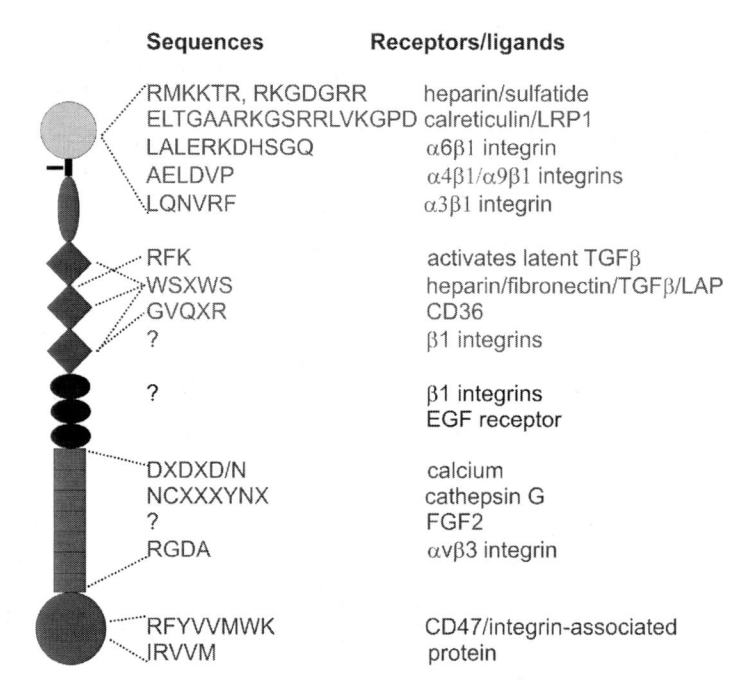

Sequences	Receptors/ligands
RMKKTR, RKGDGRR	heparin/sulfatide
ELTGAARKGSRRLVKGPD	calreticulin/LRP1
LALERKDHSGQ	α6β1 integrin
AELDVP	α4β1/α9β1 integrins
LQNVRF	α3β1 integrin
RFK	activates latent TGFβ
WSXWS	heparin/fibronectin/TGFβ/LAP
GVQXR	CD36
?	β1 integrins
?	β1 integrins
	EGF receptor
DXDXD/N	calcium
NCXXXYNX	cathepsin G
?	FGF2
RGDA	αvβ3 integrin
RFYVVMWK	CD47/integrin-associated
IRVVM	protein

Fig. 11.3 Locations of functional sequences, where known, and ligand and receptor binding sites in TSP1

additional interactions with β1 integrins and with TGF-β, latency-associated peptide, fibronectin, and the scavenger receptor CD36 (Ribeiro et al. 1999; Calzada et al. 2004a). vWC and TSRs constitute the extended central stalk region of TSP1 seen by rotary shadowing electron microscopy. These are followed by three EGF-like modules, which activate the EGF receptor (Liu et al. 2009), and seven Ca-binding modules, the last of which contains an RGD integrin-binding motif with limited accessibility. These repeats when ligated by Ca^{2+} wrap around the C-terminal lectin-like module of TSP1 to form the C-terminal globular domain of TSP1. This has also been called the "signature domain" of TSP1 because it is unique to the five TSPs. The G module mediates binding to the cell surface receptor CD47 (Isenberg et al. 2009a), and the Ca-binding modules mediate fibroblast growth factor-2 binding (Margosio et al. 2008). Two peptides derived from the G module containing a VVM motif are known to bind to CD47, but this motif is not accessible in the published crystal structure for the signature domain (Kvansakul et al. 2004). Computational modeling suggests that a conformation change could expose the VVM motif (Floquet et al. 2008), but the CD47-binding peptides also contain flanking sequences that are exposed in the crystal structure and so could be responsible for CD47 binding. One problem for resolving this issue is that the existing recombinant extracellular domain of CD47 is in a conformation that lacks TSP1 binding activity (Adams et al. 2008).

11.2.1.2 Posttranslational Modification

TSP1 is subject to several posttranslational modifications. Each subunit contains approximately four asparagine-linked complex oligosaccharides (Furukawa et al. 1989). On the basis of glycosylation of recombinant forms used for crystallography, N-glycosylation sites include Asn-584 and Asn1049 (Kvansakul et al. 2004). The majority is biantennary, but mono-, tri-, and tetra-antennary oligosaccharides are also present (Furukawa et al. 1989). The chains are further modified by core fucosylation, terminal sialylation, and addition of bisecting *N*-acetylglucosamine residues. This composition was determined for platelet TSP1; TSP1 produced by other cell types likely varies in the heterogeneity of its N-linked oligosaccharides. The TSRs contain additional novel glycosylation sites: Trp-368, -420, -423, and -480 are C-mannosylated, and Ser377, Thr432, and Thr489 are modified by the disaccharide Glcβ1-3Fuc (Hofsteenge et al. 2001; Kozma et al. 2006; Sato et al. 2006). The motifs subject to these modifications in TSP1 have been implicated in binding of TSRs to CD36, TGF-β, and sulfated glycoconjugates (Guo et al. 1992; Dawson et al. 1997; Young and Murphy-Ullrich 2004). It remains unclear whether glycosylation of the TSRs prevents interaction with any of these ligands. Potentially, alterations in the stoichiometry of C- and O-glycosylation could regulate functional activities of the TSRs. The same TSR modifications have been confirmed in F-spondin and two members of the ADAMTS family and presumably occur in other matricellular proteins that contain TSRs such as the CCNs (Fig. 11.1). Mutation of the β1-3-glucosyltransferase responsible for one of these modifications is the cause of the severe genetic disorder Peters-plus syndrome (Heinonen and Maki 2009), implying that this posttranslational modification plays an important role in the function of some TSR-containing proteins. An additional O-linked modification involving Glc or Glc-Xyl occurs at Ser555 (Nishimura et al. 1992). Finally, erythro-β-hydroxyasparagine detected in hydrolysates of purified TSP1 indicates that consensus sequences in the EGF-like modules are subject to this side chain modification (Przysiecki et al. 1987).

The disulfide bonding of TSP1 is also subject to posttranslational modification. Cys992 in the G module is unpaired and can mediate disulfide interchange in TSP1 and von Willebrand factor (vWF) (Pimanda et al. 2002). Isomerization of disulfide bonding in the Ca-binding modules by protein disulfide isomerase can influence the exposure of the cryptic RGD sequence (Hotchkiss et al. 1998).

11.2.1.3 Genetics

Inactivating mutations of the *THBS1* gene in humans have not been found, but epigenetic silencing of this gene by hypermethylation occurs in several cancers (Li et al. 1999; Yang et al. 2003; Guerrero et al. 2008; Rojas et al. 2008). A single nucleotide polymorphism (SNP) in *THBS1* has been linked to risk of familial premature myocardial infarction (Stenina et al. 2004). This nonsynonymous coding polymorphism converts Asn700 to Ser and alters Ca binding to the signature domain (Carlson et al. 2008). The mechanism by which this alters cardiovascular disease risk remains unclear.

11.2.1.4 Expression

Regulation of TSP1 expression is complex, and only a few examples can be presented here as entries into this extensive literature. The emerging understanding is that transcriptional regulation, regulation of mRNA stability, translational control, and receptor-mediated clearance are all involved in controlling TSP1 levels in the pericellular environment. TSP1 expression generally decreases with tumor progression, but not in all cancer types (reviewed in Isenberg et al. 2009b). TSP1 expression is negatively regulated by activated oncogenes including Ras and Myc and positively regulated by the tumor suppressor p53. This regulation accounts for some instances of decreased TSP1 expression in cancer. Aberrant promoter methylation also plays a role as noted above.

Conversely, TSP1 expression is elevated acutely following injury, particularly ischemia/reperfusion injuries (Sezaki et al. 2005; Thakar et al. 2005; Isenberg et al. 2008b). Tissue TSP1 levels also increase with aging and more so in association with chronic diseases of aging such as atherosclerosis, Alzheimer's disease, and type 2 diabetes (Buee et al. 1992; Riessen et al. 1998; Kang et al. 2001; Stenina et al. 2003b).

Regulation of TSP1 by hyperglycemia has been extensively investigated and illustrates the contributions of multiple mechanisms to this process. Hyperglycemia increases *THBS1* transcription in kidney mesangial cells by stimulating nuclear USF2 protein accumulation, and an NO-mediated increase in cGMP levels antagonizes this by downregulating USF2 protein levels (Wang et al. 2004). In endothelial cells, hyperglycemia activates the aryl hydrocarbon receptor, which is a transcriptional activator of the *THBS1* promoter (Dabir et al. 2008). In vascular smooth muscle cells (VSMCs), glucose regulates *THBS1* transcription via the hexosamine pathway of glucose catabolism, which may in turn modulate the activity of nuclear proteins by altering their O-GlcNAc modification (Raman et al. 2007). Glucose also controls TSP1 levels at a posttranscriptional level in vascular cells (Bhattacharyya et al. 2008). In retinal pigment epithelial cells, high glucose elevated TSP1 mRNA levels but decreased protein levels (Bhattacharyya et al. 2008).

Receptor-mediated uptake is also an important mechanism for regulating extracellular TSP1 levels. Several cell types have the ability to rapidly clear TSP1 (Murphy-Ullrich and Mosher 1987). Uptake is mediated by binding of the N-module to heparan sulfate proteoglycans (HSPGs) and LRP1/calreticulin (Elzie and Murphy-Ullrich 2004).

11.2.1.5 TSP1 Functions

TSP1 null mice are viable and show no gross abnormalities, indicating that TSP1 is not required for development (Lawler et al. 1998). Subfertility of the TSP1 null mouse is associated with altered ovarian follicle morphology and with deficient

TSP1-mediated clearance of VEGF by internalization via LRP1 (Greenaway et al. 2007). The original colony of null mice showed chronic lung inflammation that was associated with deficient latent TGF-β activation (Crawford et al. 1998; Lawler et al. 1998), but rederived TSP1 null mice in a C57Bl/6 background do not exhibit this phenotype (Isenberg et al. 2008a).

The evolution of our understanding of the function of TSP1 in platelets is illustrative of the difficulties in establishing physiological roles for matricellular proteins. The abundance of TSP1 in the α-granules of platelets led to proposals for its function in platelet aggregation, including that TSP1 served as a platelet lectin (Jaffe et al. 1982). Early studies showed that TSP1 bound to the surface of activated platelets, either directly or in a complex with fibrinogen. Furthermore, some domains and synthetic peptides derived from TSP1 could increase or decrease platelet aggregation. On this basis, several platelet receptors for TSP1 were proposed including the major platelet integrin αIIb/β3, GPIbα, CD36, CD47, and sulfatides. However, normal activation-dependent TSP1 binding to platelets from individuals lacking some of these candidate TSP1 receptors casts doubts on each model (Aiken et al. 1986; Boukerche and McGregor 1988; Kehrel et al. 1991; Tandon et al. 1991). Furthermore, when the TSP1 null mouse was created, its platelets were found to aggregate normally in response to thrombin (Lawler et al. 1998). This finding implied that this major protein in α-granules played no major role in platelet aggregation and decreased interest in this subject for some time.

Fortunately, interest was restored in the following decade by two independent discoveries. Control of vWF aggregate size by shear stress involves proteolytic cleavage by ADAMTS13 (Pimanda et al. 2004). TSP1 was found to inhibit this cleavage and thus control vWF oligomer size. A defect in the TSP1 null mouse in clot adherence in an in vivo injury model was attributed to this activity of TSP1 and provided one mechanism by which TSP1 contributes to platelet function (Bonnefoy et al. 2006). Independent of this pathway, TSP1 also regulates platelet activity via binding to its receptors CD47 and CD36. In the presence of physiological levels of the NOS substrate arginine or its product nitric oxide, platelets from TSP1 null mice became resistant to aggregation induced by thrombin (Isenberg et al. 2008e). Conversely, addition of exogenous TSP1 enhances aggregation under these conditions. Therefore, TSP1 released from platelets provides positive feedback to reinforce platelet aggregation by overcoming the thrombostatic activity of physiological NO concentrations.

The antiangiogenic activity of TSP1 was discovered in 1990 (Bagavandoss and Wilks 1990; Good et al. 1990; Taraboletti et al. 1990). TSP1 inhibits growth and migration of endothelial cells and induces their apoptosis (Guo et al. 1997a). The TSR domain was the first region of TSP1 associated with inhibiting angiogenesis, and several peptides derived from this domain have antiangiogenic activities (Tolsma et al. 1993; Vogel et al. 1993; Iruela-Arispe et al. 1999). CD36 is a necessary receptor for TSP1 to inhibit FGF2-stimulated endothelial migration in vitro and angiogenesis in the corneal assay (Dawson et al. 1997; Jimenez et al. 2000). A GVQXR motif in the third TSR probably mediates this activity. This motif is also found in the second TSR, but it is active as a synthetic peptide only if an Ile residue is epimerized to D-Ile (Dawson et al. 1999). This derivative was the starting point for creation of the mimetic ABT-510 (Haviv et al. 2005). ABT-510 is much more potent than the parent peptide for inducing endothelial cell apoptosis (Haviv et al. 2005), although this activity may be independent of its CD36 binding (Isenberg et al. 2008f). This drug candidate has completed several phase II clinical trials as a single agent for cancer patients (Ebbinghaus et al. 2007; Markovic et al. 2007; Baker et al. 2008). Studies in mice showed a synergism between TSP1 and responses to chemotherapy, which in turn led to initiation of additional cancer clinical trials using ABT-510 in combination with cytotoxic agents (Gietema et al. 2006).

Heparin-binding peptides from the TSRs and CD47-binding peptides from the G module also have potent antiangiogenic activities (Guo et al. 1997b; Iruela-Arispe et al. 1999; Kanda et al. 1999). The role of heparan sulfate binding in the antiangiogenic activity of native TSP1 remains unclear, but the increased angiogenic response and survival of full-thickness skin grafts in TSP1 null mice is shared by CD47 null but not by CD36 null mice (Isenberg et al. 2008c). This correlates with CD47 but not with CD36 being necessary for TSP1 to inhibit muscle explant angiogenesis and proangiogenic NO/cGMP signaling in endothelial cells (Isenberg et al. 2006). Thus, regulation of angiogenesis by TSP1 in the context of the cornea requires CD36, but the same activity in ischemic tissues requires only CD47.

The best characterized physiological function of NO is as the endothelium-derived vasorelaxing factor (Ignarro 2002). NO activates cGMP synthesis in VSMCs, which results in dephosphorylation of the regulatory light chain of myosin, relaxation of resistance arteries, and increased blood flow. TSP1 antagonizes this pathway by binding to CD47 on VSMC and thereby acts as a vasoconstrictor (Isenberg et al. 2007a). In TSP1 null mice, exposure to NO results in twice the local vasorelaxation response as in a wild type mouse, indicating that endogenous TSP1 physiologically limits the activity of NO (Isenberg et al. 2007a). Endogenous TSP1 also has systemic effects on blood pressure based on the decreased central pulse pressure in conscious TSP1 null mice (Isenberg et al. 2009c). Exposure to vasorelaxants, anesthesia, or autonomic blockade also results in exaggerated hypotensive responses in TSP1 null mice.

In a simple excisional wound model, the absence of TSP1 results in delayed wound closure (Agah et al. 2002). This was attributed to a deficit in macrophage recruitment into the granulation tissue of the wound. In contrast, wound healing is enhanced in the TSP1 null when the injured tissue is ischemic (Isenberg et al. 2007a, b, 2008c). In the latter wounds, TSP1 limits restoration of blood flow into the ischemic tissues by inhibiting NO-mediated vasorelaxation via its receptor CD47. Blocking antibodies that prevent interactions of TSP1 with its receptor CD47 or antisense oligonucleotides to suppress the expression of CD47 hold promise as therapeutics to enhance survival of ischemic injuries (Isenberg et al. 2007c, 2008d) and protect soft tissues and bone marrow from radiation injury while enhancing tumor ablation in mice (Maxhimer et al. 2009b).

As previously mentioned, TSP1 in tissue increases with age, and aged mice correspondingly have lower tissue cGMP levels and a further impaired ability to repair ischemic injuries (Isenberg et al. 2007b). Remarkably, aged TSP1 null mice show no decline in their ability to repair ischemic injuries. The further increase in TSP1 upon reperfusion of an ischemic tissue (Sezaki et al. 2005; Thakar et al. 2005; Isenberg et al. 2008b) creates an additional impediment to restoration and maintenance of blood flow, and TSP1 null mice show dramatically less liver damage following a warm ischemia/reperfusion injury (Isenberg et al. 2008b). Therapeutic activities of CD47 blocking antibodies have been confirmed for I/R injuries in mice and rats and, remarkably, remain effective when administered up to 30 min after reperfusion (Isenberg et al. 2008b; Maxhimer et al. 2009a).

TSP1 has both proinflammatory and anti-inflammatory activities toward many cell types in the immune system including T cells, macrophages, dendritic cells, neutrophils, and NK cells. The role of TSP1 in promoting macrophage migration into excisional skin wounds was mentioned previously, but following laser-induced retinal injury endogenous TSP1 limits migration of microglial cells in the retina (Ng et al. 2009). In addition to regulating macrophage migration, TSP1 can regulate their activation state. In ischemic limbs, endogenous TSP1 induces a proinflammatory activation state in macrophages (Brechot et al. 2008). Similarly, the percentage of M1 differentiated macrophages is increased in tumors that overexpress TSP1, and TSP1 also induces these activated macrophages to produce superoxide (Martin-Manso et al. 2008).

TSP1 inhibits T-cell receptor signaling via CD47 and an undefined HSPGs and enhances T-cell adhesion and migration via $\alpha 4\beta 1$ integrin (Li et al. 2002). The inhibitory activity in vitro is reflected by inhibition of acute cutaneous hypersensitivity reactions in vivo (Velasco et al. 2009). TSP1 also regulates T-cell Th17 differentiation (Turpie et al. 2009; Yang et al. 2009) and enhances CD4+ CD25+ T-regulatory cell differentiation (Grimbert et al. 2006).

11.2.2 Thrombospondin-2

11.2.2.1 Structure and Interactions

TSP2 was discovered based on its cDNA homology with TSP1, but the lack of a convenient source has limited the extent to which functions of this protein have been explored. TSP2 isolated from cells or tissue tends to be contaminated with TSP1, which limits its usefulness. Full-length recombinant TSP2 has been reported, but large-scale preparation has proven difficult (Chen et al. 1994). Therefore, most of the functional studies to date have employed recombinant fragments of TSP2 that can be readily purified or transfection of cells using TSP2 expression plasmids.

On the basis of their high degree of homology (Fig. 11.2), TSP1 and TSP2 were expected to have similar functions. Indeed, the N-domain of TSP2 also binds to heparin (O'Rourke et al. 1992), LRP1 (Yang et al. 2001), and the integrins $\alpha 4\beta 1$ and $\alpha 6\beta 1$ (Li et al. 2002; Calzada et al. 2003, 2004b). A similar role as in TSP1 has been ascribed to CD36 as a receptor for the TSR domain of TSP2 to mediate antiangiogenic activity (Simantov et al. 2005). As in TSP1, the EGF-like repeats from TSP2 activate EGFR (Liu et al. 2009). However, several significant differences in ligand and receptor binding should be noted (Chen et al. 1994). The N-terminal domain of TSP2 does not bind to α3β1 integrin (Calzada et al. 2003). The TSR domain of TSP2 lacks latency-associated peptide binding (Ribeiro et al. 1999) and does not activate latent TGF-β (Schultz-Cherry et al. 1995). In some contexts, TSP2 expression inhibits latent TGF-β activation (Daniel et al. 2009). Versican binds avidly to the N-module of TSP1 but only weakly to the corresponding N-terminal region of TSP2 (Kuznetsova et al. 2006). Finally, TSP2 is less active than TSP1 for binding and signaling through CD47 (Isenberg et al. 2009a).

11.2.2.2 Genetics

Like TSP1, loss of TSP2 expression in human cancers can result from hypermethylation of the gene (Whitcomb et al. 2003; Czekierdowski et al. 2008).

In a large risk association study, a t3949g SNP in the 3′-untranslated region of the TSP2 mRNA was protective for premature coronary artery disease (Topol et al. 2001). Because the disease-associated genotype has a frequency of 10%, this SNP may have a significant impact on disease risk. In a recent follow-up study of 439 cases of sudden unexpected death, the polymorphism was specifically associated with death due to plaque erosion (Burke et al. 2009). A history of cigarette smoking increased the odds ratio for the predominant TT genotype.

11.2.2.3 Expression

The promoter region for *THBS2* is quite divergent from that of *THBS1*, and the expression pattern of TSP2 during development and in adult mammals correspondingly differs from that of TSP1 (Adolph et al. 1997; Kyriakides et al. 1998b; Bornstein et al. 2000). c-Myb regulates expression of TSP2 but not TSP1, and this regulation involves altered mRNA stability (Bein et al. 1998). TSP2 is negatively regulated by ATF3 and v-Jun (Perez et al. 2001). Similar to TSP1, however, TSP2 expression in vivo increases with aging, at least in the dermis (Agah et al. 2004).

On the basis of its antiangiogenic activity, TSP2 expression has been examined as a potential prognostic marker in cancer. Downregulation of TSP2 has been reported in several human cancers (Kazuno et al. 1999; Kishi et al. 2003; Lawler and Detmar 2004). High expression of TSP2 mRNA in rectal cancer was associated with a positive response to preoperative radiotherapy (Watanabe et al. 2006). However, TSP2 was more strongly expressed in melanoma metastases versus primary melanomas (Kunz et al. 2002). A recent study of 102 pulmonary adenocarcinomas indicated that high TSP2 expression by tumor cells is a good prognostic indicator, but high expression in tumor stromal cells is associated with a poor outcome (Chijiwa et al. 2009).

11.2.2.4 TSP2 Functions

The TSP2 null mouse is viable and fertile, but connective tissue abnormalities were immediately apparent (Kyriakides et al. 1998a). The mice showed abnormal tail flexibility and skin fragility. Ultrastructural analysis showed abnormal collagen fibril assembly. Subsequent studies extended these connective tissue disorders to loss of myocardial matrix integrity, age-related dilated cardiomyopathy, and cardiac failure (Schroen et al. 2004; Swinnen et al. 2009). In addition to altering collagen fibrillogenesis, loss of matrix integrity was associated with impaired cell–matrix interactions due at least in part to the absence of TSP2-mediated clearance of MMP-2 via the LRP1 pathway in null cells (Yang et al. 2001). Excessive MMP-2 activity also results in premature softening of the uterine cervix in pregnant null mice (Kokenyesi et al. 2004).

The antiangiogenic activity of TSP2 resembles that of TSP1 (Volpert et al. 1995), although some aspects of the mechanism may differ (Armstrong et al. 2002b; Isenberg et al. 2009a). Consistent with its antiangiogenic activity, TSP2 expression is lost during progression of some human cancers as discussed above. Mice lacking TSP2 show enhanced vascularity and are more sensitive to chemical carcinogenesis (Kyriakides et al. 1998a; Hawighorst et al. 2001), and overexpression of TSP2 in cancer cells inhibits tumor growth and angiogenesis in mouse models (Streit et al. 1999). Antiangiogenic activity of TSP2 was also indicated by increased angiogenesis in a glomerulonephritis model in TSP2 null mice (Daniel et al. 2007). In contrast to the impaired excisional wound response of a TSP1 null, the TSP2 null mouse shows accelerated healing, which was attributed to the antiangiogenic activity of TSP2 (Kyriakides et al. 1999b; Agah et al. 2002). The antiangiogenic activity of TSP2 also results in increased angiogenesis in an ischemic hindlimb (Krady et al. 2008), but that advantage of the TSP2 null in ischemic injury is less than in the TSP1 null due to the absence of immediate effects of TSP2 on NO/cGMP signaling to restore blood flow (Isenberg et al. 2009a).

Because TSP1 is the only thrombospondin found in platelets, it was initially surprising that the TSP2 null mouse has a bleeding abnormality that results from a defect in platelet function (Kyriakides et al. 2003). TSP2 null platelets are present at normal numbers but exhibit defective aggregation in vivo and decreased maximal aggregation in response to ADP in vitro. The basis for these defects was traced to megakaryocytes in the bone marrow, where the absence of TSP2 limits megakaryocyte differentiation and proplatelet formation.

In addition to affecting thrombogenesis in the bone marrow, TSP2 is an autocrine inhibitor of marrow stromal cell proliferation (Hankenson and Bornstein 2002). TSP2 null mice display increased endocortical bone thickness, and in vitro studies suggested that TSP2 promotes mineralization by facilitating organization of the osteoblast-derived ECM (Alford et al. 2009). During repair of a tibial fracture, TSP2 null mice show more bone and less cartilage than wild type mice, possibly due to an altered proportion of osteoblast versus chondrocyte differentiation (Taylor et al. 2009).

TSP2 expression is induced during an inflammatory response in the skin, and TSP2 null mice exhibit an enhanced and prolonged delayed type hypersensitivity reaction (Lange-Asschenfeldt et al. 2002). Anti-inflammatory activity of TSP2 was further indicated by an increased influx of CD4+ and CD8+ T cells and monocytes/ macrophages into the kidney of mice challenged with an antiglomerular basement membrane antibody (Daniel et al. 2007). Prolonged inflammation in the TSP2 null mouse is associated with a local deficiency of T-cell apoptosis, presumed to be mediated by CD47 since a similar phenotype was observed in the CD47 null mouse (Lamy et al. 2007). However, this is not consistent with the lack of high-affinity binding of TSP2 to CD47 (Isenberg et al. 2009a). Immune responses may also be influenced by the altered foreign body response in TSP2 null mice (Kyriakides et al. 1999a).

11.2.3 Thrombospondin-3 and Thrombospondin-4

11.2.3.1 Structure, Expression, and Interactions

TSP3 was first described in 1992 and its mRNA found to be prominently expressed in lung and also expressed in bone, cartilage, and muscle (Vos et al. 1992). During development in the mouse TSP3 mRNA is expressed in brain, gut, cartilage, and lung (Iruela-Arispe et al. 1993; Qabar et al. 1994). TSP3 is expressed in cultured retinal pigment epithelial cells and cornea stromal keratocytes (Carron et al. 2000; Armstrong et al. 2002a). Osteosarcoma patients with tumors that overexpress *THBS3* have decreased relapse-free survival (Dalla-Torre et al. 2006).

TSP3 is a pentameric protein. Like other thrombospondins, TSP3 binds to heparin and calcium (Qabar et al. 1994; Chen et al. 1996). Additional receptors for TSP3 remain to be defined.

TSP4 was first described in 1994 and found to be highly expressed in heart and skeletal muscle (Lawler et al. 1993). TSP4 is also found in tendon, where it is coexpressed with cartilage oligomeric matrix protein (COMP) and can form heterooligomers (Hauser et al. 1995; Sodersten et al. 2006). Transient expression in osteogenic tissues was observed in the developing chick embryo (Tucker et al. 1995). Additional expression during development was found in the nervous system, where TSP4 may promote neurite outgrowth (Arber and Caroni 1995). Expression of TSP4 in heart is induced by pressure overload and in response to Arg8-vasopressin or angiotensin II (Mustonen et al. 2008). Notably, TSP4 mRNA and protein levels were also increased after reinnervation of the tibialis anterior muscle in rats (Zhou et al. 2006), and skeletal muscle TSP4 mRNA was similarly increased following high-intensity aerobic cycle training in human volunteers (Timmons et al. 2005).

TSP4 is a pentameric protein that binds to heparin and calcium (Lawler et al. 1995). TSP4 interacts with several other ECM proteins including collagens, laminin-1, and fibronectin (Narouz-Ott et al. 2000). As in TSP1 and TSP2, the

EGF-like repeats from TSP4 activate EGFR (Liu et al. 2009). The C-terminal signature domain has some activity to modulate NO/cGMP signaling via CD47, but less than TSP1 (Isenberg et al. 2009a). Like TSP1, TSP4 contains β-hydroxy-Asn (Stenina et al. 2005). A C-terminal peptide from TSP4 that modulates erythroid and endothelial cell proliferation was shown to bind to regulator of differentiation-1 (ROD1) and CD44, but it remains to be shown whether these are physiological ligands of TSP4 (Sadvakassova et al. 2009a, b).

11.2.3.2 Genetics

A coding polymorphism in TSP4 was associated with increased risk of familial premature myocardial infarction (Topol et al. 2001). The Ala387Pro substitution is proposed to be a gain-of-function mutation that interferes with endothelial cell adhesion and proliferation (Stenina et al. 2003a). However, several follow-up studies and a recent meta-analysis have not confirmed a statistical association with disease in different human populations (Asselbergs et al. 2006; Koch et al. 2008). A gender-specific association remains possible but needs to be confirmed (Cui et al. 2006).

11.2.3.3 Functions

TSP3 null mice are viable and fertile but show mild skeletal abnormalities as young adults that disappear by 15 weeks of age (Hankenson et al. 2005). These changes were interpreted as accelerated ossification and maturation in the long bones analyzed. One reason for the mild phenotype of TSP3 null mice may be compensation by TSP5/COMP because more dramatic changes in growth plate organization and a 20% reduction in limb length were observed when TSP3 and COMP were both deleted in the context of a collagen IX null mouse (Posey et al. 2008a).

11.2.4 Thrombospondin-5/Cartilage Oligomeric Matrix Protein

11.2.4.1 Structure and Interactions

TSP5/COMP was first described in 1992 as an acidic cartilage-specific protein (Hedbom et al. 1992). This pentameric thrombospondin is N-glycosylated at Asn-101 and Asn-721, and the heterogeneity of these oligosaccharides is developmentally regulated (Zaia et al. 1997). COMP interacts with high affinity with matrilin-1, -3, and -4 (Mann et al. 2004) and with the N-terminal NC4 domain of collagen IX (Pihlajamaa et al. 2004). COMP binds to collagens I and II and promotes fibrillogenesis (Halasz et al. 2007). The signature domain of COMP has an affinity for glycosaminoglycans, which appears to mediate COMP binding to aggrecan (Chen et al. 2007b). As noted previously, COMP can form heteropentamers with TSP4 (Sodersten et al. 2006). On the cell surface, COMP interacts with the integrin receptors α5β1 and αVβ3 to mediate cell adhesion (Chen et al. 2005). On the basis of a yeast 2 hybrid screen, granulin epithelin precursor (GEP) was identified as a COMP ligand (Xu et al. 2007). COMP potentiates the activity of this autocrine growth factor for chondrocytes.

11.2.4.2 Expression

COMP is highly expressed in tendon, and its expression is further induced in response to load (Smith et al. 1997). Expression was also induced in granulation tissue of skin following injury and was present in dermal scar tissue. Levels of circulating COMP in serum are considered to be a biomarker of cartilage breakdown, which has demonstrated utility as a diagnostic and prognostic indicator of osteoarthritis disease severity and responses to treatment (reviewed in Tseng et al. 2009). Although originally considered cartilage/skeletal specific, COMP expression was also reported in VSMC (Riessen et al. 2001) and in tumor cells of hepatocellular carcinoma (Xiao et al. 2004). Further evidence of expression in cancer was found in canine mammary tumors, mast cell tumor, and melanoma (Yamanokuchi et al. 2009).

11.2.4.3 Genetics

Point mutations in COMP cause autosomal dominant pseudoachondroplasia (PSACH) and multiple epiphyseal dysplasia, both of which give rise to dwarfism (Briggs et al. 1995; Hecht et al. 1995; Posey et al. 2008b). The mutations occur throughout the Ca-binding repeats and the C-terminal domain and interfere with secretion of the mature protein from cells, resulting in lamellar inclusions in the endoplasmic reticulum of chondrocytes.

11.2.4.4 Functions

COMP is clearly a normal component of bone, cartilage, and tendon ECM. However, it merits being considered a matricellular protein in that it shares several activities of other thrombospondins to modulate cell behavior. COMP regulates BMP-2 signaling in mesenchymal cells to modulate chondrogenesis (Kipnes et al. 2003). COMP stimulates adhesion and motility of VSMC (Riessen et al. 2001). Furthermore, COMP is not an essential structural component of ECM given that the COMP null mouse has no anatomical, histological, or ultrastructural abnormalities in skeletal development (Svensson et al. 2002).

11.3 Other TSR Proteins

In addition to several complement components, TSRs are found in a large number of secreted and cell surface proteins (Adams and Tucker 2000; Tucker 2004). Their roles in cell surface receptors such as Unc5, semaphorins, and brain angiogenesis inhibitor-1 are beyond the scope of this section, which is limited to secreted matricellular TSR proteins.

11.3.1 Neuronal TSR Proteins (Spondins)

Spondins are a family of TSR proteins that play important roles in neuronal development and physiology. SCO-spondin, R-spondin, F-spondin, mindin, and the related TSR proteins HB-GAM and midkine are expressed early in the course of neural development (Feinstein and Klar 2004).

F-spondin was first identified as a secreted protein expressed in the neural floor plate that promotes neural cell adhesion and neurite outgrowth (Klar et al. 1992). In addition to directing neuron path finding, F-spondin promotes the differentiation of primary cortical neural cells into mature neurons (Schubert et al. 2006). F-spondin interacts with at least two cell surface receptors. It binds to the extracellular domain of amyloid precursor protein (APP) and inhibits β-secretase cleavage of APP (Ho and Sudhof 2004). Proteolysis of APP also depends on the interaction of F-spondin with apoE receptor-2 via its TSR domain (Hoe et al. 2005). The N-terminal reelin domain of F-spondin mediates binding to heparin and presumably to glycosaminoglycans (Tan et al. 2008).

Function of F-spondin is probably not limited to the CNS. F-spondin expression is induced during differentiation of cementoblasts that mediate the attachment of periodontal ligament to roots of teeth and the surrounding alveolar bone (Kitagawa et al. 2006). Recently, increased expression of F-spondin was found in cartilage from patients with osteoarthritis. Similar to the activity of TSR2 in TSP1, the TSR of F-spondin was required for F-spondin to increase TGF-β activity and prostaglandin-E2 levels in cartilage explants. Recent expression and functional studies in ovarian carcinoma and neuroblastoma suggest that functions of F-spondin may extend to pathophysiology of cancer (Kobel et al. 2008; Cheng et al. 2009).

R-spondin (roof plate-specific spondin) was also discovered based on its developmental regulation in the CNS (Kamata et al. 2004). The R-spondin family quickly expanded to four members in mice and humans, which were found to be ligands for the receptors frizzled-8 and LRP6 (Nam et al. 2006). R-spondin engagement of these receptors induces canonical Wnt/β-catenin signaling and TCF-dependent gene activation. Mutations in R-spondin-4 cause anonychia, an autosomal recessive disorder characterized by the congenital absence of finger and toenails (Bergmann et al. 2006; Blaydon et al. 2006).

11.3.2 ADAMTS Family

ADAMTSs are a large family of secreted proteins composed of N-terminal protease domains followed by a disintegrin-like module, a central TSR, cysteine-rich and spacer modules, and variable numbers of C-terminal TSRs and additional modules (reviewed in Apte 2009). These matricellular proteases have a variety of functions in development, physiology, and pathophysiology. Mutations of ADAMTS13 cause thrombotic thrombocytopenic purpura, and its role in processing of vWF was addressed in Sect. 11.2.1.5. Additional inherited diseases caused by mutations in ADAMTS family members include geleophysic dysplasia (ADAMTSL2, Le Goff et al. 2008), Weill-Marchesani syndrome (ADAM-TS10, Kutz et al. 2008), and colorectal cancer (ADAMTSL3/punctin-2, Koo et al. 2007). Because of space limitations, readers should consult more comprehensive reviews for full discussion of this important family of matricellular proteases (Jones and Riley 2005; Porter et al. 2005; Bondeson et al. 2008; Apte 2009).

11.4 Tenascins

Tenascins are a family of four multimeric secreted proteins containing tenascin C (contactin/hexabrachion), tenascin R (restrictin/januscin), tenascin X, and tenascin W (tenascin N) (Jones and Jones 2000). Their four genes appear to have evolved from a common ancestor that first appeared in primitive chordates (Tucker and Chiquet-Ehrismann 2009a). The general structure of tenascin comprises an N-terminal tenascin assembly (TA) domain that mediates oligomerization into hexamers, followed by EGF-like repeats and fibronectin type-3 (FNIII) repeats (Fig. 11.1). The C-terminus has a globular domain related to fibrinogen. Binding sites for a number of cell surface receptors and other ECM components have been mapped to these domains (Jones and Jones 2000).

Typical of matricellular proteins, their expression follows complex patterns during development and in response to injury, inflammation, and malignancy in adult animals (Tucker and Chiquet-Ehrismann 2009b). Tenascin C and tenascin W show more regulation of expression, whereas tenascin R and tenascin X are more stably expressed in specific sites. On the basis of evidence for their importance in human disease, this section will focus on tenascin C and tenascin X.

11.4.1 Tenascin C

11.4.1.1 Interactions and Expression

Tenascin C is subject to both alternative splicing of exons in the fibronectin repeats and glycosylation that creates extensive heterogeneity in the mature protein.

Tenascin C interacts with the ECM via binding to fibronectin, perlecan, aggrecan, versican, and brevican and with the cell surface via receptors including integrins $\alpha 2\beta 1$, $\alpha 7\beta 1$, $\alpha 9\beta 1$, and $\alpha v\beta 3$ and EGF receptor (Orend and Chiquet-Ehrismann 2006). Through these receptors, tenascin C has context-dependent effects on cell adhesion, migration, responses to growth factors, and gene expression.

Tenascin C expression is rapidly induced following tissue injury. This results from induction by several growth factors including TGF-β, BMP2, FGF2, and PDGF (Tucker and Chiquet-Ehrismann 2009b). In addition to acute injury, tenascin C expression is induced during chronic inflammatory responses, and this extends to the inflammatory conditions characteristic of cancer. Tenascin C is overexpressed in the stroma of some cancers, where there is evidence that it promotes tumor growth, invasion, angiogenesis, and metastasis. Suppression of tenascin C expression in breast carcinoma cells suppressed their lung metastasis (Midwood and Orend 2009), whereas overexpression of tenascin C in breast carcinoma increased their proliferation and invasive behavior (Hancox et al. 2009). Consistent with these observations, high tenascin C expression is a negative prognostic factor in some cancers (Orend and Chiquet-Ehrismann 2006). Induction of tenascin C in stromal cells may play a key role because A375 human melanoma cells implanted in tenascin C null mice resulted in slower tumor growth and decreased tumor vascular density (Tanaka et al. 2004). Interestingly, VEGF induction in stroma was decreased in the absence of tenascin C.

11.4.1.2 Genetics

SNPs in tenascin C have been linked to risk of allergic diseases and asthma (Orsmark-Pietras et al. 2008). Mice lacking tenascin C are overtly normal apart from some strain-specific behavioral defects (Jones and Jones 2000). Therefore, tenascin C is dispensable for development. Consistent with its restricted expression in adult tissues at sites of inflammation or injury, the most clearly documented defects in tenascin C null mice relate to recovery from specific injuries (reviewed in Chiquet-Ehrismann and Chiquet 2003; Midwood and Orend 2009). Recently, mice lacking tenascin C were found to rapidly resolve acute joint inflammation and to be protected from erosive arthritis (Midwood et al. 2009). This was attributed to tenascin C serving as a novel endogenous activator of TLR4.

11.4.2 Tenascin X

11.4.2.1 Interactions and Expression

Tenascin X was isolated as a connective tissue protein named flexilin (Lethias et al. 1996). Tenascin X contains much longer EGF and fibronectin repeat regions than other family members, resulting in a 450-kDa subunit protein that is difficult to purify and study functionally. Little is also known about regulation of its expression apart from reported induction by BNDF in endothelial cells (Tucker and Chiquet-Ehrismann 2009b). Tenascin X is proposed to regulate spacing between collagen fibrils, and thereby control connective tissue elasticity, by binding to decorin associated with fibrils via its 10th and 11th fibronectin repeats (Elefteriou et al. 2001).

11.4.2.2 Genetics

Mutations in tenascin X cause novel recessive and dominant forms of Ehlers–Danlos syndrome (EDS) (Bristow et al. 2005). Complete deficiency in humans leads to a recessive form of EDS, and haploinsufficiency causes hypermobility type EDS. Patients with EDS have elevated risk of several complications during pregnancy, including pelvic instability, premature rupture of membranes, and postpartum hemorrhage. This has been attributed to a direct role of tenascin X in fibril cross-linking, but defects in elastin have not been adequately explained by this model (Bristow et al. 2005). Studies of wound healing in mice lacking tenascin X show its expression to occur late during wound repair, and its absence ablated late strengthening of the repaired skin (Egging et al. 2007). Therefore, tenascin X was proposed to play a specific role in the remodeling rather than initial deposition of collagen fibril matrices.

SNPs in tenascin XB are linked to risk for systemic lupus erythematosus (Kamatani et al. 2008), suggesting additional disease roles for tenascin X. Furthermore, it may play additional roles in malignancy based on its identification as an overexpressed diagnostic marker for malignant mesothelioma (Yuan et al. 2009).

11.5 SPARC

11.5.1 Structure and Interactions

Secreted protein acidic and rich in cysteine (SPARC, osteonectin, BM-40) is an evolutionarily conserved matricellular protein of multicellular eukaryotes (Bradshaw 2009). In lower animals, SPARC is essential, but it is not essential in mice, which have the additional SPARC family member SPARC-like1 (hevin) and the more distantly related SMOC-1, SMOC-2, and SPOCK/testican proteins.

SPARC contains three independently folded domains (Bradshaw and Sage 2001). The N-terminal domain is rich in acidic amino acids and binds calcium with low affinity (Fig. 11.1). The second module is cysteine rich and homologous to follistatin. The C-terminal calcium-binding domain has two EF-hand calcium-binding motifs. The C-terminal domain contains binding sites for several collagens and for PDGF. SPARC also binds to VEGF.

11.5.2 Expression

SPARC is widely expressed during murine embryonic development, but deletion of SPARC has no major effects on development (Bradshaw and Sage 2001). In adult animals, SPARC is induced in wounds and other sites of ECM turnover including in tumor stroma (Arnold and Brekken 2009). Conversely, SPARC is often down-regulated in tumor cells including non-small cell lung carcinoma, pancreatic carcinoma, and ovarian carcinoma. In human adipose tissue, SPARC expression is induced by insulin and leptin but inhibited by glucose (Kos et al. 2009).

11.5.3 Genetics

Hypermethylation of the SPARC promoter has been reported in a variety of human carcinomas (reviewed in Arnold and Brekken 2009). SNPs in SPARC have been linked to risk of idiopathic osteoporosis in men (Delany et al. 2008). An association with scleroderma was reported but could not be confirmed (Lagan et al. 2005).

11.5.4 SPARC Function

The primary described function of SPARC is in the regulation of ECM assembly (Bradshaw 2009). The skin of SPARC null mice has decreased collagen content, and collagen fibrils are smaller and more uniform than in wild type mice. *Drosophila* embryos lacking SPARC fail to deposit type IV collagen in basal lamina (Martinek et al. 2008). This phenotype may also account for cataract formation and age-dependent osteopenia in SPARC null mice (Bradshaw 2009; Delany and Hankenson 2009).

SPARC is a regulator of adipogenesis (Nie and Sage 2009). Consistent with its regulation in human adipose tissue discussed in Sect. 11.5.2, SPARC null mice show increased fat accumulation. SPARC interactions with integrins and growth factors activate ILK, which in turn inhibits GSK3β and enhances β-catenin signaling. SPARC may have additional activity to inhibit adipogenesis via its antiangiogenic activity.

There is general agreement that SPARC expression is a prognostic marker for tumor aggressiveness and patient survival in certain cancers. However, the specific role of SPARC in a given tumor type shows context dependence similar to that of other matricellular proteins (Arnold and Brekken 2009). Transgenic mouse models provide some insights into this context dependence. TRAMP mice, which develop spontaneous prostate cancers, show accelerated tumorigenesis and progression in a SPARC null background (Said et al. 2009). Expression of SPARC in the tumor cells slowed their growth and increased the cell cycle inhibitors p21 and p27. Conversely, TRAMP tumors implanted in SPARC null mice showed enhanced growth and proteolysis compared with those implanted in wild type mice. Therefore, stromal SPARC can also regulate prostate cancer growth.

11.5.5 Hevin Function

Hevin is also a collagen binding protein that regulates collagen fibrillogenesis and induces decorin production by dermal fibroblasts (Sullivan et al. 2006). Hevin null mice are viable but exhibit accelerated excisional wound repair, attributed to the inhibitory activity of hevin for fibroblast migration (Sullivan et al. 2008). Like SPARC, hevin is downregulated in some cancers and inhibited pancreatic cancer cell invasion (Esposito et al. 2007).

11.6 Small Integrin-Binding Ligand, N-Linked Glycoprotein (SIBLING) Gene Family

Osteopontin is the prototypical member of the SIBLING family that was identified as a matricellular protein. These differ from other matricellular proteins in lacking complex modular structures. Osteopontin also lacks a clear secondary structure, but like other matricellular proteins it engages different cell surface receptors to modulate cell behavior. Osteopontin polymorphisms have also been linked to systemic lupus erythematosus (Han et al. 2008). Osteopontin along with the other SIBLING family members dentin matrix protein 1, dentin sialophosphoprotein, matrix extracellular phosphoglycoprotein, and bone sialoprotein (BSP) has been implicated in the pathogenesis of several disease states. Because of space limitations, readers should refer to several excellent recent reviews (Scatena et al. 2007; Bellahcene et al. 2008; Wang and Denhardt 2008; Cho et al. 2009).

11.7 The CCN Family

The CCN family of cysteine-rich matricellular proteins consists of six highly conserved members in vertebrates, with the first three members described (*CYR61*, *CTGF*, *NOV*) providing the name of the family (Leask and Abraham 2006; Holbourn et al. 2008; Chen and Lau 2009). These secreted proteins are organized into four modular domains that follow a secretory signal peptide (Fig. 11.4). The first three domains share sequence similarities with insulin-like growth factor binding proteins (IGFBPs), von Willebrand factor type C repeat (vWC), and thrombospondin type I repeat (TSR). The fourth C-terminal domain (CT) contains a "cysteine knot" motif found in some growth factors. Each domain is encoded by a separate conserved exon, suggesting that CCN genes arose by exon shuffling through evolution (Brigstock 1999; Lau and Lam 1999). CCNs are relatively small (mostly ~40 kDa) among ECM proteins and contain 38 conserved cysteines that are distributed throughout the four domains. A polar, divergent central "hinge" region located between the vWC and TSR domains is hypersensitive to protease digestion, a process that results in proteolytic fragments of CCNs observed in some biological fluids (Brigstock 1999).

Acting in part through direct binding to various integrin receptors, CCNs regulate a broad spectrum of cellular responses, including cell adhesion and migration, differentiation and proliferation, apoptosis and survival, as well as generation of reactive oxygen species (ROS) and alteration in gene expression. CCNs can also regulate the activities of other growth factors and cytokines by modulating their bioavailability and triggering signaling cross-talk. Therefore, CCNs may function in a context-dependent manner in vivo, as they modulate the activities of other growth factors and cytokines that are coexpressed. Studies in cell culture systems and in animal models have shown that CCNs play critical roles in angiogenesis and cardiovascular development, chondrogenesis and skeletal development, wound healing and tissue repair, and pathobiology of chronic diseases such as fibrosis and cancer. Although CCNs are highly expressed in neuronal tissues during development, their functions in the neuronal system are still largely unknown.

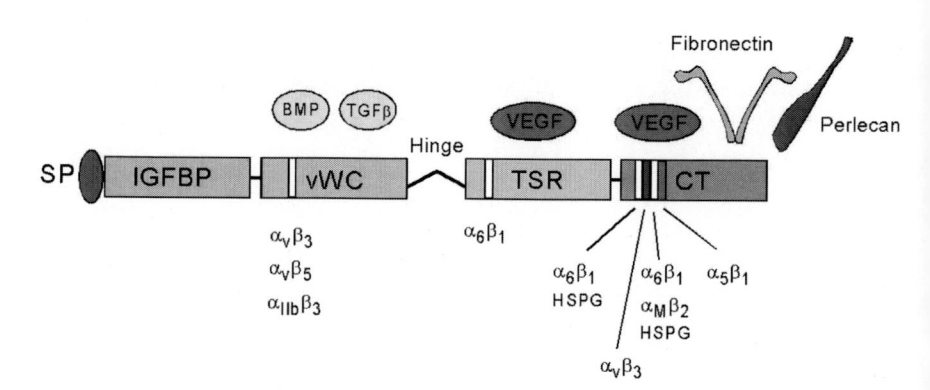

Fig. 11.4 A schematic diagram of CCN proteins. The six CCN proteins share significant structural homology, including an N-terminal secretory signal peptide (SP), followed by four modular domains – IGFBP, vWC, TSR, and CT – in which conserved cysteines are distributed throughout. CCN5 uniquely lacks the CT domain, but is otherwise conserved. A central, protease-sensitive hinge region has no sequence homology among the CCN proteins. Specific binding sites (*black and hatched bars*) for several integrins and HSPGs have been identified for CCN1 and CCN2. In addition, CCN2 interacts with BMPs and TGF-β through the vWC domain and VEGF through the TSR and CT domains and binds ECM proteins such as fibronectin and perlecan through the CT domain (Leask and Abraham 2006; Chen and Lau 2009)

11.7.1 CCN1 (CYR61)

11.7.1.1 Structure and Interactions

The first member of the CCN family described, CCN1 (*C*ysteine-*r*ich 61, Cyr61), was identified as a protein encoded by an immediate-early gene inducible by serum growth factors in fibroblasts (Lau and Lam 1999). CCN1 is secreted upon synthesis and is associated with the ECM and the cell surface, in part through its high-affinity binding to HSPGs. CCN1 induces diverse cellular responses through direct binding to integrin receptors. At least six integrins ($\alpha_2\beta_1$, $\alpha_6\beta_1$, $\alpha_v\beta_3$, $\alpha_v\beta_5$, $\alpha_{IIb}\beta_3$, and $\alpha_M\beta_2$) have been identified as signaling receptors mediating various CCN1 functions, acting in a cell type- and function-specific manner and with HSPGs as coreceptors in some contexts (Chen and Lau 2009). In fibroblasts, the HSPG syndecan-4 is critical for many CCN1 functions (Todorovic et al. 2005; Chen et al. 2007a). Although CCN proteins do not contain an RGD sequence that forms the core of binding sites for some integrins, the noncanonical binding sites in CCN1 for integrins $\alpha_v\beta_3$, $\alpha_6\beta_1$, and $\alpha_M\beta_2$, and for HSPGs have been identified (Fig. 11.4) (Chen and Lau 2009).

11.7.1.2 Genetics

Ccn1 null mice are embryonic lethal in the C57BL/6 background, with ~30% of embryos failing to form chorioallantoic fusion and die by E9.5, and the remaining embryos suffering from impaired placental vascularization, loss of vessel integrity, and severe atrioventricular septal defects (AVSD) (Mo et al. 2002; Mo and Lau 2006). *Ccn1* heterozygotes are viable but exhibit ostium primum atrial septal defects with 20% penetrance. Inactivating mutations in the human *CCN1* gene have not been found, although it is intriguing that human *CCN1* maps to a chromosomal region (1p21-31) that encodes the AVSD susceptibility gene *AVSD1*, suggesting that *CCN1* might be a candidate gene for human AVSD (Mo and Lau 2006).

11.7.1.3 Expression

During embryogenesis, CCN1 is highly expressed in the cardiovascular, neuronal, and skeletal systems (O'Brien and Lau 1992). CCN1 expression is rapidly induced by growth factors (FGF2, PDGF, TGF-β), cytokines (TNFα, IL-1), hormones (estrogen, vitamin D, angiotensin II), hypoxia, UV, mechanical stress, and bacterial and viral infections (Chen and Lau 2009). The exquisite sensitivity of CCN1 to such a broad range of environmental perturbations suggests that it is well poised to respond to insults and injuries. Indeed, CCN1 is highly expressed at sites of injury, inflammation, and tissue repair (Chen and Lau 2009).

11.7.1.4 Functions

CCN1 supports cell adhesion as an immobilized substrate and stimulates chemotaxis in many cell types through direct binding to integrin receptors. It induces adhesive signaling, including the activation of focal adhesion kinase (FAK), paxillin, Rac, actin cytoskeleton reorganization, and formation of filopodia and lamellipodia (Chen et al. 2001). CCN1 is a potent inducer of angiogenesis in vitro and in vivo and promotes vascular endothelial cell adhesion, migration, DNA synthesis, and tubule formation through integrin $\alpha_v\beta_3$. In addition, CCN1 induces chondrogenic differentiation in limb bud mesenchymal cells and stimulates osteoblast differentiation, but inhibits osteoclastogenesis (Chen and Lau 2009).

Although cell adhesion to ECM proteins generally promotes cell survival, CCN1 has the unusual ability to induce apoptosis while supporting cell adhesion in fibroblasts, acting through $\alpha_6\beta_1$ and syndecan-4 to trigger a p53-dependent death pathway (Todorovic et al. 2005). Furthermore, CCN1 synergizes with the TNF family of cytokines to induce cell death. TNFα is a proinflammatory cytokine that activates the transcription factor NFκB, which induces the expression of proinflammatory and antiapoptotic genes. However, TNFα is a strong inducer of apoptosis when de novo protein synthesis or NFκB signaling is blocked, although how it activates the apoptotic pathway in vivo is not well understood (Aggarwal 2003). Surprisingly, CCN1, CCN2, and CCN3 can each enable TNFα to induce apoptosis without inhibiting protein synthesis or NFκB signaling and enhance the cytotoxic effects of other TNF family cytokines such as FasL and TRAIL (Chen et al. 2007a; Franzen et al. 2009; Juric et al. 2009). CCN1 synergizes with TNF cytokines through integrin $\alpha_6\beta_1$ and syndecan 4 to trigger the generation of a high level of ROS via multiple pathways involving 5-lipoxygenase, neutral sphingomyelinase 1, and the mitochondria, thereby overriding the antioxidant cytoprotective effects of NFκB to enable apoptosis (Chen et al. 2007a; Juric et al. 2009). Using knockin mice in which the *Ccn1* gene is replaced by a mutant allele that is disrupted in the $\alpha_6\beta_1$-HSPG binding sites and therefore unable to synergize with TNFα or FasL, it was shown that optimal TNFα or Fas-dependent apoptosis requires CCN1 in vivo (Chen et al. 2007a; Juric et al. 2009). Thus, CCN1 is a physiologic regulator of TNF family cytokine cytotoxicity, suggesting that CCN1 may profoundly alter the functions of inflammatory cytokines such as TNFα during wound healing and injury repair.

Recent studies showed that CCN1, which is dynamically expressed at sites of wound repair, induces fibroblast senescence through its cell adhesion receptors $\alpha_6\beta_1$ and HSPGs (Jun and Lau 2010). CCN1 induces DNA damage response and p53 activation, and ROS-dependent activation of the p16^{INK4a}/pRb pathway, leading to cellular senescence and concomitant expression of antifibrotic genes characteristic of senescent cells. Moreover, CCN1 is responsible for the accumulation of senescent fibroblasts in granulation tissues of healing cutaneous wounds. Knockin mice that express a senescence-defective CCN1 mutant show few senescent cells, resulting in exacerbated fibrosis. Topical application of CCN1 protein to wounds reverses these defects. Therefore, CCN1 functions to induce cellular senescence in wound healing, thereby controlling fibrosis during tissue repair (Jun and Lau 2010).

11.7.1.5 Pathobiology

Both CCN1 and CCN2 are highly expressed in cardiomyocytes after myocardial infractions, and in VSMCs in atherosclerotic lesions and in restenosis following balloon angioplasty (Chen and Lau 2009). Knockdown of CCN1 by siRNA or by FOXO3a-mediated transcriptional repression inhibits neointimal hyperplasia after balloon angioplasty (Lee et al. 2007; Matsumae et al. 2008), indicating that blockade of CCN1 activity may ameliorate restenosis. The angiogenic activity of CCN1 may underlie its role in promoting bone fracture healing (Athanasopoulos et al. 2007) and its efficacy as safe therapeutics for lower limb ischemia in combination with FGF2 (Rayssac et al. 2009). Decreased CCN1 expression in the placenta is associated with pre-eclampsia (Gellhaus et al. 2006). CCN1 is also highly expressed in colitis and rheumatoid arthritis, consistent with a role in inflammation (Koon et al. 2008; Zhang et al. 2009).

Elevated CCN1 expression is associated with aggressive human breast cancers, ovarian carcinomas, gliomas, and esophageal squamous cell carcinomas. In these contexts, CCN1 may promote angiogenesis, survival of cancer cells, and the invasive phenotype. However, CCN1 also appears to suppress the growth of other types of tumor cells, including cells of lung cancers, endometrial cancers, melanoma, and hepatocellular carcinomas (Chen and Lau 2009). Thus, the effects of CCN1 in cancer may be cell type and context dependent, and may promote or inhibit tumor growth depending on whether angiogenic factors are limiting or whether the cancer cells are susceptible to CCN1/cytokine-induced apoptosis. In this regard, CCN1 action is a double-edge sword in prostate carcinoma cells, since it promotes the proliferation of prostate cells but also enhances the cytotoxicity of TRAIL, an immune surveillance cytokine that preferentially eliminates cancer cells (Franzen et al. 2009). Thus, prostate cancer cells may overexpress CCN proteins to promote their proliferation, although this also puts them at risk of higher susceptibility to TRAIL-mediated immune surveillance.

11.7.2 CCN2 (CTGF)

11.7.2.1 Structure and Interactions

Identified as an immediate-early gene inducible by serum and TGF-β, CCN2 (connective tissue growth factor, CTGF) shares >40% amino acid sequence identity and >60% sequence homology with CCN1, with conservation of all 38 cysteines (Brigstock 1999; Lau and Lam 1999). CCN2 was also cloned using polyclonal antibodies raised against the structurally unrelated platelet-derived growth factor prepared from platelet lysates (Bradham et al. 1991), possibly because CCN2 is stored in and released from platelet α-granules upon activation (Cicha et al. 2004). CCN2 interacts with various integrins in a cell type-dependent manner, and its binding sites for integrins $\alpha_v\beta_3$, $\alpha_5\beta_1$, and $\alpha_M\beta_2$ have been identified (Chen and

Lau 2009). CCN2 binds HSPGs and the endocytic receptor LRP-1, which serve as coreceptors with integrins in some contexts (Leask and Abraham 2006). In addition, CCN2 can act as an adaptor to other ECM proteins by its direct binding to fibronectin and perlecan (Leask and Abraham 2006).

CCN2 can modulate the signaling pathways induced by other growth factor receptors. Studies show that CCN2 can interact with the NGF receptor TrkA, which may mediate some activities of CCN2 (Wahab et al. 2005; Wang et al. 2009b). Whereas CCN2 binds TGF-β and enhances TGF-β binding to all three of its receptors, it also binds BMP-4 but inhibits BMP-4 function (Abreu et al. 2002). CCN2 binds VEGF directly and inhibit its activity, whereas proteolysis of the CCN2–VEGF complex by MMPs releases the bound VEGF in an active form (Dean et al. 2007). Thus, CCN2 may regulate the bioavailability of VEGF, releasing it for angiogenic action only when MMPs are being secreted and activated, such as during tissue remodeling and wound repair.

11.7.2.2 Genetics

Despite many similarities between the activities and patterns of expression of CCN1 and CCN2, targeted disruptions of their genes in mice show distinct phenotypes. *Ccn2* null mice are neonatal lethal due to respiratory defects as a secondary consequence of severe chondrodysplasia throughout the appendicular and axial skeleton (Ivkovic et al. 2003). *Ccn2* null embryos also suffer from pulmonary hypoplasia and defects in pancreatic islet morphogenesis (Baguma-Nibasheka and Kablar 2008; Crawford et al. 2009), but exhibit no apparent cardiovascular defects. A polymorphism in the *CCN2* promoter region (G-945C) in a British population is found to be associated with systemic sclerosis, in which the C allele creates a high affinity binding site for Sp3 and lead to repression, whereas repression is released in the G allele (Fonseca et al. 2007). Interestingly, these findings were confirmed in a Japanese population but not in a multicenter study in North America (Kawaguchi et al. 2009; Rueda et al. 2009), suggesting a potential divergence in genetic association among different populations.

11.7.2.3 Expression

Expression of CCN2 is most prominent in the embryonic cardiovascular, skeletal, and neuronal systems and in various mesenchymal tissues of the adult (Friedrichsen et al. 2003). *CCN2* is transcriptionally activated by a broad array of stimuli, including exposure to growth factors (TGF-β, VEGF, HGF), hormones (angiotensin II, endothelin-1), hypoxia, and biomechanical and shear stress. Extensive analysis of the CCN2 promoter has revealed critical promoter elements required for activation by TGF-β and endothelin-1 (Leask and Abraham 2006). CCN2 expression is negatively regulated by two major cardiac miRNAs, miR-30, and miR-133, which are downregulated during cardiac disease and associated with elevated CCN2 expression in cardiac fibrosis (Duisters et al. 2009). Similarly, mi-18a, which is downregulated in chondrocytic cells, represses CCN2 expression (Ohgawara et al. 2009).

11.7.2.4 Functions

CCN2 supports cell adhesion and stimulates cell migration in mesenchymal cell types, working through direct binding to integrin receptors with the participation of HSPGs and LRP in some contexts (Leask and Abraham 2006; Chen and Lau 2009). CCN2 can enhance DNA synthesis, although in some cell types this activity requires the presence of other mitogenic growth factors. CCN2 induces angiogenesis and promotes multiple aspects of endochondral ossification, including chondrogenesis and osteogenesis (Kubota and Takigawa 2007). However, CCN2 also inhibits Wnt signaling and antagonizes BMPs, leading to osteopenia when CCN2 is expressed in excess (Abreu et al. 2002; Smerdel-Ramoya et al. 2008). Like CCN1, CCN2 cooperates with TNFα and FasL to induce apoptosis through the generation of ROS (Chen et al. 2007a; Juric et al. 2009).

11.7.2.5 Pathobiology

Numerous studies have reported the elevated expression of CCN2 in fibrosis of a broad spectrum of tissues, including the skin, kidney, heart, liver, and lung, apparently independent of the primary etiology (Shi-wen et al. 2008). CCN2 cooperates with the pro-fibrotic growth factor TGF-β to induce a sustained level of fibrotic response that is not achieved by either factor alone (Mori et al. 1999). Thus, CCN2 has been proposed as useful marker for fibrosis and a potential target for antifibrotic therapy. Transgenic experiments showed that over expression of CCN2 by itself in the liver, kidney or heart is insufficient to induce fibrosis, but potentiates or exacerbates aspects of the fibrotic response when organisms are challenged with pro-fibrotic insults (Brigstock 2009). In the lung, however, ectopic expression of CCN2 during postnatal development can induce a fibrotic phenotype (Wu et al. 2009). CCN2 is overexpressed in human pancreatic cancer, and CCN2-specific monoclonal antibody therapy inhibits pancreatic tumor growth, lymph node metastasis, and tumor angiogenesis in rodent models (Bennewith et al. 2009; Chen and Lau 2009). In breast cancer, CCN2 is part of a gene signature that specifies osteolytic bone metastasis, and CCN2 antibody therapy suppresses breast cancer microvascularization and metastasis to bone (Chen and Lau 2009). These findings suggest that CCN2 antibody therapy holds promise as therapeutics targeting these cancers.

11.7.3 CCN3 (Nov)

CCN3 (nephroblastoma overexpressed, NOV) was identified as an aberrantly expressed gene in an avian nephroblastoma cell line (Holbourn et al. 2008). It binds to and acts through integrin receptors, mediates cell adhesion and stimulates cell migration, and induces angiogenesis (Chen and Lau 2009). CCN3 is expressed in CD34⁺ pluripotent hematopoietic stem cells of human umbilical cord blood, and plays a critical role in hematopoietic stem cell self-renewal (Gupta et al. 2007). Direct binding of CCN3 with notch leads to suppression of myogenic differentiation in vitro (Sakamoto et al. 2002). Like CCN2, CCN3 binds to and antagonizes the action of BMPs, and transgenic mice that overexpress CCN3 in osteoblasts developed osteopenia (Rydziel et al. 2007). Surprisingly, *Ccn3* null mice are viable and largely normal, but exhibit a modest, transient, and sexually dimorphic increase in bone formation and remodeling (Canalis et al. 2010). Aberrant expression of CCN3 is associated with a variety of cancers, and high CCN3 expression portends a less favorable outcome in Ewing's sarcoma and osteosarcoma (Perbal et al. 2009).

11.7.4 CCN4, CCN5, and CCN6

Three other members of the CCN family were identified as Wnt-inducible signaling proteins (WISPs) (Pennica et al. 1998). CCN4 promotes TNFα-stimulated cardiac fibroblast proliferation, and is upregulated in myocardial infraction (Venkatachalam et al. 2009), pulmonary fibrosis (Konigshoff et al. 2009) and colitis (Wang et al. 2009a). CCN5 (WISP-2) differs from other CCN proteins by lacking the CT domain while conserving the first three modular domains. CCN5 inhibits cell proliferation in smooth muscle cells and appears to suppress the invasive phenotype in breast cancer cells (Lake et al. 2003; Banerjee et al. 2008).

CCN6 is unusual among the CCN family in having a relatively divergent vWC domain, where five of the nine cysteines and a critical $\alpha_v\beta_3$ binding site are not conserved (Chen et al. 2004; Chen and Lau 2009). Inactivating mutations in human *CCN6* cause the inheritable disease progressive pseudorheumatoid dysplasia, an autosomal recessive skeletal disorder (Hurvitz et al. 1999). Like other members of the CCN family, CCN6 modulates BMP and Wnt signaling (Nakamura et al. 2007). However, both *Ccn6* null mice and mice that overexpress *Ccn6* in a broad array of tissues are apparently normal and do not display any obvious phenotypes, suggesting that the functions of CCN6 in human and mice may be significantly different (Kutz et al. 2005; Nakamura et al. 2009). CCN6 is also thought to be a tumor suppressor for inflammatory breast cancer (Kleer et al. 2007).

11.8 Future Prospects

We should expect the ranks of matricellular proteins to continue to grow. Periostin, a 90 kDa protein that binds to cell surface integrins and ECM collagen, is a recently recognized matricellular protein that plays important roles in tissue remodeling (Norris et al. 2009). A family of secreted mammalian lectins, the galectins were recently proposed to function as matricellular proteins (Elola et al. 2007). Galectins are small single domain proteins, so they do not meet the original definition of being modular proteins. Despite this deficiency, their ability to interact with a diverse set of proteins that bear suitable oligosaccharide modifications creates a similar complexity in their interactions and modulatory function. Secreted proteins that modulate cell function via enzymatic activities, such as autotaxin and ADAMs, or function as inhibitors of enzymes, such as plasminogen activator inhibitors and tissue inhibitors of metalloproteases, may also merit consideration as matricellular proteins. In common with TSP1, a number of latent TGF-β binding proteins can simultaneously interact with other ECM components to regulate the biological activity of TGF-β. These and the analogous IGFBPs clearly are modular proteins that regulate important cell functions.

Acknowledgments DDR was supported by the Intramural Research Program of the NIH, NCI, Center for Cancer Research. LFL was supported by grants from the National Institutes of Health (CA46565, GM78492, and HL81390).

References

Abreu JG, Ketpura NI, Reversade B, De Robertis EM (2002) Connective-tissue growth factor (CTGF) modulates cell signalling by BMP and TGF-beta. Nat Cell Biol 4:599–604

Adams JC, Tucker RP (2000) The thrombospondin type 1 repeat (TSR) superfamily: diverse proteins with related roles in neuronal development. Dev Dyn 218:280–299

Adams JC, Bentley AA, Kvansakul M, Hatherley D, Hohenester E (2008) Extracellular matrix retention of thrombospondin 1 is controlled by its conserved C-terminal region. J Cell Sci 121:784–795

Adolph KW, Liska DJ, Bornstein P (1997) Analysis of the promoter and transcription start sites of the human thrombospondin 2 gene (THBS2). Gene 193:5–11

Agah A, Kyriakides TR, Lawler J, Bornstein P (2002) The lack of thrombospondin-1 (TSP1) dictates the course of wound healing in double-TSP1/TSP2-null mice. Am J Pathol 161: 831–839

Agah A, Kyriakides TR, Letrondo N, Bjorkblom B, Bornstein P (2004) Thrombospondin 2 levels are increased in aged mice: consequences for cutaneous wound healing and angiogenesis. Matrix Biol 22:539–547

Aggarwal BB (2003) Signalling pathways of the TNF superfamily: a double-edged sword. Nat Rev Immunol 3:745–756

Aiken ML, Ginsberg MH, Plow EF (1986) Identification of a new class of inducible receptors on platelets. Thrombospondin interacts with platelets via a GPIIb-IIIa-independent mechanism. J Clin Invest 78:1713–1716

Alford AI, Terkhorn SP, Reddy AB, Hankenson KD (2009) Thrombospondin-2 regulates matrix mineralization in MC3T3-E1 pre-osteoblasts. Bone 46:464–471

Apte SS (2009) A disintegrin-like and metalloprotease (reprolysin-type) with thrombospondin type 1 motif (ADAMTS) superfamily-functions and mechanisms. J Biol Chem 284(46): 31493–31497

Arber S, Caroni P (1995) Thrombospondin-4, an extracellular matrix protein expressed in the developing and adult nervous system promotes neurite outgrowth. J Cell Biol 131:1083–1094

Armstrong DJ, Hiscott P, Batterbury M, Kaye S (2002a) Corneal stromal cells (keratocytes) express thrombospondins 2 and 3 in wound repair phenotype. Int J Biochem Cell Biol 34: 588–593

Armstrong LC, Bjorkblom B, Hankenson KD, Siadak AW, Stiles CE, Bornstein P (2002b) Thrombospondin 2 inhibits microvascular endothelial cell proliferation by a caspase-independent mechanism. Mol Biol Cell 13:1893–1905

Arnold SA, Brekken RA (2009) SPARC: a matricellular regulator of tumorigenesis. J Cell Commun Signal 3(3–4):255–273

Asselbergs FW, Pai JK, Pischon T, Manson JE, Rimm EB (2006) Thrombospondin-4 Ala387Pro polymorphism is not associated with vascular function and risk of coronary heart disease in US men and women. Thromb Haemost 95:589–590

Athanasopoulos AN, Schneider D, Keiper T, Alt V, Pendurthi UR, Liegibel UM, Sommer U, Nawroth PP, Kasperk C, Chavakis T (2007) Vascular endothelial growth factor (VEGF)-induced up-regulation of CCN1 in osteoblasts mediates proangiogenic activities in endothelial cells and promotes fracture healing. J Biol Chem 282:26746–26753

Bagavandoss P, Wilks JW (1990) Specific inhibition of endothelial cell proliferation by thrombospondin. Biochem Biophys Res Commun 170:867–872

Baguma-Nibasheka M, Kablar B (2008) Pulmonary hypoplasia in the connective tissue growth factor (Ctgf) null mouse. Dev Dyn 237:485–493

Baker LH, Rowinsky EK, Mendelson D, Humerickhouse RA, Knight RA, Qian J, Carr RA, Gordon GB, Demetri GD (2008) Randomized, phase II study of the thrombospondin-1-mimetic angiogenesis inhibitor ABT-510 in patients with advanced soft tissue sarcoma. J Clin Oncol 26:5583–5588

Banerjee S, Dhar G, Haque I, Kambhampati S, Mehta S, Sengupta K, Tawfik O, Phillips TA, Banerjee SK (2008) CCN5/WISP-2 expression in breast adenocarcinoma is associated with less frequent progression of the disease and suppresses the invasive phenotypes of tumor cells. Cancer Res 68:7606–7612

Bein K, Ware JA, Simons M (1998) Myb-dependent regulation of thrombospondin 2 expression. Role of mRNA stability. J Biol Chem 273:21423–21429

Bellahcene A, Castronovo V, Ogbureke KU, Fisher LW, Fedarko NS (2008) Small integrin-binding ligand N-linked glycoproteins (SIBLINGs): multifunctional proteins in cancer. Nat Rev Cancer 8:212–226

Bennewith KL, Huang X, Ham CM, Graves EE, Erler JT, Kambham N, Feazell J, Yang GP, Koong A, Giaccia AJ (2009) The role of tumor cell-derived connective tissue growth factor (CTGF/CCN2) in pancreatic tumor growth. Cancer Res 69:775–784

Bergmann C, Senderek J, Anhuf D, Thiel CT, Ekici AB, Poblete-Gutierrez P, van Steensel M, Seelow D, Nurnberg G, Schild HH, Nurnberg P, Reis A, Frank J, Zerres K (2006) Mutations in the gene encoding the Wnt-signaling component R-spondin 4 (RSPO4) cause autosomal recessive anonychia. Am J Hum Genet 79:1105–1109

Bhattacharyya S, Marinic TE, Krukovets I, Hoppe G, Stenina OI (2008) Cell type-specific post-transcriptional regulation of production of the potent antiangiogenic and proatherogenic protein thrombospondin-1 by high glucose. J Biol Chem 283:5699–5707

Blaydon DC, Ishii Y, O'Toole EA, Unsworth HC, Teh MT, Ruschendorf F, Sinclair C, Hopsu-Havu VK, Tidman N, Moss C, Watson R, de Berker D, Wajid M, Christiano AM, Kelsell DP (2006) The gene encoding R-spondin 4 (RSPO4), a secreted protein implicated in Wnt signaling, is mutated in inherited anonychia. Nat Genet 38:1245–1247

Bondeson J, Wainwright S, Hughes C, Caterson B (2008) The regulation of the ADAMTS4 and ADAMTS5 aggrecanases in osteoarthritis: a review. Clin Exp Rheumatol 26:139–145

Bonnefoy A, Daenens K, Feys HB, De Vos R, Vandervoort P, Vermylen J, Lawler J, Hoylaerts MF (2006) Thrombospondin-1 controls vascular platelet recruitment and thrombus adherence in mice by protecting (sub)endothelial VWF from cleavage by ADAMTS13. Blood 107:955–964

Bornstein P (1995) Diversity of function is inherent in matricellular proteins: an appraisal of thrombospondin 1. J Cell Biol 130:503–506

Bornstein P, Armstrong LC, Hankenson KD, Kyriakides TR, Yang Z (2000) Thrombospondin 2, a matricellular protein with diverse functions. Matrix Biol 19:557–568

Boukerche H, McGregor JL (1988) Characterization of an anti-thrombospondin monoclonal antibody (P8) that inhibits human blood platelet functions. Normal binding of P8 to thrombin-activated Glanzmann thrombasthenic platelets. Eur J Biochem 171:383–392

Bradham DM, Igarashi A, Potter RL, Grotendorst GR (1991) Connective tissue growth factor: a cysteine-rich mitogen secreted by human vascular endothelial cells is related to the SRC-induced immediate early gene product CEF-10. J Cell Biol 114:1285–1294

Bradshaw AD (2009) The role of SPARC in extracellular matrix assembly. J Cell Commun Signal 3(3–4):239–246

Bradshaw AD, Sage EH (2001) SPARC, a matricellular protein that functions in cellular differentiation and tissue response to injury. J Clin Invest 107:1049–1054

Brechot N, Gomez E, Bignon M, Khallou-Laschet J, Dussiot M, Cazes A, Alanio-Brechot C, Durand M, Philippe J, Silvestre JS, Van Rooijen N, Corvol P, Nicoletti A, Chazaud B, Germain S (2008) Modulation of macrophage activation state protects tissue from necrosis during critical limb ischemia in thrombospondin-1-deficient mice. PLoS ONE 3:e3950

Briggs MD, Hoffman SM, King LM, Olsen AS, Mohrenweiser H, Leroy JG, Mortier GR, Rimoin DL, Lachman RS, Gaines ES et al (1995) Pseudoachondroplasia and multiple epiphyseal dysplasia due to mutations in the cartilage oligomeric matrix protein gene. Nat Genet 10:330–336

Brigstock DR (1999) The connective tissue growth factor/cysteine-rich 61/nephroblastoma overexpressed (CCN) family. Endocr Rev 20:189–206

Brigstock DR (2009) Connective tissue growth factor (CCN2, CTGF) and organ fibrosis: lessons from transgenic animals. J Cell Commun Signal 4(1):1–4

Bristow J, Carey W, Egging D, Schalkwijk J (2005) Tenascin-X, collagen, elastin, and the Ehlers–Danlos syndrome. Am J Med Genet C Semin Med Genet 139C:24–30

Buee L, Hof PR, Roberts DD, Delacourte A, Morrison JH, Fillit HM (1992) Immunohistochemical identification of thrombospondin in normal human brain and in Alzheimer's disease. Am J Pathol 141:783–788

Burke A, Creighton W, Tavora F, Li L, Fowler D (2009) Decreased frequency of the 3'UTR T > G single nucleotide polymorphism of thrombospondin-2 gene in sudden death due to plaque erosion. Cardiovasc Pathol 19(3):e45–e49

Calzada MJ, Sipes JM, Krutzsch HC, Yurchenco PD, Annis DS, Mosher DF, Roberts DD (2003) Recognition of the N-terminal modules of thrombospondin-1 and thrombospondin-2 by α6β1 integrin. J Biol Chem 278:40679–40687

Calzada MJ, Annis DS, Zeng B, Marcinkiewicz C, Banas B, Lawler J, Mosher DF, Roberts DD (2004a) Identification of novel beta1 integrin binding sites in the type 1 and type 2 repeats of thrombospondin-1. J Biol Chem 279:41734–41743

Calzada MJ, Zhou L, Sipes JM, Zhang J, Krutzsch HC, Iruela-Arispe ML, Annis DS, Mosher DF, Roberts DD (2004b) α4β1 integrin mediates selective endothelial cell responses to thrombospondins in vitro and modulates angiogenesis in vivo. Circ Res 94:462–470

Canalis E, Smerdel-Ramoya A, Durant D, Economides AN, Beamer WG, Zanotti S (2010) Nephroblastoma overexpressed (Nov) inactivation sensitizes osteoblasts to bone morphogenetic protein-2, but nov is dispensable for skeletal homeostasis. Endocrinology 151:221–233

Carlson CB, Lawler J, Mosher DF (2008) Structures of thrombospondins. Cell Mol Life Sci 65:672–686

Carron JA, Hiscott P, Hagan S, Sheridan CM, Magee R, Gallagher JA (2000) Cultured human retinal pigment epithelial cells differentially express thrombospondin-1, -2, -3, and -4. Int J Biochem Cell Biol 32:1137–1142

Chanana B, Graf R, Koledachkina T, Pflanz R, Vorbruggen G (2007) AlphaPS2 integrin-mediated muscle attachment in *Drosophila* requires the ECM protein thrombospondin. Mech Dev 124:463–475

Chen CC, Lau LF (2009) Functions and mechanisms of action of CCN matricellular proteins. Int J Biochem Cell Biol 41:771–783

Chen H, Sottile J, O'Rourke KM, Dixit VM, Mosher DF (1994) Properties of recombinant mouse thrombospondin 2 expressed in *Spodoptera* cells. J Biol Chem 269:32226–32232

Chen H, Aeschlimann D, Nowlen J, Mosher DF (1996) Expression and initial characterization of recombinant mouse thrombospondin 1 and thrombospondin 3. FEBS Lett 387:36–41

Chen CC, Chen N, Lau LF (2001) The angiogenic factors Cyr61 and CTGF induce adhesive signaling in primary human skin fibroblasts. J Biol Chem 276:10443–10452

Chen N, Leu SJ, Todorovic V, Lam SCT, Lau LF (2004) Identification of a novel integrin αvβ3 binding site in CCN1 (CYR61) critical for pro-angiogenic activities in vascular endothelial cells. J Biol Chem 279:44166–44176

Chen FH, Thomas AO, Hecht JT, Goldring MB, Lawler J (2005) Cartilage oligomeric matrix protein/thrombospondin 5 supports chondrocyte attachment through interaction with integrins. J Biol Chem 280:32655–32661

Chen CC, Young JL, Monzon RI, Chen N, Todorovic V, Lau LF (2007a) Cytotoxicity of TNFα is regulated by integrin-mediated matrix signaling. EMBO J 26:1257–1267

Chen FH, Herndon ME, Patel N, Hecht JT, Tuan RS, Lawler J (2007b) Interaction of cartilage oligomeric matrix protein/thrombospondin 5 with aggrecan. J Biol Chem 282:24591–24598

Cheng YC, Liang CM, Chen YP, Tsai IH, Kuo CC, Liang SM (2009) F-spondin plays a critical role in murine neuroblastoma survival by maintaining IL-6 expression. J Neurochem 110: 947–955

Chijiwa T, Abe Y, Inoue Y, Matsumoto H, Kawai K, Matsuyama M, Miyazaki N, Inoue H, Mukai M, Ueyama Y, Nakamura M (2009) Cancerous, but not stromal, thrombospondin-2 contributes prognosis in pulmonary adenocarcinoma. Oncol Rep 22:279–283

Chiquet-Ehrismann R, Chiquet M (2003) Tenascins: regulation and putative functions during pathological stress. J Pathol 200:488–499

Cho HJ, Cho HJ, Kim HS (2009) Osteopontin: a multifunctional protein at the crossroads of inflammation, atherosclerosis, and vascular calcification. Curr Atheroscler Rep 11:206–213

Christopherson KS, Ullian EM, Stokes CC, Mullowney CE, Hell JW, Agah A, Lawler J, Mosher DF, Bornstein P, Barres BA (2005) Thrombospondins are astrocyte-secreted proteins that promote CNS synaptogenesis. Cell 120:421–433

Cicha I, Garlichs CD, Daniel WG, Goppelt-Struebe M (2004) Activated human platelets release connective tissue growth factor. Thromb Haemost 91:755–760

Crawford SE, Stellmach V, Murphy-Ullrich JE, Ribeiro SMF, Lawler J, Hynes RO, Boivin GP, Bouck N (1998) Thrombospondin-1 is a major activator of TGF-β1 in vivo. Cell 93:1159–1170

Crawford LA, Guney MA, Oh YA, Deyoung RA, Valenzuela DM, Murphy AJ, Yancopoulos GD, Lyons KM, Brigstock DR, Economides A, Gannon M (2009) Connective tissue growth factor (CTGF) inactivation leads to defects in islet cell lineage allocation and beta-cell proliferation during embryogenesis. Mol Endocrinol 23:324–336

Cui J, Randell E, Renouf J, Sun G, Green R, Han FY, Xie YG (2006) Thrombospondin-4 1186 G > C (A387P) is a sex-dependent risk factor for myocardial infarction: a large replication study with increased sample size from the same population. Am Heart J 152(543):e541–e545

Czekierdowski A, Czekierdowska S, Danilos J, Czuba B, Sodowski K, Sodowska H, Szymanski M, Kotarski J (2008) Microvessel density and CpG island methylation of the THBS2 gene in malignant ovarian tumors. J Physiol Pharmacol 59(Suppl 4):53–65

Dabir P, Marinic TE, Krukovets I, Stenina OI (2008) Aryl hydrocarbon receptor is activated by glucose and regulates the thrombospondin-1 gene promoter in endothelial cells. Circ Res 102: 1558–1565

Dalla-Torre CA, Yoshimoto M, Lee CH, Joshua AM, de Toledo SR, Petrilli AS, Andrade JA, Chilton-MacNeill S, Zielenska M, Squire JA (2006) Effects of THBS3, SPARC and SPP1 expression on biological behavior and survival in patients with osteosarcoma. BMC Cancer 6:237

Daniel C, Amann K, Hohenstein B, Bornstein P, Hugo C (2007) Thrombospondin 2 functions as an endogenous regulator of angiogenesis and inflammation in experimental glomerulonephritis in mice. J Am Soc Nephrol 18:788–798

Daniel C, Wagner A, Hohenstein B, Hugo C (2009) Thrombospondin-2 therapy ameliorates experimental glomerulonephritis via inhibition of cell proliferation, inflammation, and TGF-beta activation. Am J Physiol Renal Physiol 297:F1299–F1309

Dawson DW, Pearce SF, Zhong R, Silverstein RL, Frazier WA, Bouck NP (1997) CD36 mediates the in vitro inhibitory effects of thrombospondin-1 on endothelial cells. J Cell Biol 138: 707–717

Dawson DW, Volpert OV, Pearce SF, Schneider AJ, Silverstein RL, Henkin J, Bouck NP (1999) Three distinct D-amino acid substitutions confer potent antiangiogenic activity on an inactive peptide derived from a thrombospondin-1 type 1 repeat. Mol Pharmacol 55:332–338

Dean RA, Butler GS, Hamma-Kourbali Y, Delbe J, Brigstock DR, Courty J, Overall CM (2007) Identification of candidate angiogenic inhibitors processed by matrix metalloproteinase 2 (MMP-2) in cell-based proteomic screens: disruption of vascular endothelial growth factor (VEGF)/heparin affin regulatory peptide (pleiotrophin) and VEGF/connective tissue growth factor angiogenic inhibitory complexes by MMP-2 proteolysis. Mol Cell Biol 27:8454–8465

Delany AM, Hankenson KD (2009) Thrombospondin-2 and SPARC/osteonectin are critical regulators of bone remodeling. J Cell Commun Signal 3(3–4):227–238

Delany AM, McMahon DJ, Powell JS, Greenberg DA, Kurland ES (2008) Osteonectin/SPARC polymorphisms in Caucasian men with idiopathic osteoporosis. Osteoporos Int 19:969–978

Duisters RF, Tijsen AJ, Schroen B, Leenders JJ, Lentink V, van der Made I, Herias V, van Leeuwen RE, Schellings MW, Barenbrug P, Maessen JG, Heymans S, Pinto YM, Creemers EE (2009) miR-133 and miR-30 regulate connective tissue growth factor: implications for a role of microRNAs in myocardial matrix remodeling. Circ Res 104:170–178

Ebbinghaus S, Hussain M, Tannir N, Gordon M, Desai AA, Knight RA, Humerickhouse RA, Qian J, Gordon GB, Figlin R (2007) Phase 2 study of ABT-510 in patients with previously untreated advanced renal cell carcinoma. Clin Cancer Res 13:6689–6695

Egging D, van Vlijmen-Willems I, van Tongeren T, Schalkwijk J, Peeters A (2007) Wound healing in tenascin-X deficient mice suggests that tenascin-X is involved in matrix maturation rather than matrix deposition. Connect Tissue Res 48:93–98

Elefteriou F, Exposito JY, Garrone R, Lethias C (2001) Binding of tenascin-X to decorin. FEBS Lett 495:44–47

Elola MT, Wolfenstein-Todel C, Troncoso MF, Vasta GR, Rabinovich GA (2007) Galectins: matricellular glycan-binding proteins linking cell adhesion, migration, and survival. Cell Mol Life Sci 64:1679–1700

Elzie CA, Murphy-Ullrich JE (2004) The N-terminus of thrombospondin: the domain stands apart. Int J Biochem Cell Biol 36:1090–1101

Esposito I, Kayed H, Keleg S, Giese T, Sage EH, Schirmacher P, Friess H, Kleeff J (2007) Tumor-suppressor function of SPARC-like protein 1/Hevin in pancreatic cancer. Neoplasia 9:8–17

Feinstein Y, Klar A (2004) The neuronal class 2 TSR proteins F-spondin and mindin: a small family with divergent biological activities. Int J Biochem Cell Biol 36:975–980

Floquet N, Dedieu S, Martiny L, Dauchez M, Perahia D (2008) Human thrombospondin's (TSP-1) C-terminal domain opens to interact with the CD-47 receptor: a molecular modeling study. Arch Biochem Biophys 478:103–109

Fonseca C, Lindahl GE, Ponticos M, Sestini P, Renzoni EA, Holmes AM, Spagnolo P, Pantelidis P, Leoni P, McHugh N, Stock CJ, Shi-wen X, Denton CP, Black CM, Welsh KI, du Bois RM, Abraham DJ (2007) A polymorphism in the CTGF promoter region associated with systemic sclerosis. N Engl J Med 357:1210–1220

Franzen CA, Chen CC, Todorovic V, Juric V, Monzon RI, Lau LF (2009) The matrix protein CCN1 is critical for prostate carcinoma cell proliferation and TRAIL-induced apoptosis. Mol Cancer Res 7:1045–1055

Friedrichsen S, Heuer H, Christ S, Winckler M, Brauer D, Bauer K, Raivich G (2003) CTGF expression during mouse embryonic development. Cell Tissue Res 312:175–188

Furukawa K, Roberts DD, Endo T, Kobata A (1989) Structural study of the sugar chains of human platelet thrombospondin. Arch Biochem Biophys 270:302–312

Gellhaus A, Schmidt M, Dunk C, Lye SJ, Kimmig R, Winterhager E (2006) Decreased expression of the angiogenic regulators CYR61 (CCN1) and NOV (CCN3) in human placenta is associated with pre-eclampsia. Mol Hum Reprod 12:389–399

Gietema JA, Hoekstra R, de Vos FY, Uges DR, van der Gaast A, Groen HJ, Loos WJ, Knight RA, Carr RA, Humerickhouse RA, Eskens FA (2006) A phase I study assessing the safety and pharmacokinetics of the thrombospondin-1-mimetic angiogenesis inhibitor ABT-510 with gemcitabine and cisplatin in patients with solid tumors. Ann Oncol 17:1320–1327

Good DJ, Polverini PJ, Rastinejad F, Le BM, Lemons RS, Frazier WA, Bouck NP (1990) A tumor suppressor-dependent inhibitor of angiogenesis is immunologically and functionally indistinguishable from a fragment of thrombospondin. Proc Natl Acad Sci USA 87:6624–6628

Greenaway J, Lawler J, Moorehead R, Bornstein P, Lamarre J, Petrik J (2007) Thrombospondin-1 inhibits VEGF levels in the ovary directly by binding and internalization via the low density lipoprotein receptor-related protein-1 (LRP-1). J Cell Physiol 210:807–818

Grimbert P, Bouguermouh S, Baba N, Nakajima T, Allakhverdi Z, Braun D, Saito H, Rubio M, Delespesse G, Sarfati M (2006) Thrombospondin/CD47 interaction: a pathway to generate regulatory T cells from human CD4+ CD25-T cells in response to inflammation. J Immunol 177:3534–3541

Guerrero D, Guarch R, Ojer A, Casas JM, Ropero S, Mancha A, Pesce C, Lloveras B, Garcia-Bragado F, Puras A (2008) Hypermethylation of the thrombospondin-1 gene is associated with poor prognosis in penile squamous cell carcinoma. BJU Int 102(6):747–755

Guo NH, Krutzsch HC, Nègre E, Vogel T, Blake DA, Roberts DD (1992) Heparin- and sulfatide-binding peptides from the type I repeats of human thrombospondin promote melanoma cell adhesion. Proc Natl Acad Sci USA 89:3040–3044

Guo N, Krutzsch HC, Inman JK, Roberts DD (1997a) Thrombospondin 1 and type I repeat peptides of thrombospondin 1 specifically induce apoptosis of endothelial cells. Cancer Res 57: 1735–1742

Guo NH, Krutzsch HC, Inman JK, Shannon CS, Roberts DD (1997b) Antiproliferative and antitumor activities of D-reverse peptides derived from the second type-1 repeat of thrombospondin-1. J Pept Res 50:210–221

Gupta R, Hong D, Iborra F, Sarno S, Enver T (2007) NOV (CCN3) functions as a regulator of human hematopoietic stem or progenitor cells. Science 316:590–593

Halasz K, Kassner A, Morgelin M, Heinegard D (2007) COMP acts as a catalyst in collagen fibrillogenesis. J Biol Chem 282:31166–31173

Han S, Guthridge JM, Harley IT, Sestak AL, Kim-Howard X, Kaufman KM, Namjou B, Deshmukh H, Bruner G, Espinoza LR, Gilkeson GS, Harley JB, James JA, Nath SK (2008) Osteopontin and systemic lupus erythematosus association: a probable gene–gender interaction. PLoS ONE 3: e0001757

Hancox RA, Allen MD, Holliday DL, Edwards DR, Pennington CJ, Guttery DS, Shaw JA, Walker RA, Pringle JH, Jones JL (2009) Tumour-associated tenascin-C isoforms promote breast cancer cell invasion and growth by matrix metalloproteinase-dependent and independent mechanisms. Breast Cancer Res 11:R24

Hankenson KD, Bornstein P (2002) The secreted protein thrombospondin 2 is an autocrine inhibitor of marrow stromal cell proliferation. J Bone Miner Res 17:415–425

Hankenson KD, Hormuzdi SG, Meganck JA, Bornstein P (2005) Mice with a disruption of the thrombospondin 3 gene differ in geometric and biomechanical properties of bone and have accelerated development of the femoral head. Mol Cell Biol 25:5599–5606

Hauser N, Paulsson M, Kale AA, DiCesare PE (1995) Tendon extracellular matrix contains pentameric thrombospondin-4 (TSP-4). FEBS Lett 368:307–310

Haviv F, Bradley MF, Kalvin DM, Schneider AJ, Davidson DJ, Majest SM, McKay LM, Haskell CJ, Bell RL, Nguyen B, Marsh KC, Surber BW, Uchic JT, Ferrero J, Wang YC, Leal J, Record RD, Hodde J, Badylak SF, Lesniewski RR, Henkin J (2005) Thrombospondin-1 mimetic peptide inhibitors of angiogenesis and tumor growth: design, synthesis, and optimization of pharmacokinetics and biological activities. J Med Chem 48:2838–2846

Hawighorst T, Velasco P, Streit M, Hong YK, Kyriakides TR, Brown LF, Bornstein P, Detmar M (2001) Thrombospondin-2 plays a protective role in multistep carcinogenesis: a novel host anti-tumor defense mechanism. EMBO J 20:2631–2640

Hecht JT, Nelson LD, Crowder E, Wang Y, Elder FF, Harrison WR, Francomano CA, Prange CK, Lennon GG, Deere M et al (1995) Mutations in exon 17B of cartilage oligomeric matrix protein (COMP) cause pseudoachondroplasia. Nat Genet 10:325–329

Hedbom E, Antonsson P, Hjerpe A, Aeschlimann D, Paulsson M, Rosa-Pimentel E, Sommarin Y, Wendel M, Oldberg A, Heinegard D (1992) Cartilage matrix proteins. An acidic oligomeric protein (COMP) detected only in cartilage. J Biol Chem 267:6132–6136

Heinonen TY, Maki M (2009) Peters'-plus syndrome is a congenital disorder of glycosylation caused by a defect in the beta1, 3-glucosyltransferase that modifies thrombospondin type 1 repeats. Ann Med 41:2–10

Ho A, Sudhof TC (2004) Binding of F-spondin to amyloid-beta precursor protein: a candidate amyloid-beta precursor protein ligand that modulates amyloid-beta precursor protein cleavage. Proc Natl Acad Sci USA 101:2548–2553

Hoe HS, Wessner D, Beffert U, Becker AG, Matsuoka Y, Rebeck GW (2005) F-spondin interaction with the apolipoprotein E receptor Apoer2 affects processing of amyloid precursor protein. Mol Cell Biol 25:9259–9268

Hofsteenge J, Huwiler KG, Macek B, Hess D, Lawler J, Mosher DF, Peter-Katalinic J (2001) C-mannosylation and O-fucosylation of the thrombospondin type 1 module. J Biol Chem 276:6485–6498

Holbourn KP, Acharya KR, Perbal B (2008) The CCN family of proteins: structure-function relationships. Trends Biochem Sci 33:461–473

Hotchkiss KA, Matthias LJ, Hogg PJ (1998) Exposure of the cryptic Arg-Gly-Asp sequence in thrombospondin-1 by protein disulfide isomerase. Biochim Biophys Acta 1388:478–488

Hurvitz JR, Suwairi WM, Van HW, El-Shanti H, Superti-Furga A, Roudier J, Holderbaum D, Pauli RM, Herd JK, Van HEV, Rezai-Delui H, Legius E, Le MM, Al-Alami J, Bahabri SA, Warman ML (1999) Mutations in the CCN gene family member WISP3 cause progressive pseudorheumatoid dysplasia. Nat Genet 23:94–98

Ignarro LJ (2002) Nitric oxide as a unique signaling molecule in the vascular system: a historical overview. J Physiol Pharmacol 53:503–514

Iruela-Arispe ML, Liska DJ, Sage EH, Bornstein P (1993) Differential expression of thrombospondin 1, 2, and 3 during murine development. Dev Dyn 197:40–56

Iruela-Arispe ML, Lombardo M, Krutzsch HC, Lawler J, Roberts DD (1999) Inhibition of angiogenesis by thrombospondin-1 is mediated by two independent regions within the type 1 repeats. Circulation 100:1423–1431

Isenberg JS, Ridnour LA, Dimitry J, Frazier WA, Wink DA, Roberts DD (2006) CD47 is necessary for inhibition of nitric oxide-stimulated vascular cell responses by thrombospondin-1. J Biol Chem 281(36):26069–26080

Isenberg JS, Hyodo F, Matsumoto K, Romeo MJ, Abu-Asab M, Tsokos M, Kuppusamy P, Wink DA, Krishna MC, Roberts DD (2007a) Thrombospondin-1 limits ischemic tissue survival by inhibiting nitric oxide-mediated vascular smooth muscle relaxation. Blood 109:1945–1952

Isenberg JS, Hyodo F, Pappan LK, Abu-Asab M, Tsokos M, Krishna MC, Frazier WA, Roberts DD (2007b) Blocking thrombospondin-1/CD47 signaling alleviates deleterious effects of aging on tissue responses to ischemia. Arterioscler Thromb Vasc Biol 27:2582–2588

Isenberg JS, Romeo MJ, Abu-Asab M, Tsokos M, Oldenborg A, Pappan L, Wink DA, Frazier WA, Roberts DD (2007c) Increasing survival of ischemic tissue by targeting CD47. Circ Res 100:712–720

Isenberg JS, Maxhimer JB, Hyodo F, Pendrak ML, Ridnour LA, DeGraff WG, Tsokos M, Wink DA, Roberts DD (2008a) Thrombospondin-1 and CD47 limit cell and tissue survival of radiation injury. Am J Pathol 173:1100–1112

Isenberg JS, Maxhimer JB, Powers P, Tsokos M, Frazier WA, Roberts DD (2008b) Treatment of ischemia/reperfusion injury by limiting thrombospondin-1/CD47 signaling. Surgery 144: 752–761

Isenberg JS, Pappan LK, Romeo MJ, Abu-Asab M, Tsokos M, Wink DA, Frazier WA, Roberts DD (2008c) Blockade of thrombospondin-1-CD47 interactions prevents necrosis of full thickness skin grafts. Ann Surg 247:180–190

Isenberg JS, Romeo MJ, Maxhimer JB, Smedley J, Frazier WA, Roberts DD (2008d) Gene silencing of CD47 and antibody ligation of thrombospondin-1 enhance ischemic tissue survival in a porcine model: implications for human disease. Ann Surg 247:860–868

Isenberg JS, Romeo MJ, Yu C, Yu CK, Nghiem K, Monsale J, Rick ME, Wink DA, Frazier WA, Roberts DD (2008e) Thrombospondin-1 stimulates platelet aggregation by blocking the antithrombotic activity of nitric oxide/cGMP signaling. Blood 111:613–623

Isenberg JS, Yu C, Roberts DD (2008f) Differential effects of ABT-510 and a CD36-binding peptide derived from the type 1 repeats of thrombospondin-1 on fatty acid uptake, nitric oxide signaling, and caspase activation in vascular cells. Biochem Pharmacol 75:875–882

Isenberg JS, Annis DS, Pendrak ML, Ptaszynska M, Frazier WA, Mosher DF, Roberts DD (2009a) Differential interactions of thrombospondin-1, -2, and -4 with CD47 and effects on cGMP signaling and ischemic injury responses. J Biol Chem 284:1116–1125

Isenberg JS, Martin-Manso G, Maxhimer JB, Roberts DD (2009b) Regulation of nitric oxide signalling by thrombospondin 1: implications for anti-angiogenic therapies. Nat Rev Cancer 9:182–194

Isenberg JS, Qin Y, Maxhimer JB, Sipes JM, Despres D, Schnermann J, Frazier WA, Roberts DD (2009c) Thrombospondin-1 and CD47 regulate blood pressure and cardiac responses to vasoactive stress. Matrix Biol 28:110–119

Ivkovic S, Yoon BS, Popoff SN, Safadi FF, Libuda DE, Stephenson RC, Daluiski A, Lyons KM (2003) Connective tissue growth factor coordinates chondrogenesis and angiogenesis during skeletal development. Development 130:2779–2791

Jaffe EA, Leung LL, Nachman RL, Levin RI, Mosher DF (1982) Thrombospondin is the endogenous lectin of human platelets. Nature 295:246–248

Jimenez B, Volpert OV, Crawford SE, Febbraio M, Silverstein RL, Bouck N (2000) Signals leading to apoptosis-dependent inhibition of neovascularization by thrombospondin-1. Nat Med 6:41–48

Jones FS, Jones PL (2000) The tenascin family of ECM glycoproteins: structure, function, and regulation during embryonic development and tissue remodeling. Dev Dyn 218:235–259

Jones GC, Riley GP (2005) ADAMTS proteinases: a multi-domain, multi-functional family with roles in extracellular matrix turnover and arthritis. Arthritis Res Ther 7:160–169

Jun IL, Lau LF (2010) The matricellular protein CCN1 induces fibroblast senescence and restricts fibrosis in cutaneous wound healing. Nat Cell Biol 12(7):676–685

Juric V, Chen CC, Lau LF (2009) Fas-mediated apoptosis is regulated by the extracellular matrix protein CCN1 (CYR61) in vitro and in vivo. Mol Cell Biol 29:3266–3279

Kamata T, Katsube K, Michikawa M, Yamada M, Takada S, Mizusawa H (2004) R-spondin, a novel gene with thrombospondin type 1 domain, was expressed in the dorsal neural tube and affected in Wnts mutants. Biochim Biophys Acta 1676:51–62

Kamatani Y, Matsuda K, Ohishi T, Ohtsubo S, Yamazaki K, Iida A, Hosono N, Kubo M, Yumura W, Nitta K, Katagiri T, Kawaguchi Y, Kamatani N, Nakamura Y (2008) Identification of a significant association of a single nucleotide polymorphism in TNXB with systemic lupus erythematosus in a Japanese population. J Hum Genet 53:64–73

Kanda S, Shono T, Tomasini-Johansson B, Klint P, Saito Y (1999) Role of thrombospondin-1-derived peptide, 4N1K, in FGF-2-induced angiogenesis. Exp Cell Res 252:262–272

Kang DH, Anderson S, Kim YG, Mazzalli M, Suga S, Jefferson JA, Gordon KL, Oyama TT, Hughes J, Hugo C, Kerjaschki D, Schreiner GF, Johnson RJ (2001) Impaired angiogenesis in the aging kidney: vascular endothelial growth factor and thrombospondin-1 in renal disease. Am J Kidney Dis 37:601–611

Kawaguchi Y, Ota Y, Kawamoto M, Ito I, Tsuchiya N, Sugiura T, Katsumata Y, Soejima M, Sato S, Hasegawa M, Fujimoto M, Takehara K, Kuwana M, Yamanaka H, Hara M (2009) Association study of a polymorphism of the CTGF gene and susceptibility to systemic sclerosis in the Japanese population. Ann Rheum Dis 68:1921–1924

Kazuno M, Tokunaga T, Oshika Y, Tanaka Y, Tsugane R, Kijima H, Yamazaki H, Ueyama Y, Nakamura M (1999) Thrombospondin-2 (TSP2) expression is inversely correlated with vascularity in glioma. Eur J Cancer 35:502–506

Kehrel B, Kronenberg A, Schwippert B, Niesing-Bresch D, Niehues U, Tschope D, van de Loo J, Clemetson KJ (1991) Thrombospondin binds normally to glycoprotein IIIb deficient platelets. Biochem Biophys Res Commun 179:985–991

Kipnes J, Carlberg AL, Loredo GA, Lawler J, Tuan RS, Hall DJ (2003) Effect of cartilage oligomeric matrix protein on mesenchymal chondrogenesis in vitro. Osteoarthritis Cartilage 11:442–454

Kishi M, Nakamura M, Nishimine M, Ishida E, Shimada K, Kirita T, Konishi N (2003) Loss of heterozygosity on chromosome 6q correlates with decreased thrombospondin-2 expression in human salivary gland carcinomas. Cancer Sci 94:530–535

Kitagawa M, Kudo Y, Iizuka S, Ogawa I, Abiko Y, Miyauchi M, Takata T (2006) Effect of F-spondin on cementoblastic differentiation of human periodontal ligament cells. Biochem Biophys Res Commun 349:1050–1056

Klar A, Baldassare M, Jessell TM (1992) F-spondin: a gene expressed at high levels in the floor plate encodes a secreted protein that promotes neural cell adhesion and neurite extension. Cell 69:95–110

Kleer CG, Zhang Y, Merajver SD (2007) CCN6 (WISP3) as a new regulator of the epithelial phenotype in breast cancer. Cells Tissues Organs 185:95–99

Kobel M, Kalloger SE, Boyd N, McKinney S, Mehl E, Palmer C, Leung S, Bowen NJ, Ionescu DN, Rajput A, Prentice LM, Miller D, Santos J, Swenerton K, Gilks CB, Huntsman D (2008) Ovarian carcinoma subtypes are different diseases: implications for biomarker studies. PLoS Med 5:e232

Koch W, Hoppmann P, de Waha A, Schomig A, Kastrati A (2008) Polymorphisms in thrombospondin genes and myocardial infarction: a case-control study and a meta-analysis of available evidence. Hum Mol Genet 17:1120–1126

Kokenyesi R, Armstrong LC, Agah A, Artal R, Bornstein P (2004) Thrombospondin 2 deficiency in pregnant mice results in premature softening of the uterine cervix. Biol Reprod 70:385–390

Konigshoff M, Kramer M, Balsara N, Wilhelm J, Amarie OV, Jahn A, Rose F, Fink L, Seeger W, Schaefer L, Gunther A, Eickelberg O (2009) WNT1-inducible signaling protein-1 mediates pulmonary fibrosis in mice and is upregulated in humans with idiopathic pulmonary fibrosis. J Clin Invest 119:772–787

Koo BH, Hurskainen T, Mielke K, Aung PP, Casey G, Autio-Harmainen H, Apte SS (2007) ADAMTSL3/punctin-2, a gene frequently mutated in colorectal tumors, is widely expressed in normal and malignant epithelial cells, vascular endothelial cells and other cell types, and its mRNA is reduced in colon cancer. Int J Cancer 121:1710–1716

Koon HW, Zhao D, Xu H, Bowe C, Moss A, Moyer MP, Pothoulakis C (2008) Substance P-mediated expression of the pro-angiogenic factor CCN1 modulates the course of colitis. Am J Pathol 173:400–410

Kos K, Wong S, Tan B, Gummesson A, Jernas M, Franck N, Kerrigan D, Nystrom FH, Carlsson LM, Randeva HS, Pinkney JH, Wilding JP (2009) Regulation of the fibrosis and angiogenesis promoter SPARC/osteonectin in human adipose tissue by weight change, leptin, insulin, and glucose. Diabetes 58:1780–1788

Kozma K, Keusch JJ, Hegemann B, Luther KB, Klein D, Hess D, Haltiwanger RS, Hofsteenge J (2006) Identification and characterization of a beta 1, 3-glucosyltransferase that synthesizes the Glc-beta 1, 3-Fuc disaccharide on thrombospondin type 1 repeats. J Biol Chem 281(48): 36742–36751

Krady MM, Zeng J, Yu J, MacLauchlan S, Skokos EA, Tian W, Bornstein P, Sessa WC, Kyriakides TR (2008) Thrombospondin-2 modulates extracellular matrix remodeling during physiological angiogenesis. Am J Pathol 173:879–891

Kubota S, Takigawa M (2007) Role of CCN2/CTGF/Hcs24 in bone growth. Int Rev Cytol 257:1–41

Kunz M, Koczan D, Ibrahim SM, Gillitzer R, Gross G, Thiesen HJ (2002) Differential expression of thrombospondin 2 in primary and metastatic malignant melanoma. Acta Derm Venereol 82:163–169

Kutz WE, Gong Y, Warman ML (2005) WISP3, the gene responsible for the human skeletal disease progressive pseudorheumatoid dysplasia, is not essential for skeletal function in mice. Mol Cell Biol 25:414–421

Kutz WE, Wang LW, Dagoneau N, Odrcic KJ, Cormier-Daire V, Traboulsi EI, Apte SS (2008) Functional analysis of an ADAMTS10 signal peptide mutation in Weill-Marchesani syndrome demonstrates a long-range effect on secretion of the full-length enzyme. Hum Mutat 29:1425–1434

Kuznetsova SA, Issa P, Perruccio EM, Zeng B, Sipes JM, Ward Y, Seyfried NT, Fielder HL, Day AJ, Wight TN, Roberts DD (2006) Versican-thrombospondin-1 binding in vitro and colocalization in microfibrils induced by inflammation on vascular smooth muscle cells. J Cell Sci 119:4499–4509

Kvansakul M, Adams JC, Hohenester E (2004) Structure of a thrombospondin C-terminal fragment reveals a novel calcium core in the type 3 repeats. EMBO J 23:1223–1233

Kyriakides TR, Zhu YH, Smith LT, Bain SD, Yang Z, Lin MT, Danielson KG, Iozzo RV, LaMarca M, McKinney CE, Ginns EI, Bornstein P (1998a) Mice that lack thrombospondin 2 display connective tissue abnormalities that are associated with disordered collagen fibrillogenesis, an increased vascular density, and a bleeding diathesis. J Cell Biol 140:419–430

Kyriakides TR, Zhu YH, Yang Z, Bornstein P (1998b) The distribution of the matricellular protein thrombospondin 2 in tissues of embryonic and adult mice. J Histochem Cytochem 46: 1007–1015

Kyriakides TR, Leach KJ, Hoffman AS, Ratner BD, Bornstein P (1999a) Mice that lack the angiogenesis inhibitor, thrombospondin 2, mount an altered foreign body reaction characterized by increased vascularity. Proc Natl Acad Sci USA 96:4449–4454

Kyriakides TR, Tam JW, Bornstein P (1999b) Accelerated wound healing in mice with a disruption of the thrombospondin 2 gene. J Invest Dermatol 113:782–787

Kyriakides TR, Rojnuckarin P, Reidy MA, Hankenson KD, Papayannopoulou T, Kaushansky K, Bornstein P (2003) Megakaryocytes require thrombospondin-2 for normal platelet formation and function. Blood 101:3915–3923

Lagan AL, Pantelidis P, Renzoni EA, Fonseca C, Beirne P, Taegtmeyer AB, Denton CP, Black CM, Wells AU, du Bois RM, Welsh KI (2005) Single-nucleotide polymorphisms in the SPARC gene are not associated with susceptibility to scleroderma. Rheumatology 44:197–201

Lake AC, Bialik A, Walsh K, Castellot JJ Jr (2003) CCN5 is a growth arrest-specific gene that regulates smooth muscle cell proliferation and motility. Am J Pathol 162:219–231

Lamy L, Foussat A, Brown EJ, Bornstein P, Ticchioni M, Bernard A (2007) Interactions between CD47 and thrombospondin reduce inflammation. J Immunol 178:5930–5939

Lange-Asschenfeldt B, Weninger W, Velasco P, Kyriakides TR, von Andrian UH, Bornstein P, Detmar M (2002) Increased and prolonged inflammation and angiogenesis in delayed-type hypersensitivity reactions elicited in the skin of thrombospondin-2-deficient mice. Blood 99:538–545

Lau LF, Lam SC (1999) The CCN family of angiogenic regulators: the integrin connection. Exp Cell Res 248:44–57

Lawler J, Detmar M (2004) Tumor progression: the effects of thrombospondin-1 and -2. Int J Biochem Cell Biol 36:1038–1045

Lawler J, Duquette M, Whittaker CA, Adams JC, McHenry K, DeSimone DW (1993) Identification and characterization of thrombospondin-4, a new member of the thrombospondin gene family. J Cell Biol 120:1059–1067

Lawler J, McHenry K, Duquette M, Derick L (1995) Characterization of human thrombospondin-4. J Biol Chem 270:2809–2814

Lawler J, Sunday M, Thibert V, Duquette M, George EL, Rayburn H, Hynes RO (1998) Thrombospondin-1 is required for normal murine pulmonary homeostasis and its absence causes pneumonia. J Clin Invest 101:982–992

Le Goff C, Morice-Picard F, Dagoneau N, Wang LW, Perrot C, Crow YJ, Bauer F, Flori E, Prost-Squarcioni C, Krakow D, Ge G, Greenspan DS, Bonnet D, Le Merrer M, Munnich A, Apte SS, Cormier-Daire V (2008) ADAMTSL2 mutations in geleophysic dysplasia demonstrate a role for ADAMTS-like proteins in TGF-beta bioavailability regulation. Nat Genet 40:1119–1123

Leask A, Abraham DJ (2006) All in the CCN family: essential matricellular signaling modulators emerge from the bunker. J Cell Sci 119:4803–4810

Lee HY, Chung JW, Youn SW, Kim JY, Park KW, Koo BK, Oh BH, Park YB, Chaqour B, Walsh K, Kim HS (2007) Forkhead transcription factor FOXO3a is a negative regulator of angiogenic immediate early gene CYR61, leading to inhibition of vascular smooth muscle cell proliferation and neointimal hyperplasia. Circ Res 100:372–380

Lethias C, Descollonges Y, Boutillon MM, Garrone R (1996) Flexilin: a new extracellular matrix glycoprotein localized on collagen fibrils. Matrix Biol 15:11–19

Li Q, Ahuja N, Burger PC, Issa JP (1999) Methylation and silencing of the thrombospondin-1 promoter in human cancer. Oncogene 18:3284–3289

Li Z, Calzada MJ, Sipes JM, Cashel JA, Krutzsch HC, Annis D, Mosher DF, Roberts DD (2002) Interactions of thrombospondins with $\alpha 4\beta 1$ integrin and CD47 differentially modulate T cell behavior. J Cell Biol 157:509–519

Liu A, Garg P, Yang S, Gong P, Pallero MA, Annis DS, Liu Y, Passaniti A, Mann D, Mosher DF, Murphy-Ullrich JE, Goldblum SE (2009) Epidermal growth factor-like repeats of thrombospondins activate phospholipase Cgamma and increase epithelial cell migration through indirect epidermal growth factor receptor activation. J Biol Chem 284:6389–6402

Mann HH, Ozbek S, Engel J, Paulsson M, Wagener R (2004) Interactions between the cartilage oligomeric matrix protein and matrilins. Implications for matrix assembly and the pathogenesis of chondrodysplasias. J Biol Chem 279:25294–25298

Margosio B, Rusnati M, Bonezzi K, Cordes BL, Annis DS, Urbinati C, Giavazzi R, Presta M, Ribatti D, Mosher DF, Taraboletti G (2008) Fibroblast growth factor-2 binding to the thrombospondin-1 type III repeats, a novel antiangiogenic domain. Int J Biochem Cell Biol 40: 700–709

Markovic SN, Suman VJ, Rao RA, Ingle JN, Kaur JS, Erickson LA, Pitot HC, Croghan GA, McWilliams RR, Merchan J, Kottschade LA, Nevala WK, Uhl CB, Allred J, Creagan ET (2007) A phase II study of ABT-510 (thrombospondin-1 analog) for the treatment of metastatic melanoma. Am J Clin Oncol 30:303–309

Martinek N, Shahab J, Saathoff M, Ringuette M (2008) Haemocyte-derived SPARC is required for collagen-IV-dependent stability of basal laminae in *Drosophila* embryos. J Cell Sci 121: 1671–1680

Martin-Manso G, Galli S, Ridnour LA, Tsokos M, Wink DA, Roberts DD (2008) Thrombospondin-1 promotes tumor macrophage recruitment and enhances tumor cell cytotoxicity by differentiated U937 cells. Cancer Res 68:7090–7099

Matsumae H, Yoshida Y, Ono K, Togi K, Inoue K, Furukawa Y, Nakashima Y, Kojima Y, Nobuyoshi M, Kita T, Tanaka M (2008) CCN1 knockdown suppresses neointimal hyperplasia in a rat artery balloon injury model. Arterioscler Thromb Vasc Biol 28:1077–1083

Maxhimer JB, Shih HB, Isenberg JS, Miller TW, Roberts DD (2009a) Thrombospondin-1-CD47 blockade following ischemia reperfusion injury is tissue protective. Plast Reconstr Surg 124(6): 1880–1889

Maxhimer JB, Soto-Pantoja DR, Ridnour LA, Shih HB, DeGraff WG, Tsokos M, Wink DA, Isenberg JS, Roberts DD (2009b) Radioprotection in normal tissue and delayed tumor growth by blockade of CD47 signaling. Sci Transl Med 1:3ra7

Midwood KS, Orend G (2009) The role of tenascin-C in tissue injury and tumorigenesis. J Cell Commun Signal 3(3–4):287–310

Midwood K, Sacre S, Piccinini AM, Inglis J, Trebaul A, Chan E, Drexler S, Sofat N, Kashiwagi M, Orend G, Brennan F, Foxwell B (2009) Tenascin-C is an endogenous activator of toll-like receptor 4 that is essential for maintaining inflammation in arthritic joint disease. Nat Med 15:774–780

Mo FE, Lau LF (2006) The matricellular protein CCN1 is essential for cardiac development. Circ Res 99:961–969

Mo FE, Muntean AG, Chen CC, Stolz DB, Watkins SC, Lau LF (2002) CYR61 (CCN1) is essential for placental development and vascular integrity. Mol Cell Biol 22:8709–8720

Mori T, Kawara S, Shinozaki M, Hayashi N, Kakinuma T, Igarashi A, Takigawa M, Nakanishi T, Takehara K (1999) Role and interaction of connective tissue growth factor with transforming growth factor-beta in persistent fibrosis: a mouse fibrosis model. J Cell Physiol 181:153–159

Murphy-Ullrich JE, Mosher DF (1987) Interactions of thrombospondin with endothelial cells: receptor-mediated binding and degradation. J Cell Biol 105:1603–1611

Mustonen E, Aro J, Puhakka J, Ilves M, Soini Y, Leskinen H, Ruskoaho H, Rysa J (2008) Thrombospondin-4 expression is rapidly upregulated by cardiac overload. Biochem Biophys Res Commun 373:186–191

Nakamura Y, Weidinger G, Liang JO, Quilina-Beck A, Tamai K, Moon RT, Warman ML (2007) The CCN family member Wisp3, mutant in progressive pseudorheumatoid dysplasia, modulates BMP and Wnt signaling. J Clin Invest 117:3075–3086

Nakamura Y, Cui Y, Fernando C, Kutz WE, Warman ML (2009) Normal growth and development in mice over-expressing the CCN family member WISP3. J Cell Commun Signal 3:105–113

Nam JS, Turcotte TJ, Smith PF, Choi S, Yoon JK (2006) Mouse cristin/R-spondin family proteins are novel ligands for the Frizzled 8 and LRP6 receptors and activate beta-catenin-dependent gene expression. J Biol Chem 281:13247–13257

Narouz-Ott L, Maurer P, Nitsche DP, Smyth N, Paulsson M (2000) Thrombospondin-4 binds specifically to both collagenous and non-collagenous extracellular matrix proteins via its C-terminal domains. J Biol Chem 275:37110–37117

Ng TF, Turpie B, Masli S (2009) Thrombospondin-1-mediated regulation of microglia activation after retinal injury. Invest Ophthalmol Vis Sci 50:5472–5478

Nie J, Sage EH (2009) SPARC functions as an inhibitor of adipogenesis. J Cell Commun Signal 3(3–4):247–254

Nishimura H, Yamashita S, Zeng Z, Walz DA, Iwanaga S (1992) Evidence for the existence of O-linked sugar chains consisting of glucose and xylose in bovine thrombospondin. J Biochem 111:460–464

Norris RA, Moreno-Rodriguez R, Hoffman S, Markwald RR (2009) The many facets of the matricelluar protein periostin during cardiac development, remodeling, and pathophysiology. J Cell Commun Signal 3(3–4):275–286

O'Brien TP, Lau LF (1992) Expression of the growth factor-inducible immediate early gene cyr61 correlates with chondrogenesis during mouse embryonic development. Cell Growth Differ 3:645–654

O'Rourke KM, Laherty CD, Dixit VM (1992) Thrombospondin 1 and thrombospondin 2 are expressed as both homo- and heterotrimers. J Biol Chem 267:24921–24924

Ohgawara T, Kubota S, Kawaki H, Kondo S, Eguchi T, Kurio N, Aoyama E, Sasaki A, Takigawa M (2009) Regulation of chondrocytic phenotype by micro RNA 18a: involvement of Ccn2/Ctgf as a major target gene. FEBS Lett 583:1006–1010

Orend G, Chiquet-Ehrismann R (2006) Tenascin-C induced signaling in cancer. Cancer Lett 244:143–163

Orsmark-Pietras C, Melen E, Vendelin J, Bruce S, Laitinen A, Laitinen LA, Lauener R, Riedler J, von Mutius E, Doekes G, Wickman M, van Hage M, Pershagen G, Scheynius A, Nyberg F, Kere J (2008) Biological and genetic interaction between tenascin C and neuropeptide S receptor 1 in allergic diseases. Hum Mol Genet 17:1673–1682

Pennica D, Swanson TA, Welsh JW, Roy MA, Lawrence DA, Lee J, Brush J, Taneyhill LA, Deuel B, Lew M, Watanabe C, Cohen RL, Melhem MF, Finley GG, Quirke P, Goddard AD, Hillan KJ, Gurney AL, Botstein D, Levine AJ (1998) WISP genes are members of the connective tissue growth factor family that are up-regulated in wnt-1-transformed cells and aberrantly expressed in human colon tumors. Proc Natl Acad Sci USA 95:14717–14722

Perbal B, Lazar N, Zambelli D, Lopez-Guerrero JA, Llombart-Bosch A, Scotlandi K, Picci P (2009) Prognostic relevance of CCN3 in Ewing sarcoma. Hum Pathol 40:1479–1486

Perez S, Vial E, van Dam H, Castellazzi M (2001) Transcription factor ATF3 partially transforms chick embryo fibroblasts by promoting growth factor-independent proliferation. Oncogene 20:1135–1141

Pihlajamaa T, Lankinen H, Ylostalo J, Valmu L, Jaalinoja J, Zaucke F, Spitznagel L, Gosling S, Puustinen A, Morgelin M, Peranen J, Maurer P, Ala-Kokko L, Kilpelainen I (2004) Characterization of recombinant amino-terminal NC4 domain of human collagen IX: interaction with glycosaminoglycans and cartilage oligomeric matrix protein. J Biol Chem 279:24265–24273

Pimanda JE, Annis DS, Raftery M, Mosher DF, Chesterman CN, Hogg PJ (2002) The von Willebrand factor-reducing activity of thrombospondin-1 is located in the calcium-binding/C-terminal sequence and requires a free thiol at position 974. Blood 100:2832–2838

Pimanda JE, Ganderton T, Maekawa A, Yap CL, Lawler J, Kershaw G, Chesterman CN, Hogg PJ (2004) Role of thrombospondin-1 in control of von Willebrand factor multimer size in mice. J Biol Chem 279:21439–21448

Porter S, Clark IM, Kevorkian L, Edwards DR (2005) The ADAMTS metalloproteinases. Biochem J 386:15–27

Posey KL, Hankenson K, Veerisetty AC, Bornstein P, Lawler J, Hecht JT (2008a) Skeletal abnormalities in mice lacking extracellular matrix proteins, thrombospondin-1, thrombospondin-3, thrombospondin-5, and type IX collagen. Am J Pathol 172:1664–1674

Posey KL, Yang Y, Veerisetty AC, Sharan SK, Hecht JT (2008b) Model systems for studying skeletal dysplasias caused by TSP-5/COMP mutations. Cell Mol Life Sci 65:687–699

Przysiecki CT, Staggers JE, Ramjit HG, Musson DG, Stern AM, Bennett CD, Friedman PA (1987) Occurrence of beta-hydroxylated asparagine residues in non-vitamin K-dependent proteins containing epidermal growth factor-like domains. Proc Natl Acad Sci USA 84:7856–7860

Qabar AN, Lin Z, Wolf FW, O'Shea KS, Lawler J, Dixit VM (1994) Thrombospondin 3 is a developmentally regulated heparin binding protein. J Biol Chem 269:1262–1269

Raman P, Krukovets I, Marinic TE, Bornstein P, Stenina OI (2007) Glycosylation mediates upregulation of a potent antiangiogenic and proatherogenic protein, thrombospondin-1, by glucose in vascular smooth muscle cells. J Biol Chem 282:5704–5714

Rayssac A, Neveu C, Pucelle M, Van den Berghe L, Prado-Lourenco L, Arnal JF, Chaufour X, Prats AC (2009) IRES-based vector coexpressing FGF2 and Cyr61 provides synergistic and safe therapeutics of lower limb ischemia. Mol Ther 17:2010–2019

Ribeiro SM, Poczatek M, Schultz-Cherry S, Villain M, Murphy-Ullrich JE (1999) The activation sequence of thrombospondin-1 interacts with the latency- associated peptide to regulate activation of latent transforming growth factor-beta. J Biol Chem 274:13586–13593

Riessen R, Kearney M, Lawler J, Isner JM (1998) Immunolocalization of thrombospondin-1 in human atherosclerotic and restenotic arteries. Am Heart J 135:357–364

Riessen R, Fenchel M, Chen H, Axel DI, Karsch KR, Lawler J (2001) Cartilage oligomeric matrix protein (thrombospondin-5) is expressed by human vascular smooth muscle cells. Arterioscler Thromb Vasc Biol 21:47–54

Rojas A, Meherem S, Kim YH, Washington MK, Willis JE, Markowitz SD, Grady WM (2008) The aberrant methylation of TSP1 suppresses TGF-beta1 activation in colorectal cancer. Int J Cancer 123:14–21

Rueda B, Simeon C, Hesselstrand R, Herrick A, Worthington J, Ortego-Centeno N, Riemekasten G, Fonollosa V, Vonk MC, van den Hoogen FH, Sanchez-Roman J, Guirre-Zamorano MA, Garcia-Portales R, Pros A, Camps MT, Gonzalez-Gay MA, Gonzalez-Escribano MF, Coenen MJ, Lambert N, Nelson JL, Radstake TR, Martin J (2009) A large multicentre analysis of CTGF-945 promoter polymorphism does not confirm association with systemic sclerosis susceptibility or phenotype. Ann Rheum Dis 68:1618–1620

Rydziel S, Stadmeyer L, Zanotti S, Durant D, Smerdel-Ramoya A, Canalis E (2007) Nephroblastoma overexpressed (Nov) inhibits osteoblastogenesis and causes osteopenia. J Biol Chem 282:19762–19772

Sadvakassova G, Dobocan MC, Congote LF (2009a) Osteopontin and the C-terminal peptide of thrombospondin-4 compete for CD44 binding and have opposite effects on CD133+ cell colony formation. BMC Res Notes 2:215

Sadvakassova G, Dobocan MC, Difalco MR, Congote LF (2009b) Regulator of differentiation 1 (ROD1) binds to the amphipathic C-terminal peptide of thrombospondin-4 and is involved in its mitogenic activity. J Cell Physiol 220:672–679

Sage EH, Bornstein P (1991) Extracellular proteins that modulate cell-matrix interactions. SPARC, tenascin, and thrombospondin. J Biol Chem 266:14831–14834

Said N, Frierson HF Jr, Chernauskas D, Conaway M, Motamed K, Theodorescu D (2009) The role of SPARC in the TRAMP model of prostate carcinogenesis and progression. Oncogene 28:3487–3498

Sakamoto K, Yamaguchi S, Ando R, Miyawaki A, Kabasawa Y, Takagi M, Li CL, Perbal B, Katsube K (2002) The nephroblastoma overexpressed gene (NOV/ccn3) protein associates with Notch1 extracellular domain and inhibits myoblast differentiation via notch signaling pathway. J Biol Chem 277:29399–29405

Sato T, Sato M, Kiyohara K, Sogabe M, Shikanai T, Kikuchi N, Togayachi A, Ishida H, Ito H, Kameyama A, Gotoh M, Narimatsu H (2006) Molecular cloning and characterization of a novel human {beta}1, 3-glucosyltransferase, {beta}3Glc-T, which is localized at the endoplasmic reticulum and glucosylates O-linked fucosylglycan on thrombospondin type 1 repeat domain. Glycobiology 16(12):1194–1206

Scatena M, Liaw L, Giachelli CM (2007) Osteopontin: a multifunctional molecule regulating chronic inflammation and vascular disease. Arterioscler Thromb Vasc Biol 27:2302–2309

Schroen B, Heymans S, Sharma U, Blankesteijn WM, Pokharel S, Cleutjens JP, Porter JG, Evelo CT, Duisters R, van Leeuwen RE, Janssen BJ, Debets JJ, Smits JF, Daemen MJ, Crijns HJ, Bornstein P, Pinto YM (2004) Thrombospondin-2 is essential for myocardial matrix integrity: increased expression identifies failure-prone cardiac hypertrophy. Circ Res 95:515–522

Schubert D, Klar A, Park M, Dargusch R, Fischer WH (2006) F-spondin promotes nerve precursor differentiation. J Neurochem 96:444–453

Schultz-Cherry S, Chen H, Mosher DF, Misenheimer TM, Krutzsch HC, Roberts DD, Murphy-Ullrich JE (1995) Regulation of transforming growth factor-β activation by discrete sequences of thrombospondin 1. J Biol Chem 270:7304–7310

Sezaki S, Hirohata S, Iwabu A, Nakamura K, Toeda K, Miyoshi T, Yamawaki H, Demircan K, Kusachi S, Shiratori Y, Ninomiya Y (2005) Thrombospondin-1 is induced in rat myocardial infarction and its induction is accelerated by ischemia/reperfusion. Exp Biol Med 230:621–630

Shi-wen X, Leask A, Abraham D (2008) Regulation and function of connective tissue growth factor/CCN2 in tissue repair, scarring and fibrosis. Cytokine Growth Factor Rev 19:133–144

Simantov R, Febbraio M, Silverstein RL (2005) The antiangiogenic effect of thrombospondin-2 is mediated by CD36 and modulated by histidine-rich glycoprotein. Matrix Biol 24:27–34

Smerdel-Ramoya A, Zanotti S, Stadmeyer L, Durant D, Canalis E (2008) Skeletal overexpression of connective tissue growth factor (ctgf) impairs bone formation and causes osteopenia. Endocrinology 149:4374–4381

Smith RK, Zunino L, Webbon PM, Heinegard D (1997) The distribution of cartilage oligomeric matrix protein (COMP) in tendon and its variation with tendon site, age and load. Matrix Biol 16:255–271

Sodersten F, Ekman S, Schmitz M, Paulsson M, Zaucke F (2006) Thrombospondin-4 and cartilage oligomeric matrix protein form heterooligomers in equine tendon. Connect Tissue Res 47: 85–91

Stenina OI, Desai SY, Krukovets I, Kight K, Janigro D, Topol EJ, Plow EF (2003a) Thrombospondin-4 and its variants: expression and differential effects on endothelial cells. Circulation 108:1514–1519

Stenina OI, Krukovets I, Wang K, Zhou Z, Forudi F, Penn MS, Topol EJ, Plow EF (2003b) Increased expression of thrombospondin-1 in vessel wall of diabetic Zucker rat. Circulation 107:3209–3215

Stenina OI, Byzova TV, Adams JC, McCarthy JJ, Topol EJ, Plow EF (2004) Coronary artery disease and the thrombospondin single nucleotide polymorphisms. Int J Biochem Cell Biol 36:1013–1030

Stenina OI, Ustinov V, Krukovets I, Marinic T, Topol EJ, Plow EF (2005) Polymorphisms A387P in thrombospondin-4 and N700S in thrombospondin-1 perturb calcium binding sites. FASEB J 19:1893–1895

Streit M, Riccardi L, Velasco P, Brown LF, Hawighorst T, Bornstein P, Detmar M (1999) Thrombospondin-2: a potent endogenous inhibitor of tumor growth and angiogenesis. Proc Natl Acad Sci USA 96:14888–14893

Subramanian A, Wayburn B, Bunch T, Volk T (2007) Thrombospondin-mediated adhesion is essential for the formation of the myotendinous junction in *Drosophila*. Development 134:1269–1278

Sullivan MM, Barker TH, Funk SE, Karchin A, Seo NS, Hook M, Sanders J, Starcher B, Wight TN, Puolakkainen P, Sage EH (2006) Matricellular hevin regulates decorin production and collagen assembly. J Biol Chem 281:27621–27632

Sullivan MM, Puolakkainen PA, Barker TH, Funk SE, Sage EH (2008) Altered tissue repair in hevin-null mice: inhibition of fibroblast migration by a matricellular SPARC homolog. Wound Repair Regen 16:310–319

Svensson L, Aszodi A, Heinegard D, Hunziker EB, Reinholt FP, Fassler R, Oldberg A (2002) Cartilage oligomeric matrix protein-deficient mice have normal skeletal development. Mol Cell Biol 22:4366–4371

Swinnen M, Vanhoutte D, Van Almen GC, Hamdani N, Schellings MW, D'Hooge J, Van der Velden J, Weaver MS, Sage EH, Bornstein P, Verheyen FK, Van den Driessche T, Chuah MK, Westermann D, Paulus WJ, Van de Werf F, Schroen B, Carmeliet P, Pinto YM, Heymans S (2009) Absence of thrombospondin-2 causes age-related dilated cardiomyopathy. Circulation 120:1585–1597

Tan K, Duquette M, Liu JH, Lawler J, Wang JH (2008) The crystal structure of the heparin-binding reelin-N domain of F-spondin. J Mol Biol 381(5):1213–1223

Tanaka K, Hiraiwa N, Hashimoto H, Yamazaki Y, Kusakabe M (2004) Tenascin-C regulates angiogenesis in tumor through the regulation of vascular endothelial growth factor expression. Int J Cancer 108:31–40

Tandon NN, Ockenhouse CF, Greco NJ, Jamieson GA (1991) Adhesive functions of platelets lacking glycoprotein IV (CD36). Blood 78:2809–2813

Taraboletti G, Roberts D, Liotta LA, Giavazzi R (1990) Platelet thrombospondin modulates endothelial cell adhesion, motility, and growth: a potential angiogenesis regulatory factor. J Cell Biol 111:765–772

Taylor DK, Meganck JA, Terkhorn S, Rajani R, Naik A, O'Keefe RJ, Goldstein SA, Hankenson KD (2009) Thrombospondin-2 influences the proportion of cartilage and bone during fracture healing. J Bone Miner Res 24(6):1043–1054

Thakar CV, Zahedi K, Revelo MP, Wang Z, Burnham CE, Barone S, Bevans S, Lentsch AB, Rabb H, Soleimani M (2005) Identification of thrombospondin 1 (TSP-1) as a novel mediator of cell injury in kidney ischemia. J Clin Invest 115:3451–3459

Timmons JA, Jansson E, Fischer H, Gustafsson T, Greenhaff PL, Ridden J, Rachman J, Sundberg CJ (2005) Modulation of extracellular matrix genes reflects the magnitude of physiological adaptation to aerobic exercise training in humans. BMC Biol 3:19

Todorovic V, Chen CC, Hay N, Lau LF (2005) The matrix protein CCN1 (CYR61) induces apoptosis in fibroblasts. J Cell Biol 171:559–568

Tolsma SS, Volpert OV, Good DJ, Frazier WA, Polverini PJ, Bouck N (1993) Peptides derived from two separate domains of the matrix protein thrombospondin-1 have anti-angiogenic activity. J Cell Biol 122:497–511

Topol EJ, McCarthy J, Gabriel S, Moliterno DJ, Rogers WJ, Newby LK, Freedman M, Metivier J, Cannata R, O'Donnell CJ, Kottke-Marchant K, Murugesan G, Plow EF, Stenina O, Daley GQ (2001) Single nucleotide polymorphisms in multiple novel thrombospondin genes may be associated with familial premature myocardial infarction. Circulation 104:2641–2644

Tseng S, Reddi AH, Di Cesare PE (2009) Cartilage oligomeric matrix protein (COMP): a biomarker of arthritis. Biomark Insights 4:33–44
Tucker RP (2004) The thrombospondin type 1 repeat superfamily. Int J Biochem Cell Biol 36:969–974
Tucker RP, Chiquet-Ehrismann R (2009a) Evidence for the evolution of tenascin and fibronectin early in the chordate lineage. Int J Biochem Cell Biol 41:424–434
Tucker RP, Chiquet-Ehrismann R (2009b) The regulation of tenascin expression by tissue microenvironments. Biochim Biophys Acta 1793:888–892
Tucker RP, Adams JC, Lawler J (1995) Thrombospondin-4 is expressed by early osteogenic tissues in the chick embryo. Dev Dyn 203:477–490
Turpie B, Yoshimura T, Gulati A, Rios JD, Dartt DA, Masli S (2009) Sjogren's syndrome-like ocular surface disease in thrombospondin-1 deficient mice. Am J Pathol 175:1136–1147
Velasco P, Huegel R, Brasch J, Schroder JM, Weichenthal M, Stockfleth E, Schwarz T, Lawler J, Detmar M, Lange-Asschenfeldt B (2009) The angiogenesis inhibitor thrombospondin-1 inhibits acute cutaneous hypersensitivity reactions. J Invest Dermatol 129:2022–2030
Venkatachalam K, Venkatesan B, Valente AJ, Melby PC, Nandish S, Reusch JE, Clark RA, Chandrasekar B (2009) WISP1, a pro-mitogenic, pro-survival factor, mediates tumor necrosis factor-alpha (TNF-alpha)-stimulated cardiac fibroblast proliferation but inhibits TNF-alpha-induced cardiomyocyte death. J Biol Chem 284:14414–14427
Vogel T, Guo NH, Krutzsch HC, Blake DA, Hartman J, Mendelovitz S, Panet A, Roberts DD (1993) Modulation of endothelial cell proliferation, adhesion, and motility by recombinant heparin-binding domain and synthetic peptides from the type I repeats of thrombospondin. J Cell Biochem 53:74–84
Volpert OV, Tolsma SS, Pellerin S, Feige JJ, Chen H, Mosher DF, Bouck N (1995) Inhibition of angiogenesis by thrombospondin-2. Biochem Biophys Res Commun 217:326–332
Vos HL, Devarayalu S, de Vries Y, Bornstein P (1992) Thrombospondin 3 (Thbs3), a new member of the thrombospondin gene family. J Biol Chem 267:12192–12196
Wahab NA, Weston BS, Mason RM (2005) Connective tissue growth factor CCN2 interacts with and activates the tyrosine kinase receptor TrkA. J Am Soc Nephrol 16:340–351
Wang KX, Denhardt DT (2008) Osteopontin: role in immune regulation and stress responses. Cytokine Growth Factor Rev 19:333–345
Wang S, Skorczewski J, Feng X, Mei L, Murphy-Ullrich JE (2004) Glucose up-regulates thrombospondin 1 gene transcription and transforming growth factor-beta activity through antagonism of cGMP-dependent protein kinase repression via upstream stimulatory factor 2. J Biol Chem 279:34311–34322
Wang H, Zhang R, Wen S, McCafferty DM, Beck PL, MacNaughton WK (2009a) Nitric oxide increases Wnt-induced secreted protein-1 (WISP-1/CCN4) expression and function in colitis. J Mol Med 87:435–445
Wang XY, McLennan SV, Allen T, Tsoutsman T, Semsarian C, Twigg SM (2009b) Adverse effects of high glucose and free fatty acid on cardiomyocytes are mediated by connective tissue growth factor. Am J Physiol Cell Physiol 297:C1490–C1500
Watanabe T, Komuro Y, Kiyomatsu T, Kanazawa T, Kazama Y, Tanaka J, Tanaka T, Yamamoto Y, Shirane M, Muto T, Nagawa H (2006) Prediction of sensitivity of rectal cancer cells in response to preoperative radiotherapy by DNA microarray analysis of gene expression profiles. Cancer Res 66:3370–3374
Whitcomb BP, Mutch DG, Herzog TJ, Rader JS, Gibb RK, Goodfellow PJ (2003) Frequent HOXA11 and THBS2 promoter methylation, and a methylator phenotype in endometrial adenocarcinoma. Clin Cancer Res 9:2277–2287
Wu S, Platteau A, Chen S, McNamara G, Whitsett J, Bancalari E (2009) Conditional overexpression of connective tissue growth factor disrupts postnatal lung development. Am J Respir Cell Mol Biol 42(5):552–563
Xiao Y, Kleeff J, Guo J, Gazdhar A, Liao Q, Di Cesare PE, Buchler MW, Friess H (2004) Cartilage oligomeric matrix protein expression in hepatocellular carcinoma and the cirrhotic liver. J Gastroenterol Hepatol 19:296–302

Xu K, Zhang Y, Ilalov K, Carlson CS, Feng JQ, Di Cesare PE, Liu CJ (2007) Cartilage oligomeric matrix protein associates with granulin-epithelin precursor (GEP) and potentiates GEP-stimulated chondrocyte proliferation. J Biol Chem 282:11347–11355

Yamanokuchi K, Yabuki A, Yoshimoto Y, Arai K, Fujiki M, Misumi K (2009) Gene and protein expression of cartilage oligomeric matrix protein associated with oncogenesis in canine tumors. J Vet Med Sci 71:499–503

Yang Z, Strickland DK, Bornstein P (2001) Extracellular matrix metalloproteinase 2 levels are regulated by the low density lipoprotein-related scavenger receptor and thrombospondin 2. J Biol Chem 276:8403–8408

Yang QW, Liu S, Tian Y, Salwen HR, Chlenski A, Weinstein J, Cohn SL (2003) Methylation-associated silencing of the thrombospondin-1 gene in human neuroblastoma. Cancer Res 63: 6299–6310

Yang K, Vega JL, Hadzipasic M, Schatzmann Peron JP, Zhu B, Carrier Y, Masli S, Rizzo LV, Weiner HL (2009) Deficiency of thrombospondin-1 reduces Th17 differentiation and attenuates experimental autoimmune encephalomyelitis. J Autoimmun 32:94–103

Young GD, Murphy-Ullrich JE (2004) The tryptophan-rich motifs of the thrombospondin type 1 repeats bind VLAL motifs in the latent transforming growth factor-beta complex. J Biol Chem 279:47633–47642

Yuan Y, Nymoen DA, Stavnes HT, Rosnes AK, Bjorang O, Wu C, Nesland JM, Davidson B (2009) Tenascin-X is a novel diagnostic marker of malignant mesothelioma. Am J Surg Pathol 33:1673–1682

Zaia J, Boynton RE, McIntosh A, Marshak DR, Olsson H, Heinegard D, Barry FP (1997) Post-translational modifications in cartilage oligomeric matrix protein. Characterization of the N-linked oligosaccharides by matrix-assisted laser desorption ionization time-of-flight mass spectrometry. J Biol Chem 272:14120–14126

Zhang Q, Wu J, Cao Q, Xiao L, Wang L, He D, Ouyang G, Lin J, Shen B, Shi Y, Zhang Y, Li D, Li N (2009) A critical role of Cyr61 in interleukin-17-dependent proliferation of fibroblast-like synoviocytes in rheumatoid arthritis. Arthritis Rheum 60:3602–3612

Zhou Z, Cornelius CP, Eichner M, Bornemann A (2006) Reinnervation-induced alterations in rat skeletal muscle. Neurobiol Dis 23:595–602

Index

A
α-Aminoadipic-δ-semialdehyde, 276, 330
abductin, 268
Acetylcholine receptor, 11
ADAM, 15
ADAMTS, 15, 84–85, 164, 337, 346, 376, 385
  activation, 167
  role in versican degradation, 166–168
ADCL, See Autosomal dominant cutis laxa
Adhesion
  basic molecular toolkit, 26
  effects of versican, 168
  See Integrins
Aggregating proteoglycans, See Proteoglycans
Aggrecan, 148, 158–162, 386
  biosynthesis, 159–160
  core protein structure, 160
  electron microscopic image, 161
  in cartilage, 161–162
  interactions with HA through link protein, 160
Agrin, 11, 128, 137
  knockout mouse, 125
Allysine. See α-Aminoadipic-δ-semialdehyde
Alport syndrome, 139
Alternative exon splicing, 11, 275, 338
Alu repeats, 273, 275
AMD. See Macular degeneration, age related
Anchoring fibrils, 92–93
Aneurysms, 247–250, 287
Annelid, 9
Arachnodactyly, 247–250
ARCL, See Autosomal recessive cutis laxa
Asparagine glycosylation, 15
Asporin, 200, 212
  osteoarthritis, 205
Assembly, of ECM, 16–17
  role of integrins, 18–20

ATP6V0A2, 286
ATP7A, 286
Autosomal dominant cutis laxa (ADCL), 285–286, 290
Autosomal recessive cutis laxa (ARCL), 286–287
AVSD, 391

B
β-Aminopropionitrile, 304
Bacteria
  binding to fibulin, 344
  Interactions with fibronectin, 50
BAPN, See β-Aminopropionitrile
Basal lamina, See Basement membrane
Basement membrane, 26, 88–89, 117–140
  appearance with evolution of gastrulation, 26
  assembly, 121, 133–134
  cell surface, role in basement assembly, 133–134
  components, 119–129
  diseases, 134–140
  location in tissues, 117–118
  knockout phenotypes, table, 125
  receptors, 129–134
  ultrastructure, 118–119, 129
BCAM. See Lutheran/Basal cell adhesion molecule
Beals-Hect syndrome, 247
Biglycan, 89, 103
  binding to collagen, 200
  cardiovascular development and disease, 215–216
  knockout mouse, 103, 108, 201–205, 212, 214–215
  interactions with toll-like receptors, 211

Biglycan (*cont.*)
    role in skeletal growth and remodeling, 204, 212–215
BM-40. *See* SPARC
BMP. See Bone morphogenic protein
Bone, 23–24, 212–214
Bone morphogenic protein (BMP), 21–22
    binding to fibrillin, 22. 244, 250–252
    BMP-1/tolloid proteinases, 84
    interactions with decorin and biglycan, 208
Bone sialoprotein, 389
Brachymorphic mouse, 161
Brevican, 148, 174–178
    in brain injury, 177
    role in cancer, 178
    synthesis and turnover, 176–177

C
Cadherin domain, 5
Calcium
    binding by ECM, 15–16
    concentration in ECM, 15
Calreticulin, 373
Cambrian explosion, 26
Cancer, 155–156, 173, 178, 320–322, 352–355
Cardiac cushion formation, 154, 164, 345–348
Cardiac jelly, 154
Cartilage
    aggrecan, 161
    collagen fibril formation, 105–107
    ECM composition, 161
    hypertrophic cartilage, 93
    SLRPs in, 214–215
Cartilage-associated protein (Crtap)
    as collagen chaperone, 81
CCA. *See* Congenital contractural arachnodactyly
CCN family of matricellular proteins, 389–396
    defined, 389–390
    domain structure 370, 390
    functional roles, 390
    modulators of BMP and Wnt signaling, 394–396
CCN1 (CYR61)
    association with disease, 391
    binding to syndecan-4, 391
    expression, 391
    inducer of angiogenesis, 392
    knockout mouse, 391
    receptors, 391
    role in apoptosis, 391
CCN2 (CTGF), 393–395
    antagonist of BMP-4 signaling, 394

association with disease, 395
binding to LRP-1, 394
expression, 394–395
functions, 395
knockout mouse, 394
receptors and binding proteins, 864–865
CCN3 (Nov)
    antagonist of BMP signaling, 396
    knockout mouse, 396
CCN4, 396
CCN5, 396
CCN6, 396
    knockout mice, 396
    role in TGF-β signaling, 394
CD36, 373, 376–382
    knockout mice, 377
CD44, 149–153
    in cancer, 155–156
    knockout mouse, 150
    signaling cascade, 151
CD47, 373, 376–382
    knockout mice, 377
Chondrocytes, 105–107
Ciona fibronectin, 28–29
cho/cho mouse, 100
Chondroadherin, 198
Chondroitin sulfate, 15, 94,127, 147, 157, 162, 170–171, 177, 200, 281
Chondroplasia, 140, 394
cis-Peptide bonds in collagen, 81
cis-to-*trans* isomerization, 81
    catalyzed by cyclophilin B, 81
Coacervation, 270
Coiled-coil structure, 8–9, 13
    in Matrilin-1, 9
Cold insoluble globulin (*see* fibronectin)
Collagen, (*see* also individual collagen types), 77–109
    D-period, 85
    α chain definition, 77–79
    assembly, 95–107
    binding to fibronectin, 49
    cellular microassembly domains, 96–97
    collagen types, in vertebrates, table, 78
    crosslinking, 83
    fibril forming collagens, 83–86
    nucleators of fibril assembly, 100–101
    organizers of collagen fibril assembly, 101–102
    primordial collagen gene, 27
    proline hydroxylation, 13
    proteoglycans, 87, 94–95
    proteolytic processing, 13, 82–85

Index

protofibrils, 96
posttranslational glycosylation, 81–82
regulators of fibril assembly, 102–103
SLRP, role in collagen assembly, 103
suprastructural organization of collagens, Table, 84
telopeptides, 85
transmembrane and multiplexin collagens, 93–95
triple helix assembly and structure, 7–9, 81–82
types, table, 78
Collagen I, 27, 83–85, 382
Collagen II, 83–85, 106–107, 382
knockout mice
Collagen III, 27, 83–85
Collagen IV, 10, 84, 88–89, 124–126
7S domain, 125
α chain types, 124–126
basement membrane networks, 88–89, 124–126
carboxy-terminal domain, 5, 13
diseases, 138–139
knockout mouse, 101, 125
Collagen V, 27, 83–85, 97, 100–101, 103
knockout mouse, 100, 105
Collagen VI, 10, 84, 89–92
formation of microfibrils, 90
interactions with other ECM proteins, 89
Collagen VII, 84, 92–93, 135
Collagen VIII, 84, 93
in Descemet's membrane, 93
Collagen IX, 84, 85–88, 97, 103, 106, 382
Collagen X, 84, 93
in hypertrophic cartilage, 93
Collagen XI, 27, 83–85, 97, 100–101, 106
knockout mouse, 100
Collagen XII, 13, 32, 84–88
domain organization, 14
electron microscopic image, 12
Collagen XIII, 9, 84, 93–95
Collagen XIV, 84–88, 103
knockout mouse, 103, 107–108
Collagen XV, 84, 94–95
Collagen XVI, 84, 88, 106, 244
Collagen XVII, 9, 84, 93–95, 135
Collagen XVIII, 84, 94–95, 127, 128
endostatin, 94
knockout mouse, 125
Collagen XIX, 84, 88
Collagen XX, 84–88
Collagen XXI, 84, 88
Collagen XXII, 84, 88

Collagen XXIII, 9, 84. 93–95
Collagen XXIV, 83–85
Collagen, XXV, 9, 84, 93–95
Collagen, XXVII, 83–85
knockout mouse, 128
COMP (thrombospondin-5), 382–383
binding to thrombospondin-4, 383
domain organization, 14, 372
electron microscopic image, 12
oligomeric structure, 13
regulation of BMP2 signaling, 383
Congenital contractural arachnodactyly, 247–250
Connective tissue growth factor (CTGF), See CCN2
Cornea, 82, 97, 101, 103, 154, 204–205
collagen formation in, 104–105
Crosslinks, 83, 92, 238, 245, 272, 276–278, 303–305
CS. See Chondroitin sulfate
CSPG-2. See Versican
Crtap. See cartilage-associated protein
CTGF. See CCN2
CUB domain, 5
Cuticle collagens, 9, 26, 28
Cutis hyperelastica, 201
Cutis laxa, 285–287
ADCL, 285, 290
ARCL, 286–287
X-linked 286
Cyclophilin B, 81
Cyclosporin A, 81
CYR61. See CCN1

D

D-period of collagen, 85
DDR, See Discoidin domain receptor
Decorin, 89, 103, 107, 197–220
binding to fibrillin, 244
cardiovascular development and disease, 215–216
congenital stromal dystrophy, 205
core protein, 199–200
core protein binding to collagen, 200
elastogenesis, 216
electron microscopy of collagen fibers, 203
interactions with growth factors, 207–208
knockdown in zebrafish, 204
knockout mice, 103, 108, 200–205, 212–215
receptors, 208–211
role in cancer and angiogenesis, 217–220

Decorin (*cont.*)
  role in skeletal growth and remodeling, 204, 212–215
  signaling, 208–211
Dermatan sulfate, 147, 200–201, 208
Dermatan sulfate epimerase, 1, 200
Descemet's membrane, 93
  collagen VIII, 93
Desmosine, 272, 276–278
Diabetes, 158
Discoidin domain receptors (DDR), 18, 133
  binding to collagen I and IV, 133
  knockout mouse, 133
Domains in ECM proteins, 3–12
DS. *See* Dermatan sulfate
Dwarfism, 383
Dystroglycan, 41, 57, 131–132
Dystroglycanopathies, 137–138
Dystrophin, 131
  dystrophin-glycoprotein complex, 132

E
EB, *See* Epidermolysis bullosa
ECM. *See* Extracellular matrix
Ectopia lentis, 247
Ectodysplasin, 93
EGF. *See* Epidermal growth factor
EGFR. *See* Epidermal growth factor receptor
EF-hand domain, 5
Ehlers-Danlos syndrome, 101, 201, 212, 387
Elastase, 271
Elastic fibers, 268
  electron microscopy, 269
Elastin, 172, 267–290
  alternative splicing, 275–276
  assembly, 279–282
  binding to fibrillin, 244
  coacervation, 270
  crosslinks, 272, 276–278
  desmosine, 272, 276–278
  diseases, 283–287
  domain structure, 272–274
  developmental expression, 271
  electron microscopy, 269
  expression in aorta, 270–271
  evolution, 268
  fibulins, role in elastogenesis, 348–352
  gene structure, 274–275
  hydrophobic domains and elasticity, 278–279
  knockout mouse, 287–290
  signaling, 271
  turnover, 270–271
  physiological role, 267–268
  Xenopus genes, 273
Elastin binding protein, 172
EMT. *See* Epithelial-mesenchymal transformation
Endorepellin, 128
Endostatin, 94, 128
Endothelial-mesenchymal transformation, 154
Entactin (also called nidogen), 10, 56–57, 61, 126
  domain structure, 127
  interactions with laminin, 61
  knockout mouse, 125–126
  proteolytic processing, 15
  receptors for, 61
Epidermal basement membrane
  anchoring fibrils consisting of collagen VII, 92
Epidermal growth factor (EGF) domain, 5
  in fibrillin, 235–237
  in fibulin, 342
Epidermal growth factor receptor
  interactions with decorin, 209
  interactions with fibulin-3, 344
  interactions with tenascin-C, 386
Epidermolysis bullosa (EB and JEB), 93, 94, 134–136
Epiphycan, 202
Epithelial-mesenchymal transformation (EMT), 156, 164, 325–326
Eumetazoa, 26
  and gastrulation, 26
Evolution of ECM, 2–3, 24–33
  by exon shuffling, 25
  conserved versus variable genetic shuffling, 226–228
  role of differential splicing, 26
  of ECM toolkit, 26
  of fibronectin, 28–29
Extracellular matrix
  as nanodevices, 16–24
  assembly, 50–55
  domain structure and organization, 9–11
  evolution of, 2–3, 24–33
  general properties of, 3
  in plants, 2
  multimerization, 11–13
  oligomerization domains, 11
  posttranslational modifications, 13–15
Eye disease, 128, 173, 205–207, 247–250, 356–357

F

FAC. *See* Focal adhesion kinase
FACIT collagens, 13, 85–88, 97, 103
FACIT-like collagens, 88
FBN. *See* Fibrillin
Fibrillin
  assembly domains, 242
  binding domains, 244
  binding pro-BMP, 22, 244
  binding latent TGF-β, 22
  crosslinks, 245
  diseases, 247–250
  domain definition, 235–237
  developmental expression, 245–247
  elastic fiber assembly, 283
  interactions with fibronectin, 243
  interactions with heparin/heparan, 243
  knockout and transgenic mice, 246, 253–257, 283
    C1039G, 256
    C1163R, 256
    Fbn2-/-, 255
    Fbn1-/-/Fib2-/-, 255
    mgΔ, 253
    mgR, 253
    mgN (Fbn1-/-), 253
    Table, 255
    Tsk, 256
  Marfan syndrome, 247–250
  mechanism for disease pathogenesis, 249
  organization into microfibrils, 240–245
  processing by furins, 241
  structure, 235–237
  TGF-β-binding, 248–250
Fibrillin-1
  diseases, 247–250
  knockout mice, 246
  role in aortic wall and lung maturation, 246
Fibrillin-2
  CCA, 247–250
  expressed in early development, 246
  knockout mice, 246
Fibrillin-3, 243
  inactive in rodents, 246
Fibrin
  interactions with fibronectin, 49–50
Fibrinogen
  c-terminal domain, 5
Fibripositors, 97
Fibroblast growth factor (FGF), 21, 151, 321–322, 345
Fibromodulin, 103
  binding to collagen, 200

knockout mouse, 103, 201–205, 214–215
Fibromodulin, 103, 108
Fibronectin, 28, 41–64
  alternative splicing, 44–45
  EDA and EDB (extra domain A & B), 44–45
  V (variable) domain, 45
  assembly, 17–18, 50–55
  bacteria, binding to, 50
  basement membrane, in, 129
  cell-binding domain
  collagen-binding domains, 49
  collagen fibril organizer, 101–102
  domain structure and organization, 11
    FN1, Type 1 domain, 5, 43–44
    FN2, Type 2 domain, 5, 43–44
    FN3, Type 3, domain, 5, 43–44
  electron microscopic images, 12
  fibrin binding domain, 49–50
  gene inactivation, 42
  heparin-binding domains, 50
  in wound healing and clotting, 28
  interaction with integrins, 17, 45–49
  coupling to cytoskeleton, 49
  oligomeric structure, 13
  posttranslational modifications, 45–46
  RGD, 6, 46–49
  role of integrins and the cell, 54–55
  structure, 42–44
Fibulin, 337–359
  cardiovascular disease, 355–356
  cancer, 352–355
  cell adhesion and motility, 345–348
  domain structure, 337–342
  elastogenesis, 348–352, 358
  eye disease, 356–357
  expression, regulation, 344–345
  fibulin type module, 342–343
  gene localization, 338
  in basement membranes, 129
  in elastic fiber assembly, 282
  ligands of the fibulin family, 339–341
  receptors, 344–345
Fibulin-1
  exon splicing, 338
  in cancer, 352–355
  knockout mouse, 351–352
Fibulin-2
  in cardiovascular disease, 355
  knockout mouse, 351
Fibulin-3
  in cancer, 355
  knockout mouse, 350–351
  macular dystrophies, 357

Fibulin-4, 282
  in cardiovascular disease, 355–356
  knockout mouse, 282, 348–349
Fibulin-5, 282, 349–350
  in AMD, 357
  in cardiovascular disease, 355–356
  knockout mouse, 282
Fibulin-6
  in eye disease, 357
FGF. *See* Fibroblast growth factor
Folds in ECM proteins, 5–9
Focal adhesion kinase (FAC), 63
Furin, 84, 241, 310

G
GAG, *See* glycosaminoglycan
Galactosyl-3-sulfate ceramide, 133
Gastrulation, 26
GBM. *See* Glomerular basement membrane
Gliomedins, 93
Glycosaminoglycans
  chondroitin sulfate, 94
  heparan sulfate, 94
  in SLRPs, 197–220
  synthesis, 160
Glycosylation
  in collagen, 15, 80–82
  of fibronectin, 45–46
  of lysyl oxidase, 305–306, 315
  of thrombospondin, 374
  sites in fibrillin, 236–237
Glomerular basement membrane, 119
Glucuronate-N-acetylglucosamine, 148
Glycolipids, sulfated, 372–373
Glycosaminoglycans
  hyaluronan, 148–179
  proteoglycans, 147–179
  role in cartilage, 161–162
Glycosyl transferase, 160
Goodpasture syndrome, 139–140
GPI-Brevican, *See* Brevican
Growth Factors
  activity modulated by ECM, 20–22
  role in integrin signaling, 21

H
HA. See Hyaluronan
Hemidesmosome anchoring complex, 94, 124
Heparan sulfate proteoglycan, 126, 128, 375, 391–392
  HSPG2 mutations, 140
Heparin
  interactions with fibronectin, 50
  interactions with thrombospondin, 373

Hevin, 389
Hexabrachion, 32
IIIF1α, 326
Homocystinuria, 248
Homologous domains, function of, 9–13
Homology, defined, 4
HSPG. *See* Heparan sulfate proteoglycan
Hyaladherins, 150–153
  CD44, 149–153
  LYVE–1, 150, 153
  RHAMM, 150–153
  TSG–6, 150, 157
Hyalectins, 147
Hyaluronan, 89, 147–179
  catabolism, 149
  evolution, 149–150
  in cardiac cushions, 154
  in cancer, 155–156
  in development, 153–155
  in inflammation, 157–158
  in limb development, 154–155
  receptors, *See* Hyaladherins
  signaling, 151–153
  structure and biosynthesis, 148–149
Hyaluronan synthase
  evolution, 149–150
  Has1, 149
  Has2, 149
    in cardiac development, 154
    in limb development, 154–155
    knockout mouse, 154–155
    role in EMT, 156
  Has3, 149
Hyaluronidase, 149
Hydroxylation
  of proline, 13, 80
  of lysine, 80, 82
  of tyrosine, 312
Hydroxylysine, 80, 268
Hydroxyproline, 7, 13, 80, 268
Hypertrophic cartilage
  collagen X in, 93

I
IαI, 157
Identity, defined, 4
Ig-like domain, 5
  in perlecan, 10
IGF. *See* Insulin-like growth factor
IGF-IR, *See* Insulin-like growth factor receptor
ILK. *See* Integrin-linked kinase
Insulin-like growth factor
  interaction with decorin

Insulin-like growth factor receptor
    interactions with decorin, 211–212
Integrin, 9, 18–20, 46–49, 130–131
    as nanomachine, 19
    as stress sensors, 24
    receptors for ECM components,
        Table, 64
        collagen IV, 64, 130–131
        entactin, 61
        fibrinogen, 62
        fibronectin, 46
        laminin, 57, 130–131
        nephronectin, 61–62
        tenascin, 60
        thrombospondin, 59–60
        vitronectin, 59
    role in ECM assembly, 18–20, 54–55
    signal transduction, 19–20, 62–63
    structure, 18–20
Integrin αIIbβ3, 46, 59, 62, 344, 391
Integrin α$_M$β2, 391–392
Integrin α1β1, 57
Integrin α2β1, 57, 60, 386, 391
Integrin α3β1, 46, 57, 61, 131, 134, 373
Integrin α4β1, 46, 344, 373, 378
Integrin α4β7, 46
Integrin α5β1, 6, 46, 243–244, 344, 393
Integrin α6β1, 57, 61, 131, 134, 373,
    378, 391
Integrin α6β4, 57, 131, 134–135
Integrin α7β1, 57, 60, 131, 134, 386
Integrin α8β1, 46, 59–62
Integrin α9β1, 57, 60, 373, 386
Integrin αvβ1, 46, 59
Integrin αvβ3, 46, 57, 59–61, 244, 344, 386,
    391, 393
Integrin αvβ5, 46, 59, 344, 391
Integrin αvβ6, 46, 60, 244
Integrin αvβ8, 59
Integrin linked kinase (ILK), 63
Isolated congenital nephrotic syndrome, 138

J
JEB. See Epidermolysis bullosa
Jellyfish, 9, 28

K
Keratan sulfate chains, 160, 200–201
Keratocan, 104–105
    knockout mouse, 105, 202, 205
Knobloch syndrome, 95, 128
KS. See Keratan sulfate
Kunitz inhibitor domain, 5, 10

L
Lamina densa, 119
Lamina lucidae, 119
Lamina rara interna, 119
lamina rara externa, 119
Laminin, 27–28, 55–58, 119–124
    alternative splicing, 123
    chain types and nomenclature, 119–123
    domain structure model, 56
    electron microscopic image, 12
    entactin, interactions with, 61
    G-like domain, 5, 27, 56–57
    knockout phenotypes, 124
    laminin-type EGF-like domain, 5
    n-terminal domain, 5
    nomenclature, 122
    receptors for, 57
    structure of trimers, 123
    tissue distribution, 123–124
Laminin chains
    α1, 122
        knockout mouse, 124
    α2, 122, 124, 136–137
        in congenital muscular dystrophy,
            136–137
        knockout mouse, 137
    α3, 122, 124
        alternative splicing of, 123
        mutations causing epidermolysis
            bullosa, 134–136
    α4, 122, 137
    α5, 122, 137
        alternative splicing of, 123
        knockout mouse, 124
    β1, 122
        knockout mouse, 124
        ubiquitous in basement membranes, 124
    β2, 122, 124
        knockout mouse, 138
        mutations causing Person
            syndrome, 138
    β3, 122, 124
        mutations causing epidermolysis
            bullosa, 134–136
    β4, 122
    γ1, 123–124
        knockout mouse, 124, 133
        ubiquitous in basement membranes,
            124
    γ2, 122, 124
        mutations causing epidermolysis
            bullosa, 134–136
    γ3, 122, 124

Laminin-111, 123, 131, 137
  domain organization, 14
  electron microscopic image, 12
Laminin-113, 57
Laminin-211, 131, 136–137
Laminin-221, 122–124
Laminin-311, 123
Laminin-321, 123
Laminin-322, 123
Laminin-332, 135
Laminin-411, 123, 137
Laminin-421, 123–124
Laminin-5, *See* Laminin-332
Laminin-511, 57, 123, 135, 137
Laminin-521, 123
Latent TGF-β
  activated by thrombospondin, 373
  binding to fibrillin, 244
Latent TGF-β binding protein (LTBP), 22
  superfamily, 236–237, 250–252
  binding to fibrillin, 246, 250–252
Lecticans, 147
Leucine-rich repeats, 198
  structure, 198
Lewis, W.H., 2
Limb development, 154–155
Link protein, 148, 171
  crosslinking reaction, 304
  knockout mouse, 160
  module in hyaladherins, 150
LOX. *See* Lysyl oxidases
LOXL1–4, *See* Lysyl oxidase
LRR. *See* Leucine-rich repeats
LTBP. *See* Latent TGF-β binding protein
LTQ. *See* Lysine tyrosylquinone
Lumican, 103
  binding to collagen, 200
  interactions with toll-like receptors, 211
  knockout mice, 103, 105, 201–205
  role in cancer, 218
Lutheran/Basal cell adhesion molecule, 132–133
Lysine tyrosylquinone (LTQ), 311–312
Lysyl oxidase, 83, 281, 286, 303–328
  as a chemokine, 322–324
  biosynthesis, 305–306
  cancer, 320–322
  catalytic properties and cofactors, 312–313
  copper, 311
  domain structure, 306–311
  family members, 304
  gene localization and structure, 306
  knockout mouse, 304
  pre-prolox, 305–306
  role in fibrosis, 324–325
  structure, 317–320
  substrate specificity, 314–317
  lysyl oxidase like (LOXL), *See also* Lysyl oxidase
  domain comparison, 307
  gene structure and localization, 306
LOXL1, 325
  knockout mouse, 325
LOXL2, 325–326
  cancer, 326
  interactions with SNAIL, 326
  target of HIF1α, 326
LOXL3, 326–327
  cancer, 326–327
LOXL4, 326–327
  sequence alignment, 308–309

M
Macular degeneration, age related (AMD), 356–357
MACS
MAGP. *See* Microfibril-associated glycoprotein
Marfan syndrome, 234, 247–250
  fibrillin and TGF-β, 249–250
  mouse models, , 253–257
Matricellular proteins, 369–397
  definition, 369–371
  domain comparison, 370
  See Thrombospondin, tenascin, SPARC, spondins, CCN
Matrilin-1, 8, 382
Matrix, *See* Extracellular matrix
Matrix metalloproteinase (MMP), 15
MDC1A. *See* Merosin-deficient congenital muscular dystrophy type 1A
Mechanical stress, 22–24
  and ECM gene expression, 23
  collagen fibers and tensile stress, 23
  how cells sense stress, 24
  influences on versican synthesis, 166
Meprin, 61
Merosin-deficient congenital muscular dystrophy type 1A, 136–137
Met receptor, 209–210
Metazoa, 2–3
Microdomains for collagen assembly, 96–97
Microfibrils, 233–257
  associated with elastic fibers, 237–240
  beads-on-a-string appearance, 239

biogenesis, 240–245
consisting of collagen VI, 90
crosslinks, 245
growth factor binding, 250–252
interactions with TGF-β, 249–252
molecular organization, 240
role in elastic fiber assembly, 283
Microfibril-associated glycoprotein (MAGP), 89, 215–216, 283
 binding to fibrillin, 244
 knockout mouse, 283
miRNA-30, 394
miRNA-133, 394
MMP, See Matrix metalloproteinase
Molecular machines, 17
Monocytes, 157
Muscular dystrophy, 136–138, 205–206
Myotendinous junction, 23

N
Nanodevices, 16–24
Nephronectin, 61–62
Neurocan, 148, 174–178
 in brain injury, 177
 role in cancer, 178
 synthesis and turnover, 176–177
Nidogen, See Entactin
Nov. See CCN3
Nyctalopin, 198

O
Ortholog, 30
Osteoarthritis, 214–215, 383
 linked to asporin mutations, 205
Osteogenesis imperfecta, 81
Osteonectin. See SPARC
Osteopontin, 389
Osteoporosis, 212–214

P
PAPS. See Phosphoadenosine phosphosulfate
Paralog, 29
Parazoa, 26
PDGF. See Platelet-derived growth factor
Peptidyl-prolyl *cis* bonds, 81
Perineuronal net, 176
Perlecan, 10, 89, 95, 127–128, 140, 346, 386, 394
 binding to fibrillin, 244
 domain organization, 127
 electron microscopic image, 12
 endorepellin, 128
 homology with laminin LG and LE domains, 127

in basement membrane, 55, 117, 121
interactions with fibrillin, 238, 243
knockout mouse, 127–128
mouse with heparan sulfate attachment site deletion, 125
PG-M. See Versican
Phosphoadenosine phosphosulfate, 160
 brachymorphic mouse, 161
Phosphoserine in fibronectin, 46
Pierson syndrome, 138
Plants, ECM in, 2
Platelets, 376
Platelet-derived growth factor, 165–166
 interaction with decorin, 208
Plectin, 135
Podocan, 198
Posttranslational modifications, 13–15
Procollagen, 82–83, 86
Proline hydroxylation, See Hydroxyproline
Prolyl-3-hydroxylase, 81
Proteoglycans, 147–179, 197–219
 collagens as proteoglycans, 87, 94
 electron microscopic image, 161
 mechanism of GAG chain addition, 159–160
 See SLRPs
 structural model of hyalectin proteoglycans, 148
Protofibrils, of collagen, 96–97
Pseudorheumatoid dysplasia, 396

R
Regulators of collagen assembly, 102–103
Reichert's membrane, 123–124, 131
Resilin, 268
RGD, See fibronectin
RHAMM, 150–153
RIN2, 287

S
Schwartz-Jampel syndrome, 140
Shprintzen-Goldberg syndrome, 247
SIBLING gene family, 389
Skeletal connective tissue, 212–215
SLRP. See Small leucine-rich proteoglycan
Small leucine-rich proteoglycan (SLRP), 197–220
 angiogenesis, 218–219
 cancer and metastasis, 216–218
 cardiovascular development and disease, 215–216
 classes, 98–201
 eye disease, 205–207

Small leucine-rich proteoglycan (SLRP) (cont.)
  human diseases, 205–207
  interactions with growth factors, 207–208
  phylogenetic tree, 199
  receptors, 208–211
  role in collagen assembly, 103, 197–198
  role in skeletal development and remodeling, 212–215
  signaling, 208–211
  structure, 198–201
SPARC, 370, 387–389
  association with disease, 388–389
  evolution, 387
  hevin, 398
  knockout mice, 388
  role in collagen fibrillogenesis, 388
  structure, 387
Spondins, 370, 384
Sponge ECM, 27
  comparison of sponge and vertebrate collagens, 27
Stiff skin syndrome
Sulfatides, 133
Sulfotransferases, 160
Supravalvular aortic stenosis (SVAS), 280, 283–284
SVAS. See Supravalvular aortic stenosis
Syndecan, 18, 21, 41, 57, 391

T
T-cells, 157
TB domain, TGF-β binding domain in fibrillin, 5, 237
Telopeptides of collagen, 85
Tenascins, 385–387
  evolution of, 32–33, 385
  definition, 385
  domain structure, 370
Tenascin-C, 385–386
  alternative splicing, 385
  electron microscopic image, 12
  knockout mouse, 386
  receptors and binding proteins, 386
Tenascin-R, 32
Tenascin-W, 32
Tenascin-X, 29, 102, 386–387
  Ehlers-Danlos syndrome, 387
  knockout mouse, 102, 387
  linkage to disease, 378
Tendon
  collagen fibrils, 107–108
  role of SLRPs, 214–215

TGF-β, See Transforming growth factor-β
Thrombospondin, 59–60, 370–384
  evolution of, 29–32, 371
  domains, 6, 370, 372
  in platelets, 376
  ligands and ligand binding sites, 373
  organization of vertebrate thrombospondins, 372
  receptors and receptor binding sites, 373
Thrombospondin-1, 372–378
  antiangiogenic properties, 376–377
  effects on immune cells, 378
  expression, 375
  hypermethylation of gene in cancer, 374
  in platelets, 376
  inhibition of ADMTS13, 376
  knockout mice, 376
  nitric oxide (NO) antagonist, 377
  posttranslational modifications, 374
  receptors and ligand binding sites, 373
  wound repair, 377–378
Thrombospondin-2, 378–381
  antiangiogenic properties, 380
  bleeding abnormality, 380
  knockout mouse, 380
  receptors, 378–379
Thrombospondin-3, 381–382
  expression pattern, 381
  knockout mice, 382
Thrombospondin-4, 381–382
  tissue expression, 381–382
Thrombospondin-5 (see also COMP), 382–383
Tight skin (Tsk) mutation, 256
TIMP3, 357
Titin, 29
Toll-like receptors, 211, 305–306
Tolloid/BMP-1, 305–306
  processing of collagen C-propeptide, 84
Transforming growth factor-β (TGF-β), 21–22, 250–252, 345
  activated by thrombospondin-1, 373
  binding to fibrillin, 22 250–252
  interactions with decorin, 107
  large latent complex TGF-β (See LTBP), 22
  regulation of versican synthesis, 165–166
Transglutaminase crosslinks, 92, 245
Transmembrane collagens, 93–95
Tropoelastin, See Elastin
  definition, 272
  structure, 272
TSK. See Tight skin
TSP. See Thrombospondin
Tsukushi, 198

Index

Triple helix, collagen, 7, 81–82
Turner syndrome, 212
Tyrosine sulfation in fibronectin, 46

U
UDP-xylose, in proteoglycan synthesis, 159–160

V
Versican, 148, 162–171
    alternative splicing, 162–163
        V0, 162–163
        V1, 162–163
        V2, 162–163, 170
        V3, 162–163, 165, 170
    binding to fibrillin, 244
    CS GAG chain, 163
    degradation by MMPs, 166
    effects on cell adhesion and proliferation, 168–169
    effects on cell migration, 169–171
    in cancer, 173
    in cardiovascular disease, 174
    in eye disease, 173
    in inflammation, 172–173
    regulation by growth factors, 165–166
    role in development, 163–166
    role in heart development, 164, 170
    schematic model of versican variants, 162
    synthesis, 163–166
    turnover, 166–168
VGVAPG, 271
Vitronectin, 47, 58–59
    domain structure model, 58
    receptors for, 59
Volvox, 2
Von Willebrand factor, 6

W
Weill-Marchesani syndrome, 246
Williams-Beuren syndrome, 284–285

X
Xenopus, 273

Z
Zebrafish, 204, 273–274

Printed by Books on Demand, Germany